Dietary Habits, Beneficial Exercise and Chronic Diseases: Latest Advances and Prospects

Dietary Habits, Beneficial Exercise and Chronic Diseases: Latest Advances and Prospects

Editor

Panagiota Mitrou

MDPI • Basel • Beijing • Wuhan • Barcelona • Belgrade • Manchester • Tokyo • Cluj • Tianjin

Editor
Panagiota Mitrou
Hellenic Ministry of Health
Greece

Editorial Office
MDPI
St. Alban-Anlage 66
4052 Basel, Switzerland

This is a reprint of articles from the Special Issue published online in the open access journal *Nutrients* (ISSN 2072-6643) (available at: https://www.mdpi.com/journal/nutrients/special_issues/dietary_exercise_diseases).

For citation purposes, cite each article independently as indicated on the article page online and as indicated below:

LastName, A.A.; LastName, B.B.; LastName, C.C. Article Title. *Journal Name* **Year**, *Volume Number*, Page Range.

ISBN 978-3-0365-4991-0 (Hbk)
ISBN 978-3-0365-4992-7 (PDF)

Contents

About the Editor

Panagiota Mitrou

Dr. Mitrou is a Medical Doctor (Internist-Diabetologist) with a PhD in Glucose Metabolism. Her research activities include internal medicine, diabetes mellitus, metabolism, patient registries, and therapeutic protocols. She is a member of the "Patient Registry Monitoring Coordination Group", the "Committee on Pharmaceutical Expenditure Monitoring, Integration of Diagnostic/Therapeutic Protocols and Patient Registries", and several Working Groups on Updating the Diagnostic Protocol. Her current position is Head of the Independent Department of Therapeutic Protocols and Patient Registries in the Hellenic Ministry of Health. She is responsible for the development of diagnostic and therapeutic prescription protocols, the coordination of Scientific Working Groups, and application of therapeutic protocols in clinical practice, through their integration into the e-prescription system. Diagnostic and therapeutic protocols are useful to clinicians as a guide of good clinical practice (evidence-based medicine), and a data collection tool (providing real-world data/evidence) useful for health policies. Her work has been presented in numerous scientific meetings and has been published in scientific journals with a high Impact Factor, and it has been frequently cited in the international literature. She is a member of several conference-organizing committees, lecturer and/or chair of conferences and scientific meetings, reviewer of scientific articles in medical journals, and Editor or Section Editor-in-Chief of scientific journals.

Editorial

Is lifestyle Modification the Key to Counter Chronic Diseases?

Panagiota Mitrou

Independent Department of Therapeutic Protocols and Patient Registers, Hellenic Ministry of Health, 104 33 Athens, Greece; panagiotamitrou@gmail.com

Citation: Mitrou, P. Is lifestyle Modification the Key to Counter Chronic Diseases? *Nutrients* **2022**, *14*, 3007. https://doi.org/10.3390/nu14153007

Received: 18 July 2022
Accepted: 20 July 2022
Published: 22 July 2022

Dietary patterns, defined as the quantities, proportions, variety, or combination of different foods and drinks, as well as the frequency with which they are habitually consumed, are associated with an increased or decreased incidence of chronic diseases. Several lines of evidence indicate that healthy diet and exercise can prevent cardiovascular diseases, stroke, diabetes, and some types of cancer. Lately, an association has been found between eating habits, exercise, and psychological and/or mental disorders.

In the first article of this Special Issue, Gantenbein and Kanaka-Gantenbein reviewed the beneficial effects of the Mediterranean Diet and specifically its antioxidant and anti-inflammatory properties, enhancing metabolic, reproductive, and mental health [1].

As an alternative, the ketogenic diet, based on carbohydrate restriction, moderate protein intake, and increased fat consumption, has gained ground [2]. The restriction of carbohydrates results in an increase in the production of ketone bodies, a metabolic state in which the body utilizes fat (instead of glucose) as its primary metabolic substrate. According to Dowis and Banga, the use of the ketogenic diet might have many therapeutic effects, including weight reduction and an improvement in predisposing factors for cardiovascular diseases such as dyslipidemia and hyperglycemia [2].

However, if one finds a low-carbohydrate ketogenic diet difficult to comply with, then targeting the microbiome might be a way to improve metabolic health. According to Turroni et al., dietary products based on symbiotic agriculture could possibly modulate the human microbiome and metabolome, decreasing the risk of metabolic syndrome [3]. Moreover, as stated by van Krimpen et al., reducing gut permeability via *Lactobacillus* spp. supplementation could be a potential treatment to reduce cachexia, an inflammation-driven condition related to aging and chronic diseases [4].

Among other beneficial effects, healthy nutrition appears to have an impact on inflammatory and rheumatic diseases. As highlighted by Ratajczak et al., the type of diet followed is particularly important for patients suffering from Inflammatory Bowel Disease (IBD) who often display vitamin deficiency, such as low folate levels, resulting in anemia, neurological symptoms, and bone loss [5]. In these patients, although well-balanced dietary patterns are needed to provide all macro- and micronutrients, fiber-rich products often exacerbate gastrointestinal discomfort. As a result, the concentration of folic acid should be often evaluated in patients with IBD and given as a supplement in cases of insufficient dietary intake and/or absorption [5]. Basdeki et al. performed a systematic review and meta-analysis of previously published randomized controlled trials (RCTs) investigating the relationship of sodium intake with systemic inflammation [6]. Although the study did not verify the hypothesis that sodium intake can induce a systemic inflammatory response in humans in a dose–response manner, it highlighted major issues that could be addressed in future relevant RCTs [6]. In a systematic review, Tsiogkas et al. summarized the current evidence regarding the efficacy of Crocus sativus (Saffron) supplementation in patients with rheumatic diseases (RDs) [7]. Although current evidence indicates that saffron may have a positive effect in patients with RD, targeting inflammatory and immune responses, oxidative stress, pain, and the psychological effects of the disease, more studies are required to shed light on the efficacy and draw conclusions on the appropriate dose of this supplement [7].

In the present Special Issue, several articles focus on the effects of nutritional interventions on cardiovascular risk factors such as diabetes, hypertension, and increased body weight. According to Mahrouseh et al., eating vegetables several times a week in addition to performing physical exercise may reduce the risk of DM, although more lifestyle characteristics and socioeconomic conditions possibly underlie the increased prevalence of diabetes in Slovakia and other European countries [8]. The association between high fructose consumption and hypertension is controversial. Based on the study by Béghin et al., although the consumption of increased quantities of fructose derived from non-natural foods is associated with elevated diastolic blood pressure, the consumption of natural foods containing fructose, such as fresh fruits, does not increase blood pressure and thus remains a healthy dietary habit with pleiotropic positive metabolic effects [9]. The findings of Awoke et al. showed that women across reproductive life stages, in particular those with lower socioeconomic status, often fail to meet dietary and physical activity recommendations and should be targeted in future interventions to prevent weight gain [10]. Familial socioeconomic disadvantage was also identified as a common risk factor for obesity and early-childhood caries, which are two highly prevalent chronic diseases in childhood. In the above-mentioned study by Manohar et al., children with the highest trajectories of discretionary food intake were more likely to have increased body weight, highlighting the need for targeted health promotion interventions [11]. As reviewed by Pereira et., most dietary interventions to reduce childhood obesity are based on person-based educational approaches with modest (if any) effects on body weight, highlighting the need for multilevel interventions focused on micro- and macro-policy environmental changes and the strengthening of children, family, and community [12]. From this point of view, Scazzocchio et al., presented a new Italian educational program developed for Italian students, aiming to increase food literacy and favoring a healthier relationship with food [13].

Dietary choices may also have an impact on neurological diseases and mental health. As described by Wang et al., unhealthy dietary choices, such as the excessive intake of saturated fats and salt, can affect brain function and increase the risk for ischemic stroke [14]. On the other hand, as stated by Stoiloudis et al., healthy nutrition shows promising beneficial effects in terms of slowing down multiple sclerosis activity and progression [15]. Moreover, healthy nutrition may protect against anxiety and psychosocial maladjustment, as shown by Freret et al. in animal studies and by Khaled et al. in human studies [16,17].

Healthy nutrition can also help enhance athletic performance. In a cross-sectional study conducted by Martínez-Rodríguez et al. in Spanish female athletes, a correlation analysis highlighted the relationship of increased body weight and poor nutrition with worse results in power and endurance tests. As expected, exercise had also a beneficial effect on bone density [18]. Based on the study by Kyle et al., increased physical fitness and aerobic exercise are also important for women across the first postpartum year experiencing changes in bone mineral density and serum lipids [19].

Given the increasing prevalence of diet-related chronic diseases, in the last paper of this Special Issue, Wu et al. highlighted the need for the development of new scalable dietary monitoring techniques and presented the results of a study in Switzerland in which the nutritional quality of users' diets was estimated based on digital receipts from grocery shopping [20].

During the last few decades, there has been a tremendous rise in the incidence of chronic non-communicable diseases (NCDs) such as cardiovascular diseases, cancer, diabetes, and chronic respiratory diseases. Although population aging and other unmodifiable genetic factors are contributing to this rise, it has been recognized that there also modifiable, unhealthy lifestyle choices, such as a diet rich in saturated fats, salt, and refined carbohydrates, sugar-sweetened beverages, a lack of physical activity, and smoking, which exacerbate this phenomenon [1]. Identifying the relationship between lifestyle choices and chronic diseases can help stakeholders to design and implement programs that are oriented toward a healthy lifestyle, ensuring progress towards achieving the goal of reducing premature deaths from NCDs by one third by 2030, as outlined in the United Nation's 2030 Agenda for Sustainable Development.

Funding: This research received no external funding.

Conflicts of Interest: The author declares no conflict of interest.

References

1. Gantenbein, K.V.; Kanaka-Gantenbein, C. Mediterranean Diet as an Antioxidant: The Impact on Metabolic Health and Overall Wellbeing. *Nutrients* **2021**, *13*, 1951. [CrossRef] [PubMed]
2. Dowis, K.; Banga, S. The Potential Health Benefits of the Ketogenic Diet: A Narrative Review. *Nutrients* **2021**, *13*, 1654. [CrossRef] [PubMed]
3. Turroni, S.; Petracci, E.; Edefonti, V.; Giudetti, A.M.; D'Amico, F.; Paganelli, L.; Giovannetti, G.; Del Coco, L.; Fanizzi, F.P.; Rampelli, S.; et al. Effects of a Diet Based on Foods from Symbiotic Agriculture on the Gut Microbiota of Subjects at Risk for Metabolic Syndrome. *Nutrients* **2021**, *13*, 2081. [CrossRef] [PubMed]
4. van Krimpen, S.J.; Jansen, F.A.C.; Ottenheim, V.L.; Belzer, C.; van der Ende, M.; van Norren, K. The Effects of Pro-, Pre-, and Synbiotics on Muscle Wasting, a Systematic Review—Gut Permeability as Potential Treatment Target. *Nutrients* **2021**, *13*, 1115. [CrossRef] [PubMed]
5. Ratajczak, A.E.; Szymczak-Tomczak, A.; Rychter, A.M.; Zawada, A.; Dobrowolska, A.; Krela-Kaźmierczak, I. Does Folic Acid Protect Patients with Inflammatory Bowel Disease from Complications? *Nutrients* **2021**, *13*, 4036. [CrossRef] [PubMed]
6. Basdeki, E.D.; Kollias, A.; Mitrou, P.; Tsirimiagkou, C.; Georgakis, M.K.; Chatzigeorgiou, A.; Argyris, A.; Karatzi, K.; Manios, Y.; Sfikakis, P.P.; et al. Does Sodium Intake Induce Systemic Inflammatory Response? A Systematic Review and Meta-Analysis of Randomized Studies in Humans. *Nutrients* **2021**, *13*, 2632. [CrossRef] [PubMed]
7. Tsiogkas, S.G.; Grammatikopoulou, M.G.; Gkiouras, K.; Zafiriou, E.; Papadopoulos, I.; Liaskos, C.; Dardiotis, E.; Sakkas, L.I.; Bogdanos, D.P. Effect of *Crocus sativus* (Saffron) Intake on Top of Standard Treatment, on Disease Outcomes and Comorbidities in Patients with Rheumatic Diseases: Synthesis without Meta-Analysis (SWiM) and Level of Adherence to the CONSORT Statement for Randomized Controlled Trials Delivering Herbal Medicine Interventions. *Nutrients* **2021**, *13*, 4274. [CrossRef] [PubMed]
8. Mahrouseh, N.; Andrade, C.A.S.; Kovács, N.; Njuguna, D.W.; Varga, O. Diabetes Mellitus and Associated Factors in Slovakia: Results from the European Health Interview Survey 2009, 2014, and 2019. *Nutrients* **2021**, *13*, 2156. [CrossRef] [PubMed]
9. Béghin, L.; Huybrechts, I.; Drumez, E.; Kersting, M.; Walker, R.W.; Kafatos, A.; Molnar, D.; Manios, Y.; Moreno, L.A.; De Henauw, S.; et al. High Fructose Intake Contributes to Elevated Diastolic Blood Pressure in Adolescent Girls: Results from The HELENA Study. *Nutrients* **2021**, *13*, 3608. [CrossRef] [PubMed]
10. Awoke, M.A.; Wycherley, T.P.; Earnest, A.; Skouteris, H.; Moran, L.J. The Profiling of Diet and Physical Activity in Reproductive Age Women and Their Association with Body Mass Index. *Nutrients* **2022**, *14*, 2607. [CrossRef]
11. Manohar, N.; Hayen, A.; Scott, J.A.; Do, L.G.; Bhole, S.; Arora, A. Impact of Dietary Trajectories on Obesity and Dental Caries in Preschool Children: Findings from the Healthy Smiles Healthy Kids Study. *Nutrients* **2021**, *13*, 2240. [CrossRef] [PubMed]
12. Pereira, A.R.; Oliveira, A. Dietary Interventions to Prevent Childhood Obesity: A Literature Review. *Nutrients* **2021**, *13*, 3447. [CrossRef] [PubMed]
13. Scazzocchio, B.; Varì, R.; d'Amore, A.; Chiarotti, F.; Del Papa, S.; Silenzi, A.; Gimigliano, A.; Giovannini, C.; Masella, R. Promoting Health and Food Literacy through Nutrition Education at Schools: The Italian Experience with Maestra Natura Program. *Nutrients* **2021**, *13*, 1547. [CrossRef] [PubMed]
14. Wang, Y.; Su, X.; Chen, Y.; Wang, Y.; Zhou, J.; Liu, T.; Wang, N.; Fu, C. Unfavorable Dietary Quality Contributes to Elevated Risk of Ischemic Stroke among Residents in Southwest China: Based on the Chinese Diet Balance Index 2016 (DBI-16). *Nutrients* **2022**, *14*, 694. [CrossRef] [PubMed]
15. Stoiloudis, P.; Kesidou, E.; Bakirtzis, C.; Sintila, S.-A.; Konstantinidou, N.; Boziki, M.; Grigoriadis, N. The Role of Diet and Interventions on Multiple Sclerosis: A Review. *Nutrients* **2022**, *14*, 1150. [CrossRef] [PubMed]
16. Freret, T.; Largilliere, S.; Nee, G.; Coolzaet, M.; Corvaisier, S.; Boulouard, M. Fast Anxiolytic-Like Effect Observed in the Rat Conditioned Defensive Burying Test, after a Single Oral Dose of Natural Protein Extract Products. *Nutrients* **2021**, *13*, 2445. [CrossRef] [PubMed]
17. Khaled, K.; Hundley, V.; Tsofliou, F. Poor Dietary Quality and Patterns Are Associated with Higher Perceived Stress among Women of Reproductive Age in the UK. *Nutrients* **2021**, *13*, 2588. [CrossRef] [PubMed]
18. Martínez-Rodríguez, A.; Sánchez-Sánchez, J.; Martínez-Olcina, M.; Vicente-Martínez, M.; Miralles-Amorós, L.; Sánchez-Sáez, J.A. Study of Physical Fitness, Bone Quality, and Mediterranean Diet Adherence in Professional Female Beach Handball Players: Cross-Sectional Study. *Nutrients* **2021**, *13*, 1911. [CrossRef] [PubMed]

19. Kyle, E.M.; Miller, H.B.; Schueler, J.; Clinton, M.; Alexander, B.M.; Hart, A.M.; Larson-Meyer, D.E. Changes in Bone Mineral Density and Serum Lipids across the First Postpartum Year: Effect of Aerobic Fitness and Physical Activity. *Nutrients* **2022**, *14*, 703. [CrossRef] [PubMed]
20. Wu, J.; Fuchs, K.; Lian, J.; Haldimann, M.L.; Schneider, T.; Mayer, S.; Byun, J.; Gassmann, R.; Brombach, C.; Fleisch, E. Estimating Dietary Intake from Grocery Shopping Data—A Comparative Validation of Relevant Indicators in Switzerland. *Nutrients* **2022**, *14*, 159. [CrossRef]

Review

Mediterranean Diet as an Antioxidant: The Impact on Metabolic Health and Overall Wellbeing

Katherina V. Gantenbein and Christina Kanaka-Gantenbein *

Division of Endocrinology, Diabetes and Metabolism, First Department of Pediatrics, Aghia Sophia Children's Hospital, Medical School, National and Kapodistrian University of Athens, 11527 Athens, Greece; katherina.ganten@gmail.com
* Correspondence: chriskan@med.uoa.gr

Abstract: It has been established, worldwide, that non-communicable diseases such as obesity, diabetes, metabolic syndrome, and cardiovascular events account for a high percentage of morbidity and mortality in contemporary societies. Several modifiable risk factors, such as sedentary activities, sleep deprivation, smoking, and unhealthy dietary habits have contributed to this increase. Healthy nutrition in terms of adherence to the Mediterranean diet (MD), rich in fruits, legumes, vegetables, olive oil, herbs, spices, and high fiber intake may contribute to the decrease in this pandemic. The beneficial effects of the MD can be mainly attributed to its numerous components rich in anti-inflammatory and antioxidant properties. Moreover, the MD may further contribute to the improvement of reproductive health, modify the risk for neurodegenerative diseases, and protect against depression and psychosocial maladjustment. There is also evidence highlighting the impact of healthy nutrition in female people on the composition of the gut microbiota and future metabolic and overall health of their offspring. It is therefore important to highlight the beneficial effects of the MD on metabolic, reproductive, and mental health, while shaping the overall health of future generations. The beneficial effects of MD can be further enhanced by increased physical activity in the context of a well-balanced healthy lifestyle.

Keywords: Mediterranean diet; antioxidation; metabolic health; reproductive health; gut microbiota; non-communicable diseases; flavonoids; polyphenols; resveratrol; olive oil

Citation: Gantenbein, K.V.; Kanaka-Gantenbein, C. Mediterranean Diet as an Antioxidant: The Impact on Metabolic Health and Overall Wellbeing. *Nutrients* **2021**, *13*, 1951. https://doi.org/10.3390/nu13061951

Academic Editor: Panagiota Mitrou

Received: 17 April 2021
Accepted: 31 May 2021
Published: 6 June 2021

Publisher's Note: MDPI stays neutral with regard to jurisdictional claims in published maps and institutional affiliations.

1. The Burden of Non-Communicable Diseases

In recent decades, there has been a striking rise in the incidence of non-communicable diseases such as metabolic syndrome, obesity, type 2 diabetes, and non-alcoholic fatty liver disease in modern westernized societies and several countries under development, all predisposing to increased cardiovascular risk. The main factors contributing to the emergence of these diseases are related to people's diet and behavior, such as the increase in food rich in saturated fat and carbohydrates, including sweetened beverages, lack of physical exercise and smoking. All these factors lead to obesity, hypertension, hyperlipidemia, and insulin resistance [1].

According to the World Health Organization (WHO) non-communicable diseases such as cardiovascular diseases, cancer, respiratory diseases, and diabetes account for 70% of all deaths worldwide. Among the six designated WHO regions, the WHO European region is most severely affected by non-communicable diseases [1].

Several modifiable and non-modifiable factors account for this phenomenon. However, the human genome has not been significantly modified during these last decades and there is growing evidence that epigenetic alterations, mainly due to both intrauterine but also extrauterine insults, may significantly contribute to this increase. Moreover, Western societies have moved from everyday physical activity for walking, jogging, and outdoor leisure to mostly indoor activities, progressively inducing a change in lifestyle with more time spent on sedentary activities. Even the mobility of the modern citizen has shifted from walking, biking, jogging, etc., to sedentary ways of transportation such as driving or using public transport. Furthermore, the widespread use of technology has enabled human communication through all forms of telecommunication, reducing thus the need for face-to-face interaction or meeting outdoors.

This lifestyle change was already established before the outbreak of the COVID-19 pandemic. However, during this pandemic, screen time has experienced a significant increase, since it has become evident that technology has a major contribution in all forms of everyday life activities and has significantly modified many aspects of human behavior. For example, technology has enabled medical consultation through telemedicine, entertainment through all forms of screen time, such as Playstation-based or other technology-based games, f.ex. serious games, or even time spent to watch a ballet or concert or theater action at home. Even the learning process during the lockdown at all levels of education, i.e., primary, in kindergartens and primary schools, secondary, i.e., gymnasium or lyceum or even tertiary level such as university activities has shifted from the classic well-known learning process through teaching in classes, or in amphitheaters in higher education, to a screen-based teaching/educational process. Overall, there has therefore been a significant reduction in the time spent on physical activity and a huge increase in screen time and overall sedentary activities. Therefore, due to this lack of physical activity that the lockdown strategy has imposed during this very recent pandemic, it is of upmost importance to highlight other modifiable factors that may contribute to overall health, such as healthy nutrition.

In addition to the tremendous lack of physical activity during recent years, there is growing evidence that other modifiable risk factors such as smoking, but most importantly, modern nutrition, play a crucial role in the increasing incidence of these non-communicable diseases such as obesity, type 2 diabetes, metabolic syndrome, or non-alcoholic fatty liver disease development, across a wide range of age groups.

Western type diet is characterized by big meals rich in fat and carbohydrate content with a low content in fibers, while sweet beverages significantly contribute to the obesogenic environment, facilitating the development of visceral obesity and insulin resistance. On the contrary, dietary fibers, which have several anti-inflammatory and anti-proliferative actions, are not often represented in the Western diet. When ingested, dietary fibers are not digested in the gastrointestinal tract, but they are fermented by the gut microbiota, leading to the generation of short-chain fatty acids (SCFAs) with anti-inflammatory and anti-aging actions. Furthermore, dietary fibers positively shape the composition of gut microbiota and immunity [2,3].

2. The Benefits of the Mediterranean Diet

The important role of a healthy nutrition to combat insulin resistance, obesity, and their deleterious consequences, such as non-alcoholic fatty liver disease (NAFLD), polycystic ovarian syndrome (PCOS), sleep apnea, type 2 Diabetes (T2D) and cardiovascular events, all through the same pathogenetic mechanism of insulin resistance, has become evident through the beneficial effect of the so-called Mediterranean diet (MD). The MD, originally the typical Cretan diet, is mainly a plant-based diet, rich in fruits, vegetables, legumes, and nuts, in association with a moderate consumption of fish and dairy products and a low consumption of red meat and red wine. In addition, herbs, teas, and spices are also highly represented in the MD [4–6].

The MD mainly consists of carbohydrates, proteins, and fibers, while it is low in fat content. The most important source of fat in the MD is olive oil, which contains mainly unsaturated fatty acids. As Calabrese et al. mentioned, olive oil also contains 3,3-dimethyl-1-butanol, which prevents the formation of trimethylamine-1-oxide. High levels of trimethylamine-1-oxide are associated with an increased likelihood of cardiovascular events [3]. Furthermore, the consumption of fiber-rich food alters the gut microbiome and enriches the microbiome diversity, which is important for the immune system, possessing anti-inflammatory capacities [7,8]. According to Anderson et al., a diet rich in fibers, such as the MD, is associated with a lower prevalence of cardiovascular diseases, diabetes, metabolic syndrome, and gastrointestinal diseases like gastroesophageal reflux disease, gastric ulcer or diverticulitis [9].

These foods which have beneficial health effects have been called, in the recent literature, functional foods, i.e., foods that have a great contribution in overall health. They contain biologically active nutrients, such as polyphenols which have a significant impact on the prevention and management of chronic non-communicable diseases due to their beneficial anti-oxidative, anti-bacterial or anti-inflammatory effects [10]. Systemic oxidative stress is a hallmark of obesity and metabolic syndrome. Oxidative stress at the cellular level originates from an imbalance between endogenous reactive oxygen species (ROS) production and the natural anti-oxidation system [4]. The molecular structure of polyphenols, important in counteracting oxidative stress, consists of at least two aromatic rings with a hydroxyl group and a carbon bridge between the aromatic rings. They are classified in two groups, the flavonoids, including flavonols, flavones, isoflavones, flavanones, flavan-3-ols, anthocyanidins and dihydrochalcones and the nonflavonoids including phenolic acid, lignans and stilbenes [11].

Moreover, the MD consists of many components rich in mono-unsaturated fatty acid such as oleic acid, in olive oil, omega-3-polyunsaturated fatty acid, such as alpha- linolenic acid in tree nuts, like walnuts, high amounts of flavonoids and antioxidants found in fruits and vegetables and high amounts of dietary fibers that have a great impact on the composition of gut microbiota—all considered to have health-promoting properties and to enhance longevity [12,13]. As already reported above, the significant lack of physical activity that emerged during the lockdown strategy to combat the recent COVID-19 pandemic should be counteracted by a healthier nutrition, such as the MD, Therefore, a review article highlighting the many beneficial effects of MD is especially relevant during this period that is characterized by this shift towards sedentary activities.

3. The Impact of Mediterranean Diet on Metabolic Health

3.1. The Impact on Obesity Prevention and Management

The increasing prevalence of obesity in westernized societies has become a major public health problem, reaching epidemic dimensions in both adult but also adolescent populations [14,15]. Obesity increase is multifactorial, being the result of lifestyle modifications such as a shift towards sedentary activity and a change from the MD towards a high-fat, carbohydrate-rich Western diet. As highlighted above, oxidative stress is crucially implicated in the chronic inflammation process that is an important component in obesity pathogenesis. Therefore, this chronic inflammation state contributes both to the increased rates of obesity and its complications, such as metabolic syndrome, nonalcoholic fatty liver disease, dyslipidemia, diabetes, all conferring an increased cardiovascular risk [4,14,15]

Therefore, obesity is generally considered as a low grade inflammation state affecting the whole body [4]. An increased intake of high energy food rich in animal-saturated fat leads to oxidative stress at the cellular level and inflammation by increasing the expression of nicotinamide adenine dinucleuotide phosphate (NADPH) oxidase (NOX) in adipocytes. This triggers adipogenesis and the production of reactive oxygen species. Through the activation of the nuclear factor-kappa B (NF-κB) pathway, it results in an increased expression of the proinflammatory mediators such as tumor necrosis factor alpha (TNFα), interleukin 6 (IL6), leptin while decreasing the secretion of the protective adipokine, adiponectin [4,16]. The increased expression of proinflammatory cytokines leads to downregulation of the anti-inflammatory molecule, AMP-activated protein kinase (AMPK), and the induction of the acute-phase-protein C-reactive protein (CRP) [4].

It has been shown that polyphenols have an anti-inflammatory function. They inhibit proinflammatory molecules and modulate the inflammatory pathways like NF-κB, MAPK and the arachidonic acid pathway [17]. As Li et al. [18] proved, the administration of the flavonoid hesperidin can reduce the secretion of cytokines like interleukin 1 (Il-1) and TNFα and restricts the inflammation in a murine rheumatic arthritis model [18]. Spices are a rich source of polyphenols as well and have anti-inflammatory, anti-carcinogenic and antioxidant capacities. Curcumin, for example, is known for its protective properties against DNA-mutations [19]. Special interest has been laid to resveratrol, a component of grapes, that has been shown to promote intracellular glucose transport and reduce insulin secretion. It has also been shown to inhibit the NADPH-induced oxidative stress and activate the anti-inflammatory molecule peroxisome proliferator-activated receptor gamma (PPARγ) [4]. Moreover, the ingestion of herbal components such as green tea, rich in polyphenols such as epigallocatechin-3-O-gallate (EGCG), contributes to diabetes management and has a lipid-lowering effect, while increasing energy expenditure and decreasing inflammation and oxidative stress through the inhibition of the NF-κB pathway [4]. Furthermore, coffee consumption has been shown to reduce the risk of several diseases such as diabetes mellitus and non-alcoholic fatty liver disease [20]. The mechanism behind this observation could be explained by the polyphenols contained in coffee such as caffeoylquinic acids, trigonelline and N-methylpyridinium. These lead to the upregulation of antioxidant and detoxifying enzymes and the downregulation of proinflammatory mediators via the activation of nuclear factor erythroid 2-related factor 2 (Nrf2) [20]. Obesity is also related with changes in lipid profile [14] with a reduction in high density lipoprotein (HDL) and an increase in triglycerides. Increased visceral adipose tissue is further associated with oxidative stress and it has been shown that telomeres are shorter especially in obesity that is accompanied by metabolic abnormalities leading to metabolic syndrome [21]. Faster shortening of the telomeres and an inhibition of the telomerase activity has therefore been observed in unhealthy obesity [22,23]. In both "healthy" and "unhealthy" obesity, a strategy aiming to reduce oxidative stress at the cellular level can modify morbidity and promote overall health. MD can thus be considered as a natural antioxidant, able to reduce mortality, by reducing the incidence of cardiovascular, metabolic, endocrine, and neurodegenerative diseases through the beneficial effects of its components, mainly polyphenols [24].

Therefore, the effects of MD on the prevention of obesity have simultaneously protective effects on its complications, and it is difficult to discern the specific effects of the different components of the MD on the different expression patterns of inflammation and whether the reduction in insulin resistance may rather contribute to the reduction in metabolic syndrome or merely of type 2 Diabetes, or of nonalcoholic fatty liver disease, since all these non-communicable diseases listed above are interrelated, with many common pathways [4].

3.2. The Impact on Diabetes Mellitus

Diabetes mellitus, especially Type 2 Diabetes mellitus (T2DM), is one of the leading diseases in Western societies, in addition to cardiovascular diseases. One of the reasons for the increase in the incidence of diabetes is the high-fat, high-carbohydrate, and low-fiber diet. Consequently, the MD is an important component in the prevention and combat of its pathogenesis. According to Esposito et al., the adherence to a MD has beneficial effects in both the prevention and treatment of diabetes. It has been demonstrated that people following MD had lower hemoglobin A1c (HbA1c) levels compared to the control group. Furthermore, it was observed that fasting glucose could also be reduced by the MD [25]. However, it was not investigated which components of the MD are responsible for this observation. One example could be the fat content. In the MD, more unsaturated fatty acids are consumed, especially as a component of olive oil or table olives, and less saturated fatty acids from animal fat. This has an inhibitory effect on the pathogenesis of diabetes mellitus. Increased monounsaturated fatty acids and reduced saturated fatty acids in the diet have been associated with increased insulin sensitivity and improved beta cell function [26,27]. In addition, the unsaturated fatty acids of olive oil have been shown to increase glucagon-like-peptide-1-(GLP1) expression [28]. GLP1 stimulates insulin secretion and inhibits glucagon secretion. An increased GLP1 expression is also achieved by the SCFAs. As MD is rich in dietary fiber, and the fermentation of this fiber leads to the formation of

SCFAs, these then, in turn, stimulate GLP1 expression [26]. Further consideration of the beneficial effects of MD in the pathogenesis of diabetes mellitus could be the widespread use of spices and herbs, also important components of the MD. Zare et al. have been able to show through a randomized control trial (RCT) that cinnamon can improve the glycemic status in patients with diabetes, especially in those with a higher body mass index (BMI) [29]. Cinnamaldehyde, the main component of cinnamon, has also been shown to have lipid-lowering capacities [30]. Furthermore, tea consumption, that is also highly represented in the MD, has health-promoting benefits in the combat against diabetes. Thus, according to Alkhatib et al., the tea has beneficial effects on the pathogenesis of diabetes mellitus by improving glucose metabolism and consequently lowering the fasting glucose level. Moreover, further studies have demonstrated that the consumption of tea is associated with a reduced waist-to-hip ratio and lower levels of blood pressure [10]. The pathomechanism behind the positive effects of these products is still unclear. It could be due to their rich content in polyphenols [4]. There is evidence that glucose uptake is prevented by the inhibition of the enzymes α-amylase and α-glycosidase. Glucose transport by sodium–glucose-linked transporter 1 (SGLT1) could also act as a target for the polyphenols. In addition, there is evidence that polyphenols increase glucose uptake in muscle cells, reduce gluconeogenesis in the liver, and prevent the inflammation-induced destruction of the pancreatic β-cells [31]. An example of these beneficial molecules is the polyphenol quercetin, which reduces the metabolic stress in mitochondria and acts like the widely used antidiabetic drug metformin [32]. Moreover, it can increase glucose uptake through GLUT4 transporter [33]. Another example is rosemary. According to Naimi et al. [34], rosemary contains a large amount of polyphenols, especially carnosic acid, rosmarinic acid and carnosol. These molecules show important antidiabetic effects. They inhibit a-glucosidase and thus intestinal glucose uptake. They also inhibit dipeptidyl peptidase 4, which inactivates the enzyme GLP1. In addition, they inhibit hepatic gluconeogenesis and increase glucose uptake in muscle cells, thus counteracting hyperglycemia. The beneficial properties of rosemary and its polyphenol components have also been confirmed in human studies. As shown in the review by Naimi et al. [34], rosemary also has lipid-lowering and anti-inflammatory effects [34].

3.3. The Impact of MD on Non-Alcoholic Fatty Liver Disease

Non-alcoholic fatty liver disease (NAFLD) is one of the leading non-communicable diseases affecting around 20–30% of the general population in Western societies. This increase in incidence occurs in parallel with the increasing incidence of obesity and insulin resistance [35]. According to the "multiple hits" hypothesis, oxidative stress, hypovitaminosis E, low-grade inflammation, and gut microbiota dysbiosis, all contribute to the pathogenesis of NAFLD [36]. A nutraceutical approach in the prevention and management of the disease is the adherence to the MD. MD, rich in nuts with antioxidant properties, may contribute to the decrease in the burden of NAFLD. Although the exact role of the different components of the MD on the prevention and management of NAFLD is not fully elucidated, it became clear that many components have lipid-lowering properties and reduce insulin resistance at the hepatic level, while they have anti-inflammatory and antioxidant properties, further reducing endoplasmic stress. In a randomized control trial (RCT), Yaskolka et al. were able to demonstrate that an MD enriched in green plants with reduced red meat intake was able to enhance intrahepatic fat loss in comparison to the control diet, proving the beneficial effects of MD on reducing hepatic steatosis [37]. Furthermore, in another RCT, Franco et al. were able to demonstrate that a low-glycemic index MD in combination with physical activity was able to reduce the NAFLD score, highlighting the importance of combined lifestyle modifications, including both healthy diet and exercise [38].

3.4. The Impact of MD on the Metabolic Syndrome

Metabolic syndrome (MetS) is defined as the simultaneous occurrence of abdominal obesity and at least two of the following: dyslipidemia, mainly characterized by low HDL levels and high triglycerides, and hyperglycemia and/or arterial hypertension. MetS can be further accompanied by hyperuricemia [39], NAFLD, as well as microalbuminuria [40,41]. The presence of the MetS confers a higher risk of cardiovascular events than the addition of these risk factors separately. Furthermore, the MetS is associated with numerous other complications as well as with an increased risk of neoplasia [42,43]. Nowadays, the incidence of MetS is steadily increasing, mainly as a cause of modern lifestyle with increasing obesity rates, and the combination of increased sedentary activities and non-adherence to healthy nutrition. Pathophysiologically, the excess of adipocytes leads to a constant inflammatory state since adipocytes secrete numerous proinflammatory cytokines. This inflammatory state is characterized by the ectopic deposition of lipids, which leads to insulin resistance in muscle cells, liver steatosis with altered metabolism in the liver and beta-cell dysfunction in pancreatic cells and consequently decreased insulin secretory activity that progressively cannot compensate for the increased insulin demands necessary to counteract the peripheral insulin resistance [24].

Kastorini et al. investigated the effects of the MD on metabolism through a meta-analysis. They found that adherence to the MD has a positive effect on metabolism. MD is associated with a lower rate of MetS and its components such as abdominal obesity, low HDL levels and hypertension [44]. The inflammatory state could therefore be a target for the treatment of metabolic syndrome. As already mentioned, the MD has anti-inflammatory effects. An example for these beneficial effects is the tomato, which plays an essential role in the MD. According to Ghavipour et al., the tomato has anti-inflammatory and antioxidant effects [45]. Tomato consumption has been associated with reduced body weight [46]. Olive oil, in turn, as already reported above, can also suppress the inflammatory response. The inflammatory reaction can be further attenuated by the polyphenol quercetin since it induces the secretion of the anti-inflammatory factor, adiponectin, in adipocytes. This leads to a reduction in body weight in obese mice in vivo. Furthermore, it was shown that the administration of quercetin improves the components of the MetS such as dyslipidemia, hypertension, and insulin resistance [47]. In addition, the polyphenol resveratrol was able to inhibit adipocyte proliferation [48,49].

4. The Impact on Cardiovascular Disease

Cardiovascular events are mainly due to atherosclerosis and intravascular plaques formation. These plaques can either lead to vascular stenosis and occlusion or break off with the blood flow inducing the formation of a thrombus. Both events subsequently lead to tissue blood underperfusion, tissue ischemia and final necrosis. Cardiovascular disease (CVD) is one of the leading causes of mortality in Western societies. Predisposing factors can be divided into the non-modifiable ones, such as positive family history, age, or male gender, and the modifiable ones, such as smoking, lack of physical exercise and unhealthy diet [50].

An important parameter responsible for the aggravation of atherosclerosis is hyperlipidemia with the combination of high low density lipoprotein (LDL) and low HDL levels. In addition to the high LDL levels, conferring an atherosclerotic cardiovascular risk, it has recently become evident that the susceptibility of LDL particles to aggregate constitutes a further prothrombotic factor and a risk factor for future cardiovascular death. In a very elegant study using data from the Healthy Nordic diet group, Ruuth et al. were able to demonstrate that decreased aggregation susceptibility could be achieved through a diet rich in vitamin E, such as the MD [51]. Furthermore, a study on the effect of a short-term adherence to MD versus fast food (FF) diet was able to achieve a change of HDL lipidome towards a healthier HDL status, further highlighting the cardioprotective effect of MD [52]. Moreover, in vitro and in vivo studies have shown that the consumption of garlic can improve the lipid status, by lowering the level of LDL and increasing the level of HDL. This is achieved by inhibiting enzymes in the pathway of cholesterol synthesis, for example, monooxygenase and 3-hydroxy-3-methyl-glutaryl-coenzyme A reductase (HMG-CoA-reductase). Furthermore, this lowers the blood pressure and inhibits platelet aggregation by inhibiting cyclooxygenase (COX) activity [53]. Thus, the complications of atherosclerosis, such as heart attack or stroke, can be prevented by garlic consumption. An antihypertensive and cardioprotective effect has also been associated with the consumption of olive oil [54]. A diet rich in olive oil consumption, especially of extra-virgin olive oil, allows the intake of phytosterols and polyphenols that have a protective function on the endothelium preventing arteriosclerosis and cardiovascular events. In addition, it increases the expression of anti-inflammatory molecules like PPARγ and reduces the expression of proinflammatory molecules such as IL1β and cyclooxygenase (COX) 2 [4]. The unsaturated fatty acids that are highly represented in the MD and especially in the extra virgin olive oil have been demonstrated, according to the PREDIMED Trial, to contribute to diminished ceramide accumulation and therefore to confer a cardioprotective effect. It is well known that aberrant accumulation of ceramides, that frequently occurs because of excess saturated fat consumption, may lead to impaired cellular function, including impaired insulin function [55], while the MD has a favorable impact on CVD by diminishing ceramide concentration. As mentioned previously, MD contains many biologically active molecules called polyphenols with anti-inflammatory and antioxidant effects [19]. These molecules have also antiplatelet activity [56,57]. Their entire mechanism of action is still unclear [58]. Studies showed that polyphenols, such as oleuropein found in olive oil, reduce oxidative stress, and inhibit the COX1 pathway, which is involved in platelet aggregation [59]. In vitro and in vivo studies showed that polyphenols such as chlorogenic acid or the anthocyanin cyanidin-3-glucoside interrupt thrombin-induced platelet aggregation [60,61]. Another described mechanism of action of polyphenols is the increase in nitric monoxide (NO) levels after consumption of quercetin and catechin [62]. NO has a vasodilating effect and inhibits platelet aggregation [63]. Resveratrol, a polyphenoid that is present in berries, grapes, and grape products such as red wine, has also been proven to have protective effects on the cardiovascular system. It reduces blood pressure, has antiapoptotic effects on cardiomyocytes and prevents thrombus formation [64]. An important complication of hypertension and atherosclerosis is myocardial infarction, in which the occlusion of a blood vessel leads to the reduced perfusion of the myocardium. This leads to remodeling mechanisms that affect both the structure and viability of the myocardium. Nowadays, attempts are being made to prevent these remodeling mechanisms by administering angiotensin converting enzyme (ACE) inhibitors. However, it would be interesting to have a similar effect through natural substances, the so-called neutraceuticals. According to Suzuki et al., catechins, the phenol component of tea, protect against remodeling after myocardial infarction by suppressing proinflammatory cytokine secretion [65]. This anti-inflammatory effect has also been demonstrated in a murine autoimmune myocarditis model, where catechin administration was shown to improve ventricular contractility [66].

5. The Impact of MD on Reproductive Health

Healthy nutrition is also important in other areas of overall health, such as reproductive health. It is well known that obesity and its complications are associated with reduced fertility in both women and men [67,68]. Therefore, a healthy low-fat diet such as the MD may have important beneficial effects on reproductive capacity [69]. According to a study by Einarsson et al., overweight women who lost weight before insemination had an increased spontaneous pregnancy rate compared to overweight women who did not lose weight. However, in the group of overweight women who underwent assisted reproduction technologies (ARTs), such as in vitro fertilization (IVF), there was no significant difference in pregnancy rates between women who achieved weight loss and the control group [70]. Therefore, it is important to have a healthy lifestyle from an early age so that obesity can be prevented long before issues such as infertility arise.

5.1. The Impact of MD on Female Reproductive Health

Polycystic ovarian syndrome (PCOS) also belongs to the group of metabolic diseases. PCOS is considered as the expression of the metabolic syndrome and insulin resistance at the reproductive axis of the female, although its exact pathomechanism is still unclear. However, obesity and insulin resistance are the main components of its pathogenesis [71]. Women with PCOS have an increased risk of developing type 2 diabetes mellitus. The definition of PCOS includes the fulfillment of the following criteria: hyperandrogenism, either clinical or biochemical, chronic anovulation, expressed as menstrual irregularities, and polycystic appearance of the ovaries on pelvic ultrasound in around 50% of the cases [72]. It is claimed that low-grade inflammation is present in PCOS, which drives the clinical picture [73]. Another parameter that is important in the pathogenesis of PCOS is the presence of advanced glycation end products (AGEs). These substances are either endogenously produced or exogenously supplied through food ingestion [74]. They are mainly the result of the ingestion of high sugar diet and may also be found in animal fat- and protein-rich products and less represented in fruits and vegetables [75]. High AGE levels are associated with increased insulin resistance, that in turn, drives sex-hormone binding globulin (SHBG) reduction and androgen excess. AGEs bind to their receptors (receptor of advanced glycation end products, RAGE), which are also expressed in the ovarian tissue. By binding to RAGE, they induce an inflammatory response that stimulates oxidative stress. AGEs thus accumulate in the ovarian tissue, leading to ovarian dysfunction [74]. Considering all these observations, a dietary modification with a reduced intake of AGEs would be beneficial in PCOS pathology.

Barrea et al. studied how diet differs between women with PCOS and healthy women. They found that women with PCOS follow unhealthier diets with the ingestion of simple carbohydrates, saturated fat, and little fiber content. The constellation of these diets contributes to the inflammatory state and induces oxidative stress [76]. The oxidative stress, in turn, stimulates androgen synthesis in the ovarian tissue. This leads to an increased inflammatory response, so that a vicious circle occurs. In contrast, a healthy high fiber diet with mainly unsaturated and few saturated fatty acids can reduce glucosemia and lipidemia and thus limit the inflammatory response [77]. This oxidative stress can be counteracted through the numerous anti-inflammatory substances contained in healthy diets, as the MD. As Amini et al. have described, polyunsaturated fatty acids have positive effects on insulin sensitivity and can limit hyperandrogenism [78]. Furthermore, Zhang et al. used an Lipopolysaccharide (LPS)-induced sepsis model to investigate the effect of the ω3 polyunsaturated fatty acid-derived metabolite resolvin. It was observed that resolvin leads to a downregulation of proinflammatory mediators and thus to a limitation of the inflammatory response [79]. The lipid-lowering and anti-inflammatory effects of ω3 fatty acids can hence be used as a treatment method in PCOS patients. These molecules are found in fish and olive oil, nutrients abundantly consumed in MD [78]. In addition to vegetables, fish and olive oil, herbs and spices are also frequently used in MD. According to Ashkar et al., these products also have beneficial effects in limiting PCOS pathology. Studies have shown that cinnamon, for example, reduces insulin resistance and fasting insulin levels. Green tea stimulates metabolism and has a lipolytic effect. Chamomile has positive effects on weight control and lowers insulin levels. In murine PCOS models, green tea and chamomile were shown to reduce the number of ovarian cysts observed in pelvic ultrasound. Mint and licorice counteract hirsutism, a common sign of PCOS due to hyperandrogenism. Fennel lowers testosterone levels. Finally, Marrubium vulgare lowers cholesterol, glucose, and oxidative stress [80].

Furthermore, in PCOS, insulin has a stimulatory and anti-apoptotic effect on the theca cells, resulting in the hyperplasia of these cells [81]. According to Wong, the polyphenol resveratrol has proapoptotic properties on the theca cells and can thus antagonize the insulin effect [82]. In addition, resveratrol has been shown to decrease the level of proinflammatory cytokines, such as Interleukin 6, Interleukin 18, TNFα and the acute-phase-protein CRP, which are all markers of the inflammatory state in PCOS. Furthermore,

Brenjian et al. described that resveratrol can modulate the stress of the endoplasmic reticulum, also involved in the pathogenesis of PCOS [83]. In addition, this leads to an induction of the expression of SIRT1, which also seems to have protective effects against oxidative stress in the ovarian tissue [84]. It has also been shown that resveratrol decelerates ovarian ageing in rodent experiments and thus increases fertility [85]. A further polyphenol with beneficial properties in the pathogenesis of PCOS seems to be quercetin. It is known that PCOS patients have lower levels of adiponectin and adiponectin receptors. As widely known, adiponectin is beneficial for metabolism as it improves glucose homeostasis and has lipid-lowering effects. Rezvan et al. investigated how the expression levels of adiponectin and its receptors change after quercetin supplementation in PCOS patients. They found that quercetin induces an upregulation of these proteins and could therefore be used as a treatment for PCOS [86,87].

5.2. The Impact on Male Reproductive Health

Male fertility is also influenced by lifestyle. Oxidative stress, for example, can contribute to male infertility. A high amount of reactive oxygen species is associated with reduced sperm quality, vitality, and concentration as well as DNA damage in spermatocytes. Oxidative stress is induced by both internal and external factors. Internal factors can be the result of cryptorchidism or testicular inflammation. External factors that lead to enhanced oxidative stress include obesity, smoking and an unhealthy diet [88]. According to Salas-Huetos et al., oxidative stress can be reduced by the antioxidant properties of fruits and vegetables, since these antioxidants act as scavengers and reduce reactive oxygen species [89,90]. Another parameter associated with impaired male fertility is the presence of xenoestrogens, which are found nowadays in increased amounts in meat products [90]. Xenoestrogents are shown to reduce the ejaculate volume, the sperm motility and vitality [91]. Furthermore, it has been observed that increased consumption of trans lipids is associated with reduced sperm quality and lower testosterone levels. In contrast, the polyunsaturated fatty acids found in fish and olive oil or nuts are associated with better sperm quality and morphology [92]. The effects of dyslipidemia on fertility were studied in several animal experimental models. Saez Lancellotti et al. studied how a diet high in cholesterol alters sperm quality. The results showed that the high cholesterol diet led to the formation of abnormal sperm and reduced sperm motility [93]. These observations were significantly reversed after olive oil supplementation [94]. Altogether, these observations support the notion that adherence to the MD would be beneficial in male reproductive issues. The mechanism behind this is most likely due to the polyphenols. According to Simas et al. the polyphenol resveratrol improves male fertility, by ameliorating sperm mobility and functionality. It has been further demonstrated that the diabetes complications on the reproductive system such as DNA fragmentation and disturbed sperm acrosome integrity can be reduced by the ingestion of resveratrol [95]. Resveratrol can also improve male fertility rates in the procedure of cryoconservation since it protects the sperm from oxidative stress [96]. The polyphenols, mainly catechins, contained in green tea also have a protective effect on sperm quality. They protect against spontaneous mutations and chromosomal aberrations. According to Rahman et al., polyphenols may therefore improve fertility through their anti-inflammatory and antioxidant effects [88].

6. Transgenerational Effects on the Offspring

Healthy diet already before pregnancy, but mainly during pregnancy and lactation, has a major impact on the metabolic and overall health of the offspring [97–99]. As described in the review by Amati et al., maternal nutrition before and during pregnancy has a major impact on fetal health greatly contributing to the gut microbiota diversity of the offspring [97,100]. It was observed that children whose mothers followed MD had a lower risk of developing congenital heart defects. Moreover, intrauterine development is also positively influenced by the MD [97]. Following an MD during pregnancy is associated with a lower rate of small for gestational age (SGA) neonates [98]. In addition, newborns whose mothers followed MD during pregnancy showed a lower probability of developing neural tube defects. Furthermore, it has been demonstrated that maternal diet during pregnancy also affects the rate of congenital malformations, like gastroschisis [99]. Adherence to the MD is further associated with a reduced likelihood of gestational diabetes and hypertensive pregnancy diseases [100]. In contrast, the risk of developing gestational diabetes is increased in overweight women, which also entails many complications for the newborn, including both adverse metabolic health [100] and future neurodevelopment [99,101]. Furthermore, the high body weight of the mother is associated with lower fertilization rate and miscarriage [99]. Moreover, the healthy diet of a pregnant woman rich in dietary fiber shapes the gut microbiota of her offspring [100]. The gut microbiota composition is further affected by the diet of the newborn, since babies fed by infant formulas have a higher number of pathogenic bacteria in comparison to breast-fed babies [102]. However, the positive effects of MD are not limited to the pregnancy and early postnatal period. It has been shown that children whose mothers had a high adherence to the MD are less likely to develop atopic diseases and behavioral problems [97].

7. Impact on Autoimmune Disease

Autoimmune diseases are another type of diseases for which a positive impact can be achieved through dietary changes. As an example, we focus on Type 1 diabetes mellitus (T1DM), an autoimmune disease where the autoimmune destruction of the pancreatic beta cells take place, so that insulin synthesis and secretion progressively diminish, up to complete exhaustion. The disease is mainly manifested in early childhood but can also be diagnosed in later ages, and if not appropriately managed through the years, is associated with severe complications, especially micro- and macrovascular events. The pathogenesis of T1DM has not been fully elucidated yet. It is considered that a complex pathogenesis is responsible for its occurrence, notably the combination of inherent genetic predisposing factors with external environmental triggers such as viral infections or dietary components contributing to the autoimmune pathogenesis of the insulitis [103]. Studies have shown that patients with T1DM have an altered intestinal flora compared to healthy individuals [104]. It can begin in infancy or even earlier in life, where it has been shown that breastfed children of mothers who eat a low meat diet have a lower risk of developing T1DM [105]. A high fat diet leads to the colonization of the intestine of mainly Bacteroides species. In contrast, a high fiber diet induces a microbiome rich in Prevotella species [106]. A gut microbiome rich in Bacteroides is correlated with a T1DM-associated intestinal dysbiosis. The type of dominant bacterial species has an influence on the synthesized SCFAs. Beneficial SCFAs, such as butyrates, have immunomodulatory effects and suppress inflammatory responses. In contrast, Bacteroides-dominant gut flora results in the formation of proprionate, succinate and acetate, which have proinflammatory effects and promote autoimmunity by increasing the epithelial permeability of the gut [102]. As already described, this intestinal flora is strongly influenced by nutrition. According to Yang et al., the consumption of a polyphenol-rich diet, such as the MD, increases the amount of beneficial bacterial species and suppresses the growth of pathogenic bacterial species [107].

There is also evidence for the positive effects of the MD in other autoimmune diseases. One example is multiple sclerosis, an autoimmunological disease characterized by immune cell recruitment, mainly T cells, in the central nervous system, and subsequent demyelination. It has been observed that the consumption of saturated fatty acids and especially long-chain fatty acids is associated with an aggravation of the symptomatology and an increase in immune cell recruitment [108]. In contrast, SCFAs have been shown to have an immunomodulatory effect [109]. It has been further suggested that flavonoids have a remyelination-promoting effect [110,111].

8. Impact on Neurodegeneration and Mental Health

8.1. Polyphenols against Neurodegeneration

Unhealthy diets also have effects on neurodegeneration. The high-fat, low-fiber Western diet has been associated with an increased risk of Parkinson's disease. The lack of dietary fiber leads to an excess of lipopolysaccharide-producing bacteria in the microbiome, resulting in damage to the intestinal barrier and mitochondrial function [112]. Furthermore, fats are thought to increase the permeability of the blood–brain barrier and increase the quantity of β-amyloid plaques deposition. Polyphenols instead lead to an improvement in cognition, which can be shown by both cognitive tests and functional MRI (fMRI) studies. According to Rajaram et al., the consumption of fruits, green tea, and nuts, which are all important components of the Mediterranean diet, has neuroprotective effects [113]. These effects could be explained by the antioxidant effect of polyphenols [114]. Hesperidin, for example, which is found in citrus fruits, scavenges free oxygen radicals. Furthermore, it suppresses the formation of proinflammatory cytokines and thus also has an anti-inflammatory effect. In addition, according to several studies, it has been shown that hesperidin increases cerebral blood flow, which further prevents cognitive decline. The neuroprotective effect of hesperidin was observed in both animal experiments and in human studies [115]. Hesperidin is just one example of the multitude of polyphenols contained in the MD with neuroprotective properties. Other bioactive compounds are anthocyanins and carotenoids, which are abundantly found in fruits and vegetables. These molecules reduce oxidative stress, by lowering the number of free radicals. Further studies have shown that anthocyanins and the compound of olive oil, oleacanthal, prevent the aggregation of Aβ amyloids and tau protein [116]. Isocyanates found in cruciferous vegetables and oleacanthal act as natural COX inhibitors and thus limit inflammatory responses [117–119]. These findings are based both on in vitro and in vivo studies. The research group of Yammine et al. investigated the effect of phenols on murine N2 neuronal cells after treatment with 7-ketocholesterol. 7-ketocholesterol is an oxidation product that is found in high concentration in neurodegenerative diseases. It promotes oxidative stress, apoptosis, autophagy, mitochondrial and peroxisomal dysfunction. These effects were reduced after the treatment of N2 cells with the polyphenols, resveratrol, quercetin and apigenin [120]. In summary, it can be said that polyphenols, and thus also the MD, have a preventive effect against neurodegenerative diseases such as Alzheimer's or Parkinson's disease.

8.2. The Impact on Mental Health

The current COVID-19 pandemic has exacerbated neuropsychological disorders, since the lockdown strategy led to significant lifestyle changes that were accompanied by increased social isolation, uncertainty for the future and unemployment. The main consequence of all these changes is depression [121]. Depression is nowadays one of the leading causes of disability [122]. Therefore, it is extremely important to find methods to relieve the symptoms, especially in a natural way, since antidepressant medication has often multiple side effects. One example could be healthy food intake. As already described, a healthy diet helps to combat the pathogenesis of many diseases. The same holds true for depression. As Riera-Sampol et al. have shown, both adherence to MD and a low BMI are beneficial against depression. These positive properties of MD on depression are probably due to nutrients such as unsaturated fatty acids and polyphenols [123]. In an RCT, Parletta et al. investigated how symptoms of depression change based on one's diet. They showed that the group that followed MD had less stress and negative emotions and scored better on depression scores. To determine the beneficial component of MD, they also studied how the amount of omega-3 and omega-6 polyunsaturated fatty acids was related to the clinical picture. They found that high omega-3 and low omega-6 polyunsaturated fatty acids were associated with a more favorable outcome [124]. The importance of n3-polyunsaturated fatty acids in the pathology of depression was also confirmed by further studies. In addition, comparing n3-polyunsaturated fatty acids to placebo as a treatment of depression also demonstrated the beneficial effects of n3-polyunsaturated fatty acids [125].

These results can probably be attributed to the anti-inflammatory and immunomodulatory properties of n3-polyunsaturated fatty acids since depression is associated with an inflammatory state [126]. Further substances that are associated with a reduced risk of developing depression are polyphenols [127]. The uptake of polyphenols mainly occurs through the intestinal flora, a process that creates an interaction between polyphenols and intestinal flora. Polyphenols regulate the composition of gut bacteria by stimulating the growth of beneficial bacteria and inhibiting the growth of harmful bacteria. The bacteria break down the polyphenols and produce polyphenol metabolites and polyphenol-related bacteria metabolites. SCFAs are an example of bacterial metabolites. The formation of SCFAs induces the secretion of hormones, including leptin, which showed beneficial effects on depressive symptoms in animal experiments [11]. Furthermore, polyphenols can have a positive impact on the homeostasis of neurotransmitters. In patients suffering from depression, a reduced level of dopamine and serotonin has been observed. According to Gu et al., the polyphenol resveratrol modulates this, by increasing the levels of both dopamine and serotonin in a dose-dependent manner. In addition, it increases the level of the neuroprotective molecules brain-derived nerve growth factor (BDNF) and neuropeptide Y (NPY), which are also found to be decreased in patients with depression [128].

9. Possible Adverse Effects of Mediterranean Diet

In addition to the beneficial effects of MD and the polyphenols it contains, adverse effects have also been described. One example is the polyphenol resveratrol. According to Shaito et al., there are indications that the anti-inflammatory effect is dose dependent. A high-dose administration of resveratrol in animal experiments showed proinflammatory and cytotoxic effects. In addition, pharmacokinetic interactions have been described, mainly due to CYP450 induction, which may affect the pharmacokinetics of other drugs [129]. EGCG present in tea has also been associated with adverse effects when administered at high concentrations. A hepatotoxic effect with an increase in hepatic transaminase has been reported in animal experiments as well as in humans. This led to a market withdrawal of high dose-EGCG-supplements [130]. In conclusion, the adverse health effects of polyphenols have mainly been described in the context of an excess supraphysiological intake that does not correspond to the natural diet. Thus, in a healthy balanced diet, the numerous beneficial effects outweigh by far the putative negative effects. In Figure 1, the numerous beneficial effects of MD and its components are illustrated, affecting most aspects of human life and health.

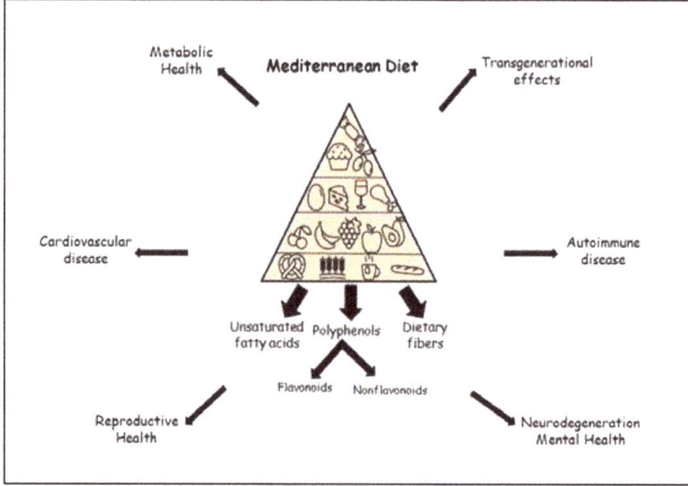

Figure 1. Beneficial effects of the Mediterranean diet.

10. Conclusions

It has been established, worldwide, that non-communicable diseases, such as obesity, diabetes, non-alcoholic fatty liver disease, metabolic syndrome, and cardiovascular events account for a high percentage of morbidity and mortality in modern westernized societies. Several modifiable factors, such as increase in sedentary activities, sleep deprivation, smoking, and unhealthy dietary habits have significantly contributed to this increase. Healthy nutrition in terms of adherence to the Mediterranean diet, rich in fruits, legumes, vegetables, olive oil, nuts, herbs, spices, and high fiber intake with a reduction in processed red meat intake may contribute to the decrease in this pandemic. Moreover, it seems that MD may further contribute to reproductive health in both men and women, since higher rates of subfertility have been noted nowadays in westernized societies and may further contribute to the overall health of future generations. Moreover, healthy nutrition may modify the risk for neurodegenerative diseases such as Alzheimer disease, or Parkinson's disease, or even protect against depression and psychosocial maladjustment. There is a lot of evidence highlighting the impact of healthy nutrition of the woman on the composition of gut microbiota and metabolic health of her offspring and may further protect from congenital malformations or even adverse neurodevelopment in the offspring. It is important to highlight all these beneficial effects of the MD and specifically the protective components of the MD such as polyphenols, poly-unsaturated fatty acids, flavonoids, terpenoids, all with anti-inflammatory and antioxidant properties, enhancing metabolic, reproductive, and mental health, and shaping the overall health of future generations. The beneficial effects of a healthy nutrition can be further enhanced by increased physical activity and the avoidance of sleep deprivation and excess psychosocial stress in the context of a healthy lifestyle modification.

Author Contributions: C.K.-G. conceptualized this review. K.V.G. searched the literature and drafted the manuscript. C.K.-G. revised and finalized the manuscript. All authors have read and agreed to the published version of the manuscript.

Funding: The research received no external funding.

Institutional Review Board Statement: Not applicable.

Informed Consent Statement: Not applicable.

Data Availability Statement: This is a review article, not presenting any unpublished data. All articles used are properly cited in the references.

Conflicts of Interest: The authors declare no conflict of interest.

References

1. Iriti, M.; Varoni, E.M.; Vitalini, S. Healthy Diets and Modifiable Risk Factors for Non-Communicable Diseases-The European Perspective. *Foods* **2020**, *9*, 940. [CrossRef] [PubMed]
2. De Filippo, C.; Di Paola, M.; Ramazzotti, M.; Albanese, D.; Pieraccini, G.; Banci, E.; Miglietta, F.; Cavalieri, D.; Lionetti, P. Diet, Environments, and Gut Microbiota. A Preliminary Investigation in Children Living in Rural and Urban Burkina Faso and Italy. *Front. Microbiol.* **2017**, *8*, 1979. [CrossRef] [PubMed]
3. Calabrese, C.M.; Valentini, A.; Calabrese, G. Gut Microbiota and Type 1 Diabetes Mellitus: The Effect of Mediterranean Diet. *Front. Nutr.* **2020**, *7*, 612773. [CrossRef] [PubMed]
4. Nani, A.; Murtaza, B.; Sayed Khan, A.; Khan, N.A.; Hichami, A. Antioxidant and Anti-Inflammatory Potential of Polyphenols Contained in Mediterranean Diet in Obesity: Molecular Mechanisms. *Molecules* **2021**, *26*, 985. [CrossRef] [PubMed]
5. Davis, C.; Bryan, J.; Hodgson, J.; Murphy, K. Definition of the Mediterranean Diet; a Literature Review. *Nutrients* **2015**, *7*, 9139–9153. [CrossRef] [PubMed]
6. Saura-Calixto, F.; Goñi, I. Definition of the Mediterranean diet based on bioactive compounds. *Crit. Rev. Food. Sci. Nutr.* **2009**, *49*, 145–152. [CrossRef] [PubMed]
7. Hooper, L.V.; Littman, D.R.; Macpherson, A.J. Interactions between the microbiota and the immune system. *Science* **2012**, *336*, 1268–1273. [CrossRef] [PubMed]
8. Bailey, M.A.; Holscher, H.D. Microbiome-Mediated Effects of the Mediterranean Diet on Inflammation. *Adv. Nutr.* **2018**, *9*, 193–206. [CrossRef]

9. Anderson, J.W.; Baird, P.; Davis, R.H., Jr.; Ferreri, S.; Knudtson, M.; Koraym, A.; Waters, V.; Williams, C.L. Health benefits of dietary fiber. *Nutr. Rev.* **2009**, *67*, 188–205. [CrossRef] [PubMed]

10. Alkhatib, A.; Tsang, C.; Tiss, A.; Bahorun, T.; Arefanian, H.; Barake, R.; Khadir, A.; Tuomilehto, J. Functional Foods and Lifestyle Approaches for Diabetes Prevention and Management. *Nutrients* **2017**, *9*, 1310. [CrossRef] [PubMed]

11. Zhou, N.; Gu, X.; Zhuang, T.; Xu, Y.; Yang, L.; Zhou, M. Gut Microbiota: A Pivotal Hub for Polyphenols as Antidepressants. *J. Agric. Food. Chem.* **2020**, *68*, 6007–6020. [CrossRef]

12. Widmer, R.J.; Flammer, A.J.; Lerman, L.O.; Lerman, A. The Mediterranean diet, its components, and cardiovascular disease. *Am. J. Med.* **2015**, *128*, 229–238. [CrossRef] [PubMed]

13. de Lorgeril, M.; Renaud, S.; Mamelle, N.; Salen, P.; Martin, J.L.; Monjaud, I.; Guidollet, J.; Touboul, P.; Delaye, J. Mediterranean alpha-linolenic acid-rich diet in secondary prevention of coronary heart disease. *Lancet* **1994**, *343*, 1454–1459. [CrossRef]

14. Stadler, J.T.; Marsche, G. Obesity-Related Changes in High-Density Lipoprotein Metabolism and Function. *Int. J. Mol. Sci.* **2020**, *21*, 8985. [CrossRef] [PubMed]

15. Knight, J.A. Diseases and disorders associated with excess body weight. *Ann. Clin. Lab. Sci.* **2011**, *41*, 107–121. [PubMed]

16. Dandona, P.; Aljada, A.; Chaudhuri, A.; Mohanty, P.; Garg, R. Metabolic syndrome: A comprehensive perspective based on interactions between obesity, diabetes, and inflammation. *Circulation* **2005**, *111*, 1448–1454. [CrossRef] [PubMed]

17. Yahfoufi, N.; Alsadi, N.; Jambi, M.; Matar, C. The Immunomodulatory and Anti-Inflammatory Role of Polyphenols. *Nutrients* **2018**, *10*, 1618. [CrossRef]

18. Li, R.; Li, J.; Cai, L.; Hu, C.M.; Zhang, L. Suppression of adjuvant arthritis by hesperidin in rats and its mechanisms. *J. Pharm. Pharmacol.* **2008**, *60*, 221–228. [CrossRef] [PubMed]

19. Griffiths, K.; Aggarwal, B.B.; Singh, R.B.; Buttar, H.S.; Wilson, D.; De Meester, F. Food Antioxidants and Their Anti-Inflammatory Properties: A Potential Role in Cardiovascular Diseases and Cancer Prevention. *Diseases* **2016**, *4*, 28. [CrossRef]

20. Kolb, H.; Kempf, K.; Martin, S. Health Effects of Coffee: Mechanism Unraveled? *Nutrients* **2020**, *12*, 1842. [CrossRef]

21. Lejawa, M.; Osadnik, K.; Osadnik, T.; Pawlas, N. Association of Metabolically Healthy and Unhealthy Obesity Phenotypes with Oxidative Stress Parameters and Telomere Length in Healthy Young Adult Men. Analysis of the MAGNETIC Study. *Antioxidants* **2021**, *10*, 93. [CrossRef] [PubMed]

22. Cheng, F.; Carroll, L.; Joglekar, M.V.; Januszewski, A.S.; Wong, K.K.; Hardikar, A.A.; Jenkins, A.J.; Ma, R.C.W. Diabetes, metabolic disease, and telomere length. *Lancet Diabetes Endocrinol.* **2021**, *9*, 117–126. [CrossRef]

23. Canudas, S.; Becerra-Tomás, N.; Hernández-Alonso, P.; Galié, S.; Leung, C.; Crous-Bou, M.; De Vivo, I.; Gao, Y.; Gu, Y.; Meinilä, J.; et al. Mediterranean Diet and Telomere Length: A Systematic Review and Meta-Analysis. *Adv. Nutr.* **2020**, *11*, 1544–1554. [CrossRef] [PubMed]

24. Di Daniele, N.; Noce, A.; Vidiri, M.F.; Moriconi, E.; Marrone, G.; Annicchiarico-Petruzzelli, M.; D'Urso, G.; Tesauro, M.; Rovella, V.; De Lorenzo, A.; et al. Impact of Mediterranean diet on metabolic syndrome, cancer and longevity. *Oncotarget* **2017**, *8*, 8947–8979. [CrossRef] [PubMed]

25. Esposito, K.; Giugliano, D. Mediterranean diet and type 2 diabetes. *Diabetes Metab. Res. Rev.* **2014**, *30*, 34–40. [CrossRef] [PubMed]

26. Martín-Peláez, S.; Fito, M.; Castaner, O. Mediterranean Diet Effects on Type 2 Diabetes Prevention, Disease Progression, and Related Mechanisms. A Review. *Nutrients* **2020**, *12*, 2236. [CrossRef]

27. Bhaswant, M.; Poudyal, H.; Brown, L. Mechanisms of enhanced insulin secretion and sensitivity with n-3 unsaturated fatty acids. *J. Nutr. Biochem.* **2015**, *26*, 571–584. [CrossRef] [PubMed]

28. Thombare, K.; Ntika, S.; Wang, X.; Krizhanovskii, C. Long chain saturated and unsaturated fatty acids exert opposing effects on viability and function of GLP-1-producing cells: Mechanisms of lipotoxicity. *PLoS ONE* **2017**, *12*, e0177605. [CrossRef]

29. Zare, R.; Nadjarzadeh, A.; Zarshenas, M.M.; Shams, M.; Heydari, M. Efficacy of cinnamon in patients with type II diabetes mellitus: A randomized controlled clinical trial. *Clin. Nutr.* **2019**, *38*, 549–556. [CrossRef]

30. Subash Babu, P.; Prabuseenivasan, S.; Ignacimuthu, S. Cinnamaldehyde–a potential antidiabetic agent. *Phytomedicine* **2007**, *14*, 15–22. [CrossRef]

31. Kim, Y.; Keogh, J.B.; Clifton, P.M. Polyphenols and Glycemic Control. *Nutrients* **2016**, *8*, 17. [CrossRef] [PubMed]

32. Eid, H.M.; Martineau, L.C.; Saleem, A.; Muhammad, A.; Vallerand, D.; Benhaddou-Andaloussi, A.; Nistor, L.; Afshar, A.; Arnason, J.T.; Haddad, P.S. Stimulation of AMP-activated protein kinase and enhancement of basal glucose uptake in muscle cells by quercetin and quercetin glycosides, active principles of the antidiabetic medicinal plant Vaccinium vitis-idaea. *Mol. Nutr. Food Res.* **2010**, *54*, 991–1003. [CrossRef] [PubMed]

33. Dhanya, R.; Arya, A.D.; Nisha, P.; Jayamurthy, P. Quercetin, a Lead Compound against Type 2 Diabetes Ameliorates Glucose Uptake via AMPK Pathway in Skeletal Muscle Cell Line. *Front. Pharmacol.* **2017**, *8*, 336. [CrossRef] [PubMed]

34. Naimi, M.; Vlavcheski, F.; Shamshoum, H.; Tsiani, E. Rosemary Extract as a Potential Anti-Hyperglycemic Agent: Current Evidence and Future Perspectives. *Nutrients* **2017**, *9*, 968. [CrossRef] [PubMed]

35. Angelico, F.; Ferro, D.; Baratta, F. Is the Mediterranean Diet the Best Approach to NAFLD Treatment Today? *Nutrients* **2021**, *13*, 739. [CrossRef] [PubMed]

36. Tilg, H.; Moschen, A.R. Evolution of inflammation in nonalcoholic fatty liver disease: The multiple parallel hits hypothesis. *Hepatology* **2010**, *52*, 1836–1846. [CrossRef] [PubMed]

37. Yaskolka Meir, A.; Rinott, E.; Tsaban, G.; Zelicha, H.; Kaplan, A.; Rosen, P.; Shelef, I.; Youngster, I.; Shalev, A.; Blüher, M.; et al. Effect of green-Mediterranean diet on intrahepatic fat: The DIRECT PLUS randomised controlled trial. *Gut* **2021**. [CrossRef]

38. Franco, I.; Bianco, A.; Mirizzi, A.; Campanella, A.; Bonfiglio, C.; Sorino, P.; Notarnicola, M.; Tutino, V.; Cozzolongo, R.; Giannuzzi, V.; et al. Physical Activity and Low Glycemic Index Mediterranean Diet: Main and Modification Effects on NAFLD Score. Results from a Randomized Clinical Trial. *Nutrients* **2020**, *13*, 66. [CrossRef]
39. Pugliese, N.R.; Mengozzi, A.; Virdis, A.; Casiglia, E.; Tikhonoff, V.; Cicero, A.F.G.; Ungar, A.; Rivasi, G.; Salvetti, M.; Barbagallo, C.M.; et al. The importance of including uric acid in the definition of metabolic syndrome when assessing the mortality risk. *Clin. Res. Cardiol.* **2021**, 1–10. [CrossRef]
40. Huang, P.L. A comprehensive definition for metabolic syndrome. *Dis. Model. Mech.* **2009**, *2*, 231–237. [CrossRef] [PubMed]
41. Kim, C.H.; Younossi, Z.M. Nonalcoholic fatty liver disease: A manifestation of the metabolic syndrome. *Clevel. Clin. J. Med.* **2008**, *75*, 721–728. [CrossRef] [PubMed]
42. Esposito, K.; Chiodini, P.; Colao, A.; Lenzi, A.; Giugliano, D. Metabolic syndrome and risk of cancer: A systematic review and meta-analysis. *Diabetes Care* **2012**, *35*, 2402–2411. [CrossRef] [PubMed]
43. Uzunlulu, M.; Caklili, O.T.; Oguz, A. Association between Metabolic Syndrome and Cancer. *Ann. Nutr. Metab.* **2016**, *68*, 173–179. [CrossRef] [PubMed]
44. Kastorini, C.M.; Milionis, H.J.; Esposito, K.; Giugliano, D.; Goudevenos, J.A.; Panagiotakos, D.B. The effect of Mediterranean diet on metabolic syndrome and its components: A meta-analysis of 50 studies and 534,906 individuals. *J. Am. Coll. Cardiol.* **2011**, *57*, 1299–1313. [CrossRef] [PubMed]
45. Ghavipour, M.; Saedisomeolia, A.; Djalali, M.; Sotoudeh, G.; Eshraghyan, M.R.; Moghadam, A.M.; Wood, L.G. Tomato juice consumption reduces systemic inflammation in overweight and obese females. *Br. J. Nutr.* **2013**, *109*, 2031–2035. [CrossRef] [PubMed]
46. Li, Y.F.; Chang, Y.Y.; Huang, H.C.; Wu, Y.C.; Yang, M.D.; Chao, P.M. Tomato juice supplementation in young women reduces inflammatory adipokine levels independently of body fat reduction. *Nutrition* **2015**, *31*, 691–696. [CrossRef]
47. Rivera, L.; Morón, R.; Sánchez, M.; Zarzuelo, A.; Galisteo, M. Quercetin ameliorates metabolic syndrome and improves the inflammatory status in obese Zucker rats. *Obesity* **2008**, *16*, 2081–2087. [CrossRef]
48. Sun, W.; Yu, S.; Han, H.; Yuan, Q.; Chen, J.; Xu, G. Resveratrol Inhibits Human Visceral Preadipocyte Proliferation and Differentiation in vitro. *Lipids* **2019**, *54*, 679–686. [CrossRef]
49. Fischer-Posovszky, P.; Kukulus, V.; Tews, D.; Unterkircher, T.; Debatin, K.M.; Fulda, S.; Wabitsch, M. Resveratrol regulates human adipocyte number and function in a Sirt1-dependent manner. *Am. J. Clin. Nutr.* **2010**, *92*, 5–15. [CrossRef]
50. Balakumar, P.; Maung, U.K.; Jagadeesh, G. Prevalence and prevention of cardiovascular disease and diabetes mellitus. *Pharmacol. Res.* **2016**, *113*, 600–609. [CrossRef]
51. Ruuth, M.; Nguyen, S.D.; Vihervaara, T.; Hilvo, M.; Laajala, T.D.; Kondadi, P.K.; Gisterå, A.; Lähteenmäki, H.; Kittilä, T.; Huusko, J. Susceptibility of low-density lipoprotein particles to aggregate depends on particle lipidome, is modifiable, and associates with future cardiovascular deaths. *Eur. Heart J.* **2018**, *39*, 2562–2573. [CrossRef] [PubMed]
52. Zhu, C.; Sawrey-Kubicek, L.; Beals, E.; Hughes, R.L.; Rhodes, C.H.; Sacchi, R.; Zivkovic, A.M. The HDL lipidome is widely remodeled by fast food versus Mediterranean diet in 4 days. *Metabolomics* **2019**, *15*, 114. [CrossRef] [PubMed]
53. Rahman, K.; Lowe, G.M. Garlic and cardiovascular disease: A critical review. *J. Nutr.* **2006**, *136*, 736s–740s. [CrossRef] [PubMed]
54. Gorzynik-Debicka, M.; Przychodzen, P.; Cappello, F.; Kuban-Jankowska, A.; Marino Gammazza, A.; Knap, N.; Wozniak, M.; Gorska-Ponikowska, M. Potential Health Benefits of Olive Oil and Plant Polyphenols. *Int. J. Mol. Sci.* **2018**, *19*, 686. [CrossRef] [PubMed]
55. Wang, D.D.; Toledo, E.; Hruby, A.; Rosner, B.A.; Willett, W.C.; Sun, Q.; Razquin, C.; Zheng, Y.; Ruiz-Canela, M.; Guasch-Ferré, M.; et al. Plasma Ceramides, Mediterranean Diet, and Incident Cardiovascular Disease In the PREDIMED Trial (Prevención con Dieta Mediterránea). *Circulation* **2017**, *135*, 2028–2040. [CrossRef] [PubMed]
56. Hirsch, G.E.; Viecili, P.R.N.; de Almeida, A.S.; Nascimento, S.; Porto, F.G.; Otero, J.; Schmidt, A.; da Silva, B.; Parisi, M.M.; Klafke, J.Z. Natural Products with Antiplatelet Action. *Curr. Pharm. Des.* **2017**, *23*, 1228–1246. [CrossRef] [PubMed]
57. El Haouari, M.; Rosado, J.A. Medicinal Plants with Antiplatelet Activity. *Phytother. Res.* **2016**, *30*, 1059–1071. [CrossRef] [PubMed]
58. Viuda-Martos, M.; Ruiz-Navajas, Y.; Fernández-López, J.; Pérez-Alvarez, J.A. Spices as functional foods. *Crit. Rev. Food Sci. Nutr.* **2011**, *51*, 13–28. [CrossRef] [PubMed]
59. Ed Nignpense, B.; Chinkwo, K.A.; Blanchard, C.L.; Santhakumar, A.B. Polyphenols: Modulators of Platelet Function and Platelet Microparticle Generation? *Int. J. Mol. Sci.* **2019**, *21*, 146. [CrossRef] [PubMed]
60. Fuentes, E.; Caballero, J.; Alarcón, M.; Rojas, A.; Palomo, I. Chlorogenic acid inhibits human platelet activation and thrombus formation. *PLoS ONE* **2014**, *9*, e90699. [CrossRef] [PubMed]
61. Zhou, F.H.; Deng, X.J.; Chen, Y.Q.; Ya, F.L.; Zhang, X.D.; Song, F.; Li, D.; Yang, Y. Anthocyanin Cyanidin-3-Glucoside Attenuates Platelet Granule Release in Mice Fed High-Fat Diets. *J. Nutr. Sci. Vitaminol.* **2017**, *63*, 237–243. [CrossRef] [PubMed]
62. Ludovici, V.; Barthelmes, J.; Nagele, M.P.; Flammer, A.J.; Sudano, I. Polyphenols: Anti-Platelet Nutraceutical? *Curr. Pharm. Des.* **2018**, *24*, 146–157. [CrossRef] [PubMed]
63. Loscalzo, J. Nitric oxide insufficiency, platelet activation, and arterial thrombosis. *Circ. Res.* **2001**, *88*, 756–762. [CrossRef] [PubMed]
64. Mohar, D.S.; Malik, S. The Sirtuin System: The Holy Grail of Resveratrol? *J. Clin. Exp. Cardiol.* **2012**, *3*, 216. [CrossRef]
65. Suzuki, J.; Ogawa, M.; Maejima, Y.; Isobe, K.; Tanaka, H.; Sagesaka, Y.M.; Isobe, M. Tea catechins attenuate chronic ventricular remodeling after myocardial ischemia in rats. *J. Mol. Cell. Cardiol.* **2007**, *42*, 432–440. [CrossRef]

66. Suzuki, J.; Ogawa, M.; Futamatsu, H.; Kosuge, H.; Sagesaka, Y.M.; Isobe, M. Tea catechins improve left ventricular dysfunction, suppress myocardial inflammation and fibrosis, and alter cytokine expression in rat autoimmune myocarditis. *Eur. J. Heart Fail.* **2007**, *9*, 152–159. [CrossRef] [PubMed]

67. Schagdarsurengin, U.; Steger, K. Epigenetics in male reproduction: Effect of paternal diet on sperm quality and offspring health. *Nat. Rev. Urol.* **2016**, *13*, 584–595. [CrossRef] [PubMed]

68. Best, D.; Bhattacharya, S. Obesity and fertility. *Horm. Mol. Biol. Clin. Investig.* **2015**, *24*, 5–10. [CrossRef] [PubMed]

69. Chavarro, J.E.; Schlaff, W.D. Introduction: Impact of nutrition on reproduction: An overview. *Fertil. Steril.* **2018**, *110*, 557–559. [CrossRef] [PubMed]

70. Einarsson, S.; Bergh, C.; Friberg, B.; Pinborg, A.; Klajnbard, A.; Karlström, P.O.; Kluge, L.; Larsson, I.; Loft, A.; Mikkelsen-Englund, A.L.; et al. Weight reduction intervention for obese infertile women prior to IVF: A randomized controlled trial. *Hum. Reprod.* **2017**, *32*, 1621–1630. [CrossRef]

71. Delitala, A.P.; Capobianco, G.; Delitala, G.; Cherchi, P.L.; Dessole, S. Polycystic ovary syndrome, adipose tissue and metabolic syndrome. *Arch. Gynecol. Obstet.* **2017**, *296*, 405–419. [CrossRef] [PubMed]

72. Norman, R.J.; Dewailly, D.; Legro, R.S.; Hickey, T.E. Polycystic ovary syndrome. *Lancet* **2007**, *370*, 685–697. [CrossRef]

73. González, F. Inflammation in Polycystic Ovary Syndrome: Underpinning of insulin resistance and ovarian dysfunction. *Steroids* **2012**, *77*, 300–305. [CrossRef] [PubMed]

74. Garg, D.; Merhi, Z. Advanced Glycation End Products: Link between Diet and Ovulatory Dysfunction in PCOS? *Nutrients* **2015**, *7*, 10129–10144. [CrossRef] [PubMed]

75. Uribarri, J.; Woodruff, S.; Goodman, S.; Cai, W.; Chen, X.; Pyzik, R.; Yong, A.; Striker, G.E.; Vlassara, H. Advanced glycation end products in foods and a practical guide to their reduction in the diet. *J. Am. Diet. Assoc.* **2010**, *110*, 911–916. [CrossRef] [PubMed]

76. Barrea, L.; Arnone, A.; Annunziata, G.; Muscogiuri, G.; Laudisio, D.; Salzano, C.; Pugliese, G.; Colao, A.; Savastano, S. Adherence to the Mediterranean Diet, Dietary Patterns and Body Composition in Women with Polycystic Ovary Syndrome (PCOS). *Nutrients* **2019**, *11*, 2278. [CrossRef] [PubMed]

77. Barrea, L.; Marzullo, P.; Muscogiuri, G.; Di Somma, C.; Scacchi, M.; Orio, F.; Aimaretti, G.; Colao, A.; Savastano, S. Source and amount of carbohydrate in the diet and inflammation in women with polycystic ovary syndrome. *Nutr. Res. Rev.* **2018**, *31*, 291–301. [CrossRef] [PubMed]

78. Amini, L.; Tehranian, N.; Movahedin, M.; Ramezani Tehrani, F.; Ziaee, S. Antioxidants and management of polycystic ovary syndrome in Iran: A systematic review of clinical trials. *Iran J. Reprod. Med.* **2015**, *13*, 1–8. [PubMed]

79. Zhang, J.; Wang, M.; Ye, J.; Liu, J.; Xu, Y.; Wang, Z.; Ye, D.; Zhao, M.; Wan, J. The Anti-inflammatory Mediator Resolvin E1 Protects Mice Against Lipopolysaccharide-Induced Heart Injury. *Front. Pharmacol.* **2020**, *11*, 203. [CrossRef]

80. Ashkar, F.; Rezaei, S.; Salahshoornezhad, S.; Vahid, F.; Gholamalizadeh, M.; Dahka, S.M.; Doaei, S. The Role of medicinal herbs in treatment of insulin resistance in patients with Polycystic Ovary Syndrome: A literature review. *Biomol. Concepts* **2020**, *11*, 57–75. [CrossRef] [PubMed]

81. Spaczynski, R.Z.; Tilly, J.L.; Mansour, A.; Duleba, A.J. Insulin and insulin-like growth factors inhibit and luteinizing hormone augments ovarian theca-interstitial cell apoptosis. *Mol. Hum. Reprod.* **2005**, *11*, 319–324. [CrossRef] [PubMed]

82. Wong, D.H.; Villanueva, J.A.; Cress, A.B.; Duleba, A.J. Effects of resveratrol on proliferation and apoptosis in rat ovarian theca-interstitial cells. *Mol. Hum. Reprod.* **2010**, *16*, 251–259. [CrossRef] [PubMed]

83. Brenjian, S.; Moini, A.; Yamini, N.; Kashani, L.; Faridmojtahedi, M.; Bahramrezaie, M.; Khodarahmian, M.; Amidi, F. Resveratrol treatment in patients with polycystic ovary syndrome decreased pro-inflammatory and endoplasmic reticulum stress markers. *Am. J. Reprod. Immunol.* **2020**, *83*, e13186. [CrossRef] [PubMed]

84. Ochiai, A.; Kuroda, K. Preconception resveratrol intake against infertility: Friend or foe? *Reprod. Med. Biol.* **2020**, *19*, 107–113. [CrossRef] [PubMed]

85. Chen, Z.G.; Luo, L.L.; Xu, J.J.; Zhuang, X.L.; Kong, X.X.; Fu, Y.C. Effects of plant polyphenols on ovarian follicular reserve in aging rats. *Biochem. Cell Biol.* **2010**, *88*, 737–745. [CrossRef] [PubMed]

86. Rezvan, N.; Moini, A.; Gorgani-Firuzjaee, S.; Hosseinzadeh-Attar, M.J. Oral Quercetin Supplementation Enhances Adiponectin Receptor Transcript Expression in Polycystic Ovary Syndrome Patients: A Randomized Placebo-Controlled Double-Blind Clinical Trial. *Cell J.* **2018**, *19*, 627–633. [PubMed]

87. Rezvan, N.; Moini, A.; Janani, L.; Mohammad, K.; Saedisomeolia, A.; Nourbakhsh, M.; Gorgani-Firuzjaee, S.; Mazaherioun, M.; Hosseinzadeh-Attar, M.J. Effects of Quercetin on Adiponectin-Mediated Insulin Sensitivity in Polycystic Ovary Syndrome: A Randomized Placebo-Controlled Double-Blind Clinical Trial. *Horm. Metab. Res.* **2017**, *49*, 115–121. [CrossRef] [PubMed]

88. Rahman, S.U.; Huang, Y.; Zhu, L.; Feng, S.; Khan, I.M.; Wu, J.; Li, Y.; Wang, X. Therapeutic Role of Green Tea Polyphenols in Improving Fertility: A Review. *Nutrients* **2018**, *10*, 834. [CrossRef]

89. Kefer, J.C.; Agarwal, A.; Sabanegh, E. Role of antioxidants in the treatment of male infertility. *Int. J. Urol.* **2009**, *16*, 449–457. [CrossRef]

90. Salas-Huetos, A.; Bulló, M.; Salas-Salvadó, J. Dietary patterns, foods and nutrients in male fertility parameters and fecundability: A systematic review of observational studies. *Hum. Reprod. Update* **2017**, *23*, 371–389. [CrossRef] [PubMed]

91. Rozati, R.; Reddy, P.P.; Reddanna, P.; Mujtaba, R. Role of environmental estrogens in the deterioration of male factor fertility. *Fertil. Steril.* **2002**, *78*, 1187–1194. [CrossRef]

92. Nassan, F.L.; Chavarro, J.E.; Tanrikut, C. Diet and men's fertility: Does diet affect sperm quality? *Fertil. Steril.* **2018**, *110*, 570–577. [CrossRef]

93. Saez Lancellotti, T.E.; Boarelli, P.V.; Monclus, M.A.; Cabrillana, M.E.; Clementi, M.A.; Espínola, L.S.; Cid Barría, J.L.; Vincenti, A.E.; Santi, A.G.; Fornés, M.W. Hypercholesterolemia impaired sperm functionality in rabbits. *PLoS ONE* **2010**, *5*, e13457. [CrossRef] [PubMed]

94. Saez Lancellotti, T.E.; Boarelli, P.V.; Romero, A.A.; Funes, A.K.; Cid-Barria, M.; Cabrillana, M.E.; Monclus, M.A.; Simón, L.; Vicenti, A.E.; Fornés, M.W. Semen quality and sperm function loss by hypercholesterolemic diet was recovered by addition of olive oil to diet in rabbit. *PLoS ONE* **2013**, *8*, e52386. [CrossRef]

95. Simas, J.N.; Mendes, T.B.; Fischer, L.W.; Vendramini, V.; Miraglia, S.M. Resveratrol improves sperm DNA quality and reproductive capacity in type 1 diabetes. *Andrology* **2021**, *9*, 384–399. [CrossRef] [PubMed]

96. Garcez, M.E.; dos Santos Branco, C.; Lara, L.V.; Pasqualotto, F.F.; Salvador, M. Effects of resveratrol supplementation on cryopreservation medium of human semen. *Fertil. Steril.* **2010**, *94*, 2118–2121. [CrossRef] [PubMed]

97. Amati, F.; Hassounah, S.; Swaka, A. The Impact of Mediterranean Dietary Patterns During Pregnancy on Maternal and Offspring Health. *Nutrients* **2019**, *11*, 1098. [CrossRef] [PubMed]

98. Biagi, C.; Nunzio, M.D.; Bordoni, A.; Gori, D.; Lanari, M. Effect of Adherence to Mediterranean Diet during Pregnancy on Children's Health: A Systematic Review. *Nutrients* **2019**, *11*, 997. [CrossRef] [PubMed]

99. Mehta, S.H. Nutrition and pregnancy. *Clin. Obstet. Gynecol.* **2008**, *51*, 409–418. [CrossRef] [PubMed]

100. Mulligan, C.M.; Friedman, J.E. Maternal modifiers of the infant gut microbiota: Metabolic consequences. *J. Endocrinol.* **2017**, *235*, R1–R12. [CrossRef] [PubMed]

101. Friedman, J.E. Developmental Programming of Obesity and Diabetes in Mouse, Monkey, and Man in 2018: Where Are We Headed? *Diabetes* **2018**, *67*, 2137–2151. [CrossRef] [PubMed]

102. Mejía-León, M.E.; Barca, A.M. Diet, Microbiota and Immune System in Type 1 Diabetes Development and Evolution. *Nutrients* **2015**, *7*, 9171–9184. [CrossRef] [PubMed]

103. Ilonen, J.; Lempainen, J.; Veijola, R. The heterogeneous pathogenesis of type 1 diabetes mellitus. *Nat. Rev. Endocrinol.* **2019**, *15*, 635–650. [CrossRef] [PubMed]

104. Han, H.; Li, Y.; Fang, J.; Liu, G.; Yin, J.; Li, T.; Yin, Y. Gut Microbiota and Type 1 Diabetes. *Int. J. Mol. Sci.* **2018**, *19*, 995. [CrossRef] [PubMed]

105. Niinistö, S.; Takkinen, H.M.; Uusitalo, L.; Rautanen, J.; Vainio, N.; Ahonen, S.; Nevalainen, J.; Kenward, M.G.; Lumia, M.; Simell, O.; et al. Maternal intake of fatty acids and their food sources during lactation and the risk of preclinical and clinical type 1 diabetes in the offspring. *Acta Diabetol.* **2015**, *52*, 763–772. [CrossRef] [PubMed]

106. Wu, G.D.; Chen, J.; Hoffmann, C.; Bittinger, K.; Chen, Y.Y.; Keilbaugh, S.A.; Bewtra, M.; Knights, D.; Walters, W.A.; Knight, R.; et al. Linking long-term dietary patterns with gut microbial enterotypes. *Science* **2011**, *334*, 105–108. [CrossRef] [PubMed]

107. Yang, Q.; Liang, Q.; Balakrishnan, B.; Belobrajdic, D.P.; Feng, Q.J.; Zhang, W. Role of Dietary Nutrients in the Modulation of Gut Microbiota: A Narrative Review. *Nutrients* **2020**, *12*, 381. [CrossRef] [PubMed]

108. Timmermans, S.; Bogie, J.F.; Vanmierlo, T.; Lütjohann, D.; Stinissen, P.; Hellings, N.; Hendriks, J.J. High fat diet exacerbates neuroinflammation in an animal model of multiple sclerosis by activation of the Renin Angiotensin system. *J. Neuroimmune Pharmacol.* **2014**, *9*, 209–217. [CrossRef]

109. Vinolo, M.A.; Rodrigues, H.G.; Nachbar, R.T.; Curi, R. Regulation of inflammation by short chain fatty acids. *Nutrients* **2011**, *3*, 858–876. [CrossRef] [PubMed]

110. Zhang, Y.; Yin, L.; Zheng, N.; Zhang, L.; Liu, J.; Liang, W.; Wang, Q. Icariin enhances remyelination process after acute demyelination induced by cuprizone exposure. *Brain. Res. Bull.* **2017**, *130*, 180–187. [CrossRef] [PubMed]

111. Katz Sand, I. The Role of Diet in Multiple Sclerosis: Mechanistic Connections and Current Evidence. *Curr. Nutr. Rep.* **2018**, *7*, 150–160. [CrossRef]

112. Jackson, A.; Forsyth, C.B.; Shaikh, M.; Voigt, R.M.; Engen, P.A.; Ramirez, V.; Keshavarzian, A. Diet in Parkinson's Disease: Critical Role for the Microbiome. *Front. Neurol.* **2019**, *10*, 1245. [CrossRef] [PubMed]

113. Rajaram, S.; Jones, J.; Lee, G.J. Plant-Based Dietary Patterns, Plant Foods, and Age-Related Cognitive Decline. *Adv. Nutr.* **2019**, *10*, S422–S436. [CrossRef]

114. Gildawie, K.R.; Galli, R.L.; Shukitt-Hale, B.; Carey, A.N. Protective Effects of Foods Containing Flavonoids on Age-Related Cognitive Decline. *Curr. Nutr. Rep.* **2018**, *7*, 39–48. [CrossRef] [PubMed]

115. Hajialyani, M.; Hosein Farzaei, M.; Echeverría, J.; Nabavi, S.M.; Uriarte, E.; Sobarzo-Sánchez, E. Hesperidin as a Neuroprotective Agent: A Review of Animal and Clinical Evidence. *Molecules* **2019**, *24*, 648. [CrossRef] [PubMed]

116. Li, D.; Wang, P.; Luo, Y.; Zhao, M.; Chen, F. Health benefits of anthocyanins and molecular mechanisms: Update from recent decade. *Crit. Rev. Food Sci. Nutr.* **2017**, *57*, 1729–1741. [CrossRef] [PubMed]

117. Grodzicki, W.; Dziendzikowska, K. The Role of Selected Bioactive Compounds in the Prevention of Alzheimer's Disease. *Antioxidants* **2020**, *9*, 229. [CrossRef] [PubMed]

118. Li, W.; Sperry, J.B.; Crowe, A.; Trojanowski, J.Q.; Smith, A.B., 3rd; Lee, V.M. Inhibition of tau fibrillization by oleocanthal via reaction with the amino groups of tau. *J. Neurochem.* **2009**, *110*, 1339–1351. [CrossRef]

119. Burčul, F.; Generalić Mekinić, I.; Radan, M.; Rollin, P.; Blažević, I. Isothiocyanates: Cholinesterase inhibiting, antioxidant, and anti-inflammatory activity. *J. Enzyme Inhib. Med. Chem.* **2018**, *33*, 577–582. [CrossRef] [PubMed]

120. Yammine, A.; Zarrouk, A.; Nury, T.; Vejux, A.; Latruffe, N.; Vervandier-Fasseur, D.; Samadi, M.; Mackrill, J.J.; Greige-Gerges, H.; Auezova, L.; et al. Prevention by Dietary Polyphenols (Resveratrol, Quercetin, Apigenin) Against 7-Ketocholesterol-Induced Oxiapoptophagy in Neuronal N2a Cells: Potential Interest for the Treatment of Neurodegenerative and Age-Related Diseases. *Cells* **2020**, *9*, 2346. [CrossRef] [PubMed]
121. Ettman, C.K.; Abdalla, S.M.; Cohen, G.H.; Sampson, L.; Vivier, P.M.; Galea, S. Prevalence of Depression Symptoms in US Adults Before and During the COVID-19 Pandemic. *JAMA Netw. Open* **2020**, *3*, e2019686. [CrossRef] [PubMed]
122. James, S.L.; Abate, D.; Abate, K.H.; Abay, S.M.; Abbafati, C.; Abbasi, N.; Abbastabar, H.; Abd-Allah, F.; Abdela, J.; Abdelalim, A.; et al. Global, regional, and national incidence, prevalence, and years lived with disability for 354 diseases and injuries for 195 countries and territories, 1990–2017: A systematic analysis for the Global Burden of Disease Study 2017. *Lancet* **2018**, *392*, 1789–1858. [CrossRef]
123. Riera-Sampol, A.; Bennasar-Veny, M.; Tauler, P.; Nafría, M.; Colom, M.; Aguilo, A. Association between Depression, Lifestyles, Sleep Quality and Sense of Coherence in a Population with Cardiovascular Risk. *Nutrients* **2021**, *13*, 585. [CrossRef]
124. Parletta, N.; Zarnowiecki, D.; Cho, J.; Wilson, A.; Bogomolova, S.; Villani, A.; Itsiopoulos, C.; Niyonsenga, T.; Blunden, S.; Meyer, B.; et al. A Mediterranean-style dietary intervention supplemented with fish oil improves diet quality and mental health in people with depression: A randomized controlled trial (HELFIMED). *Nutr. Neurosci.* **2019**, *22*, 474–487. [CrossRef] [PubMed]
125. Huang, Q.; Liu, H.; Suzuki, K.; Ma, S.; Liu, C. Linking What We Eat to Our Mood: A Review of Diet, Dietary Antioxidants, and Depression. *Antioxidants* **2019**, *8*, 376. [CrossRef] [PubMed]
126. Simopoulos, A.P. Omega-3 fatty acids in inflammation and autoimmune diseases. *J. Am. Coll. Nutr.* **2002**, *21*, 495–505. [CrossRef] [PubMed]
127. Bayes, J.; Schloss, J.; Sibbritt, D. Effects of Polyphenols in a Mediterranean Diet on Symptoms of Depression: A Systematic Literature Review. *Adv. Nutr.* **2020**, *11*, 602–615. [CrossRef] [PubMed]
128. Gu, Z.; Chu, L.; Han, Y. Therapeutic effect of resveratrol on mice with depression. *Exp. Ther. Med.* **2019**, *17*, 3061–3064. [CrossRef]
129. Shaito, A.; Posadino, A.M.; Younes, N.; Hasan, H.; Halabi, S.; Alhababi, D.; Al-Mohannadi, A.; Abdel-Rahman, W.M.; Eid, A.H.; Nasrallah, G.K.; et al. Potential Adverse Effects of Resveratrol: A Literature Review. *Int. J. Mol. Sci.* **2020**, *21*, 2084. [CrossRef] [PubMed]
130. Hayat, K.; Iqbal, H.; Malik, U.; Bilal, U.; Mushtaq, S. Tea and its consumption: Benefits and risks. *Crit. Rev. Food Sci. Nutr.* **2015**, *55*, 939–954. [CrossRef] [PubMed]

nutrients

Review

The Potential Health Benefits of the Ketogenic Diet: A Narrative Review

Kathryn Dowis and Simran Banga *

Department of Biology, Western Kentucky University, Bowling Green, KY 42101, USA;
kathryn.dowis803@topper.wku.edu
* Correspondence: simran.banga@wku.edu; Tel.: +270-745-4748

Abstract: Considering the lack of a comprehensive, multi-faceted overview of the ketogenic diet (KD) in relation to health issues, we compiled the evidence related to the use of the ketogenic diet in relation to its impact on the microbiome, the epigenome, diabetes, weight loss, cardiovascular health, and cancer. The KD diet could potentially increase genetic diversity of the microbiome and increase the ratio of *Bacteroidetes* to *Firmicutes*. The epigenome might be positively affected by the KD since it creates a signaling molecule known as β-hydroxybutyrate (BHB). KD has helped patients with diabetes reduce their HbA1c and reduce the need for insulin. There is evidence to suggest that a KD can help with weight loss, visceral adiposity, and appetite control. The evidence also suggests that eating a high-fat diet improves lipid profiles by lowering low-density lipoprotein (LDL), increasing high-density lipoprotein (HDL), and lowering triglycerides (TG). Due to the Warburg effect, the KD is used as an adjuvant treatment to starve cancer cells, making them more vulnerable to chemotherapy and radiation. The potential positive impacts of a KD on each of these areas warrant further analysis, improved studies, and well-designed randomized controlled trials to further illuminate the therapeutic possibilities provided by this dietary intervention.

Keywords: β-hydroxybutyrate (BHB); body mass index (BMI), type 1 diabetes; type 2 diabetes (T2D); hemoglobin A1c (HbA1c); visceral adipose tissue (VAT); cardiovascular disease (CVD); high-density lipoprotein (HDL); low-density lipoprotein (LDL); Apolipoprotein B (ApoB)

Citation: Dowis, K.; Banga, S. The Potential Health Benefits of the Ketogenic Diet: A Narrative Review. *Nutrients* **2021**, *13*, 1654. https://doi.org/10.3390/nu13051654

Academic Editor: Marcellino Monda

Received: 7 April 2021
Accepted: 9 May 2021
Published: 13 May 2021

Publisher's Note: MDPI stays neutral with regard to jurisdictional claims in published maps and institutional affiliations.

1. Introduction

Ketogenic diets have started to increase in popularity as doctors and researchers investigate the potential benefits. Nutritional ketosis, the aspirational endpoint of ketogenic diets, is achieved by restricting carbohydrate intake, moderating protein consumption, and increasing the number of calories obtained from fat [1]. Theoretically, this restriction of carbohydrates causes the body to switch from glucose metabolism as a primary means of energy production. This results in the use of ketone bodies from fat metabolism, a metabolic state where the body prefers to utilize fat as its primary fuel source. Recent studies utilizing Low-carbohydrate, High-fat (LCHF) diets, such as the ketogenic diet, show promise in helping patients lose weight, reverse the signs of metabolic syndrome, reduce, or eliminate insulin requirements for type II diabetics [2], reduce inflammation, improve epigenetic profiles, alter the microbiome, improve lipid profiles, supplement cancer treatments, and potentially increase longevity [3] and brain function.

The number of Americans suffering from obesity, diabetes, and metabolic syndrome is on the rise. The markers of metabolic syndrome include an increase in abdominal adiposity, insulin resistance, elevated triglycerides, and hypertension [4,5]. All of these negative health markers increase the risk of cardiovascular disease, diabetes, stroke, and Alzheimer's disease. According to WebMD, there are currently 27 million people with Type 2 diabetes and 86 million with pre-diabetes. In addition, the Centers of Disease Control and Prevention (CDC) also estimates that almost 40% of adults and around 20% of American children are obese [6,7]. Many researchers believe these diseases are a result of carbohydrate intolerance and insulin resistance. Thus, a diet that reduces the exposure to carbohydrates, including whole grains, might become a more logical recommendation for improving health [8]. In line with this, two dietary regimens, the standard ketogenic diet, and the therapeutic ketogenic diet (Figure 1), which restrict carbohydrate consumption to varying degrees are being studied for their health impacts. The therapeutic ketogenic diet, which severely restricts both carbohydrates and protein, is typically used in the treatment of epilepsy and cancer. However, the Dietary Guidelines for Americans suggests that between 45 and 65% of caloric intake should come from carbohydrates (Figure 1). If a person consumed 2000 calories per day that would equate to an average of 225–325 g of carbohydrates each day [9].

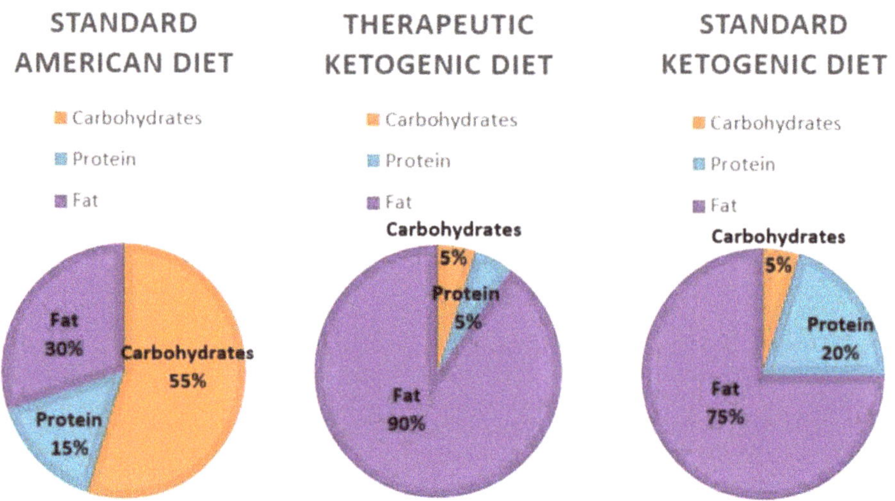

Figure 1. A comparison between the macronutrient breakdown of the standard American diet, therapeutic ketogenic diet, and the typical ketogenic diet. The therapeutic ketogenic diet is typically used in epilepsy and cancer treatments.

One emerging diet that is becoming mainstream is a low-carb/high-fat diet. However, there is a difference between a low-carb and a low-carb ketogenic diet (LCKD). Ketosis is normally achieved through either fasting or carbohydrate restriction. It is important to clarify that a low-carb diet typically refers to a diet with an intake of 50 to 150 g of carbohydrate per day. However, although this is a lower amount of carbohydrates than the standard American diet, it is not low enough to enter nutritional ketosis. Only when a patient restricts carbohydrates to less than 50 g/day will the body be incapable of fueling the body by glucose and will switch to burning fat [10]. The ketogenic diet is a reversal of the current food pyramid supported by the dietary guidelines. Thus, instead of a diet rich in carbohydrates, it is high in fat (Figure 2). The resulting carbohydrate restriction lowers blood glucose levels, and the subsequent insulin changes will instruct the body to change from a state of storing fat to a state of fat oxidation [10]. Once fats are utilized as the primary fuel source in the liver, the production of ketone bodies begins, a process known as ketogenesis. During ketosis, three major ketone bodies are formed and utilized by the body for energy: acetone, acetoacetate, and β-hydroxybutyrate [11]. All cells that contain mitochondria can meet their energy demands with ketone bodies, including the brain and muscle. In addition, research suggests that β-hydroxybutyrate acts as a signal molecule and may play a role in suppressing appetite [12].

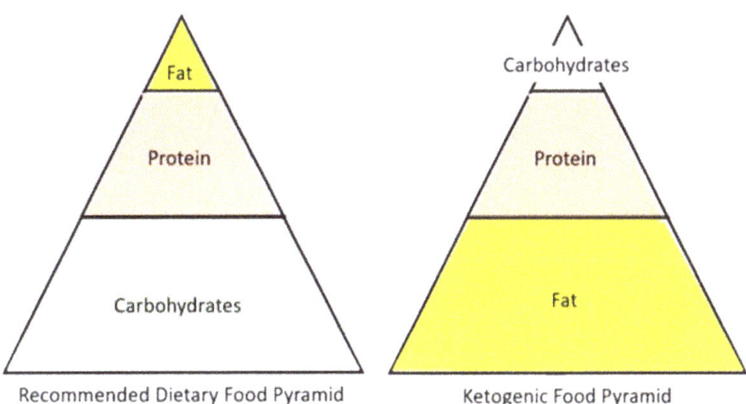

Figure 2. A visual comparison of the recommended dietary food pyramid, including major macromolecule components, to the ketogenic diet food pyramid.

However, there is some heterogeneity in the available data. Thus, the aim of this review is to highlight the role the ketogenic diet has in altering the microbiome, epigenetics, weight loss, diabetes, cardiovascular disease, and cancer as summarized below (Figure 3).

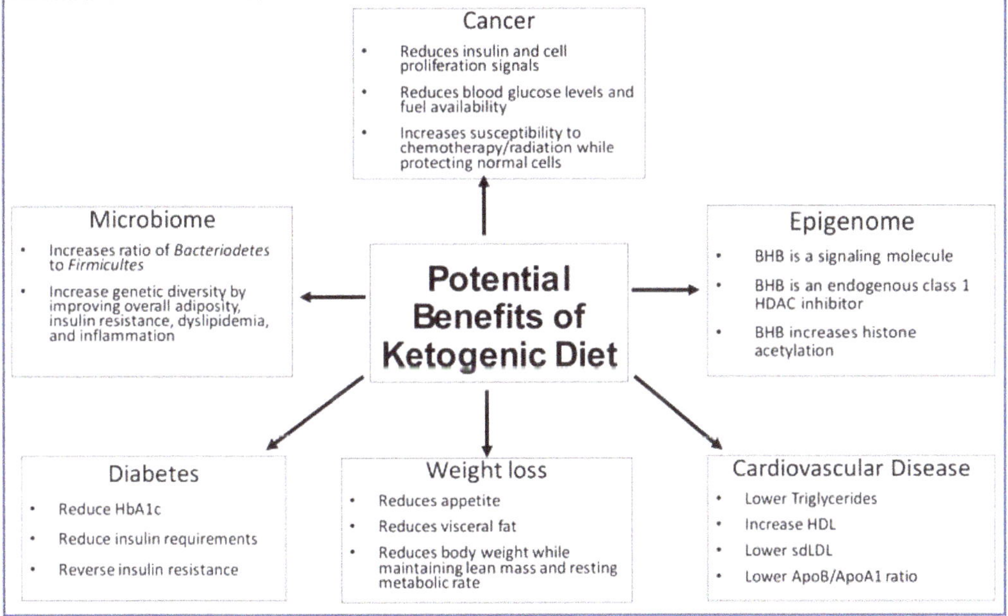

Figure 3. The potential therapeutic impacts of the ketogenic diet on the microbiome, epigenome, diabetes, weight loss and cardiovascular disease.

2. The Effect of the Ketogenic Diet on the Microbiome

The microbiome consists of trillions of microscopic organisms in the human gastrointestinal tract. It comprises over 8000 different types of bacteria, viruses, and fungi living in a complex ecosystem [13]. Recent research suggests that the genetic make-up of a microbiome can be affected by lifestyle factors which include but are not limited to sleep, exercise, antibiotic use, and even diet. These bacteria can alter our response to different food sources because they differ in their ability to harvest energy from food, affecting the postprandial glucose response (PPGR) [13]. Since the controlling of glucose levels in the blood seems to reduce the risk of metabolic disease, diabetes, and obesity, this might be an innovative way to help reduce disease risk. A study conducted at the Weizmann Institute demonstrated that a mathematical algorithm could be used to determine an individual's microbiome profile and predict their glycemic response to different types of foods [14]. Thus, the patients were able to change from stable blood glucose to unstable levels by simply eating the foods that the program predicted as good or bad based on their microbiome. Their initial results were confirmed by a repeat study at the Mayo Clinic with a different population [13]. It is important to note that the composition of the microbiome, which is believed to have a fundamental role in human health, is shaped predominantly by environmental factors. According to a study conducted by Rothschild et al. [15], the average heritability of the gut microbiome taxa is only 1.9%, while over 20% of variability was associated with diet and lifestyle.

Thus, research into the complex interactions that exist between diet, the microbiome, and host metabolic rates have increased. A study exploring the benefits of prebiotic foods, such as inulin and oligosaccharides, observed an increase in the number of *Bifidobacteria* in the colon and the presence of other critical butyrate-producing bacteria [16]. Another study determined that the diversity of the gut microbiota was influenced more by a Westernized diet than by the body mass index of the subjects [17]. The patients who followed the Westernized diet showed an increase in *Firmicutes* and a decrease in *Bacteroidetes* in their microbiome, which are negative changes. A review article also reported positive changes in the gut microbiome and overall health in energy-restrictive diets or diets rich in fiber and vegetables [18]. Thus, people eating processed and bland food had reduced diversity of their microbiota, while people eating a diet rich in fruit and vegetables had increased diversity in their gut microbiota [19]. Moreover, gut biomes that lacked genetic diversity were related to overall adiposity, insulin resistance, dyslipidemia, and an inflammatory phenotype [20].

Discovering how the gut microbiota and diet interact and how this interaction is connected to overall health, is critical. It is important to determine whether new dietary changes, such as a ketogenic diet, will positively or negatively affect overall microbiome diversity and species make-up. Some research has found that whole grains play an important role in the development of a healthy microbiome and are necessary for good health [21]. Thus, a person consuming a ketogenic diet might not consume enough whole grains to maintain a healthy microbiome [12]. According to Adam-Perrot et al. [12] low-carb diets are at greater risk of being nutritionally inadequate by lacking in fiber, necessary vitamins, minerals, and iron. This idea is based on analysis of popular diets and food surveys conducted to determine nutrient intake while consuming varying levels of carbohydrates [22]. Thus, it is even more critical that people on a LCKD choose desirable low carbohydrate foods that are rich in fiber. In addition, a ketogenic diet should maintain moderate protein intake of around 1.5 g/day per kg of respective body weight [23]. If people consume red meat and organ meats, then they should be able to obtain adequate amounts of iron as well. Additionally, the consumption of small amounts of leafy greens, nuts, berries, and resistant starchy vegetables, all of which are optional ketogenic foods, could potentially maintain healthy gut microbiota [23].

Currently, scientists do not have any data on the long-term effects of the ketogenic diet on the gut microbiome. Based on various studies, many predict that the diet will positively affect the microbiome by increasing the *Bacteroidetes* and *Bifidobacteria* species associated with improved health and decreasing microbial species known to increase health risks. In fact, a study found that the disrupted gut microbiota of epileptic infants was improved with a one-week ketogenic diet, which managed to increase their *Bacteroides* amount by ~24% [24]. Another 6-month study on children with refractory epilepsy found a significant decrease in *Firmicutes* and an increase in *Bacteroides* although the overall diversity decreased [25].

Studies have shown that a low ratio of *Firmicutes* to *Bacteroidetes* is an indicator of a healthy microbiome [26]. A few studies found that obese patients were more likely to have a higher *Firmicutes* to *Bacteroidetes* ratio [26–28] and higher levels of short chain fatty acids (SCFAs) in their stool [5]. However, another study found that obese patients showed an increase in *Bacteroidetes*, while *Firmicutes* remained the same [29]. Therefore, it appears that reducing obesity with the KD may result in positive changes in the microbiome. A study by Basciani et al. [30] recently analyzed the changes in the gut microbiota in obese, insulin-resistant patients who followed isocaloric ketogenic diets which varied in their source of proteins. The very low-calorie ketogenic diets (VLCKDs) contained either whey, vegetable, or animal proteins. The data indicated all groups had a decrease in relative abundance of *Firmicutes* and an increase in *Bacteroidetes* after 45 days. However, the positive changes were less pronounced in the group that consumed animal protein sources.

Recently, a few short-term studies tested the impact of the KD on patient microbiomes. A study by Nagpal et al. [31] analyzed the effect of a modified Mediterranean Ketogenic Diet (MMKD) vs. the American Heart Association Diet (AHAD) on the microbiome of patients with normal cognition or mild cognitive impairment. They found that the MMKD did not show significant changes in the *Firmicutes* or *Bacteroides* phyla at 6 weeks. However, they did see a decrease in the family *Bifidobacteriaceae* and an increase in family *Verrucomicrobiaceae*, which was considered a positive change. Furthermore, the beneficial SCFA, butyrate, increased in the MMKD. The presence of butyrate has been known to increase gut health [31].

3. The Effect of the Ketogenic Diet on the Epigenome

Epigenetics refers specifically to changes "on top" of the genome that can modify and alter levels of gene expression. These epigenetic markers are heritable, yet recent research suggests that some changes can be reversed or occur through environmental changes [20]. The modifications of the genome involve DNA methylation, changes to chromatin structure, histone modification, and noncoding RNAs. Most notable are histone modifications. For example, the N-terminal of histone tails can be acetylated, methylated, phosphorylated, ubiquitinated, or SUMOylated. Histone deacetylases (HDACs) are enzymes that can remove acetyl groups and condense the chromatin. Similarly, sirtuins (SIRTs) are also capable of deacetylating histones. Histone lysine methylation can either activate or repress a gene's activity based on the exact location and number of methyl groups added to the histone tail [32]. Research has found that most epigenetic modification occur during early embryogenesis, but the genome can acquire changes later in life. Some of the later epigenetic modifications are caused or modified because of diet [32].

Some ketogenic food sources that positively regulate epigenetic activity are cruciferous vegetables, dietary fiber, foods rich in long-chain fatty acids, and berries, such as raspberries [20]. The benefits of some of these food sources have a multitude of positive effects. For instance, black raspberries not only positively affect methylation patterns in the WNT-signaling pathway, but they also profoundly impact the microbiome make-up (increased *Lactobacillus*, *Bacteroidaceae*, and anti-inflammatory bacterial species), and increased production of butyrate by fermentation in the gut [20]. Thus, it appears that diets rich in certain foods can positively modify genes that increase overall cell health.

The benefits of the ketogenic diet might also go beyond treating existing disease, and instead help prevent chronic and degenerative disease [23]. A literature review by Miller et al. [23] argued that a state of nutritional ketosis will positively affect mitochondrial function and enhance resistance to oxidative stress and noted that the ketones directly up-regulate bioenergetic proteins that influence antioxidant defenses [23]. According to Boison [33], "Ketone bodies, such as β-hydroxybutyrate (BHB), and their derivatives have received the most attention as mediators of the anti-seizure, neuroprotective, and anti-inflammatory effects of KD therapy" [34–36]. The ketogenic diet's mechanism of action might be due to increased levels of adenosine [37,38], which blocks DNA methylation and, thus, exerts an epigenetic change. A study in epileptic rats subjected to the KD therapy found ameliorated DNA methylation mediated changes in gene expression by increasing adenosine [39], which blocks DNA methylation [40]. It is also being studied for its role in the aging process since it is linked to the positive regulation of epigenetic modifications, such as nuclear lamin architecture [41], reduced telomere length [42,43], DNA methylation, and chromatin structure [44].

The effect of the ketogenic diet on brain health appears to be well supported and is due specifically to the production of BHB [23]. They found that BHB is more than a fuel molecule; it plays important roles in cell signaling. The signaling functions of BHB link the effects of environmental factors on epigenetic regulation and cellular processes since it is an endogenous class 1 HDAC inhibitor [45]. Thus, a ketogenic diet has been linked to increased global histone acetylation, with a specific increase in the expression of protective genes, such as Foxo3a [46].

Evidence also suggests that BHB can have a direct epigenetic effect via a novel histone modification known as β-hydroxybutyrlation of H3K9, which results in improved gene regulation in the hypothalamus and improved overall aging [47]. Furthermore, the energy carrier molecule, nicotinamide adenine dinucleotide (NAD) is important in oxidative respiration. In its oxidative state (NAD+), NAD also acts as a cofactor for sirtuin enzymes and poly-ADP-ribose polymerase (PARP). Sirtuins and PARP play roles in gene expression, DNA damage repair, and fatty acid metabolism [46]. The energy available to a cell is measured by the NAD+/NADH ratio, which is modified by the utilization of glucose versus BHB as a fuel source [48]. During a ketogenic state, more NAD is found in the oxidative state which allows sirtuins and PARP to be more active. Additionally, catabolism of BHB into acetyl-CoA, another energy carrier molecule, raises acetyl-CoA levels. It has been found that the production of two moles of acetyl-CoA using BHB as the precursor reduces only one mole of NAD+ to NADH. However, four moles of NAD+ are produced by glucose metabolism. Thus, the ketogenic diet creates excess NAD+ for the cell and has a positive impact on the redox state of the cell [48]. This might have positive impacts on the activity of NAD+ dependent enzymes, such as sirtuins. Newman et al. [49] found that increased acetyl-CoA favors both enzymatic and nonenzymatic protein acetylation, specifically in the mitochondria, which improves overall mitochondrial function.

BHB produced by a ketogenic diet may also increase the efficiency of ATP production in the mitochondria and reduce the number of free radicals. As a result of the positive impacts of BHB, one study found that BHB precursor molecules improved cognition and disease progression in an Alzheimer's mouse model [50]. Additionally, the presence of BHB showed improvement in a case study of a patient with Alzheimer's disease [51]. The presence of D-β-hydroxybutyrate protect neurons from oxidative damage by reducing the cytosolic NAD+/NADPH ratio, resulting in an increase in the antioxidant agent known as reduced glutathione [52]. BHB also inhibits NF-kB, which is known to regulate the expression of multiple pro-inflammatory genes. This results in a diminished pro-inflammatory response [52]. Similarly, the BHB precursor, 1,3 butanediol, also modulates the expression of the inflammasome via histone β-hydroxybutyrlation. Thus, it reduces the expression of caspase-1, IL-1B, and IL-18 [53], which are inflammation markers. A study in *C. elegans* found that BHB alone could extend their life span [3]. Thus, the endogenous effects of BHB produced by a ketogenic diet might enhance health and increase longevity.

4. The Effect of the Ketogenic Diet on Weight Loss

According to recent Harvard models, 50% of the children today are likely to be obese by the age of 35 years [9]. As scientists try to determine the most effective strategies to combat the obesity epidemic, many studies have emerged that compare the health outcomes of different diets. A recent meta-analysis of seven random-controlled trials using diazoxide or octreotide for suppressing insulin secretion in obese patients found that it led to reduced body weight, fat mass, while maintaining lean mass [54]. However, the cost of artificially reducing insulin levels was an increase in blood glucose levels. While these studies seem promising as an indicator of biomarkers that can stimulate weight loss, it seems more logical to help patients achieve lower insulin levels via changes to their diet. The reduction of carbohydrate intake naturally reduces blood glucose levels, thus reducing insulin as a result. Many studies have now demonstrated that the ketogenic diet reduces both blood glucose and insulin levels [55–57].

A study conducted by Fumagalli et al. [58] analyzed the genetic profiles of patients and looked at the impacts on metabolism. They specifically looked at human CHC22 clathrin, which plays a central role in intracellular traffic of insulin-responsive glucose transporter 4 (GLUT4). The GLUT4 pathway is the dominant mechanism used by humans to remove glucose from the circulating blood after a meal. They found two major gene variants, one which is more frequent in farming populations than in hunter-gatherers. Hunter-gatherers have the gene that allows GLUT4 to be sequestered more effectively and thus have an inherent increased risk of insulin resistance. It is hypothesized that as humans became farmers and increased glucose in the diet, it was beneficial for the blood sugar to be lowered more easily with the newer form of CHC22. Thus, people with different forms of CHC22 are likely to differ in their ability to clear blood sugar after a meal. The people with the form that allows blood sugar levels to remain elevated could eventually lead to diabetes in the face of a high-carbohydrate load in the diet. This new finding might explain why some patients are successful on a high-carbohydrate low-fat diet, while others prefer to maintain weight with a low-carbohydrate, high-fat diet [58].

The importance of dietary adherence is of great concern for the success of any diet study. The study conducted by Shai et al. [59] that was able to control for the feeding of at least one meal a day (cafeteria meal), might better reveal the true effects of a sustained ketogenic diet. The Shai study [59] compared a low-fat, restricted-calorie diet (LFD), a Mediterranean, restricted-calorie diet (MD), and a low-carbohydrate, non-restricted calorie diet (LC) on 322 moderately obese subjects over a period of two years. The dietary adherence was >85% at the end of two years. This study instructed the LC group to be ketogenic for the first 2 months (<20 g/day) and gradually increase to 120 g/day of carbohydrates. The results found that the greatest weight loss occurred in the low-carb group and both the LC and MD were more effective than the LFD. Although, the weight loss during the first 3 months in the LC group was significantly greater than either of the other two groups, as carbohydrates were added back into their diet, their weight rebounded back to a level close to the MD group. Shai et al. [59] found that one of the benefits of the LC group was the similar calorie deficit achieved even though it was not a calorie-restricted diet. The researchers propose that a LC diet may be the optimal choice for individuals that cannot follow a calorie restricted diet since these subjects will be permitted to eat until satiated but will still most likely end up lowering their total caloric intake.

A similar long-term (56 week) ketogenic study was conducted on 66 obese people with a BMI >30 [60]. All patients were instructed to eat <20 g of carbohydrates in the form of green vegetables and salads for 12 weeks and then they could increase the carbohydrates to 40 g/day for the remainder of the study. The weight and body mass index of all patients decreased significantly. More interestingly, the patients were advised to maintain a state of nutritional ketosis and they were able to show continued decreases in both BW and BMI throughout the study. Consequently, this study did not show the plateau and gradual increases seen in the Shai study [59] which allowed the reintroduction of carbohydrates after the initial weight loss period. A similar study by Samaha et al. [61] also found that patients lost significantly more weight on a 30 g/carbohydrate per day diet for six months compared to a LFD. Another possible benefit from the ketogenic diet is that there is a measurable biomarker that signifies dietary adherence, which is β-hydroxybutyrate (BHB). When an individual is in ketosis, the body will begin ketone production and the level of BHB in the blood will be over 0.5 mmol. Studies that include this measurement can therefore confirm dietary adherence and determine the true effects of the diet on health outcomes, like weight loss. Mohorko et al. [57] conducted a 12-week ketogenic diet study on obese patients who were calorie restricted (1200–1500kcal) for the first two weeks and then were instructed to eat ad-libitum for hunger for the remaining weeks while eating the macronutrient composition necessary to remain in a state of nutritional ketosis. BHB was measured throughout the study and patients maintained levels above 0.5 mmol throughout the 12 weeks. Patients showed significant weight loss in both the men and women groups (average of (-)18 kg for men and (-)11 kg for women). Interestingly, as the diet progressed, the patients Fat Mass (FM) became the largest component of weight loss and it significantly correlated with BHB. Another valuable outcome in this study was the reduction of the hunger hormone, leptin, as well as a slight increase in energy expenditure, even while weight decreased throughout all 12 weeks. Another long-term study was done by Hallberg et al. [2] which followed diabetic patients on a ketogenic diet for one year. At the beginning of this study, 92% of the patients in the ketogenic group were obese. These patients were instructed to eat less than 30 g of total carbohydrates per day and the goal was to maintain BHB blood levels of 0.5–3.0 mmol/L. These patients had an average of 12% decrease in body weight, with some patients achieving as high as ~40% change. The patients who were in the standard care diet group (American Diabetic Association recommended diet) did not see any significant change in body weight [2].

A short-term, 4-week ketogenic diet (KD) on 20 obese Chinese females had profound outcomes [62]. In this study, compliance to the diet was measured with urinary ketone strips. These participants were given a monitored 4-week normal diet which was followed up with a 4-week KD with the same daily caloric intake but a drastic reduction in carbohydrates to <10% of calories. The effect was a significant decrease in body weight, body mass index, waist circumference, hip circumference, body fat %, and decreased fasting leptin levels. Similar positive outcomes were seen in other KD diet studies [56,63,64]. Similarly, a recent meta-analysis concluded that very low-calorie ketogenic diets are a very effective strategy for treating obesity [65]. An 8-week study conducted by Goss et al. [66] compared the very low carbohydrate diet (VLCD) (<10% carbohydrates) to a low-fat diet in older obese adults with BMI between 30 and 40. This study precisely measured fat loss with DXA and MRI measurements. Both groups exhibited decrease in total fat, but the VLCD experienced ~3 fold greater decrease in visceral adipose tissue and a significant decrease in intermuscular adipose tissue with a 5-fold greater reduction in total body fat mass.

Another long-term study monitored weight loss as well as changes in visceral fat mass using DEXA. The study by Moreno et al. [67] compared a very low-calorie ketogenic diet (VLCK) to a low-calorie (LC) diet as a treatment for obesity over two years. Participants in the active stage consumed 600–800 kcal/day and <50 g of carbohydrates per day until they were 80% of target weight loss goals (stage 1). Urinary ketone strips were used during stage 1 to confirm a state of ketosis. Then they used a standard low-calorie diet (10% below total metabolic expenditure) during stage 2 until they achieved another 20% weight loss, followed by long-term maintenance of weight loss in stage 3. The comparison control group used the low-calorie diet throughout the study to achieve weight loss. The weight loss in kilograms in the VLCK diet was double that of the LC diet throughout most of the study and remained significant. The amount of visceral fat loss in the VLCK diet group was 3X greater than the control group while preserving lean body and skeletal bone mass. The main side effects recorded in the VLCK were fatigue, headache, constipation, and nausea. However, none of these side effects were severe enough to cause the patients to drop out of the study and most subsided within the first month [67].

A meta-analysis conducted by Bueno et al. [68] compared randomized controlled trials of very low carb ketogenic diets (VLCKD) with low fat diets for 1 year. This study found a significant difference in decreased body weight for the VLCKD group. Another study compared a KD (<30 g carbohydrates/day) with two control groups (standard American diet (SAD) without exercise and SAD with 3-5 days of exercise for 30 minutes) over ten weeks [69]. The KD outperformed the other control groups in all variables tested, with 5 out of 7 being statistically significant. The patients showed significant decreases in body mass index (BMI), body fat mass (BFM), and weight while their resting metabolic rate

(RMR) increased. The RMR in the experimental group produced a positive, sizeable change with a magnitude of slope that was more than 10X the two control SAD groups. These results reveal that diet plays a more significant role in outcomes than exercise [69].

The ability to control hunger is also a key component to weight loss success. Castro et al. [70] evaluated patients from the very low-calorie ketogenic diet (VLCK) study and found a negative correlation between BHB levels and the urge to eat and feelings of hunger during the phase of maximum ketosis, even though there was no significant change in ghrelin hormone. This result is supported by other large investigations in overweight and obese adults which also found that low-carbohydrate diets were more effective in controlling hunger than low-fat diets [71,72]. A 2-week study conducted by Choi et al. [73] compared varying nutrition drinks on weight loss in obese adults. There were three groups: 4:1 fat to protein and carbohydrate ratio, 1.7:1 ratio with increased protein, and a balanced nutrition drink with similar carbohydrates to recommended dietary advice. All groups decreased body weight and body fat mass, but only the 1.7:1 KD-group maintained protein mass. Furthermore, only the KD groups improved blood lipid levels with appetite reduction. Since this was a nutritional drink feeding study, all the groups had similar caloric reduction; thus, results were due to macronutrient composition. In addition, levels of ketosis were strongly related to positive differences in food cravings, alcohol cravings, physical activity, sleep patterns, and sexual activity [73]. This outcome might also be supported by a recent finding that postprandial glycemic dips were the best predictor of appetite and energy intake following a meal and large glycemic dips are usually associated with high carbohydrate consumption [74]. Furthermore, a study showed that high carbohydrate meals had a greater impact on brain reward and homeostatic activity in ways that could impede weight loss maintenance [75]. Interestingly, the increased brain activity findings were partially associated with higher insulin levels, too. Thus, the ability of the KD to reduce hunger, lower glycemic fluctuations, and reduce influences on areas of the brain associated with addiction are all positive signs that a ketogenic diet should be considered as a treatment option for obesity.

One of the major concerns for rapid weight loss is the lowering of the resting metabolic rate (RMR). This bodily change can lead to weight regain, which is known as adaptive thermogenesis. Thus, it is typical for hunger to increase and energy expenditure to decrease during weight loss, which is a hindrance to long-term weight loss maintenance. Gomez-Arbelaez et al. [76] tested this outcome in subjects on the very low-calorie ketogenic (VLCK) diet study and followed them for 2 years. In this study, twenty obese patients lost 20.2 kg of body weight after four months and sustained this weight loss without the expected reduction in RMR. Authors of the study hypothesize that RMR did not drop because the subjects maintained their lean body mass. DEXA scans revealed that although they lost ~20 kg of fat mass, they only lost 1 kg of muscle mass. This conclusion was also supported by normal renal activity and positive nitrogen balance while subjects maintained their fat loss upon follow-up [76].

A study by Hall et al. [77] hypothesized that the development of obesity is "a consequence of the insulin-driven shift in fat partitioning toward storage and away from oxidation resulting from an increased proportion of dietary carbohydrates." To test this hypothesis, they tested seventeen obese men in metabolic wards with a four-week high-carbohydrate diet followed by a four week, isocaloric ketogenic diet. The results showed that a state of ketosis increased energy expenditure (~100 kcal/d), most likely due to beta oxidation and the partitioning of fuel towards ATP production rather than fat storage [77]. However, this level of energy expenditure change due to a ketogenic diet is not as high as measured in another study. In the study by Ebbeling et al. [78], it was noted that short-term feeding studies do not consider the body's process of fat adaptation, which takes at least 2–3 weeks, if not longer. Thus, the Framingham study by Ebbeling et al. [78] conducted a randomized trial on 164 patients where they lost weight and were then placed on varying diets of carbohydrate content for twenty weeks to measure changes in energy expenditure. The difference in total energy expenditure was 209–278 kcal/d or around 60 kcal/d increase for every 10% decrease in the carbohydrate percentage of total energy intake. This study concluded that dietary quality could affect energy expenditure independently of body weight. In accordance, Mobbs et al. [79] has suggested that ketogenic diets "reverse obesity by preventing the inhibitory effects of lipids on glycolysis, thus maintaining relatively elevated post-prandial thermogenesis." Further studies will need to be conducted to evaluate and confirm the exact mechanisms of action.

More recent studies on the KD are analyzing the outcomes of the diet in conjunction with other comorbidities related to obesity. A small study was conducted by Carmen et al. [80] that followed three obese participants on a 10% carbohydrate KD for 6–7 months that exhibited comorbid binge eating and food addiction symptoms. No adverse effects were found, and participants had reductions in binge eating episodes and food addiction symptoms. All three lost 10–24% BW and maintained treatment outcomes 9–17 months after initiating the diet and continued adherence to the diet [80]. Another study looked at the outcomes for male and female severely obese patients who also suffered from non-alcoholic fatty liver syndrome (NAFLD) [81]. They used a very low-calorie ketogenic diet of <50 g of carbohydrates and <800 kcal/day. Both males and females showed significant losses in body weight. However, males lost significantly more weight and had greater reductions in waist circumference. The patients also improved their biomarker for NAFLD, which was a reduction in gamma-glutamyl transferase [81]. To determine if the ketogenic diet negatively affects kidney function, Bruci et al. [82] conducted a 3-month very low-calorie ketogenic diet (VLCKD) study for weight loss in obese patients with and without mild kidney failure. All patients were advised to consume <20 g carbohydrates and 500–800 calories per day. The average mean weight loss from initial weight was nearly 20%, participants had significant reduction in fat mass, and 27.7% of the patients with mild kidney failure acquired normalized glomerular filtrate rate. It was, therefore, concluded that a KD not only leads to weight loss but also improvement in kidney function.

Please refer to Table S1 in the Supplementary Materials for a comparison of studies evaluating the KD in relation to weight loss outcomes.

5. The Effect of the Ketogenic Diet on Diabetes

According to the latest CDC report, an estimated 30 million people have diabetes and around 84 million have pre-diabetes. That statistic predicts that ~45% of Americans are either diabetic or pre-diabetic. Diabetes is a major health concern that is accompanied by a long list of secondary complications and diabetics are at increased risk of microvascular pathology of the retina, renal glomerulus, peripheral neuropathy, and atherosclerotic disease affecting arteries [83]. Many of these diabetic complications have been linked to elevated levels of glucose over long periods of time, which is measured as hemoglobin A1c (HbA1c) [83].

Type 2 diabetes is caused by hyperinsulinemia and insulin levels are directly affected by carbohydrate consumption. Protein intake can cause slight increases in blood glucose and subsequent insulin secretion, but fat consumption has no major effect on either [84]. If hyperinsulinemia is directly affected by nutrient intake, then it could be argued that these blood markers could be controlled by the conscious control of food choices. Of further note, the American Diabetes Association (ADA) recommends a goal of an HbA1c less than 7%, and the American College of Endocrinology sets a target level of 6.5%, even though few patients ever obtain that goal. Thus, Brownlee et al. [83] argued that patients should increase efforts to minimize glycemic variability since it can reduce risks of diabetic complications, independent of HbA1c. The DiRECT study by Lean et al. [85] found that weight loss alone could result in almost 46% of patients achieving diabetes remission at 12 months. Yet, this does not address the issue of diabetic patients who are not overweight. Thus, many scientists are now examining the potential benefits related to diabetes and improved blood markers that can result from eating a ketogenic diet. Although no professional organization in endocrinology or diabetology has focused on the rational use of ketogenic diet for either diabetes or obesity conditions, Kalra et al. [86]

argues nutrition should be considered as an integral part of metabolic management of diabetes, and the ketogenic diet should at least be offered as a treatment option.

Interestingly, the use of a diet low in carbohydrates for the treatment of diabetes is not a new or novel idea. In fact, prior to the invention of insulin, diet was the main intervention used by diabetic patients. The physicians, Dr. Elliot Joslin and Dr. Frederick Allen, were both recommending their patients in the 1920s to eat foods without carbohydrate content, and it highly resembled the current ketogenic recommendations [87]. According to Feinman et al. [88] the number one goal of both type 1 and type 2 diabetics should be glycemic control. It is argued that carbohydrate restriction can benefit diabetic patient blood markers even in the absence of weight loss [88]. This is important since many diabetics are not overweight yet still need to manage their blood glucose levels. The benefits of carbohydrate restriction in type 1 diabetics reduces the error in determining insulin amount to match the increased blood glucose since dramatic spikes are less likely [88].

A recent study compared the use of a low-calorie (LC) diet vs. a very low-carbohydrate ketogenic diet (VLCKD) on health outcomes for type 2 diabetics. The VLCKD group approached normal blood sugar level in just 24 weeks unlike the LC group [87]. The VLCKD group reduced insulin doses by half, on average, and sulfonylurea doses were halved or discontinued. The HbA1c levels dropped significantly in the VLCKD to 6.2% vs. 7.5% in the LC group. Thus, the VLCKD group managed to reach both the ADA and American College of Endocrinology target level for HbA1c. According to Hussain et al. [87] the VLCKD was not found to have an adverse effect on glucose metabolism, insulin resistance, or cause chronic dehydration. However, they did caution that diabetic patients should only attempt this nutritional therapy while being closely monitored by a physician to reduce the risk of hypoglycemia since drugs will need to be quickly reduced to match changes in blood markers elicited by the diet [87]. A study by Webster et al. [89] found that type 2 diabetic patients who self-selected to follow a KD reduced their mean HbA1c from 7.5% to 5.9% at the 15-month follow-up. As a result, their HbA1c levels reached the normal range (which is under 6.0%), and they had achieved partial or full type 2 diabetes remission.

A study conducted by Westman et al. [8] compared the effects of a low-carbohydrate ketogenic diet (LCKD) versus a low-glycemic index diet (LGID) on glycemic control in type 2 diabetic patients which was measured by hemoglobin A1C (HbA1c). They enrolled forty nine patients and randomly assigned them to the different diets. Both groups followed group meetings, nutritional advice, and an exercise recommendation. Both interventions showed improvements in hemoglobin A1c, fasting glucose, fasting insulin, and weight loss. However, the LCKD had greater improvements, including a reduction or elimination of diabetes medications in 95% of patients vs. 62% in the LGID group [8]. As mentioned previously, the study by Dashti et al. [60] compared the health outcomes of a ketogenic diet on obese diabetics with high blood glucose levels to non-diabetic obese patients over 56 weeks. This study concluded that all markers, such as body weight, body mass index, blood glucose, total cholesterol, LDL, triglycerides, and urea all showed a significant decrease in both groups throughout the study, with more positive outcomes seen in the diabetic group [60]. The kidney tests also showed normal function. This study demonstrated that the diet is safe to use for longer periods of time in obese diabetic subjects.

A year-long randomized study compared the effects of a very low-carbohydrate ketogenic diet (LCK) versus a moderate-carbohydrate, calorie-restricted, low-fat diet (MCCR) in pre-diabetic or type 2 diabetic patients [90]. The results showed that the LCK exhibited greater improvements in their HbA1c, weight loss, and medication use than those assigned to the MCCR diet [90]. Another randomized controlled study by the same researcher compared the LCK against the diet program based on the online American Diabetes Associations' "Create Your Plate" diet. The purpose of this study was twofold. The researcher had already seen the benefits of the LCK in a previous study that had personalized intervention. They wanted to see if an online program could be just as successful at helping overweight individuals with type 2 diabetes. The results indicated that the online ketogenic program was more successful in helping patients manage their diabetes by reducing their HbA1c, lowering triglycerides, increasing weight loss, and retention rates were higher than in the control group [91]. In addition, a previous study has discovered that a carbohydrate restricted diet was more successful than a low-fat diet in improving diabetic markers for metabolic syndrome in forty subjects with atherogenic dyslipidemia [92].

A recent study recently conducted at Indiana University was one of the first long-term studies that required use of routine blood tests to determine the patients' state of nutritional ketosis while maintaining a KD diet. Patients were highly compliant, and experienced improved diabetic conditions [2]. The diet intervention also reversed the diabetic status of some patients, whose HbA1cs became normal. The 2-year follow-up to this study revealed that 74% of KD group remained enrolled [93]. This group had a significant improvement in HbA1c, fasting glucose, and fasting insulin while the usual care group had no changes from baseline. The mean dose of prescribed insulin decreased by 81% and the diabetes reversal increased to 53.5%. Diabetes remission was 17.6% and diabetes complete remission was 6.7% [93]. The long-term success in diabetes treatment for this digitally monitored continuous care intervention group is evidence of the feasibility and adherence of the KD in type 2 diabetes treatment [2,93].

Additionally, the study by Shai et al. [59] showed that patients were able to reduce their fasting blood glucose on a low carbohydrate or a Mediterranean diet, while the low-fat group saw the opposite effect. The patients in the low carbohydrate group were also able to significantly decrease their HbA1c [59]. Another meta-analysis that compared very low-carbohydrate ketogenic diets (VLCKDs) to low-fat diets (LFDs) found that the VLCKD showed greater improvements in fasting glucose, insulin analysis, HbA1c, and C-reactive protein [68]. Additionally, a recent meta-analysis of low-carbohydrate or very-low carbohydrate diets found that patients adhering to the diet for 6 months can have diabetes remission without severe complications [94]. Several recent studies on the KD show positive improvements in glycemic profiles [56,66,82,89,95].

Currently, the ADA recommends that type 1 diabetics eat a low-fat diet rich in whole grain carbohydrates. One study showed the low-fat diet has not been found to improve HbA1c in all patients, regardless of diabetes state [96]. It looked at the HbA1c outcomes for type 1 diabetics (T1D) who were advised to reduce carbohydrate intake (<75 g of carbs/day) to reduce the need for insulin. The patients in this study had a 50% adherence rate, and those who strictly adhered to the diet reduced their HbA1c by 1.8%. Another randomized trial [97] determined the feasibility of a LC diet (<75 g/day) versus standard carb counting in adults with T1D. Of the ten people in the 12-week study, the LC group exhibited significant decreases in HbA1c, decreased daily insulin use, and reduction in body weight. All of the outcomes in the carb counting group were unchanged. Thus, these T1D patients had positive outcomes without meeting the KD threshold of <50 g/day while consuming significantly less carbohydrates than the typical diet.

Interestingly, some type 1 diabetes patients have taken it upon themselves to treat and control their diabetes with the very low–carbohydrate diet (VLCD), against the advice of current medical professionals. Lennerz et al. [98] evaluated the results of this choice by recruiting type 1 diabetics who self-selected to follow a VLCD (<30 g/day). They found these patients on a social media site and then asked for permission to contact physicians and confirm health outcomes. Shockingly, 97% of the patients were able to achieve the ADA glycemic targets for HbA1c with an average of 5.6% and a mean daily insulin dosage of 0.40 U/kg per day. Participants in this group reported increased levels of overall health, increased satisfaction with diabetes management, and decreased number of adverse events. These results are unprecedented in type 1 diabetic patients. If these outcomes are confirmed in clinical trials, the chronic health issues associated with type 1 diabetes could be prevented or significantly reduced by diet alone. Almost one-fourth of these patients did not discuss their VLCD with their care providers, which means they were making these changes without the support of their physicians. Even in an intensively treated group in the Diabetes Control and Complication Trial, the best HbA1c achieved was 7.2%, but that was coupled with increased rates of hypoglycemia [98].

Although there are only a few randomized controlled trials evaluating the effects of the KD on diabetes, there are some recent case studies and qualitative studies that shed some light on the issue [55,64,99,100]. The positive outcomes in these studies might reflect the motivation of these patients who opted or volunteered to ensue KD diets. A paper by Walton et al. [64] presented 11 case studies on women with T2D that volunteered to eat a KD with <30 g of carbohydrate per day. Their HbA1c was > 6.5% and dropped to 5.6% with diabetes reversal. Another case study by Lichtash et al. [99] involved a women patient with T2D and normal weight. After failed glycemic control with standard care, she voluntarily began a KD with intermittent fasting. Her HbA1c dropped from 9.3% to 5.8% after 14 months while maintaining her weight. Similarly, Wong et al. [100] examined type 1 and type 2 diabetics who opted to do a KD for >3 months. Participants reported better glycemic control, decreased medicine use, weight loss, and satiety. Most of these patients expressed the KD as a normal way of eating and plan to continue for the rest of their lives. A similar T2D cohort was recruited [55] for a retrospective study on 49 patients who followed KD for >3 months and compared their outcomes to 75 patients who followed usual care (UC). 100% of the KD cohort either discontinued or reduced insulin dosage while only 23% of UC did. The KD cohort had a greater reduction in fasting plasma glucose, weight loss, as well as a superior reduction in HbA1c compared to UC. Thus, it seems that those patients who opt to follow the diet are having positive outcomes.

Please refer to Table S2 in the Supplementary Materials for a comparison of studies evaluating the KD in relation to diabetic outcomes.

6. The Effect of the Ketogenic Diet on Lipidology and Cardiovascular Risk

Cardiovascular disease (CVD) and its risk factors are a major health issue in industrialized nations. Moreover, large epidemiological studies are starting to show that CVD is becoming a larger problem in developing or low-income countries as well [101]. There has been a long-standing viewpoint that a diet high in saturated fat is unhealthy and will eventually lead to cardiovascular disease. Many hypothesized that a diet rich in saturated fat will increase LDL, and thus more fat in the blood leads to fat deposits in the vessels, resulting in increased risk of cardiovascular disease [102]. This idea was fortified by Ancel Keys in his 7-country study and eventually led to the diet-heart hypothesis [102]. Moreover, the United States accepted the idea proposed by Keys and adopted the low-fat diet (LFD) as the optimal diet to fight the increasing levels of CVD in the U.S. Additionally, the mainstream view for decades was that high total cholesterol also leads to atherosclerosis and cardiovascular disease [103]. As a result, the prescription of the LFD that consisted of ~60% energy from carbohydrates became the standard of care for physicians starting in the 1980s [98]. According to the 2015 Dietary Guidelines for Americans, people are still recommended to consume a diet that limits saturated fat intake to less than 10%, with some organizations placing even stricter limitations and advising around 7% [104]. However, randomized controlled trials have started to question the validity that saturated fat intake and a single blood marker, LDL, can accurately predict risk. Many scientists now argue for the need to analyze specifically how different types of macronutrients that replace saturated fat in the diet are impacting risk [105]. It is also important to consider the data regarding LDL as the single biomarker chosen to monitor and determine cardiovascular risk [105].

New research is also starting to question that mindset set forth by Ancel Keys, and many scientists have argued that the global dietary recommendations should be revisited and updated [106]. For example, a recent analysis of the literature done by Ravnskov et al. [103] compiled all the data on PubMed from initial to 2015 on over 68,000 patients. Ravnskov et al. [103] argued that that if the main goal of prevention of disease is prolonging life, then all-cause mortality should be the measurement used for determining health outcomes. Interestingly, they found that 30% of patients showed no association between LDL and all-cause mortality, while 70% showed a statistically significant inverse relationship. Contradicting the diet-heart hypothesis, they also found that the 4-year mortality among patients with the highest levels of LDL were almost 36% lower than those patients with the lowest LDL levels. Furthermore, the patients placed on statins had higher rates of mortality risk than those with the highest LDL. The results of these studies question the standardized method of using total cholesterol and LDL as the biomarkers of coronary heart disease.

Thus, if total cholesterol and LDL are not true indicators of cardiovascular risk, then one must ask what other blood markers could serve as better indicators of coronary heart disease. In a review by Feinman et al. [88] they argue that the best indicators of CVD risk are ApoB [107], the ratio of TC/HDL, increased levels of small dense LDL particles (sdLDL) [108,109], and the ratio of ApoB to ApoA1 [88]. If these markers are, in fact, a better indicator of disease risk, then understanding the effect of diet on these other biomarkers is of great importance. One study by Krauss et al. [110] compared patients who consumed diets of varying carbohydrate intake (54%, 39% or 26%) with the amount of saturated fat varying between 7% or 15%. This study showed that a high saturated fat intake, combined with carbohydrate restriction (26%) did raise total LDL. However, the higher total LDL levels were due to an increase in the larger sized LDL particles, which are less atherogenic than the sdLDL, and the patients saw a subsequent lowering of the sdLDL particles [9,110].

A large prospective study called the European and Prospective Investigation into Cancer and Nutrition Study (EPIC) also found that diets high in glycemic load (GL) and glycemic index (GI) were associated with a greater risk in Cardiovascular Heart Disease (CHD) [111]. Glycemic index is a measurement of the ability of carbohydrates to increase blood glucose levels. The glycemic load is the product of the GI of a particular food and its available carbohydrate. This study included around 520,000 men and women between the ages of 35 and 70 over a period of 8 years [111]. The study found a greater risk of CHD with higher sugar consumption. Their findings supported other observational studies that suggest that replacing saturated fat with sugar or refined carbohydrates might increase cardiovascular risk, rather than lower it [112,113]. Additionally, the very large PURE study recently showed that a diet higher in saturated fat did increase LDL, but also increased HDL, lowered triglycerides (TG), lowered the TC/HDL ratio, and lowered the ApoB/ApoA1 ratio [106]. They also found that the diets high in carbohydrate intake had the complete opposite effect on these atherogenic biomarkers. The benefit of the PURE study is that it revealed the risk associated with varying macronutrient composition in diets from over 5 continents in 18 countries, regardless of cultural food trends. Thus, it was a global look at the effect of dietary patterns on health regardless of background and ethnicity. The PURE study concluded their findings do not support the current recommendations to limit total fat intake to 30% of energy and saturated fat to less than 10%, and the recommended amount of <7% saturated fat might even be harmful. Instead, they argue that individuals who eat a diet high in carbohydrates might benefit by replacing some of those carbs with fat [106]. According to the PURE study, the ApoB to ApoA1 ratio was the strongest lipid predictor of myocardial infarction and ischemic stroke. Since this biomarker has been found to increase with carbohydrate intake, they concluded that this factor could provide the mechanistic explanation for higher risks seen in people with the highest carbohydrate intake [106]. This idea was supported by a recent article on Medscape, which argued that the predictive power of the ApoB to ApoA1 ratio was superior to other biomarkers to assess CV risk [114]. It also mentioned adding other lipid parameters to the ApoB/ApoA1 ratio did not improve the predictive power.

A study done by Lu et al. [115] compared the ability of either the ApoB/ApoA1 ratio or LDL to predict coronary heart disease (CHD) in normal and overweight patients. They found every quartile increase in the ApoB/ApoA1 ratio showed an increase in CHD prevalence. Meanwhile, the increases in LDL quartiles did not predict the highest percentages of CHD [115]. The ratio had an even stronger predictive capability in the overweight subjects. Furthermore, other studies have also supported the findings of the PURE study. One study conducted on postmenopausal women found an inverse relationship between dietary saturated fat intake and atherogenic disease progression [116]. Another study previously mentioned even found a positive association between plasma phospholipids and CHD mortality [117]. According to another study conducted by Dreon et al. [108], a decrease in saturated fat intake did lower total LDL, but it appeared to only reduce the amount of the large, buoyant LDL particles. They argue that more emphasis on CVD risk should be placed on high levels of triglycerides (TG), decreased concentration of HDL, and increased amounts of sdLDL particles. If these biomarkers are potentially more effective predictors of coronary heart disease, then the analysis of a diet's effect on these lipid markers is of great importance [109].

Only a few studies have looked at the health impact of very high fat consumption (VLCKD) on overall health (which could include analysis of weight maintenance, lipid profiles, and inflammation markers [69]. To accurately determine the effect of a KD on cardiovascular risk markers, it is important to only look at studies that restricted carbohydrates below 50 g/day to ensure the patients would be in a state of nutritional ketosis. One study compared a KD to the standard American diet (SAD) and the SAD plus exercise. Not only did the KD outperform the other groups in multiple health outcomes, but it also showed a much more significant decline in triglycerides [69]. Another study compared a LC diet group (<30 g/day) to a LF diet in obese patients after 6 months [61]. Once again, the LC group had a drastic decrease in TG, while no significant difference was seen in total cholesterol (TC), HDL or LDL. This led investigators to conclude that the LC diet did not have adverse effects on serum lipid levels.

The impact of the prescribed low-fat diet versus diets higher in fat on cardiovascular lipids levels are beginning to emerge. One 2-year diet study compared the effect of a low-fat diet (LFD), low carbohydrate diet (LC), and a Mediterranean diet (MD) on lipid profiles of overweight patients [59]. The LC group had a significant decrease in triglycerides and the total cholesterol/HDL ratio decreased the most in the LC group. Their ratio decreased by 20% compared to a 12% decrease in the LF group [59]. The beneficial biomarker, HDL, increased in all groups, while the LDL changes were similar in all groups, which has also been noted in other studies [118]. A metabolic ward study of shorter duration conducted by Hall et al. [77] also found that triglycerides decreased in the reduced carbohydrate group. However, they saw the LDL levels increase in the LC group. Meanwhile, the Choi et al. [73] study mentioned earlier did not find an increase in LDL. It was conducted on obese patients with tightly controlled nutrition drinks, which had similar calorie reduction. Only the KD groups improved blood lipid profiles while reducing appetite. The KD groups saw a decrease in triglycerides and LDL, and no significant change in HDL [73].

Another 6-month study compared a low-calorie KD to a low-calorie diet in obese patients; some were diabetic. They found that both the diabetic and non-diabetic patients in the KD group showed the best lipid outcomes [87]. They found a significant decrease in triglycerides, a decrease in total cholesterol, a decrease in LDL, and an increase in HDL. A study conducted by Walton et al. [64] followed 11 women with type 2 diabetes for 90 days on a KD. The women in this study had increased HDL, a significant decrease in TG, and a significant decrease in the TG: HDL ratio, although LDL levels were not significantly changed. Another cardiovascular benefit was the lowering of the patient's systolic and diastolic blood pressure. When evaluating type 1 diabetic patients who self-selected to be on a LCD, they found that these patients showed a decrease in TG, while having increases in HDL, TC, and LDL [98]. The researchers hypothesized that the total LDL elevation on the KD, if associated with a low TG, may reflect an increase in the large, buoyant lipoprotein particles which are considered a lower risk subtype. When the KD was followed for one year in type 2 diabetics and adherence was confirmed with BHB, it was noted that TG decreased by 24%, HDL increased 18%, LDL increased 10%, while ApoB was unchanged [2]. Although these lipid changes are considered favorable, the increase in LDL seen in some groups is still an area of concern. One analysis suggested that the risk from a slight increase in LDL might be offset by emphasizing the consumption of unsaturated fatty acids rather than saturated fatty acids [9].

The DIETFITS study also concluded that the increase in saturated fat intake may improve overall lipid profiles if they are adhering to a high-quality, whole-food based, low carbohydrate diet [104]. One major area of concern would be whether the KD would have these same beneficial changes in patients with dyslipidemia. A 56-week study tested the effect of the KD on obese patients with and without high cholesterol levels [119]. It is important to note that these patients were instructed to include 5 tablespoons of olive oil into the diet, which is a form of unsaturated fatty acids. Throughout the experiment, the patients saw continuous improvements in their lipid markers. Not only did both groups have decreased LDL, decreased TG, and increased HDL levels, but the patients with high cholesterol levels also ended the study with blood profiles that were more like normal subjects.

A more recently published case study on a young man who used a Mediterranean KD diet for treating his IBS had some interesting findings [120]. The doctors looked at more detailed lipid subfractions to determine the lipid outcomes of cardiovascular risk, which was unique. First, the authors mention that a typical lipid profile analysis would suggest the diet was having adverse effects on the patient. His total cholesterol changed from 160 to 450 mg/dL, even though a portion of that was due to increased HDL levels. Many argue that HDL-P is a superior predictive measure of good cardiovascular health. The HLD-P in this patient increased from 5699 nmol/L to 12,080 nmol/L. The current association between LDL-C and cardiovascular risk is driven by atherogenic small dense and/or oxidized LDL. It is believed that these two components can penetrate the endothelium of blood vessels and contribute to plaque formation [121,122]. Yet, large LDL are not associated with cardiovascular risk and may provide a protective effect. This patient saw an increase of LDL from 90 to 321 mg/dL. The LDL subfraction revealed that almost the entire increase in his LDL-C was caused by an increase in large LDL, while his small and medium LDL decreased by almost 10%. Thus, these authors argued that the typical analysis of lipid profiles from patients on ketogenic diets may not accurately reveal risk unless more detailed lipid subfraction tests are conducted.

Please refer to Table S3 in the Supplementary Materials for a comparison of studies evaluating the KD in relation to lipidology outcomes.

7. The Effect of the Ketogenic Diet on Cancer

Cancer currently remains the second leading cause of death (~22%) in the United States and is second only to heart disease [123]. Typically, cancer occurs in adults because of multiple mutations in numerous genes, genes that usually regulate cell growth and proliferation [124–126]. Today it is the accepted model that as many as six mutations need to occur to produce cancer (usually to oncogenes and tumor suppressor genes). Oncogenes are genes that regulate cellular pathways that can increase cellular growth, while tumor suppressor genes regulate pathways that inhibit abnormal cell growth. As the mutated cell population expands, it accrues the necessary changes to ignore growth control signals, avoid apoptosis, escape immune surveillance, and creates an environment to thrive (using mechanisms like angiogenesis and tolerance for anoxic environments) and eventually the capability to metastasize [124]. These mutations can result from many causes, such as DNA replication errors, failed DNA repair mechanisms, mutagen exposures, or increased reactive oxygen species [127].

Consequently, preventative mechanisms that could lower cancer incidence would be related to reducing these external causes or activating internal pathways to reduce cellular error. Additionally, epidemiologic evidence linking obesity to elevated cancer incidence found that 14% and 20% of all cancer deaths in men and women, respectively, are due to being overweight and obese [128]. As a result, the Annual Report to the Nation on Cancer emphasized the increasing contribution of obesity on cancer incidence [129]. One mechanism believed to contribute to obesity's role in cancer is the increase in adipocytes in the body, which can increase circulating levels of insulin and Insulin Growth Factor 1 (IGF1) hormones. These hormones bind receptors in many cell types and activate P13K/AKT signaling pathways that increase cell survival and upregulate transcription factors that promote cell proliferation [130]. Both hormones also increase glucose uptake into cells, resulting in increased energy molecules being available for cell growth. Insulin is an anabolic hormone that promotes glucose uptake into cells, reduces the release of fatty acids from adipocytes, prevents ketone production in the liver, and stimulates fat and glycogen storage [131]. Additionally, many recent publications support the idea that prolonged, increased levels of serum insulin is likely to promote cancer growth [132–134].

The alterations in the metabolism of cancer cells were first described by Warburg et al. in 1927 [135] It was discovered that cancer cells acquire mutations in critical genes that change the way cancer cells acquire energy. First, cancer cells use glycolysis for ATP production and reduce their dependency on the oxidative cellular respiration in the mitochondria. This results in the cancer cells gaining only 2 ATP per glucose molecule instead of the average 36 ATP from typical cellular respiration processes, resulting in an enormous demand for glucose. Secondly, it allows the cancers cells to rapidly divide even in the absence of oxygen, since glycolysis is an anaerobic process that occurs in the cytosol. Currently, altered metabolism has been described as a primary signature of cancer [125,136,137]. Since this discovery, the use of metabolic therapies for dealing with cancer have been overshadowed by discoveries in the genetics and molecular signatures of cancer [138].

Therefore, it seems reasonable to hypothesize that diet could have profound effects on reducing cancer risk, especially if that diet is known to decrease body weight, lower insulin levels, and target the metabolic weaknesses of cancer cells. Some researchers hypothesize that the ketogenic diet might reduce cancer risk because it capitalizes on the reduced expression of ketolytic enzymes in cancer cells [48]. The diet would starve the cancer cells by reducing their ability to utilize glucose, while normal cells can adapt and begin utilizing ketone bodies for their energy demands. Another potential benefit could be the decrease in insulin that results from being in nutritional ketosis, which would reduce insulin-like growth factors that support cancer proliferation [48]. Especially given the fact that 20% of all cancer cases in North America can be attributed to obesity and 38% of all attributable cancer cases are linked to the increase in BMI since 1982 [139]. There have also been numerous studies that have linked cancer risk to hyperinsulinemia [140–144]. It is suggested that insulin resistance leads to hyperinsulinemia, and insulin has both pro-mitotic and antiapoptotic activity that may assist in tumor progression. Thus, any diet that can reduce obesity and lower insulin levels, such as the ketogenic diet, might reduce cancer risk.

Support for a KD as a mono-therapeutic approach for treating cancer is demonstrated in many mouse models. However, due to the heterogeneity of these studies (types of cancers, KD protocol, length of study, etc.), we discuss them separately. Poff et al. [145] tested a KD on systematic metastatic cancer in mice. They found that KD alone significantly decreased blood glucose levels, reduced tumor growth, and improved mean survival time by 56.7%. A similar study looked at the effect of the KD on mice with gastric tumor cells. Both tumor growth and mean survival time were improved [146]. In one study, Allen et al. [147] found that a KD reduced tumor growth in lung cancer xenografts.

In another study, they tested the use of a calorie-restricted KD on the growth and vascularity of malignant mouse astrocytoma (CT-2A) and human malignant glioma (U87-MG). When compared to an unrestricted high carbohydrate standard diet, they found that tumor growth decreased by 65% for CT-2A and 35% for U87-MG tumors [148]. They also found that signs of angiogenesis were reduced in the calorie restricted KD group. It is important to note that the mice in this study were fed KetoCal, a new nutritionally balanced high fat/low carbohydrate ketogenic diet for children with epilepsy. This finding suggests that the use of KetoCal should be considered not only for epilepsy, but as an alternative therapeutic option for malignant brain cancer. Another study found that a KetoCal KD diet also increased mean survival time and slowed tumor growth in mice with brain cancer [149]. Additionally, one study on mice by Morsher et al. [150] compared a KD and SD on neuroblastoma, with or without calorie restriction. It was found that the best results were in the calorie restricted KD group, with reduced tumor growth and survival time.

Meanwhile, a few studies have tried to compare the effect of a KD (with varying levels of carbohydrate amounts) on prostate cancer, with differing results. Caso et al. [151]

studied mice that were either randomized into a standard Western diet, non-carbohydrate KD (NCKD) with 0% carbs, 10% carbohydrate KD, or 20% carbohydrate KD. The group with the slowest tumor growth was the 20% carbohydrate KD, while the WD had the most rapid growth. However, they did not find a significant improvement in survival among any of the carbohydrate restricted groups when compared to the WD. This result is different than a similar study done by Masko et al. [152], which compared a NCKD, 10% carbohydrate, and 20% carbohydrate diet in mice with prostate cancer. They concluded that none of these diet groups differed greatly in their tumor size throughout most of the study, and the diet did not affect survival. However, another study conducted on mice with prostate cancer compared a WD with a NCKD and found that the NCKD was significantly associated with lower tumor volumes at the end of the 53-day experiment [153]. Regardless of the varying results, a meta-analysis done by Klement et al. [154] analyzed a total of 29 animal studies and found that the majority (72%) found evidence of reduced tumor growth because of KDs.

The data of the effect of KD in human patients is limited mostly to case studies and cohort studies. A meta-analysis of 24 human studies, found that 42% found that the KD can reduce tumor growth [154]. In addition, it has been found that most human studies had positive impacts [154,155], with many other studies found it stabilized disease [154,155] and one study found a pro-tumorigenic effect of the KD [154,155]. However, another review of 14 studies of the use of KD in cancer found mixed results [154]. It was found that people responded differently to the diet, with some cancers being reduced, some neutral in effect, and some cancers getting progressively worse. This finding could be related to a recent publication by Chang et al. [156] that tested relative expression of several key enzymes in ketolytic and glycolytic metabolism in human anaplastic glioma and glioblastoma. They found genetically heterogeneous tumors with varying expressions of key enzymes. However, they found most cells had an enzyme profile with decreased levels of mitochondrial ketolytic enzymes and increased expression of glycolytic enzymes, suggesting that human brain tumors are more dependent on glucose and have defects in ketone metabolism.

The prognosis of patients with gliomas is extremely poor, with an average survival duration of 1.5 years [138]. Due to the poor outcomes with brain cancer, many studies using KD have been aimed at helping brain cancer patients. A small study by van der Louw et al. [157] followed three patients with recurrent diffuse intrinsic pontine glioma (DIPG). Although all three patients succumbed to the disease, it was determined that the use of KD is safe and feasible, but its effect on survival was not clear. Another 12-week randomized, controlled study also found that the use of KD in women with ovarian and endometrial cancer had favorable effects on physical function, perceived energy, and diminished food cravings for starchy and fast-food fats [158].

One of the most intriguing studies was a case study of a 38-year-old man with glioblastoma multiforme was treated with standard of care (SOC) along with a calorie-restricted ketogenic metabolic therapy, hyperbaric oxygen therapy, and other metabolic therapies [159]. The patient remains in excellent health with no neurological issues after 24 months of treatment. Thus, it seems that the ketogenic diet might be best utilized as an adjuvant therapy and should be started when the disease is first diagnosed. Recently the KEATING study [160] used either the modified ketogenic diet (MKD) or the medium chain triglyceride ketogenic diet (MCTKD) as an adjuvant therapy for glioblastoma. The Global Health Status (GHS) increased for patients in MKD cohort and decreased for the MCTKD patients. They had a low retainment with only 3 of 12 patients completing the 12-month intervention. The three patients who did complete the study chose to continue doing the KD. The researchers of the KEATING study suggested that the KD intervention should be reduced to six weeks and only be utilized during the time of chemo and radiation therapy.

Yet, another study by Panhans et al. [161] had greater compliance. This study recruited patients with a diversity of CNS malignancies (GBM, astrocytoma, and oligodendroglioma). These patients were asked to do a more standard KD of 3:1 for 120 days and aimed to keep carbohydrates under 20 g/day. One cohort was provided KD meals by Epigenix Foundation for the first 30 days, while the others were given only meal plans. Adherence to the diet was confirmed with ketone and glucose levels measured with Precision Xtra meters. The six patients with the highest ketones were alive at the end of the study. The two patients with the lowest ketones succumbed to their disease. Five patients were able to maintain 100% adherence for the duration of the study. Overall, patients' symptoms improved, which included higher energy levels, increased physical activity, increased cognitive function, decreased appetite, and reduced seizure. It is important to note that one patient had increased seizures. The researchers stated that the KD was well tolerated and discussed its feasibility for future experiments. This cancer clinic also stated that as interest in the KD grows, they now openly discuss the risks and potential benefits on a regular basis with patients and emphasize the lack of robust clinical evidence.

The ketogenic diet is also now being tested as an adjuvant therapy for other cancers as well. For example, Clinicaltrials.gov currently lists over 100 trials looking at the ketogenic diet and 12 of those were related to CNS malignancies [161]. Therefore, data is starting to emerge on the impacts of KD on other cancer types. For instance, a study compared the typical diet with 55% of calories from carbohydrate (CHO) against a KD with around 6% from CHO in breast cancer patients in a 6-week trial [162]. The KD group's global quality of life was higher at the 6-week mark and no adverse effects were seen in either group. Interestingly, the KD group lowered caloric intake without any restrictions, which may have been due to the satiating effects of fat. The KD diet was found to have no adverse effects on thyroid hormones, electrolytes, LDH, urea, or albumin. Yet, the KD diet was found to have potential beneficial effects, such as significantly reduced levels of lactate and ALP. Decreased lactate levels might slow metastases by reducing the acidity of the tumor microenvironment while reducing its ability to use it as a substrate for increasing biomass. Furthermore, it is believed that increased levels of ALP in breast cancer is a negative prognostic marker.

Another 12-week study in ovarian and endometrial cancer patients found an adherence level of 57–80% [163]. The focus of this study was to determine if the diet negatively affected lipid profiles since that is a current concern of many doctors and may restrict their decision on whether to suggest the KD diet for their cancer patients. They compared the KD versus the American Cancer Society (ACS) high-fiber, low fat diet. No changes were seen in lipid profiles to TC, TG, HDL-C, LDL-C, TC:HDL-C ratio or TG:HDL-C ratio after adjusting for baseline levels and weight loss. Another recent study looked at the effects of the diet on the body composition of KD patients while receiving radiation therapy. Klement et al. [164] compared a nonKD vs a KD with supplemental essential amino acids (KETOCOMP study). The KD had significantly associated with loss of 0.5 kg of fat mass and 0.4 kg of body weight per week, while showing no change in fat free mass or skeletal muscle mass. Thus, KD with ample amino acid intake could improve body composition during radiotherapy. Finally, a recent study conducted by Hagihara et al. [165] analyzed the effects of a 3-month KD as an adjuvant therapy for patients with advanced cancers of many types. They found that the diet was well tolerated, did not have any major negative outcomes, and improved life expectancy. Researchers were also able to stratify survival outcomes with three factors: albumin, blood sugar, and CRP levels. Thus, it was argued that stable adherence and highly reproducible results should be in favor of using the ketogenic diet as a standard for therapeutic treatment during chemotherapy with advanced cancer diagnoses.

Please refer to Table S4 in the Supplementary Materials for a comparison of studies evaluating the KD in relation to cancer outcomes.

8. Discussion

A well-formulated ketogenic diet can provide low carbohydrate intake, while providing adequate fiber sources such as seeds, nuts, coconut, avocado, spinach, broccoli, cauliflower, and berries. Together, all these rich pre-biotic foods would lead to an increase in *Bacteroides* and *Bifidobacterium* and a subsequent decrease in *Firmicutes*. With disease rates increasing rapidly in the United States and other modern nations, it is increasingly important that we determine the safety, efficacy, and potential life-saving benefits of alternative diets. What might be discovered is that patients should be given individualized diets based on the species comprising their microbiome. This might enable patients to eat certain foods that maximize their ability to remain in a state of nutritional ketosis and optimize their overall health outcomes. The regular monitoring of the microbiome might be necessary to continually moderate and change dietary needs for diversity. It might also be determined that fecal microbiota transplants might be necessary to fully alter and change the microbiome at the onset of a new diet which could then be further modified and enhanced through diet. Regardless, much more research is needed in this area to determine the effect of the ketogenic diet on the microbiome.

Even though the ketogenic diet shows promise in helping patients lose weight, obesity is more than excess adipose tissue being stored on the body. It has been linked to many other metabolic issues, such as diabetes, cardiovascular disease, neurological disorders, and cancer. The ability to improve glycemic control in diabetics is critical for long-term health especially since some would argue that the biggest indicator of cardiovascular disease risk was HbA1c [88]. Surprisingly, the United Kingdom Prospective Diabetes Study (UKPDS) examined 5102 newly diagnosed type 2 diabetics and found that patients showed a 14% decrease in myocardial infarction for every 1% reduction in their HbA1c [166–168]. The ability to have tight glycemic control is even more challenging in type 1 diabetics since they are unable to make insulin and must inject it in response to glucose spikes induced by diet. Thus, their greatest challenge is controlling postprandial glycemia [98]. Some scientists argue that reducing carbohydrate intake is the easiest way for a type 1 diabetic to control their blood sugar levels since it will reduce the error in determining the insulin amount needed to match their increased blood glucose levels [88]. However, the benefits from the low carbohydrate diet might also improve other health markers in diabetics, such as abdominal fat and health-related quality of life factors as shown in other studies [169,170]. Type 2 diabetics have also improved or eliminated their diabetic state through diet, specifically a diet that restricts carbohydrate consumption. Type 2 diabetes results in insulin resistant cells and this has been linked to other complications and atherosclerotic processes such as inflammation, decreased size of LDL particles, and endothelial dysfunction [171]. Thus, the benefits of a healthy, low carbohydrate diet on diabetes might also improve the markers for cardiovascular disease as well.

Although the debate about diet and heart health continues, many new studies are revealing that the picture is much more complicated than the diet-heart hypothesis suggested. The need for more randomized, controlled studies of long-term duration are necessary to determine the true effect of dietary macronutrients on cardiovascular risks. It appears from preliminary studies that a ketogenic diet might have favorable outcomes on CVD, but some still view the idea with great skepticism. In medicine, randomized controlled trials are considered the gold standard and many physicians feel that there is not enough of these studies to consider changing their medical advice. It is interesting that while current scientists are unwilling to consider these dietary recommendations due to the lack of long-term evidence, the entire United States adopted the current dietary guidelines based mainly on an epidemiological study done by Ancel Keys [102]. Additionally, when the available randomized controlled studies and prospective cohort studies of that time were analyzed, they did not support the recommendation of dietary fat and coronary heart disease [172,173]. Regardless, the necessity in discovering a healthy diet for most people is an important endeavor, especially since we are currently seeing an epidemic of diabetes and obesity, both of which are linked to cardiovascular disease risk.

The potential of the ketogenic diet to aid in cancer treatment is still up for debate. However, the positive results seen in mice warrant that this metabolic therapy should be evaluated further. From the studies presented, it appears that in mice and humans, the diet seems to be most beneficial when used as an adjuvant with other therapies and when administered as soon as possible. It might also be critical to genetically analyze each tumor and determine its metabolic profile to determine if it is exhibiting the Warburg effect. If so, then the KD diet might be a useful addition to the treatment protocol. In summary, a ketogenic diet may have positive impacts on the pathogenesis of cancer, although the determination of its use as a monotherapy or adjuvant therapy in humans need further study.

In conclusion, it is becoming more and more apparent that a "systems biology" approach to human health might be the way of the future. Future studies might need to consider numerous factors such as lifestyle, dietary intake, genotype, gut microbiome composition, and genome-wide information on the epigenome to create a successful plan for maximizing good health. According to Gerhauser et al. [20], "this ambitious goal can only be reached in large interdisciplinary research projects, combining expertise of food technologists, nutritionists, food chemists, molecular biologists, epigeneticists, clinicians, nutritional epidemiologists, bioinformaticians, and statisticians to achieve an integrated view of the influence of diet on human health." Others argue that a diagnosis of high-risk epigenetic states may lead to a better understanding of the links between nutrition, the epigenome, and cancer risk [32]. If these markers can be identified and better understood, then new interventions can be created. New research suggests that long-term dietary choices affect diversity and gene expression of the gut microbiome. One such path might be the use of the ketogenic diet to increase beneficial metabolites which can have positive impacts on the genome. Additionally, a recent study analyzed the genetic variants for personalized management of ketogenic diets [174] and it suggested that certain genetic and dynamic markers of KD response may help identify individuals that will benefit the most from a KD diet. Thus, the use of the ketogenic diet might have a multitude of therapeutic effects, including but not limited to, helping with weight loss, improving lipid markers for cardiovascular health, healing a disrupted microbiome, improving epigenetic markers, reversing diabetes, or reducing the need for medication, and improving responses to cancer treatments. However, if a high fat/low carbohydrate KD diet seems too restrictive, then the use of personalized nutritional advice using microbiome sequencing might be the way of the future for stabilizing many of these diseases and improving metabolic health.

Supplementary Materials: The following are available online at https://www.mdpi.com/article/10.3390/nu13051654/s1, Table S1: Main studies reporting the effects of low-carbohydrate diets, including ketogenic diets, on weight loss, Table S2: Main studies reporting the effects of low-carbohydrate diets, including ketogenic diets, on diabetes health markers, Table S3: Main studies reporting the effects of low-carbohydrate diets, including ketogenic diets, on lipidology health markers, Table S4: Main studies reporting the effects of low-carbohydrate diets, including ketogenic diets, on cancer.

Author Contributions: Both authors contributed to conceptualization, writing, and editing of this review. Both authors have read and agreed to the published version of the manuscript.

Funding: No funding was utilized in writing this review.

Institutional Review Board Statement: Not applicable.

Informed Consent Statement: Not applicable.

Data Availability Statement: Not applicable.

Acknowledgments: We thank Rozanne Wille and Charles Larssen for critical reading of the manuscript. This work is supported by the Office of Research & Creative Activity and the Department of Biology, Western Kentucky University.

Conflicts of Interest: The authors declare no conflict of interest.

References

1. Moore, J.; Westman, E.C. *Keto Clarity: Your Definitive Guide to the Benefits of a Low-Carb, High-Fat Diet*; Victory Belt Publishing Inc.: Las Vegas, NV, USA, 2020.
2. Hallberg, S.J.; McKenzie, A.L.; Williams, P.T.; Bhanpuri, N.H.; Peters, A.L.; Campbell, W.W.; Hazbun, T.L.; Volk, B.M.; McCarter, J.P.; Phinney, S.D.; et al. Effectiveness and Safety of a Novel Care Model for the Management of Type 2 Diabetes at 1 Year: An Open-Label, Non-Randomized, Controlled Study. *Diabetes Ther.* **2018**, *9*, 583–612. [CrossRef] [PubMed]
3. Edwards, C.; Canfield, J.; Copes, N.; Rehan, M.; Lipps, D.; Bradshaw, P.C. D-beta-hydroxybutyrate extends lifespan in C. elegans. *Aging* **2014**, *6*, 621–644. [CrossRef] [PubMed]
4. Longo, V.D.; Mattson, M.P. Fasting: Molecular Mechanisms and Clinical Applications. *Cell Metab.* **2014**, *19*, 181–192. [CrossRef] [PubMed]
5. Den Besten, G.; van Eunen, K.; Groen, A.K.; Venema, K.; Reijngoud, D.J.; Bakker, B.M. The role of short-chain fatty acids in the interplay between diet, gut microbiota, and host energy metabolism. *J. Lipid Res.* **2013**, *54*, 2325–2340. [CrossRef]
6. Fryar, C.D.; Carroll, M.D.; Afful, J. *Prevalence of Overweight, Obesity, and Severe Obesity among Adults Aged 20 and over: United States, 1960–1962 through 2017–2018*; E-Stats; NCHS Health: Hyattsville, MD, USA, 2020. Available online: https://www.cdc.gov/nchs/data/hestat/obesity-adult-17-18/obesity-adult.htm (accessed on 12 May 2021).
7. Fryar, C.D.; Carroll, M.D.; Afful, J. *Prevalence of Overweight, Obesity, and Severe Obesity among Children and Adolescents Aged 2–19 Years: United States, 1963–1965 through 2015–2016*; E-Stats; NCHS Health: Hyattsville, MD, USA, 2020. Available online: https://www.cdc.gov/nchs/data/hestat/obesity_child_15_16/obesity_child_15_16.htm (accessed on 12 May 2021).
8. Westman, E.C.; Yancy, W.S.; Mavropoulos, J.C.; Marquart, M.; McDuffie, J.R. The effect of a low-carbohydrate, ketogenic diet versus a low-glycemic index diet on glycemic control in type 2 diabetes mellitus. *Nutr. Metab.* **2008**, *5*, 36. [CrossRef]
9. Abbasi, J. Interest in the Ketogenic Diet Grows for Weight Loss and Type 2 Diabetes. *JAMA* **2018**, *319*, 215–217. [CrossRef]
10. Westman, E.C.; Feinman, R.D.; Mavropoulos, J.C.; Vernon, M.C.; Volek, J.S.; Wortman, J.A.; Yancy, W.S.; Phinney, S.D. Low-carbohydrate nutrition and metabolism. *Am. J. Clin. Nutr.* **2007**, *86*, 276–284. [CrossRef]
11. Paoli, A. Ketogenic Diet for Obesity: Friend or Foe? *Int. J. Environ. Res. Public Health* **2014**, *11*, 2092–2107. [CrossRef]
12. Adam-Perrot, A.; Clifton, P.; Brouns, F. Low-carbohydrate diets: Nutritional and physiological aspects. *Obes. Rev.* **2006**, *7*, 49–58. [CrossRef]
13. Wallace, C. Dietary advice based on the bacteria in your gut. 2018 February 25. Wall Street Journal. Available online: https://www.wsj.com/articles/dietary-advice-based-on-the-bacteria-in-your-gut-1519614301 (accessed on 12 May 2021).
14. Zeevi, D.; Korem, T.; Zmora, N.; Israeli, D.; Rothschild, D.; Weinberger, A.; Ben-Yacov, O.; Lador, D.; Avnit-Sagi, T.; Lotan-Pompan, M.; et al. Personalized Nutrition by Prediction of Glycemic Responses. *Cell* **2015**, *163*, 1079–1094. [CrossRef]
15. Rothschild, D.; Weissbrod, O.; Barkan, E.; Kurilshikov, A.; Korem, T.; Zeevi, D.; Costea, P.I.; Godneva, A.; Kalka, I.N.; Bar, N.; et al. Environment dominates over host genetics in shaping human gut microbiota. *Nature* **2018**, *555*, 210–215. [CrossRef]
16. Rivière, A.; Selak, M.; Lantin, D.; Leroy, F.; De Vuyst, L. Bifidobacteria and Butyrate-Producing Colon Bacteria: Importance and Strategies for Their Stimulation in the Human Gut. *Front. Microbiol.* **2016**, *7*, 979. [CrossRef]
17. Davis, S.C.; Yadav, J.S.; Barrow, S.D.; Robertson, B.K. Gut microbiome diversity influenced more by the Westernized dietary regime than the body mass index as assessed using effect size statistic. *Microbiologyopen* **2017**, *6*, e00476. [CrossRef]
18. Lynch, S.V.; Pedersen, O. The Human Intestinal Microbiome in Health and Disease. *N. Engl. J. Med.* **2016**, *375*, 2369–2379. [CrossRef]
19. Claesson, M.J.; Jeffery, I.B.; Conde, S.; Power, S.E.; O'Connor, E.M.; Cusack, S.; Harris, H.M.B.; Coakley, M.; Lakshminarayanan, B.; O'Sullivan, O.; et al. Gut microbiota composition correlates with diet and health in the elderly. *Nature* **2012**, *488*, 178–184. [CrossRef]
20. Gerhauser, C. Impact of dietary gut microbial metabolites on the epigenome. *Philos. Trans. R. Soc. B Biol. Sci.* **2018**, *373*, 20170359. [CrossRef]
21. Gong, L.; Cao, W.; Chi, H.; Wang, J.; Zhang, H.; Liu, J.; Sun, B. Whole cereal grains and potential health effects: Involvement of the gut microbiota. *Food Res. Int.* **2018**, *103*, 84–102. [CrossRef]
22. Kennedy, E.T.; A Bowman, S.; Spence, J.T.; Freedman, M.; King, J. Popular Diets. *J. Am. Diet. Assoc.* **2001**, *101*, 411–420. [CrossRef]
23. Miller, V.J.; Villamena, F.A.; Volek, J.S. Nutritional Ketosis and Mitohormesis: Potential Implications for Mitochondrial Function and Human Health. *J. Nutr. Metab.* **2018**, *2018*, 5157645. [CrossRef]
24. Xie, G.; Zhou, Q.; Qiu, C.-Z.; Dai, W.-K.; Wang, H.-P.; Li, Y.-H.; Liao, J.-X.; Lu, X.-G.; Lin, S.-F.; Ye, J.-H.; et al. Ketogenic diet poses a significant effect on imbalanced gut microbiota in infants with refractory epilepsy. *World J. Gastroenterol.* **2017**, *23*, 6164–6171. [CrossRef]
25. Zhang, Y.; Zhou, S.; Zhou, Y.; Yu, L.; Zhang, L.; Wang, Y. Altered gut microbiome composition in children with refractory epilepsy after ketogenic diet. *Epilepsy Res.* **2018**, *145*, 163–168. [CrossRef]
26. Ley, R.E.; Turnbaugh, P.J.; Klein, S.; Gordon, J.I. Microbial ecology: human gut microbes associated with obesity. *Nature* **2006**, *444*, 1022–1023. [CrossRef]
27. Turnbaugh, P.J.; Ley, R.E.; Mahowald, M.A.; Magrini, V.; Mardis, E.R.; Gordon, J.I. An obesity-associated gut microbiome with increased capacity for energy harvest. *Nat. Cell Biol.* **2006**, *444*, 1027–1031. [CrossRef]

28. Ley, R.E.; Bäckhed, F.; Turnbaugh, P.; Lozupone, C.A.; Knight, R.D.; Gordon, J.I. Obesity alters gut microbial ecology. *Proc. Natl. Acad. Sci. USA* **2005**, *102*, 11070–11075. [CrossRef]

29. Schwiertz, A.; Taras, D.; Schaefer, K.; Beijer, S.; Bos, N.A.; Donus, C.; Hardt, P.D. Microbiota and SCFA in Lean and Overweight Healthy Subjects. *Obesity* **2010**, *18*, 190–195. [CrossRef]

30. Basciani, S.; Camajani, E.; Contini, S.; Persichetti, A.; Risi, R.; Bertoldi, L.; Strigari, L.; Prossomariti, G.; Watanabe, M.; Mariani, S.; et al. Very-Low-Calorie Ketogenic Diets with Whey, Vegetable, or Animal Protein in Patients With Obesity: A Randomized Pilot Study. *J. Clin. Endocrinol. Metab.* **2020**, *105*, 2939–2949. [CrossRef]

31. Nagpal, R.; Neth, B.J.; Wang, S.; Craft, S.; Yadav, H. Modified Mediterranean-ketogenic diet modulates gut microbiome and short-chain fatty acids in association with Alzheimer's disease markers in subjects with mild cognitive impairment. *EBioMedicine* **2019**, *47*, 529–542. [CrossRef]

32. Bishop, K.S.; Ferguson, L.R. The Interaction between Epigenetics, Nutrition and the Development of Cancer. *Nutrients* **2015**, *7*, 922–947. [CrossRef]

33. Boison, D. New insights into the mechanisms of the ketogenic diet. *Curr. Opin. Neurol.* **2017**, *30*, 187–192. [CrossRef]

34. Freeman, J.M.; Kossoff, E.H. Ketosis and the Ketogenic Diet, 2010: Advances in Treating Epilepsy and Other Disorders. *Adv. Pediatr.* **2010**, *57*, 315–329. [CrossRef]

35. Youm, Y.-H.; Nguyen, K.Y.; Grant, R.W.; Goldberg, E.L.; Bodogai, M.; Kim, D.; D'Agostino, D.; Planavsky, N.J.; Lupfer, C.; Kanneganti, T.D.; et al. The ketone metabolite β-hydroxybutyrate blocks NLRP3 inflammasome–mediated inflammatory disease. *Nat. Med.* **2015**, *21*, 263–269. [CrossRef] [PubMed]

36. Rahman, M.; Muhammad, S.; Khan, M.A.; Chen, H.; Ridder, D.A.; Müller-Fielitz, H.; Pokorná, B.; Vollbrandt, T.; Stölting, I.; Nadrowitz, R.; et al. The β-hydroxybutyrate receptor HCA2 activates a neuroprotective subset of macrophages. *Nat. Commun.* **2014**, *5*, 3944. [CrossRef] [PubMed]

37. Lusardi, T.A.; Akula, K.K.; Coffman, S.Q.; Ruskin, D.N.; Masino, S.A.; Boison, D. Ketogenic diet prevents epileptogenesis and disease progression in adult mice and rats. *Neuropharmacology* **2015**, *99*, 500–509. [CrossRef] [PubMed]

38. Masino, S.A.; Li, T.; Theofilas, P.; Sandau, U.S.; Ruskin, D.N.; Fredholm, B.B.; Geiger, J.D.; Aronica, E.; Boison, D. A ketogenic diet suppresses seizures in mice through adenosine A1 receptors. *J. Clin. Investig.* **2011**, *121*, 2679–2683. [CrossRef]

39. Kobow, K.; Kaspi, A.; Harikrishnan, K.N.; Kiese, K.; Ziemann, M.; Khurana, I.; Fritzsche, I.; Hauke, J.; Hahnen, E.; Coras, R.; et al. Deep sequencing reveals increased DNA methylation in chronic rat epilepsy. *Acta Neuropathol.* **2013**, *126*, 741–756. [CrossRef]

40. Williams-Karnesky, R.L.; Sandau, U.S.; Lusardi, T.A.; Lytle, N.K.; Farrell, J.M.; Pritchard, E.M.; Kaplan, D.L.; Boison, D. Epigenetic changes induced by adenosine augmentation therapy prevent epileptogenesis. *J. Clin. Investig.* **2013**, *123*, 3552–3563. [CrossRef]

41. Dechat, T.; Pfleghaar, K.; Sengupta, K.; Shimi, T.; Shumaker, D.K.; Solimando, L.; Goldman, R.D. Nuclear lamins: Major factors in the structural organization and function of the nucleus and chromatin. *Genes Dev.* **2008**, *22*, 832–853. [CrossRef]

42. Armanios, M.; Alder, J.K.; Parry, E.M.; Karim, B.; Strong, M.A.; Greider, C.W. Short Telomeres are Sufficient to Cause the Degenerative Defects Associated with Aging. *Am. J. Hum. Genet.* **2009**, *85*, 823–832. [CrossRef]

43. Hewitt, G.M.; Jurk, D.; Marques, F.D.; Correia-Melo, C.; Hardy, T.L.D.; Gackowska, A.; Anderson, R.; Taschuk, M.T.; Mann, J.; Passos, J.F. Telomeres are favoured targets of a persistent DNA damage response in ageing and stress-induced senescence. *Nat. Commun.* **2012**, *3*, 708. [CrossRef]

44. Sun, D.; Yi, S.V. Impacts of Chromatin States and Long-Range Genomic Segments on Aging and DNA Methylation. *PLoS ONE* **2015**, *10*, e0128517. [CrossRef]

45. Shimazu, T.; Hirschey, M.D.; Newman, J.; He, W.; Shirakawa, K.; Le Moan, N.; Grueter, C.A.; Lim, H.; Saunders, L.R.; Stevens, R.D.; et al. Suppression of Oxidative Stress by -Hydroxybutyrate, an Endogenous Histone Deacetylase Inhibitor. *Science* **2013**, *339*, 211–214. [CrossRef]

46. Moreno, C.L.; Mobbs, C.V. Epigenetic mechanisms underlying lifespan and age-related effects of dietary restriction and the ketogenic diet. *Mol. Cell. Endocrinol.* **2017**, *455*, 33–40. [CrossRef]

47. Xie, Z.; Zhang, D.; Chung, D.; Tang, Z.; Huang, H.; Dai, L.; Qi, S.; Li, J.; Colak, G.; Chen, Y.; et al. Metabolic Regulation of Gene Expression by Histone Lysine β-Hydroxybutyrylation. *Mol. Cell* **2016**, *62*, 194–206. [CrossRef]

48. Dąbek, A.; Wojtala, M.; Pirola, L.; Balcerczyk, A. Modulation of Cellular Biochemistry, Epigenetics and Metabolomics by Ketone Bodies. Implications of the Ketogenic Diet in the Physiology of the Organism and Pathological States. *Nutrients* **2020**, *12*, 788. [CrossRef]

49. Newman, J.C.; Verdin, E. β-Hydroxybutyrate: A Signaling Metabolite. *Annu. Rev. Nutr.* **2017**, *37*, 51–76. [CrossRef]

50. Kashiwaya, Y.; Bergman, C.; Lee, J.-H.; Wan, R.; King, M.T.; Mughal, M.R.; Okun, E.; Clarke, K.; Mattson, M.P.; Veech, R.L. A ketone ester diet exhibits anxiolytic and cognition-sparing properties, and lessens amyloid and tau pathologies in a mouse model of Alzheimer's disease. *Neurobiol. Aging* **2013**, *34*, 1530–1539. [CrossRef]

51. Newport, M.T.; VanItallie, T.B.; Kashiwaya, Y.; King, M.T.; Veech, R.L. A new way to produce hyperketonemia: Use of ketone ester in a case of Alzheimer's disease. *Alzheimer Dement.* **2015**, *11*, 99–103. [CrossRef]

52. Kashiwaya, Y.; Takeshima, T.; Mori, N.; Nakashima, K.; Clarke, K.; Veech, R.L. D-beta-hydroxybutyrate protects neurons in models of Alzheimer's and Parkinson's disease. *Proc. Natl. Acad. Sci. USA* **2000**, *97*, 5440–5444. [CrossRef]

53. Chakraborty, S.; Galla, S.; Cheng, X.; Yeo, J.-Y.; Mell, B.; Singh, V.; Yeoh, B.; Saha, P.; Mathew, A.V.; Vijay-Kumar, M.; et al. Salt-Responsive Metabolite, β-Hydroxybutyrate, Attenuates Hypertension. *Cell Rep.* **2018**, *25*, 677–689. [CrossRef]

54. Huang, Z.; Wang, W.; Huang, L.; Guo, L.; Chen, C. Suppression of Insulin Secretion in the Treatment of Obesity: A Systematic Review and Meta-Analysis. *Obesity* **2020**, *28*, 2098–2106. [CrossRef]

55. Ahmed, S.R.; Bellamkonda, S.; Zilbermint, M.; Wang, J.; Kalyani, R.R. Effects of the low carbohydrate, high fat diet on glycemic control and body weight in patients with type 2 diabetes: Experience from a community-based cohort. *BMJ Open Diabetes Res. Care* **2020**, *8*, e000980. [CrossRef] [PubMed]

56. Michalczyk, M.M.; Klonek, G.; Maszczyk, A.; Zajac, A. The Effects of a Low Calorie Ketogenic Diet on Glycaemic Control Variables in Hyperinsulinemic Overweight/Obese Females. *Nutrients* **2020**, *12*, 1854. [CrossRef] [PubMed]

57. Mohorko, N.; Černelič-Bizjak, M.; Poklar-Vatovec, T.; Grom, G.; Kenig, S.; Petelin, A.; Jenko-Pražnikar, Z. Weight loss, improved physical performance, cognitive function, eating behavior, and metabolic profile in a 12-week ketogenic diet in obese adults. *Nutr. Res.* **2019**, *62*, 64–77. [CrossRef] [PubMed]

58. Fumagalli, M.; Camus, S.M.; Diekmann, Y.; Burke, A.; Camus, M.D.; Norman, P.J.; Joseph, A.; Abi-Rached, L.; Benazzo, A.; Rasteiro, R.; et al. Genetic diversity of CHC22 clathrin impacts its function in glucose metabolism. *eLife* **2019**, *8*, e41517. [CrossRef]

59. Shai, I.; Schwarzfuchs, D.; Henkin, Y.; Shahar, D.R.; Witkow, S.; Greenberg, I.; Golan, R.; Fraser, D.; Bolotin, A.; Vardi, H.; et al. Weight Loss with a Low-Carbohydrate, Mediterranean, or Low-Fat Diet. *N. Engl. J. Med.* **2008**, *359*, 229–241, Erratum in: *N. Engl. J. Med.* **2009**, *361*, 2681. [CrossRef]

60. Dashti, H.M.; Mathew, T.C.; Khadada, M.; Al-Mousawi, M.; Talib, H.; Asfar, S.K.; Behbahani, A.I.; Al-Zaid, N.S. Beneficial effects of ketogenic diet in obese diabetic subjects. *Mol. Cell. Biochem.* **2007**, *302*, 249–256. [CrossRef]

61. Samaha, F.F.; Iqbal, N.; Seshadri, P.; Chicano, K.L.; Daily, D.A.; McGrory, J.; Williams, T.; Williams, M.; Gracely, E.J.; Stern, L. A Low-Carbohydrate as Compared with a Low-Fat Diet in Severe Obesity. *N. Engl. J. Med.* **2003**, *348*, 2074–2081. [CrossRef]

62. Kong, Z.; Sun, S.; Shi, Q.; Zhang, H.; Tong, T.K.; Nie, J. Short-Term Ketogenic Diet Improves Abdominal Obesity in Overweight/Obese Chinese Young Females. *Front. Physiol.* **2020**, *11*, 856. [CrossRef]

63. Monda, V.; Polito, R.; Lovino, A.; Finaldi, A.; Valenzano, A.; Nigro, E.; Corso, G.; Sessa, F.; Asmundo, A.; Di Nunno, N.; et al. Short-Term Physiological Effects of a Very Low-Calorie Ketogenic Diet: Effects on Adiponectin Levels and Inflammatory States. *Int. J. Mol. Sci.* **2020**, *21*, 3228. [CrossRef]

64. Walton, C.M.; Perry, K.; Hart, R.H.; Berry, S.L.; Bikman, B.T. Improvement in Glycemic and Lipid Profiles in Type 2 Diabetics with a 90-Day Ketogenic Diet. *J. Diabetes Res.* **2019**, *2019*, 8681959. [CrossRef]

65. Castellana, M.; Conte, E.; Cignarelli, A.; Perrini, S.; Giustina, A.; Giovanella, L.; Giorgino, F.; Trimboli, P. Efficacy and safety of very low calorie ketogenic diet (VLCKD) in patients with overweight and obesity: A systematic review and meta-analysis. *Rev. Endocr. Metab. Disord.* **2020**, *21*, 5–16. [CrossRef]

66. Goss, A.M.; Gower, B.; Soleymani, T.; Stewart, M.; Pendergrass, M.; Lockhart, M.; Krantz, O.; Dowla, S.; Bush, N.; Barry, V.G.; et al. Effects of weight loss during a very low carbohydrate diet on specific adipose tissue depots and insulin sensitivity in older adults with obesity: A randomized clinical trial. *Nutr. Metab. (Lond.)* **2020**, *17*, 1–12. [CrossRef]

67. Moreno, B.; Crujeiras, A.B.; Bellido, D.; Sajoux, I.; Casanueva, F.F. Obesity treatment by very low-calorie-ketogenic diet at two years: Reduction in visceral fat and on the burden of disease. *Endocrine* **2016**, *54*, 681–690. [CrossRef]

68. Bueno, N.B.; De Melo, I.S.V.; De Oliveira, S.L.; da Rocha Ataide, T. Very-low-carbohydrate ketogenic diet v. low-fat diet for long-term weight loss: A meta-analysis of randomised controlled trials. *Br. J. Nutr.* **2013**, *110*, 1178–1187. [CrossRef]

69. Gibas, M.K.; Gibas, K.J. Induced and controlled dietary ketosis as a regulator of obesity and metabolic syndrome pathologies. *Diabetes Metab. Syndr. Clin. Res. Rev.* **2017**, *11* (Suppl. S1), S385–S390. [CrossRef]

70. Castro, A.I.; Gomez-Arbelaez, D.; Crujeiras, A.B.; Granero, R.; Aguera, Z.; Jimenez-Murcia, S.; Sajoux, I.; Lopez-Jaramillo, P.; Fernandez-Aranda, F.; Casanueva, F.F. Effect of A Very Low-Calorie Ketogenic Diet on Food and Alcohol Cravings, Physical and Sexual Activity, Sleep Disturbances, and Quality of Life in Obese Patients. *Nutrients* **2018**, *10*, 1348. [CrossRef]

71. McClernon, F.J.; Yancy, W.S.; Eberstein, J.A.; Atkins, R.C.; Westman, E.C. The Effects of a Low-Carbohydrate Ketogenic Diet and a Low-Fat Diet on Mood, Hunger, and Other Self-Reported Symptoms. *Obesity* **2007**, *15*, 182. [CrossRef]

72. Martin, C.K.; Rosenbaum, D.; Han, H.; Geiselman, P.J.; Wyatt, H.R.; Hill, J.O.; Brill, C.; Bailer, B.; Miller-Iii, B.V.; Stein, R.; et al. Change in Food Cravings, Food Preferences, and Appetite During a Low-Carbohydrate and Low-Fat Diet. *Obesity* **2011**, *19*, 1963–1970. [CrossRef]

73. Choi, H.-R.; Kim, J.; Lim, H.; Park, Y.K. Two-Week Exclusive Supplementation of Modified Ketogenic Nutrition Drink Reserves Lean Body Mass and Improves Blood Lipid Profile in Obese Adults: A Randomized Clinical Trial. *Nutrients* **2018**, *10*, 1895. [CrossRef]

74. Wyatt, P.; Berry, S.E.; Finlayson, G.; O'Driscoll, R.; Hadjigeorgiou, G.; Drew, D.A.; Al Khatib, H.; Nguyen, L.H.; Linenberg, I.; Chan, A.T.; et al. Postprandial glycaemic dips predict appetite and energy intake in healthy individuals. *Nat. Metab.* **2021**, *3*, 523–529. [CrossRef]

75. Holsen, L.M.; Hoge, W.S.; Lennerz, B.S.; Cerit, H.; Hye, T.; Moondra, P.; Goldstein, J.M.; Ebbeling, C.B.; Ludwig, D.S. Diets Varying in Carbohydrate Content Differentially Alter Brain Activity in Homeostatic and Reward Regions in Adults. *J. Nutr.* **2021**. [CrossRef]

76. Gomez-Arbelaez, D.; Crujeiras, A.B.; Castro, A.I.; Martinez-Olmos, M.A.; Canton, A.; Ordoñez-Mayan, L.; Sajoux, I.; Galban, C.; Bellido, D.; Casanueva, F.F. Resting metabolic rate of obese patients under very low calorie ketogenic diet. *Nutr. Metab.* **2018**, *15*, 18. [CrossRef]

77. Hall, K.D.; Chen, K.Y.; Guo, J.; Lam, Y.Y.; Leibel, R.L.; Mayer, L.E.; Reitman, M.L.; Rosenbaum, M.; Smith, S.R.; Walsh, B.T.; et al. Energy expenditure and body composition changes after an isocaloric ketogenic diet in overweight and obese men. *Am. J. Clin. Nutr.* **2016**, *104*, 324–333. [CrossRef]

78. Ebbeling, C.B.; A Feldman, H.; Klein, G.L.; Wong, J.M.W.; Bielak, L.; Steltz, S.K.; Luoto, P.K.; Wolfe, R.R.; Wong, W.W.; Ludwig, D.S. Effects of a low carbohydrate diet on energy expenditure during weight loss maintenance: Randomized trial. *BMJ* **2018**, *363*, k4583, Erratum in: *BMJ* **2020**, *371*, m4264. [CrossRef]

79. Mobbs, C.V.; Mastaitis, J.; Yen, K.; Schwartz, J.; Mohan, V.; Poplawski, M.; Isoda, F. Low-carbohydrate diets cause obesity, low-carbohydrate diets reverse obesity: A metabolic mechanism resolving the paradox. *Appetite* **2007**, *48*, 135–138. [CrossRef]

80. Carmen, M.; Safer, D.L.; Saslow, L.R.; Kalayjian, T.; Mason, A.E.; Westman, E.C.; Dalai, S.S. Treating binge eating and food addiction symptoms with low-carbohydrate Ketogenic diets: A case series. *J. Eat. Disord.* **2020**, *8*, 1–7. [CrossRef]

81. D'Abbondanza, M.; Ministrini, S.; Pucci, G.; Migliola, E.N.; Martorelli, E.-E.; Gandolfo, V.; Siepi, D.; Lupattelli, G.; Vaudo, G. Very Low-Carbohydrate Ketogenic Diet for the Treatment of Severe Obesity and Associated Non-Alcoholic Fatty Liver Disease: The Role of Sex Differences. *Nutrients* **2020**, *12*, 2748. [CrossRef]

82. Bruci, A.; Tuccinardi, D.; Tozzi, R.; Balena, A.; Santucci, S.; Frontani, R.; Mariani, S.; Basciani, S.; Spera, G.; Gnessi, L.; et al. Very Low-Calorie Ketogenic Diet: A Safe and Effective Tool for Weight Loss in Patients with Obesity and Mild Kidney Failure. *Nutrients* **2020**, *12*, 333. [CrossRef]

83. Brownlee, M.; Hirsch, I.B. Glycemic Variability: A Hemoglobin A1c–Independent Risk Factor for Diabetic Complications. *JAMA* **2006**, *295*, 1707–1708. [CrossRef]

84. Nuttall, F.Q.; Gannon, M.C. Plasma Glucose and Insulin Response to Macronutrients in Nondiabetic and NIDDM Subjects. *Diabetes Care* **1991**, *14*, 824–838. [CrossRef]

85. Lean, M.E.; Leslie, W.S.; Barnes, A.C.; Brosnahan, N.; Thom, G.; McCombie, L.; Peters, C.; Zhyzhneuskaya, S.; Al-Mrabeh, A.; Hollingsworth, K.G.; et al. Primary care-led weight management for remission of type 2 diabetes (DiRECT): An open-label, cluster-randomised trial. *Lancet* **2018**, *391*, 541–551. [CrossRef]

86. Kalra, S.; Singla, R.; Rosha, R.; Dhawan, M. Ketogenic diet: Situational analysis of current nutrition guidelines. *J. Pak. Med. Assoc.* **2018**, *68*, 1836–1839. [PubMed]

87. Hussain, T.A.; Mathew, T.C.; Dashti, A.A.; Asfar, S.; Al-Zaid, N.; Dashti, H.M. Effect of low-calorie versus low-carbohydrate ketogenic diet in type 2 diabetes. *Nutrition* **2012**, *28*, 1016–1021. [CrossRef] [PubMed]

88. Feinman, R.D.; Pogozelski, W.K.; Astrup, A.; Bernstein, R.K.; Fine, E.J.; Westman, E.C.; Accurso, A.; Frassetto, L.; Gower, B.A.; McFarlane, S.I.; et al. Dietary carbohydrate restriction as the first approach in diabetes management: Critical review and evidence base. *Nutrition* **2015**, *31*, 1–13, Erratum in: *Nutrition* **2019**, *62*, 213. [CrossRef]

89. Webster, C.C.; Murphy, T.E.; Larmuth, K.M.; Noakes, T.D.O.; Smith, J.A.H. Diet, Diabetes Status, and Personal Experiences of Individuals with Type 2 diabetes Who Self-Selected and Followed a Low Carbohydrate High Fat diet. *Diabetes Metab. Syndr. Obes. Targets Ther.* **2019**, *12*, 2567–2582. [CrossRef]

90. Saslow, L.R.; Daubenmier, J.J.; Moskowitz, J.T.; Kim, S.; Murphy, E.J.; Phinney, S.D.; Ploutz-Snyder, R.; Goldman, V.; Cox, R.M.; Mason, A.E.; et al. Twelve-month outcomes of a randomized trial of a moderate-carbohydrate versus very low-carbohydrate diet in overweight adults with type 2 diabetes mellitus or prediabetes. *Nutr. Diabetes* **2017**, *7*, 1–6. [CrossRef]

91. Saslow, L.R.; E Mason, A.; Kim, S.; Goldman, V.; Ploutz-Snyder, R.; Bayandorian, H.; Daubenmier, J.; Hecht, F.M.; Moskowitz, J.T. An Online Intervention Comparing a Very Low-Carbohydrate Ketogenic Diet and Lifestyle Recommendations Versus a Plate Method Diet in Overweight Individuals With Type 2 Diabetes: A Randomized Controlled Trial. *J. Med. Internet Res.* **2017**, *19*, e36. [CrossRef]

92. Volek, J.S.; Phinney, S.D.; Forsythe, C.E.; Quann, E.E.; Wood, R.J.; Puglisi, M.J.; Kraemer, W.J.; Bibus, D.M.; Fernandez, M.L.; Feinman, R.D. Carbohydrate Restriction has a More Favorable Impact on the Metabolic Syndrome than a Low Fat Diet. *Lipids* **2009**, *44*, 297–309. [CrossRef]

93. Athinarayanan, S.J.; Adams, R.N.; Hallberg, S.J.; McKenzie, A.L.; Bhanpuri, N.H.; Campbell, W.W.; Volek, J.S.; Phinney, S.D.; McCarter, J.P. Long-Term Effects of a Novel Continuous Remote Care Intervention Including Nutritional Ketosis for the Management of Type 2 Diabetes: A 2-Year Non-randomized Clinical Trial. *Front. Endocrinol.* **2019**, *10*, 348. [CrossRef]

94. Goldenberg, J.Z.; Day, A.; Brinkworth, G.D.; Sato, J.; Yamada, S.; Jönsson, T.; Beardsley, J.; A Johnson, J.; Thabane, L.; Johnston, B.C. Efficacy and safety of low and very low carbohydrate diets for type 2 diabetes remission: Systematic review and meta-analysis of published and unpublished randomized trial data. *BMJ* **2021**, *372*, m4743. [CrossRef]

95. Tay, J.; Luscombe-Marsh, N.D.; Thompson, C.H.; Noakes, M.; Buckley, J.D.; Wittert, A.G.; Yancy, W.S.Y., Jr.; Brinkworth, G.D. Comparison of low- and high-carbohydrate diets for type 2 diabetes management: A randomized trial. *Am. J. Clin. Nutr.* **2015**, *102*, 780–790. [CrossRef]

96. Nielsen, J.V.; Gando, C.; Joensson, E.; Paulsson, C. Low carbohydrate diet in type 1 diabetes, long-term improvement and adherence: A clinical audit. *Diabetol. Metab. Syndr.* **2012**, *4*, 23. [CrossRef]

97. Krebs, J.D.; Strong, A.P.; Cresswell, P.; Reynolds, A.N.; Hanna, A.; Haeusler, S. A randomised trial of the feasibility of a low carbohydrate diet vs standard carbohydrate counting in adults with type 1 diabetes taking body weight into account. *Asia Pac. J. Clin. Nutr.* **2016**, *25*, 78–84. [CrossRef]

98. Lennerz, B.S.; Barton, A.; Bernstein, R.K.; Dikeman, R.D.; Diulus, C.; Hallberg, S.; Rhodes, E.T.; Ebbeling, C.B.; Westman, E.C.; Yancy, W.S.; et al. Management of Type 1 Diabetes With a Very Low–Carbohydrate Diet. *Pediatrics* **2018**, *141*, e20173349. [CrossRef]
99. Lichtash, C.; Fung, J.; Ostoich, K.C.; Ramos, M. Therapeutic use of intermittent fasting and ketogenic diet as an alternative treatment for type 2 diabetes in a normal weight woman: A 14-month case study. *BMJ Case Rep.* **2020**, *13*, e234223. [CrossRef]
100. Wong, K.; Raffray, M.; Roy-Fleming, A.; Blunden, S.; Brazeau, A.-S. Ketogenic Diet as a Normal Way of Eating in Adults with Type 1 and Type 2 Diabetes: A Qualitative Study. *Can. J. Diabetes* **2021**, *45*, 137–143. [CrossRef]
101. Teo, K.K.; Dokainish, H. The Emerging Epidemic of Cardiovascular Risk Factors and Atherosclerotic Disease in Developing Countries. *Can. J. Cardiol.* **2017**, *33*, 358–365. [CrossRef]
102. Forouhi, N.G.; Krauss, R.M.; Taubes, G.; Willett, W. Dietary fat and cardiometabolic health: Evidence, controversies, and consensus for guidance. *BMJ* **2018**, *361*, k2139. [CrossRef]
103. Ravnskov, U.; Diamond, D.M.; Hama, R.; Hamazaki, T.; Hammarskjöld, B.; Hynes, N.; Kendrick, M.; Langsjoen, P.H.; Malhotra, A.; Mascitelli, L.; et al. Lack of an association or an inverse association between low-density-lipoprotein cholesterol and mortality in the elderly: A systematic review. *BMJ Open* **2016**, *6*, e010401. [CrossRef]
104. Shih, C.W.; E Hauser, M.; Aronica, L.; Rigdon, J.; Gardner, C.D. Changes in blood lipid concentrations associated with changes in intake of dietary saturated fat in the context of a healthy low-carbohydrate weight-loss diet: A secondary analysis of the Diet Intervention Examining The Factors Interacting with Treatment Success (DIETFITS) trial. *Am. J. Clin. Nutr.* **2019**, *109*, 433–441, Erratum in: *Am. J. Clin. Nutr.* **2020**, *111*, 490. [CrossRef]
105. Siri-Tarino, P.W.; Sun, Q.; Hu, F.B.; Krauss, R.M. Saturated fat, carbohydrate, and cardiovascular disease. *Am. J. Clin. Nutr.* **2010**, *91*, 502–509. [CrossRef] [PubMed]
106. Dehghan, M.; Mente, A.; Zhang, X.; Swaminathan, S.; Li, W.; Mohan, V.; Iqbal, R.; Kumar, R.; Wentzel-Viljoen, E.; Rosengren, A.; et al. Associations of fats and carbohydrate intake with cardiovascular disease and mortality in 18 countries from five continents (PURE): A prospective cohort study. *Lancet* **2017**, *390*, 2050–2062. [CrossRef]
107. Barter, P.J.; Ballantyne, C.M.; Carmena, R.; Cabezas, M.C.; Chapman, M.J.; Couture, P.; De Graaf, J.; Durrington, P.N.; Faergeman, O.; Frohlich, J.; et al. Apo B versus cholesterol in estimating cardiovascular risk and in guiding therapy: Report of the thirty-person/ten-country panel. *J. Intern. Med.* **2006**, *259*, 247–258. [CrossRef] [PubMed]
108. Dreon, D.M.; A Fernstrom, H.; Campos, H.; Blanche, P.; Williams, P.T.; Krauss, R.M. Change in dietary saturated fat intake is correlated with change in mass of large low-density-lipoprotein particles in men. *Am. J. Clin. Nutr.* **1998**, *67*, 828–836. [CrossRef] [PubMed]
109. Dreon, D.M.; A Fernstrom, H.; Williams, P.T.; Krauss, R.M. A very-low-fat diet is not associated with improved lipoprotein profiles in men with a predominance of large, low-density lipoproteins. *Am. J. Clin. Nutr.* **1999**, *69*, 411–418. [CrossRef] [PubMed]
110. Krauss, R.M.; Blanche, P.J.; Rawlings, R.S.; Fernstrom, H.S.; Williams, P.T. Separate effects of reduced carbohydrate intake and weight loss on atherogenic dyslipidemia. *Am. J. Clin. Nutr.* **2006**, *83*, 1025–1031. [CrossRef] [PubMed]
111. Sieri, S.; Agnoli, C.; Grioni, S.; Weiderpass, E.; Mattiello, A.; Sluijs, I.; Sanchez, M.J.; Jakobsen, M.U.; Sweeting, M.; Van Der Schouw, Y.T.; et al. Glycemic index, glycemic load, and risk of coronary heart disease: A pan-European cohort study. *Am. J. Clin. Nutr.* **2020**, *112*, 631–643. [CrossRef]
112. Sacks, F.M.; Lichtenstein, A.H.; Wu, J.H.; Appel, L.J.; Creager, M.A.; Kris-Etherton, P.M.; Miller, M.; Rimm, E.B.; Rudel, L.L.; Robinson, J.G.; et al. Dietary Fats and Cardiovascular Disease: A Presidential Advisory From the American Heart Association. *Circulation* **2017**, *136*, e1–e23. [CrossRef]
113. Hu, T.; Bazzano, L. The low-carbohydrate diet and cardiovascular risk factors: Evidence from epidemiologic studies. *Nutr. Metab. Cardiovasc. Dis.* **2014**, *24*, 337–343. [CrossRef]
114. Brookes, L. The apo B/A-I Ratio-A Stronger Predictor of Cardiovascular Events Than LDL, HDL, or Total Cholesterol, Triglycerides, or Lipid Ratios. Medscape. 2006. Available online: http://www.medscape.com/viewarticle/541799 (accessed on 12 May 2021).
115. Lu, M.; Lu, Q.; Zhang, Y.; Tian, G. ApoB/apoA1 is an effective predictor of coronary heart disease risk in overweight and obesity. *J. Biomed. Res.* **2011**, *25*, 266–273. [CrossRef]
116. Mozaffarian, D.; Rimm, E.B.; Herrington, D.M. Dietary fats, carbohydrate, and progression of coronary atherosclerosis in postmenopausal women. *Am. J. Clin. Nutr.* **2004**, *80*, 1175–1184. [CrossRef]
117. Clarke, R.; Shipley, M.; Armitage, J.; Collins, R.; Harris, W. Plasma phospholipid fatty acids and CHD in older men: Whitehall study of London civil servants. *Br. J. Nutr.* **2009**, *102*, 279–284. [CrossRef]
118. Guldbrand, H.; Dizdar, B.; Bunjaku, B.; Lindström, T.; Bachrach-Lindström, M.; Fredrikson, M.; Östgren, C.J.; Nystrom, F.H. In type 2 diabetes, randomisation to advice to follow a low-carbohydrate diet transiently improves glycaemic control compared with advice to follow a low-fat diet producing a similar weight loss. *Diabetologia* **2012**, *55*, 2118–2127. [CrossRef]
119. Dashti, H.M.; Al-Zaid, N.S.; Mathew, T.C.; Al-Mousawi, M.; Talib, H.; Asfar, S.K.; Behbahani, A.I. Long Term Effects of Ketogenic Diet in Obese Subjects with High Cholesterol Level. *Mol. Cell. Biochem.* **2006**, *286*, 1–9. [CrossRef]
120. Norwitz, N.G.; Loh, V. A Standard Lipid Panel Is Insufficient for the Care of a Patient on a High-Fat, Low-Carbohydrate Ketogenic Diet. *Front. Med.* **2020**, *7*, 97. [CrossRef]

121. Boullier, A.; Bird, D.A.; Chang, M.-K.; Dennis, E.A.; Friedman, P.; Gillotte-Taylor, K.; Hörkkö, S.; Palinski, W.; Quehenberger, O.; Shaw, P.; et al. Scavenger Receptors, Oxidized LDL, and Atherosclerosis. *Ann. N. Y. Acad. Sci.* **2006**, *947*, 214–223. [CrossRef]
122. Dhaliwal, B.S.; Steinbrecher, U.P. Scavenger receptors and oxidized low density lipoproteins. *Clin. Chim. Acta* **1999**, *286*, 191–205. [CrossRef]
123. Johnson, N.B.; Hayes, L.D.; Brown, K.; Hoo, E.C.; Ethier, A.K. Centers for Disease Control and Prevention (CDC) CDC National Health Report: Leading causes of morbidity and mortality and associated behavioral risk and protective factors—United States, 2005–2013. *MMWR Suppl.* **2014**, *63*, 25356673.
124. Golemis, E.A.; Scheet, P.; Beck, T.N.; Scolnick, E.M.; Hunter, D.J.; Hawk, E.; Hopkins, N. Molecular mechanisms of the preventable causes of cancer in the United States. *Genes Dev.* **2018**, *32*, 868–902. [CrossRef]
125. Hanahan, D.; Weinberg, R.A. Hallmarks of Cancer: The Next Generation. *Cell* **2011**, *144*, 646–674. [CrossRef]
126. Pickup, M.W.; Mouw, J.K.; Weaver, V.M. The extracellular matrix modulates the hallmarks of cancer. *EMBO Rep.* **2014**, *15*, 1243–1253. [CrossRef]
127. Sabharwal, S.S.; Schumacker, P.T. Mitochondrial ROS in cancer: Initiators, amplifiers or an Achilles' heel? *Nat. Rev. Cancer* **2014**, *14*, 709–721. [CrossRef]
128. Calle, E.E.; Rodriguez, C.; Walker-Thurmond, K.; Thun, M.J. Overweight, Obesity, and Mortality from Cancer in a Prospectively Studied Cohort of U.S. Adults. *N. Engl. J. Med.* **2003**, *348*, 1625–1638. [CrossRef]
129. Eheman, C.; Henley, S.J.; Ballard-Barbash, R.; Jacobs, E.J.; Schymura, M.J.; Noone, A.-M.; Pan, L.; Anderson, R.N.; Fulton, J.E.; Kohler, B.A.; et al. Annual Report to the Nation on the status of cancer, 1975–2008, featuring cancers associated with excess weight and lack of sufficient physical activity. *Cancer* **2012**, *118*, 2338–2366. [CrossRef]
130. Font-Burgada, J.; Sun, B.; Karin, M. Obesity and Cancer: The Oil that Feeds the Flame. *Cell Metab.* **2016**, *23*, 48–62. [CrossRef]
131. Bolla, A.M.; Caretto, A.; Laurenzi, A.; Scavini, M.; Piemonti, L. Low-Carb and Ketogenic Diets in Type 1 and Type 2 Diabetes. *Nutrients* **2019**, *11*, 962. [CrossRef]
132. DeCensi, A.; Puntoni, M.; Gandini, S.; Guerrieri-Gonzaga, A.; Johansson, H.A.; Cazzaniga, M.; Pruneri, G.; Serrano, D.; Schwab, M.; Hofmann, U.; et al. Differential effects of metformin on breast cancer proliferation according to markers of insulin resistance and tumor subtype in a randomized presurgical trial. *Breast Cancer Res. Treat.* **2014**, *148*, 81–90. [CrossRef] [PubMed]
133. Dowling, R.J.; Niraula, S.; Chang, M.C.; Done, S.J.; Ennis, M.; McCready, D.R.; Leong, W.L.; Escallon, J.M.; Reedijk, M.; Goodwin, P.J.; et al. Changes in insulin receptor signaling underlie neoadjuvant metformin administration in breast cancer: A prospective window of opportunity neoadjuvant study. *Breast Cancer Res.* **2015**, *17*, 1–12. [CrossRef] [PubMed]
134. Gunter, M.J.; Xie, X.; Xue, X.; Kabat, G.C.; Rohan, T.E.; Wassertheil-Smoller, S.; Ho, G.Y.; Wylie-Rosett, J.; Greco, T.; Yu, H.; et al. Breast Cancer Risk in Metabolically Healthy but Overweight Postmenopausal Women. *Cancer Res.* **2015**, *75*, 270–274. [CrossRef] [PubMed]
135. Warburg, O.; Wind, F.; Negelein, E. The Metabolism of Tumors in the Body. *J Gen Physiol.* **1927**, *7–8*, 519–530. [CrossRef]
136. Cantor, J.R.; Sabatini, D.M. Cancer Cell Metabolism: One Hallmark, Many Faces. *Cancer Discov.* **2012**, *2*, 881–898. [CrossRef]
137. Ward, P.S.; Thompson, C.B. Metabolic Reprogramming: A Cancer Hallmark Even Warburg Did Not Anticipate. *Cancer Cell* **2012**, *21*, 297–308. [CrossRef]
138. Woolf, E.C.; Syed, N.; Scheck, A.C. Tumor Metabolism, the Ketogenic Diet and β-Hydroxybutyrate: Novel Approaches to Adjuvant Brain Tumor Therapy. *Front. Mol. Neurosci.* **2016**, *9*, 122. [CrossRef]
139. Arnold, M.; Pandeya, N.; Byrnes, G.; Renehan, A.G.; Stevens, G.A.; Ezzati, M.; Ferlay, J.; Miranda, J.J.; Romieu, I.; Dikshit, R.; et al. Global burden of cancer attributable to high body-mass index in 2012: A population-based study. *Lancet Oncol.* **2015**, *16*, 36–46. [CrossRef]
140. Gunter, M.J.; Hoover, D.R.; Yu, H.; Wassertheil-Smoller, S.; Manson, J.E.; Li, J.; Harris, T.G.; Rohan, T.E.; Xue, X.; Ho, G.Y.F.; et al. A Prospective Evaluation of Insulin and Insulin-like Growth Factor-I as Risk Factors for Endometrial Cancer. *Cancer Epidemiol. Biomark. Prev.* **2008**, *17*, 921–929. [CrossRef]
141. Gunter, M.J.; Hoover, D.R.; Yu, H.; Wassertheil-Smoller, S.; Rohan, T.E.; Manson, J.E.; Howard, B.V.; Wylie-Rosett, J.; Anderson, G.L.; Ho, G.Y.F.; et al. Insulin, Insulin-like Growth Factor-I, Endogenous Estradiol, and Risk of Colorectal Cancer in Postmenopausal Women. *Cancer Res.* **2008**, *68*, 329–337. [CrossRef]
142. Kaaks, R. Nutrition, Insulin, IGF-1 Metabolism and Cancer Risk: A Summary of Epidemiological Evidence. *Novartis Found. Symp.* **2008**, *262*, 247–264. [CrossRef]
143. Cohen, D.H.; Leroith, D. Obesity, type 2 diabetes, and cancer: The insulin and IGF connection. *Endocr. Relat. Cancer* **2012**, *19*, F27–F45. [CrossRef]
144. Renehan, A.G.; Frystyk, J.; Flyvbjerg, A. Obesity and cancer risk: The role of the insulin–IGF axis. *Trends Endocrinol. Metab.* **2006**, *17*, 328–336. [CrossRef]
145. Poff, A.M.; Ari, C.; Seyfried, T.N.; D'Agostino, D.P. The Ketogenic Diet and Hyperbaric Oxygen Therapy Prolong Survival in Mice with Systemic Metastatic Cancer. *PLoS ONE* **2013**, *8*, e65522. [CrossRef]
146. Otto, C.; Kaemmerer, U.; Illert, B.; Muehling, B.; Pfetzer, N.; Wittig, R.; Voelker, H.U.; Thiede, A.; Coy, J.F. Growth of human gastric cancer cells in nude mice is delayed by a ketogenic diet supplemented with omega-3 fatty acids and medium-chain triglycerides. *BMC Cancer* **2008**, *8*, 122. [CrossRef] [PubMed]

147. Allen, B.G.; Bhatia, S.K.; Buatti, J.M.; Brandt, K.E.; Lindholm, K.E.; Button, A.M.; Szweda, L.I.; Smith, B.J.; Spitz, D.R.; Fath, M.A. Ketogenic Diets Enhance Oxidative Stress and Radio-Chemo-Therapy Responses in Lung Cancer Xenografts. *Clin. Cancer Res.* **2013**, *19*, 3905–3913. [CrossRef] [PubMed]
148. Zhou, W.; Mukherjee, P.; A Kiebish, M.; Markis, W.T.; Mantis, J.G.; Seyfried, T.N. The calorically restricted ketogenic diet, an effective alternative therapy for malignant brain cancer. *Nutr. Metab.* **2007**, *4*, 5. [CrossRef] [PubMed]
149. Abdelwahab, M.G.; Fenton, K.E.; Preul, M.C.; Rho, J.M.; Lynch, A.; Stafford, P.; Scheck, A.C. The Ketogenic Diet Is an Effective Adjuvant to Radiation Therapy for the Treatment of Malignant Glioma. *PLoS ONE* **2012**, *7*, e36197. [CrossRef]
150. Morscher, R.J.; Aminzadeh-Gohari, S.; Feichtinger, R.G.; Mayr, J.A.; Lang, R.; Neureiter, D.; Sperl, W.; Kofler, B. Inhibition of Neuroblastoma Tumor Growth by Ketogenic Diet and/or Calorie Restriction in a CD1-Nu Mouse Model. *PLoS ONE* **2015**, *10*, e0129802. [CrossRef]
151. Caso, J.; Masko, E.M.; Ii, J.A.T.; Poulton, S.H.; Dewhirst, M.; Pizzo, S.V.; Freedland, S.J.; Thomas, J.A. The effect of carbohydrate restriction on prostate cancer tumor growth in a castrate mouse xenograft model. *Prostate* **2012**, *73*, 449–454. [CrossRef]
152. Masko, E.M.; Thomas, J.A.; Antonelli, J.A.; Lloyd, J.C.; Phillips, T.E.; Poulton, S.H.; Dewhirst, M.W.; Pizzo, S.V.; Freedland, S.J. Low-Carbohydrate Diets and Prostate Cancer: How Low Is "Low Enough"? *Cancer Prev. Res.* **2010**, *3*, 1124–1131. [CrossRef]
153. Kim, H.S.; Masko, E.M.; Poulton, S.L.; Kennedy, K.M.; Pizzo, S.V.; Dewhirst, M.W.; Freedland, S.J. Carbohydrate restriction and lactate transporter inhibition in a mouse xenograft model of human prostate cancer. *BJU Int.* **2012**, *110*, 1062–1069. [CrossRef]
154. Klement, R.J. Beneficial effects of ketogenic diets for cancer patients: A realist review with focus on evidence and confirmation. *Med. Oncol.* **2017**, *34*, 132. [CrossRef]
155. Oliveira, C.L.; Mattingly, S.; Schirrmacher, R.; Sawyer, M.B.; Fine, E.J.; Prado, C.M. A Nutritional Perspective of Ketogenic Diet in Cancer: A Narrative Review. *J. Acad. Nutr. Diet.* **2018**, *118*, 668–688. [CrossRef]
156. Chang, H.T.; Olson, L.K.; A Schwartz, K. Ketolytic and glycolytic enzymatic expression profiles in malignant gliomas: Implication for ketogenic diet therapy. *Nutr. Metab.* **2013**, *10*, 47. [CrossRef]
157. Van Der Louw, E.J.; Reddingius, R.E.; Olieman, J.F.; Neuteboom, R.F.; Catsman-Berrevoets, C.E. Ketogenic diet treatment in recurrent diffuse intrinsic pontine glioma in children: A safety and feasibility study. *Pediatr. Blood Cancer* **2018**, *66*, e27561. [CrossRef]
158. Cohen, C.W.; Fontaine, K.R.; Arend, R.C.; Soleymani, T.; Gower, B.A. Favorable Effects of a Ketogenic Diet on Physical Function, Perceived Energy, and Food Cravings in Women with Ovarian or Endometrial Cancer: A Randomized, Controlled Trial. *Nutrients* **2018**, *10*, 1187. [CrossRef]
159. Elsakka, A.M.A.; Bary, M.A.; Abdelzaher, E.; Elnaggar, M.; Kalamian, M.; Mukherjee, P.; Seyfried, T.N. Management of Glioblastoma Multiforme in a Patient Treated With Ketogenic Metabolic Therapy and Modified Standard of Care: A 24-Month Follow-Up. *Front. Nutr.* **2018**, *5*, 20. [CrossRef]
160. Martin-McGill, K.J.; Marson, A.G.; Smith, C.T.; Young, B.; Mills, S.J.; Cherry, M.G.; Jenkinson, M.D. Ketogenic diets as an adjuvant therapy for glioblastoma (KEATING): A randomized, mixed methods, feasibility study. *J. Neuro Oncol.* **2020**, *147*, 213–227. [CrossRef]
161. Panhans, C.M.; Gresham, G.; Amaral, L.J.; Hu, J. Exploring the Feasibility and Effects of a Ketogenic Diet in Patients with CNS Malignancies: A Retrospective Case Series. *Front. Neurosci.* **2020**, *14*, 390. [CrossRef]
162. Khodabakhshi, A.; Seyfried, T.N.; Kalamian, M.; Beheshti, M.; Davoodi, S.H. Does a ketogenic diet have beneficial effects on quality of life, physical activity or biomarkers in patients with breast cancer: A randomized controlled clinical trial. *Nutr. J.* **2020**, *19*, 1–10. [CrossRef]
163. Cohen, C.W.; Fontaine, K.R.; Arend, R.C.; Gower, B.A. A Ketogenic Diet Is Acceptable in Women with Ovarian and Endometrial Cancer and Has No Adverse Effects on Blood Lipids: A Randomized, Controlled Trial. *Nutr. Cancer* **2019**, *72*, 584–594. [CrossRef]
164. Klement, R.J.; Schäfer, G.; Sweeney, R.A. A ketogenic diet exerts beneficial effects on body composition of cancer patients during radiotherapy: An interim analysis of the KETOCOMP study. *J. Tradit. Complement. Med.* **2020**, *10*, 180–187. [CrossRef]
165. Hagihara, K.; Kajimoto, K.; Osaga, S.; Nagai, N.; Shimosegawa, E.; Nakata, H.; Saito, H.; Nakano, M.; Takeuchi, M.; Kanki, H.; et al. Promising Effect of a New Ketogenic Diet Regimen in Patients with Advanced Cancer. *Nutrients* **2020**, *12*, 1473. [CrossRef]
166. Stratton, I.M.; I Adler, A.; Neil, H.A.W.; Matthews, D.R.; E Manley, S.; A Cull, C.; Hadden, D.; Turner, R.C.; Holman, R.R. Association of glycaemia with macrovascular and microvascular complications of type 2 diabetes (UKPDS 35): Prospective observational study. *BMJ* **2000**, *321*, 405–412. [CrossRef] [PubMed]
167. Turner, R.C. The U.K. Prospective Diabetes Study: A review. *Diabetes Care* **1998**, *21*, C35–C38. [CrossRef]
168. Turner, R.C.; Millns, H.; Neil, H.A.W.; Stratton, I.M.; E Manley, S.; Matthews, D.R.; Holman, R.R. Risk factors for coronary artery disease in non-insulin dependent diabetes mellitus: United Kingdom prospective diabetes study (UKPDS: 23). *BMJ* **1998**, *316*, 823–828. [CrossRef]
169. Sasakabe, T.; Haimoto, H.; Umegaki, H.; Wakai, K. Association of decrease in carbohydrate intake with reduction in abdominal fat during 3-month moderate low-carbohydrate diet among non-obese Japanese patients with type 2 diabetes. *Metabolism* **2015**, *64*, 618–625. [CrossRef]
170. Guldbrand, H.; Lindström, T.; Dizdar, B.; Bunjaku, B.; Östgren, C.; Nystrom, F.; Bachrach-Lindström, M. Randomization to a low-carbohydrate diet advice improves health related quality of life compared with a low-fat diet at similar weight-loss in Type 2 diabetes mellitus. *Diabetes Res. Clin. Pract.* **2014**, *106*, 221–227. [CrossRef]

171. Watson, K.E.; Harmel, A.L.P.; Matson, G. Atherosclerosis in Type 2 Diabetes Mellitus: The Role of Insulin Resistance. *J. Cardiovasc. Pharmacol. Ther.* **2003**, *8*, 253–260. [CrossRef]
172. Harcombe, Z.; Baker, J.S.; Cooper, S.M.; Davies, B.; Sculthorpe, N.; DiNicolantonio, J.J.; Grace, F. Evidence from randomised controlled trials did not support the introduction of dietary fat guidelines in 1977 and 1983: A systematic review and meta-analysis. *Open Heart* **2015**, *2*, e000196, Erratum in: *Open Heart* **2015**, *2*, pii: openhrt-2014-000196corr1. [CrossRef] [PubMed]
173. Harcombe, Z.; Baker, J.S.; Davies, B. Evidence from prospective cohort studies did not support the introduction of dietary fat guidelines in 1977 and 1983: A systematic review. *Br. J. Sports Med.* **2016**, *51*, 1737–1742. [CrossRef]
174. Aronica, L.; Volek, J.; Poff, A.; D'Agostino, D.P. Genetic variants for personalised management of very low carbohydrate ketogenic diets. *BMJ Nutr. Prev. Health* **2020**, *3*, 363–373. [CrossRef] [PubMed]

Article

Effects of a Diet Based on Foods from Symbiotic Agriculture on the Gut Microbiota of Subjects at Risk for Metabolic Syndrome

Silvia Turroni [1,†], Elisabetta Petracci [2,†], Valeria Edefonti [3,*], Anna M. Giudetti [4], Federica D'Amico [1,5], Lisa Paganelli [6], Giusto Giovannetti [7], Laura Del Coco [4], Francesco P. Fanizzi [4], Simone Rampelli [1], Debora Guerra [8], Claudia Rengucci [9], Jenny Bulgarelli [10], Marcella Tazzari [10], Nicoletta Pellegrini [11], Monica Ferraroni [3], Oriana Nanni [2] and Patrizia Serra [2,‡]

[1] Department of Pharmacy and Biotechnology, University of Bologna, via Belmeloro, 6, 40126 Bologna, Italy; silvia.turroni@unibo.it (S.T.); federica.damico8@unibo.it (F.D.); simone.rampelli@unibo.it (S.R.)

[2] Biostatistics and Clinical Trial Unit, Istituto Romagnolo per lo Studio dei Tumori "Dino Amadori"—IRST S.r.l., IRCCS, via P. Maroncelli, 40, 47014 Meldola, Italy; elisabetta.petracci@irst.emr.it (E.P.); oriana.nanni@irst.emr.it (O.N.); patrizia.serra@irst.emr.it (P.S.)

[3] Branch of Medical Statistics, Biometry, and Epidemiology "G.A. Maccacaro", Department of Clinical Sciences and Community Health, Università degli Studi di Milano, via A. Vanzetti, 5, 20133 Milano, Italy; monica.ferraroni@unimi.it

[4] Department of Biological and Environmental Sciences and Technologies, University of Salento, via Province of Lecce—Monteroni, 73100 Lecce, Italy; anna.giudetti@unisalento.it (A.M.G.); laura.delcoco@unisalento.it (L.D.C.); fp.fanizzi@unisalento.it (F.P.F.)

[5] Department of Medical and Surgical Sciences, University of Bologna, via Massarenti, 9, 40138 Bologna, Italy

[6] Donna Impresa Coldiretti Forlì-Cesena e Rimini, via E. Forlanini, 11, 47121 Forlì, Italy; l.paganelli@hotmail.com

[7] Centro Colture Sperimentali, CCS Aosta S.r.l., Frazione Olleyes 9, 11020 Quart, Italy; giustogiovannetti@hotmail.com

[8] Department of Medical Oncology, Istituto Romagnolo per lo Studio dei Tumori "Dino Amadori"—IRST S.r.l., IRCCS, via P. Maroncelli, 40, 47014 Meldola, Italy; debora.guerra@irst.emr.it

[9] Bioscience Laboratory, Istituto Romagnolo per lo Studio dei Tumori "Dino Amadori"—IRST S.r.l., IRCCS, via P. Maroncelli, 40, 47014 Meldola, Italy; claudia.rengucci@irst.emr.it

[10] Immunotherapy, Cell Therapy and Biobank Unit, Istituto Romagnolo per lo Studio dei Tumori "Dino Amadori"—IRST S.r.l., IRCCS, via P. Maroncelli, 40, 47014 Meldola, Italy; jenny.bulgarelli@irst.emr.it (J.B.); marcella.tazzari@irst.emr.it (M.T.)

[11] Department of Agricultural, Food, Environmental and Animal Sciences, Università di Udine, via delle Scienze 206, 33100 Udine, Italy; nicoletta.pellegrini@uniud.it

[*] Correspondence: valeria.edefonti@unimi.it; Tel.: +39-02-5032-0853; Fax: +39-02-5032-0866

[†] Both authors contributed equally to this work.

[‡] This paper is dedicated to the memory of Prof. Dino Amadori, founder of our institute (Istituto Romagnolo per lo Studio dei Tumori "Dino Amadori"—IRST S.r.l., IRCCS), who devoted his life to the study and treatment of cancer.

Citation: Turroni, S.; Petracci, E.; Edefonti, V.; Giudetti, A.M.; D'Amico, F.; Paganelli, L.; Giovannetti, G.; Del Coco, L.; Fanizzi, F.P.; Rampelli, S.; et al. Effects of a Diet Based on Foods from Symbiotic Agriculture on the Gut Microbiota of Subjects at Risk for Metabolic Syndrome. *Nutrients* **2021**, *13*, 2081. https://doi.org/10.3390/nu13062081

Academic Editor: Panagiota Mitrou

Received: 25 May 2021
Accepted: 14 June 2021
Published: 17 June 2021

Publisher's Note: MDPI stays neutral with regard to jurisdictional claims in published maps and institutional affiliations.

Abstract: Diet is a major driver of gut microbiota variation and plays a role in metabolic disorders, including metabolic syndrome (MS). Mycorrhized foods from symbiotic agriculture (SA) exhibit improved nutritional properties, but potential benefits have never been investigated in humans. We conducted a pilot interventional study on 60 adults with $\geq$ 1 risk factors for MS, of whom 33 consumed SA-derived fresh foods and 27 received probiotics over 30 days, with a 15-day follow-up. Stool, urine and blood were collected over time to explore changes in gut microbiota, metabolome, and biochemical, inflammatory and immunologic parameters; previous dietary habits were investigated through a validated food-frequency questionnaire. The baseline microbiota showed alterations typical of metabolic disorders, mainly an increase in *Coriobacteriaceae* and a decrease in health-associated taxa, which were partly reversed after the SA-based diet. Improvements were observed in metabolome, MS presence (two out of six subjects no longer had MS) or components. Changes were more pronounced with less healthy baseline diets. Probiotics had a marginal, not entirely favorable, effect, although one out of three subjects no longer suffered from MS. These findings suggest that improved dietary patterns can modulate the host microbiota and metabolome, counteracting the risk of developing MS.

Keywords: adult volunteers; dietary intervention; dietary patterns; gut microbiota; pilot study; symbiotic agriculture; metabolic dysfunction; metabolic syndrome; metabolome

1. Introduction

Metabolic syndrome (MS) is characterized by several metabolic abnormalities, including abdominal obesity, elevated values of triglycerides, blood pressure, or fasting glucose, or reduced high-density lipoprotein (HDL) cholesterol [1]. This pathological condition has been increasing over recent years, mainly due to changes in lifestyle and unbalanced diets, with a prevalence of 10–40% in the European population, depending on age and gender [2,3]. Subjects with untreated MS can easily develop cardiovascular and cerebrovascular disease, with an increased risk of mortality [4]. Moreover, several epidemiological and clinical studies support the hypothesis that MS may also be an important etiologic factor for the development and progression of certain types of cancer and for overall cancer mortality [5–9].

An increasing number of studies have shown that the gut microbiota (i.e., the vast and diverse set of microorganisms that populate our intestine) may play a role in the pathogenesis and progression of MS [10–14]. In particular, the studies are consistent in highlighting a dysbiotic (i.e., unbalanced) profile, comparable to that observed in other metabolic disorders, characterized by: (i) reduced diversity (a well-recognized hallmark of healthy gut and overall health); (ii) reduced proportions of beneficial commensals, mostly short-chain fatty acid (SCFA) producers; and (iii) increased amounts of opportunistic pathogens or pathobionts, including proteobacteria and other taxa whose pathogenic potential has only recently been revealed, for example *Coriobacteriaceae*. This layout could contribute in various ways to MS, e.g., by affecting satiety, favoring fat storage, altering intestinal cholesterol absorption, reducing hepatic glycogenesis and increasing triglyceride synthesis, exerting prothrombotic and hypertensive effects and, not least, disrupting the integrity of the epithelial barrier, thus consolidating a chronic low-grade inflammatory state [15–19]. It is, therefore, not surprising that the gut microbiota has been proposed as a target in interventions aimed at mitigating the risk of MS. In particular, given its sensitivity to variations in the amount of food and especially to the composition of the diet itself [20,21], dietary interventions for microbiota modulation, including the supplementation of pre- and probiotics, have been and still are the subject of numerous and recent studies [2,15,22].

Symbiotic agriculture (SA) is an agricultural production process aimed at restoring, safeguarding and employing in agro-ecosystems the natural symbiosis present between soil microorganisms (mainly fungi and bacteria) and the plant systems of cultivated species. The objectives that animate this new vision of agriculture are to: (i) increase the sustainability of agricultural practices by favoring the mechanisms to restore the biological fertility of soils and bio-sequestration of carbon, increasing the efficiency of crop fertilization interventions and reducing greenhouse gas emissions from the soil; (ii) increase the resistance of crops to adversity; (iii) increase the yield of crops (over-yielding) and plantations with off-land and under-earth luxuration by re-functionalizing N-organic; (iv) produce food and feed with greater shelf-life and greater transferable antioxidant and secondary metabolites relevant to human health; and (v) increase and improve the nutritional properties and the natural content of vitamins and metabolites produced exclusively by associated microorganisms, such as cobalamin (B12) and menaquinone (K2) [23–25]. In particular, SA systems make extensive use of mycorrhizal fungi and bacteria as ecologically and economically relevant fertilizers, which contribute to ecosystem functioning and crop productivity. The ultimate impact of mycorrhized farming on the nutritional and nutraceutical value of derived foods, such as fruits, vegetables, legumes and cereals, has recently been proven, especially in terms of antioxidant capacity, phenolic content and secondary metabolites levels [24–26]. However, as far as we know, to date, no SA-derived foods have been tested in a dietary intervention on humans.

To investigate whether a diet based on SA-derived products could impact the gut microbiota of subjects at risk for MS, we designed a pilot interventional study where subjects with at least one predisposing MS factor were provided with fresh food products from local certified organic SA production for 30 days. In parallel, a second group of subjects with similar characteristics followed their usual diet, except for probiotic supplementation. After a baseline interview, including an assessment of usual dietary habits, subjects were asked to collect fecal samples at baseline and during intervention until the end of follow-up for microbiota profiling.

In addition, through a comprehensive characterization of the enrolled subjects, the study aimed to provide preliminary evidence for other potential effects of a SA-based intervention, such as those on anthropometric and biochemical factors, urine metabolome profile, and the inflammatory and immunological status of the included subjects. All these aspects were integrated and analyzed with respect to the baseline dietary profile of study participants.

2. Materials and Methods

2.1. Study Setting and Participants

This pilot intervention study was promoted and conducted by the Istituto Romagnolo per lo Studio dei Tumori "Dino Amadori" (IRST) between October 2018 and September 2019 in the catchment area of Romagna, Italy. Subjects aged 18–65 years and with at least one of the following conditions were eligible for enrollment: abdominal obesity, hypertension, dyslipidemia, impaired fasting glucose or insulin resistance (for the formal definition of MS see Table 1 of Alberti et al., 2009 [27]). Subjects with severe or uncontrolled conditions or under treatment with antibiotics or following a specific diet regimen, such as vegan or celiac individuals, were excluded from the study.

To promote and facilitate study recruitment, an illustrative brochure was distributed in the IRST area. Moreover, a press conference was organized, and the study was promoted in local newspapers. In all cases, information on the study and reference staff was provided.

The protocol was approved by the CEROM Ethical Committee (Study ID: IRST B086 L4P1755, Ethical approval ID: 6759/2018). All participants provided written informed consent or assent.

2.2. Study Intervention

The study was designed to enroll two groups of subjects: those receiving fresh foods from organic symbiotic crops with mycorrhizae (SA-group) and those integrating their habitual diet with receiving probiotics with sachet formulation (PROB-group). Subjects from both groups were asked not to change their usual diet during the study period. The dietary intervention lasted 30 days. For a preliminary evaluation of the durability of potential changes in the gut microbiota after intervention, a follow up of 15 days was planned. Although comparison between the two groups was not an objective of the present study, subjects were randomly allocated to one of the two groups using the nQuery Advisor®, Version 7.0 (Statsols, Statistical Solutions Limited, Cork, Ireland) mixed block non-stratified randomization procedure.

Subjects in the SA-group substituted their habitual foods with fresh, baked or steamed products chosen from a ~130-item menu-like list based on seasonal availability, including: fruit and vegetables in season (e.g., apples, pears, kiwis, beetroot, broccoli, cabbage, carrots, cauliflower, celery chickpeas, beans, fennel, salad, lentils, potatoes, spinach, tomatoes, and pumpkin); whole wheat or spelt-based products (bread, pasta, bakery products, including focaccia, breadsticks, cakes, and biscuits, as well as flour to be directly used in recipes); dairy products (e.g., milk, cheese, and yoghurt) from cattle, sheep, and goats; different cuts of poultry and red meat; extra virgin olive oil; tomato sauce and pesto; jam and fruit juices. All the products were derived from local certified organic SA production and prepared by farmers; therefore, they were free from colorants, preservatives or other additives normally present in preserved foods; they were delivered twice a week directly to each volunteer's home. The diet also included the use of aromatic herbs.

Table 1. Distribution at baseline of anthropometric, biochemical, and immunological characteristics for all study participants and separately for the dietary intervention groups. Italy, 2018–2019.

	All (*n* = 60)		SA-Group (*n* = 33)		PROB-Group (*n* = 27)		*p*
	Median	[Min–Max]	Median	[Min–Max]	Median	[Min–Max]	
Gender, *n* (%)							0.176
Male	13	(21.7)	5	(15.2)	8	(29.6)	
Female	47	(78.3)	28	(84.8)	19	(70.4)	
Smoking habit, *n* (%) [1]							0.860
Never smoker	26	(48.1)	14	(46.7)	12	(50.0)	
Ex-smoker	22	(40.7)	12	(40.0)	10	(41.7)	
Current smoker	6	(11.1)	4	(13.3)	2	(8.3)	
Age at enrollment, years	46.9	[18.3–86.4]	52.7	[34.6–86.4]	45.3	[18.3–64.2]	0.015
Weight, kg	70.5	[44.0–103.0]	70.0	[44.0–103.0]	72.0	[47–94.5]	0.953
Height, m	1.65	[1.4–1.8]	1.7	[1.4–1.8]	1.7	[1.5–1.8]	0.183
BMI, kg/m^2	25.7	[19.2–36.8]	26.1	[19.2–36.8]	25.3	[19.8–33.3]	0.427
Waist circumference, cm	85.0	[64.0–113.0]	85.0	[67.0–113.0]	84.0	[64.0–102.0]	0.639
Hip circumference, cm	105.0	[89.0–123.0]	105.0	[89.0–123]	104.0	[90.0–116.0]	0.312
WHR	0.8	[0.7–1.0]	0.8	[0.7–1.0]	0.8	[0.7–1.0]	0.783
Abdomen circumference, cm	98.5	[69.0–120]	98.0	[78.0–120.0]	99.0	[69.0–111.0]	0.582
Glucose, mg/dL	82.5	[66.0–212.0]	83.0	[66.0–212.0]	82.0	[72.0–103.0]	0.271
Cholesterol, mg/dL [1]	193.0	[136.0–269.0]	190.0	[136.0–269.0]	195.0	[139.0–269.0]	0.345
HDL, mg/dL [1]	59.0	[31.0–94.0]	65.0	[31.0–94.0]	55.0	[34.0–86.0]	0.061
LDL, mg/dL [1]	113.0	[55.0–171.0]	106.0	[67.0–171.0]	119.0	[55.0–167.0]	0.064
Triglycerides, mg/dL [1]	90.0	[43.0–365.0]	91.5	[43.0–365.0]	90.0	[44.0–243.0]	0.879
Cortisol, µg/L [1]	125.0	[61.0–268.0]	124.5	[68.0–206.0]	129.0	[61.0–268.0]	0.744
Insulin, mU/L [1]	9.2	[3.0–93.3]	8.9	[3.0–28.2]	10.1	[5.1–93.3]	0.169
Systolic BP, mmHg [1]	120.0	[97.0–155.0]	120.0	[100.0–155.0]	115.0	[97.0–150.0]	0.072
Diastolic BP, mmHg [1]	70.0	[55.0–90.0]	70.0	[60.0–90.0]	70.0	[55.0–90.0]	1.000
MS, *n* (%) [1]							0.488
No	50	(84.7)	26	(81.2)	24	(88.9)	
Yes	9	(15.3)	6	(18.8)	3	(11.1)	
INF-γ [1]	0	[0.0–7.5]	0	[0–2.8]	0	[0–7.5]	0.646
IL-6 [1]	1.3	[0.0–254.4]	1.3	[0–55.5]	1.6	[0–254.4]	0.613
IL-10 [1]	0.3	[0.0–15.0]	0	[0–15.0]	0.6	[0–4.5]	0.087
IL-17A [1]	0	[0.0–18.8]	0	[0–2.6]	0.8	[0–18.8]	0.004
TNFα [1]	0.2	[0.0–67.9]	0	[0–67.9]	0.3	[0–11.9]	0.419
IMI categories, *n* (%) [1]							0.265
0–3	24	(40.7)	16	(50.0)	8	(29.6)	
4–5	23	(39.0)	10	(31.3)	13	(48.2)	
6–8	12	(20.3)	6	(18.7)	6	(22.2)	

BMI: body mass index; WHR: waist-to-hip ratio; HDL: high-density lipoprotein; LDL: low-density lipoprotein; BP: blood pressure; MS: metabolic syndrome; IMI: Italian Mediterranean Index. [1] With the exception of smoking habit, missing values were present only for one patient.

The probiotics provided to the subjects in the PROB-group were manufactured by Probiotical S.p.A., Novara (Italy). Each sachet included LF08 (*Lactobacillus fermentum*), LP09 (*Lactobacillus plantarum*), and LS01 (*Lactobacillus salivarius*) at 3.33 billion CFUs each. Maltodextrin was used as the excipient. This mixture was chosen based on previous studies that showed a beneficial effect of these *Lactobacillus* species on markers of MS [28–30]. Each individual in PROB-group was asked to take one 2.5-g sachet every day.

2.3. Collection of Participants' Information and Samples

At baseline, a nutritional visit by trained personnel allowed to collect demographic and anamnestic information and to measure anthropometry. Dietary habits over the past year were also assessed through the self-administration of the European Prospective Investigation into Cancer and Nutrition (EPIC) food-frequency questionnaire (FFQ) [31,32]. No nutritional counselling was provided during the visit. In addition, a blood sample for the determination of biochemical parameters and cytokine levels, and one sample of stool and urine for the characterization of the microbiota and metabolome, respectively, were obtained from all subjects.

The biochemical parameters were immediately assessed, whereas serum was separated by centrifuging the blood samples and stored at $-80\,^{\circ}$C until use. The serum levels of the human inflammatory cytokines IFN-γ, IL-6, IL-10, IL-17A and TNF-α were measured by a multiplexed bead-based immunoassay (Flex set Cytometric Bead Array (CBA), BD Bioscience, San Jose, CA, USA). Samples were acquired with the FACSCanto flow cytometer (BD Bioscience) and the data were analyzed by Diva software and CBA software (BD Bioscience). Fecal samples were collected in sterile containers and stored at $-80\,^{\circ}$C at IRST Bioscience Laboratory before being shipped on dry ice to the Microbiology Laboratory at the Department of Pharmacy and Biotechnology, University of Bologna (Bologna, Italy) for gut microbiota analysis. Urine samples were collected in sterile containers on the same day of blood collection and stored at $-80\,^{\circ}$C at IRST Bioscience Laboratory before being shipped on dry ice to the General and Inorganic Chemistry Laboratory at the Department of Biological and Environmental Sciences and Technologies, University of Salento (Lecce, Italy). Anthropometric information and biological samples were collected at multiple time points before and during intervention as well as in the follow-up (see Figure 1 for more details).

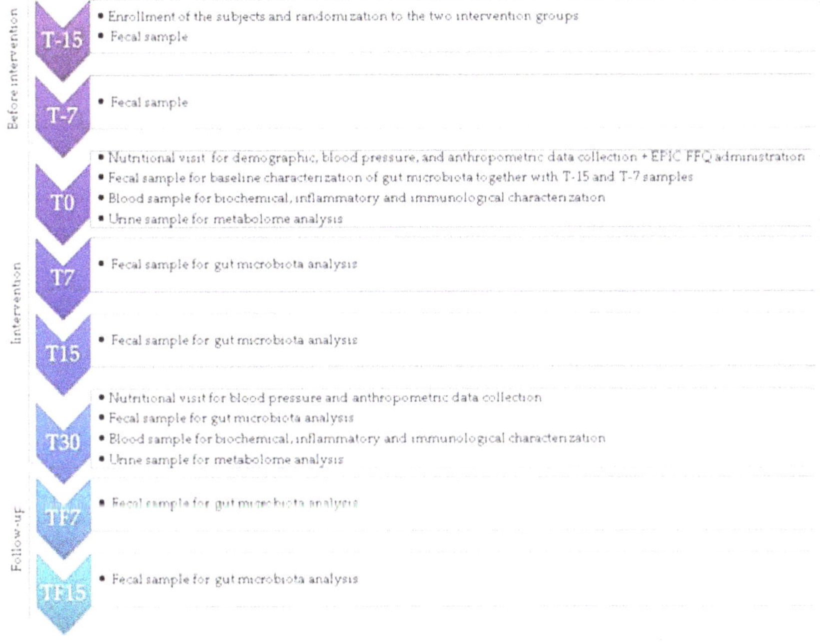

Figure 1. Study timeline. Italy, 2018–2019. The time points are grouped as follows: (i) T-15 and T-7: 15 and 7 days before the intervention (Before intervention); (ii) T0: start of the intervention; T7, T15 and T30: 7, 15 and 30 days from the beginning of the intervention (Intervention); and (iii) TF7 and TF15: 7 and 15 days after the end of the intervention (Follow-up). The validated semi-quantitative European Prospective Investigation into Cancer and Nutrition (EPIC) Food Frequency Questionnaire (FFQ) was administered to collect information on consumption frequency of food items.

The study lasted 30 days, with a 15-day follow-up (TF). Fecal samples were collected weekly in the two-week run-in period for a more reliable depiction of the basal microbiota configuration. Biological sample collection and anthropometric, biochemical and immunological measurements were performed at multiple time points as shown. Dietary habits were assessed at baseline using a validated food frequency questionnaire (see Figure 1 for more details).

2.4. Dietary Habits at Baseline

The current study collected information on consumption frequency of food items as derived at baseline from the 188-item validated semi-quantitative EPIC FFQ [33], designed to capture local dietary habits for the Varese, Turin, and Florence centers [32]. Estimates of daily intakes of energy, minerals, macro- and micro-nutrients (altogether indicated as "nutrients", hereafter) were derived by linking the food items with the Italian Food Composition Tables [34] through a dedicated software package [31].

2.5. Gut Microbiota Analysis through Illumina Sequencing

Microbial DNA was extracted from fecal samples using the repeated bead-beating plus column method [35] with a few modifications [36]. For the baseline, the feces of three replicates (collected weekly, i.e., at T-15, T-7, and T0, see Figure 1) were pooled together. A parallel sensitivity analysis explored the baseline variation at the separate T-15, T-7, and T0 time-points, with an additional focus on genera that changed significantly during SA-based diet. Feces processing was performed as described below. Briefly, approximately 250 mg of each sample was suspended in 1 mL of lysis buffer with four 3-mm glass beads and 0.5 g of 0.1-mm zirconia beads (BioSpec Products, Bartlesville, OK, USA), and bead-beaten in a FastPrep homogenizer (MP Biomedicals, Irvine, CA, USA) at 5.5 movements/s for 1 min three times. The samples were then incubated for 15 min at 95 °C and centrifuged at 13,000 rpm for 5 min. The supernatants were added with 260 μL of 10 M ammonium acetate, and incubated for 30 min with isopropanol (one volume). After washing with 70% ethanol, the nucleic acid pellet was suspended in 100 μL of TE buffer. RNA was removed by 15-min incubation with 2 μL of DNase-free RNase (10 mg/mL) at 37 °C. For the subsequent DNA purification steps, the DNeasy Blood and Tissue Kit (QIAGEN, Hilden, Germany) was used. DNA was assessed for concentration and quality using the NanoDrop ND-1000 spectrophotometer (NanoDrop Technologies, Wilmington, DE, USA).

The V3–V4 hypervariable region of the 16S rRNA gene was amplified using primers 341F and 785R [37], including overhang adapter sequences for Illumina sequencing. For amplification, KAPA HiFi HotStart ReadyMix (Roche, Basel, Switzerland) was used with the following thermal cycle: 95 °C for 3 min, 25 cycles of 95 °C for 30 s, 55 °C for 30 s, and 72 °C for 30 s, and 72 °C for 5 min. Amplicons were purified using magnetic beads (Agencourt AMPure XP, Beckman Coulter, Brea, CA, USA). A limited-cycle PCR was used to add Illumina sequencing adapters and barcodes. After another purification step, samples were pooled at equimolar concentration of 4 nM, denatured and diluted to 5 pM. The final library was sequenced on an Illumina MiSeq platform following a 2 × 250 bp paired-end protocol per manufacturer's instructions (Illumina, San Diego, CA, USA). Raw sequencing reads were deposited in the National Center for Biotechnology Information Sequence Read Archive (Bioproject ID PRJNA726866).

For sequence processing, PANDASeq [38] and QIIME 2 [39] were used. Reads were filtered for length and quality, and subsequently binned into amplicon sequence variants (ASVs) using DADA2 [40]. The VSEARCH algorithm [41] and the Greengenes database (May 2013 release) were used for taxonomic assignment. Chimeras were discarded during the analysis. Different alpha diversity metrics, such as the inverse Simpson index, Faith's Phylogenetic Diversity (PD whole tree) and the number of observed ASVs, were used. For beta diversity, weighted and unweighted UniFrac distances and Bray-Curtis dissimilarity were used to construct Principal Coordinates Analysis (PCoA) graphs. Publicly available sequences of the gut microbiota from age- and sex-matched healthy Italians were downloaded and processed as above. Specifically, we recovered sequences from De Filippis et al. (45 Italian adults; NCBI SRA SRP042234) [42], Schnorr et al. (2 Italian adults; MG-RAST mgp12183) [43] and Biagi et al. (13 Italian adults; MG-RAST mgp17761) [44].

2.6. Urine Metabolomics by Nuclear Magnetic Resonance Spectroscopy

For the Nuclear Magnetic Resonance (NMR) Spectroscopy analysis, 540 μL of urine, thawed at room temperature and mixed, was added to 60 μL of saline buffer solution

(KH_2PO_4, in 100% D_2O containing 0.03% w/w TSP as chemical shift reference and 2 mM sodium azide, pH 7.4), and transferred into a 5mm NMR tube. ^{1}H-NMR spectra were acquired using a Bruker Avance III 600 Ascend NMR spectrometer (Bruker, Milan, Italy), operating at 600.13 MHz for 1H observation, equipped with a TCI cryoprobe (Triple Resonance inverse Cryoprobe) incorporating a z-axis gradient coil and automatic tuning-matching (ATM). Samples were loaded on a Bruker Automatic Sample Changer, interfaced with the IconNMR software (Bruker, Milan, Italy), and analyzed in automatic mode, setting a time delay of 5 min between sample injection and pre-acquisition calibrations for complete temperature equilibration (300 K). Measurements were repeated once in random order after the completion of the first entire set. For each sample, a standard 1D ^{1}H-NMR (ZGCPPR Bruker standard pulse sequence) spectrum, with pre-saturation and composite pulse for selection, was recorded, with 64 transients, 16 dummy scans, 5s relaxation delay, size of FID (free induction decay) of 64 K data points, spectral width of 12,019.230 Hz (20.0276 ppm), acquisition time of 2.73 s and saturation of the solvent signal during the relaxation delay. The resulting FIDs were multiplied by an exponential weighting function equivalent to a line broadening of 0.3 Hz prior to Fourier transformation, automated phasing and baseline correction. Molecular constituent identification was performed by analysis of several spectroscopic NMR data. The compounds were identified by correspondence with literature data [45], according to their chemical shifts, multiplicity and homonuclear and heteronuclear coupling, exhibited in the 1D and 2D NMR spectra. In particular, ^{1}H-^{1}H J-resolved, ^{1}H-^{1}H COSY Correlation Spectroscopy, ^{1}H-^{13}C HSQC Heteronuclear Single Quantum Correlation, ^{1}H-^{13}C HMBC, Heteronuclear Multiple Bond Correlation NMR experiments and a freely available electronic database containing detailed information about metabolites were used (see https://hmdb.ca/, last accessed on 17 May 2021, and reference [46]). NMR data were processed using TopSpin 3.6.1 and Analysis of Mixture, Amix 3.9.13 (Bruker, Biospin, Milan, Italy), for both simultaneous visual inspection and the successive bucketing process.

2.7. Statistical Analyses

Participant characteristics were summarized by means of descriptive statistics such as median, minimum and maximum values or interquartile range (IQR) for continuous variables, and frequencies and percentages for categorical ones. Student's *t*-test or the Mann Whitney U test and the Chi-square or the Fisher's exact test, as appropriate, were used to compare baseline characteristics (i.e., demographic, anthropometric, biochemical parameters, cytokines, and actual adherence to a Mediterranean-style diet) between SA-group and PROB-group. As cytokines presented with highly skewed distributions, some preliminary data transformations were attempted within the Box-Cox family. However, none of them significantly improved the original skewness, given the presence of several zeros, and therefore the analyses were performed on the untransformed data. To compare the above-mentioned data over time (at baseline, T0, and after intervention, T30), the paired *t*-test, the Wilcoxon signed rank test, or the McNemar test was used, as appropriate.

Adherence to a Mediterranean-style diet was assessed by the calculation of the Italian Mediterranean Index (IMI), which was designed to specifically target dietary habits of the Italian population [47], measured by the EPIC FFQ as in our study population. Briefly, this score considered intakes of 11 items, including 6 typical Mediterranean foods (pasta; typical Mediterranean vegetables such as raw tomatoes, leafy vegetables, onion, and garlic, salad, and fruiting vegetables; fruit; legumes; olive oil; and fish), 4 non-Mediterranean foods (soft drinks, butter, red meat, and potatoes) and alcohol. Subjects received 1 point if consumption of typical Mediterranean foods was in the 3rd tertile of the distribution, and 0 points otherwise; when consumption of non-Mediterranean foods was in the first tertile of the distribution, the study participant received 1 point and 0 points otherwise. Ethanol intakes up to 12 g d^{-1} received 1 point, while abstainers and persons who consumed <12 g d^{-1} scored 0. Possible scores ranged from 0 to 11. Details on component definition and standard portions for optimal scoring were provided in [47]. The final index was then divided based on the final categories provided in [47] to improve comparability. Comparisons across the three IMI categories were conducted by referring to the Kruskal Wallis test and the Chi-square or the Fisher's exact test, as appropriate.

An exploratory factor analysis (EFA) was carried out on a selected list of 27 nutrients to summarize overall dietary behavior at baseline in terms of a smaller number of underlying unobservable and randomly varying factors, which can be interpreted as dietary patterns (DPs) derived from EFA (EFA-based DPs). After factorability checks on the nutrient-based correlation matrix (visual inspection, Bartlett's test of sphericity, overall and individual measures of sampling adequacy), the main analysis was based on: (i) principal component method; (ii) eigenvalue >1 and scree-plot criteria, to choose how many factors to retain; (iii) varimax rotation, to make factor naming easier; and (iv) 0.63 cut-off criterion for factor labeling. To quantify the adherence of each subject's diet to each EFA-based DP, we estimated the factor scores for each subject and DP, following the weighted least squares method. We further calculated the Pearson correlation coefficients between the EFA-based DP scores and the daily amount of 37 selected food groups and condiments, derived from the original food items on the same subjects (see for example [48,49] for a more detailed description of the methodology). A cluster analysis (CLU) was carried out on the EFA-based DP scores to further classify subjects according to one (and only one) indicator of similarity in dietary habits among subjects at baseline (CLU-based DP or dietary cluster) [50]. We adopted the Partitioning Around Medoids (PAM) CLU algorithm: as compared to k-means, the PAM algorithm is less sensitive to outliers, and it is integrated with the average silhouette method to choose the optimal number of clusters [50]. Either Euclidean or Manhattan distances were considered, with similar results; the Euclidean distance was selected for the final analysis. The results of the average silhouette method were integrated with model parsimony and cluster interpretation, for the final decision on the optimal number of clusters to retain. Cluster labeling was qualitative and relied on the position of each cluster center within the ranges of the factor scores used as input data. A sensitivity analysis was also conducted considering other clustering methods, including hierarchical clustering and Gaussian mixture models.

For microbiota analysis, the significance of separation in PCoA of beta diversity between study subjects and age- and sex-matched healthy Italians, as well as within each intervention group over time, was tested by a permutation test with pseudo-F ratio (function "adonis" in the R vegan package). To assess differences in alpha diversity and microbiota composition among groups, Kruskal–Wallis or Friedman tests followed by post hoc Wilcoxon tests (paired or unpaired as needed) were performed. Kendall rank correlation test was used to assess the associations between genus-level relative abundances and anthropometric, biochemical, immunological and metabolomic variables. Only statistically significant correlations with absolute Kendall's tau $\geq$0.2 were considered. As for the integration with dietary information, differences in beta diversity and composition at various phylogenetic levels were evaluated across the different clusters defined at baseline. In addition, the food groups and condiments most contributing to the ordination space were identified using the function "envfit" of the R vegan package. When appropriate, p-values were corrected for multiple comparisons using the Benjamini–Hochberg or false discovery rate (FDR) method. An FDR-adjusted p-value ≤ 0.05 was considered as statistically significant. A p-value between 0.05 and 0.1 was considered a tendency.

To investigate, within a unified framework, whether changes over time at genus level were associated with any temporal improvement in the components of MS, the nonparametric rank-based longitudinal methodology proposed by Noguchi et al., 2012 [51] was applied. This method is robust to outliers, heavily skewed data, and has competitive performance for small sample sizes compared to its parametric counterpart. The factorial design chosen considered one whole-plot factor, stratifying subjects in independent groups, and one sub-plot factor, a time variable for the six repeated measures (T0, T7, T15, T30, TF7, and TF15). All analyses were performed for each genus separately and for each intervention group. Two alternative versions of the whole-plot factor were proposed.

The former solution considered a variable given by the difference between T0 and T30 in the number of (altered) dichotomous MS components and then categorized in: -1 if the subject had a worsening (at T30) in at least one factor, 0 if nothing changed at T30, and 1 if the subject experienced an improvement (at T30) in at least one factor. The latter solution considered the relative variation in each of the 5 MS factors (e.g., (triglycerides(T0)-triglycerides(T30))/triglycerides(T0), continuous variable). Such new variables were then categorized as follows: -1 if the subject experienced a worsening $\geq 5\%$, 1 if there was an improvement $\geq 5\%$, and 0 if there was no change or it was <5% in both directions. The 5% threshold was considered as the minimally relevant expected change given the study intervention. Given the small number of subjects generally having a worsening over time, in all the analyses, these were considered with those not experiencing any change or a very small (<5%) one. All the fitted models included the main effects for time and for the MS component change variable, as well as an interaction term between them. In this way, we could assess whether a different temporal trajectory in genus relative abundances was present between the subjects with and without any improvement in metabolic disorders. The ANOVA-type statistics (ATS) were considered for the interpretation of the results.

The ^{1}H NMR spectra of urine (ZGCPPR Bucker standard pulse sequence) were data-reduced to equal length integral segments of 0.02 ppm bucket width considering the NMR spectral range 9.5–0.5 ppm for the bucketing process and multivariate analyses. Resonances of residual water (4.95–4.60 ppm) and urea (6.00–5.60 ppm) were discarded because of the variability (though limited) of urea signal and variations in the suppression of the water signal. Moreover, NMR signals of creatinine (4.08–4.03 and 3.07–3.03 ppm) and citrate (2.70–2.65 and 2.57–2.51 ppm) were combined to account for shifting signals [52] and the remaining buckets were then normalized to the total area to minimize differences in urine concentration between samples and subsequently mean-centered. For statistical analyses, all the imported data were mean-centered and divided by the square root of the standard deviation of each variable using the Pareto scaling algorithm. Unsupervised (blinded) investigation of the data was performed by Principal Component Analysis (PCA) and subsequently analyzed using Orthogonal Projections to Latent Structure Discriminant Analysis (OPLS-DA). In particular, using the NMR buckets as input variables, the PCA was preliminarily used to explore the potential differences in the metabolome profile at baseline and/or presence of outliers (95% confidence ellipse using Hotelling's T^2 statistics). OPLS-DA analysis was also performed on NMR bucket-reduced data, in which results were clearly discriminated in the first predictive t [1] component. The parameters calculated to assess the validity of the established models were the total amount of variation between and within the groups (R^2Y and R^2X) and the predictive ability of the models as determined by permutation test and seven-fold cross-validation (Q^2). NMR discriminant variables were evaluated by the S-line Plots, identified with the loading scaled as a correlation coefficient value (p(corr)) of the OPLS-DA models.

Most of the calculations were performed using the open-source statistical computing environment R [53]. The dietary data were analyzed with libraries psych [54], cluster [55], cclust [56], and mclust [57]; the microbiome data were analyzed with libraries vegan (http://www.cran.r-project.org/package=vegan/, last accessed on 17 May 2021), Made4 [58] and nparLD [51]. Metabolome data were analyzed using SIMCA-14 software (Sartorius Stedim Biotech, Umeå, Sweden).

3. Results

3.1. Description of the Study Participants at Baseline

3.1.1. Anthropometric, Biochemical and Immunological Characteristics

Participants were recruited between October 2018 and September 2019. Of the 67 subjects assessed for eligibility, 7 were excluded because did not meet the inclusion criteria. The analyses were therefore performed on 60 subjects, if not otherwise indicated.

Table 1 shows the baseline characteristics of the recruited subjects, altogether and separately for the two study groups. Most of the study subjects were females (78.3%)

and the median age was 47 years [IQR: 12]. Nine subjects suffered from MS at baseline, (with 6 of them belonging to the SA-group and 3 of them belonging to the PROB-group). Percentages of participants in each adherence category (from the lowest to the highest one) to the IMI were equal to 40.7%, 39.0% and 20.3%, respectively. This is in line with previous literature on dietary patterns of Italian subjects from the EPICOR study, a prospective collaborative investigation of the causes of cardiovascular diseases in Italian volunteers recruited in 1993–1998 within the Italian section of EPIC (47,021 Italian men and women in total) [47]. No substantial differences were observed across SA- and PROB-groups in any examined variable, with the exception of age (higher in SA-group) and IL-17A (lower in SA-group), as compared to PROB-group ($p = 0.015$ and $p = 0.004$, respectively, Mann Whitney U test). However, the distribution of IL-17A was quite extreme: only 36% of the data were different from zero, with one subject showing a value higher than the 90th percentile of the overall distribution.

3.1.2. Dietary Habits

The analysis of dietary data at baseline was based on 59 subjects, as 1 subject did not fill in most of the FFQ items, thus resulting in a total energy intake <500 kcal.

The distribution of study participants and of their baseline characteristics according to categories of adherence to Mediterranean diet, as measured by the IMI, are shown in Table 2. The distribution of baseline characteristics was similar across categories of adherence to the IMI (all p-values were nonsignificant).

Table 2. Distribution at baseline of anthropometric, biochemical, and immunological characteristics by Italian Mediterranean Index tertiles ($n = 59$). Italy, 2018–2019.

	Low Adherence (Index Range: 0–3) ($n = 24$, 40.7%)		Medium Adherence (Index Range: 4–5) ($n = 23$, 39.0%)		High Adherence (Index Range 6–8) [2] ($n = 12$, 20.3%)		p
	Median	[Min–Max]	Median	[Min–Max]	Median	[Min–Max]	
Gender, n (%)							0.848
Male	5	(20.8)	6	(26.1)	2	(16.7)	
Female	19	(79.0)	17	(73.9)	10	(83.3)	
Smoking habit, n (%) [1]							0.797
Never smoker	11	(50.0)	10	(47.6)	5	(45.5)	
Ex-smoker	8	(36.4)	10	(47.6)	4	(36.4)	
Current smoker	3	(13.6)	1	(4.8)	2	(18.2)	
Age at enrollment, years	46.2	[18.3–86.4]	53.7	[35.4–84.8]	46.1	[40.5–55.0]	0.389
Weight, kg	73.5	[47.0–103.0]	67.0	[44.0–94.5]	66.5	[56.0–84.5]	0.431
Height, m	1.7	[1.43–1.8]	1.7	[1.4–1.8]	1.7	[1.58–1.8]	0.949
BMI, kg/m^2	26.5	[19.2–36.8]	25.4	[20.0–31.9]	24.7	[20.3–31.8]	0.380
Waist circumference, cm	85.5	[64.0–113.0]	83.0	[70.0–105.0]	84.0	[68.0–101.0]	0.543
Hip circumference, cm	107.0	[90.0–123.0]	104.0	[89.0–115.0]	101.5	[90.0–122.0]	0.391
WHR	0.8	[0.7–1.0]	0.8	[0.7–1.0]	0.8	[0.7–1.0]	0.952
Abdomen circumference, cm	99.5	[69.0–120.0]	98.0	[81.0–110.0]	97.0	[75–115.0]	0.501
Glucose, mg/dL	86.5	[72.0–212.0]	82.0	[72.0–123.0]	79.0	[66.0–88.0]	0.051
Cholesterol, mg/dL [1]	193.5	[136.0–269]	193.5	[139.0–228.0]	185.0	[145–269.0]	0.486
HDL, mg/dL [1]	55.5	[31.0–94.0]	58.0	[34.0–79.0]	63.0	[41.0–82.0]	0.176
LDL, mg/dL [1]	117.0	[67.0–171.0]	111.0	[67.0–157.0]	110.5	[55.0–167.0]	0.535
Triglycerides, mg/dL [1]	96.0	[44.0–365.0]	90.0	[50.0–267.0]	90.5	[43.0–164.0]	0.740
Cortisol, µg/L [1]	119.5	[61.0–268.0]	141.5	[85.0–226.0]	121.0	[68.0–254.0]	0.471
Insulin, mU/L [1]	8.9	[3.0–93.3]	10.0	[5.4–35.0]	8.1	[3.2–27.7]	0.334
Systolic BP, mmHg [1]	120.0	[100.0–140.0]	118.0	[97.0–155.0]	120.0	[107.0–130.0]	0.984
Diastolic BP, mmHg [1]	72.5	[60.0–90]	70.0	[55.0–90.0]	70.0	[60.0–90.0]	0.514
MS, n (%) [1]							0.719
No	19	(79.2)	19	(86.4)	11	(91.7)	
Yes	5	(20.8)	3	(13.6)	1	(8.3)	
INF-γ [1]	0	[0–2.3]	0	[0–2.8]	0.2	[0–7.5]	0.151
IL-6 [1]	1.0	[0–55.5]	1.6	[0–5.7]	1.3	[0–254.4]	0.489
IL-10 [1]	0.1	[0–15.0]	0.4	[0–1.9]	0.7	[0–4.5]	0.520
IL-17A [1]	0	[0–3.6]	0	[0–3.3]	0.5	[0–18.8]	0.411
TNFα [1]	0	[0–11.0]	0.345	[0–5.3]	0.6	[0–11.9]	

BMI: body mass index; WHR: waist-to-hip ratio; HDL: high-density lipoprotein; LDL: low-density lipoprotein; BP: blood pressure; MS: metabolic syndrome. [1] With the exception of smoking habit and MS, missing values were present only for one patient. [2] No subjects in our study sample reached the maximum IMI score of 11.

Visual inspection, Bartlett's test of sphericity (making it possible to reject the null hypothesis that the correlation matrix is the identity matrix with a $p < 0.001$), overall

(0.84) and individual measures of sampling adequacy (20 nutrients with measures $\geq$0.90) suggested that the nutrient-based correlation matrix was adequate for EFA (Supplementary Table S1). Table 3 gives the factor-loading matrix for the three retained DPs.

Table 3. Factor loading matrix [1] and explained variances for the three major dietary patterns identified by principal component factor analysis on baseline nutrient information. Italy, 2018–2019.

Nutrient	Dietary Pattern		
	Animal Products	Vitamins and Fiber	Regional
Animal protein	**0.96**	-	-
Vegetable protein	0.34	0.48	**0.73**
Cholesterol	**0.88**	0.15	0.13
Saturated fatty acids	**0.80**	0.43	0.13
Monounsaturated fatty acids	0.48	**0.66**	0.42
Linoleic acid	**0.64**	0.40	0.44
Linolenic acid	0.49	0.61	0.40
Other polyunsaturated fatty acids	-	-	**0.70**
Soluble carbohydrates	0.43	**0.66**	0.29
Starch	0.46	0.19	**0.70**
Sodium	**0.78**	0.30	0.27
Calcium	**0.66**	0.54	-
Potassium	0.61	**0.72**	0.25
Phosphorus	**0.80**	0.44	0.31
Iron	0.54	0.57	0.57
Zinc	**0.82**	0.29	0.42
Thiamin (vitamin B1)	**0.69**	0.44	0.41
Riboflavin (vitamin B2)	**0.70**	0.43	-
Vitamin B6	**0.73**	0.46	0.32
Total folate	0.33	**0.77**	0.44
Niacin	**0.83**	0.29	0.27
Vitamin C	0.21	**0.88**	-
Retinol	**0.72**	-	0.17
Beta-carotene	-	**0.87**	0.24
Vitamin D	**0.79**	0.10	-
Vitamin E	0.28	**0.76**	0.47
Total fiber	0.22	**0.80**	0.47
Proportion of explained variance (%)	37.99	27.28	15.09
Cumulative explained variance (%)	37.99	65.27	80.36

[1] Estimates from a principal component factor analysis on 27 nutrients. For each factor, loadings greater or equal to 0.63 indicated important or "dominant nutrients" in the current paper and were shown in bold typeface; loadings smaller than 0.1 were suppressed.

The selected DPs explained ~80% of the total variance. Any nutrient had one or more factor loadings $\geq$0.30, thus suggesting that all the selected nutrients were relevant in this analysis. The greater the loading of a given nutrient to a factor was, the higher the contribution of that nutrient to the factor. The first DP was named "Animal products", as it was characterized by high loadings on animal protein, cholesterol, niacin, zinc, saturated fatty acids, phosphorus, vitamin D, sodium, vitamin B6, retinol, riboflavin, thiamin, calcium, and linoleic acid. The second DP, named "Vitamins and fiber", was characterized by high loadings on vitamin C, beta-carotene, total fiber, total folate, vitamin E, potassium, monounsaturated fatty acids, and soluble carbohydrates. The third DP, named "Regional", had high loadings on vegetable protein, other polyunsaturated fatty acids, and starch. The communalities—measuring the proportion of each nutrient's variance explained by the retained DPs altogether—were generally satisfactory, being greater or equal to 0.70, except for five nutrients (other polyunsaturated fatty acids, retinol, vitamin D, riboflavin and soluble carbohydrates). In addition, when considering Pearson correlation coefficients >0.45 with the amount of selected food groups on the same subjects, the "Animal products" DP score was positively correlated with (in order from the highest to the lowest coefficients) consumption of red meat (especially, beef and pork), offal, processed meat, fish, eggs, coffee, cheese, and olive oil; the "Vitamins and fiber" DP score had positive correlation coefficients with root vegetables, other (than citrus) fruit, olive oil, leafy vegetables (raw and cooked), cabbages, soups and bouillon, whereas the "Regional" DP was positively correlated with the consumption of grains (wholemeal), tea (including herbal tea), and leafy vegetables (raw and cooked).

Table 4 provides a description of the CLU-based DPs or clusters identified at baseline using the PAM CLU method on the EFA-based DP scores.

Table 4. Description of the dietary patterns identified at baseline from cluster analysis [1]: cluster size (i.e., number of subjects included in each cluster) and cluster centers. Italy, 2018–2019.

Cluster Name [2]	Cluster Size	Cluster Center (Medoid)		
		Animal Products	Vitamins and Fiber	Regional
C1-High consumers	11	0.11	0.24	**0.88** [3]
C2-Low consumers	19	−0.72	−0.59	−0.41
C3-Omnivorous with meat prevalence	18	**0.81** [3]	−0.30	−0.30
C4-Omnivorous with plant-based foods prevalence	11	−0.69	**0.70** [3]	−0.70

[1] Estimates from the Partitioning Around Medoids clustering algorithm carried out on the factor scores derived from a previous Principal Component Factor Analysis on nutrient information at baseline. The optimal number of clusters was equal to four, as derived from a combination of criteria, including results of the average silhouette method, parsimony and cluster interpretation. [2] Cluster names were based on the position of center coordinate within the range of the factor scores used as input data. Specifically, coordinates exceeding the third quartile (in absolute value) indicated extreme dietary behavior. Quartiles (Q) of the factor scores at baseline were as follows: "Animal products" pattern: Q1: −0.69; Q2: −0.24; Q3: 0.48; "Vitamins and fiber" pattern: Q1: −0.53; Q2: −0.19; Q3: 0.37; "Regional" pattern: Q1: −0.60; Q2: −0.30; Q3: 0.32. [3] For each cluster, center coordinates greater than or equal to the third quartile score are shown in bold typeface.

The optimal number of clusters was equal to four. Each cluster showed an extreme behavior (exceeding the third score quartile) in one of its center coordinates, except for cluster number 2 (C2) (19 subjects). Specifically, the C1 center was extreme on the "Regional" factor (11 subjects), and the C3 center was extreme on the "Animal products" factor (18 subjects), whereas the C4 center was extreme on the "Vitamins and fiber" pattern (11 subjects). The C2 coordinates were all lower than the corresponding factor medians, being close to the first quartile for the "Animal products" and "Vitamins and fiber" factors and being between the first quartile and the median of the "Regional" factor score: we therefore named C2 as "Low consumers". Similarly, higher-than-median score coordinates described C1 for the remaining "Animal products" and "Vitamins and fiber" patterns; we indicated it as the "High consumers" cluster, especially extreme on the "Regional" DP. The extreme coordinate of the C3 center on the "Animal products" factor was balanced with approximately median score coordinates on the "Vitamins and fiber" and "Regional" factors, thus pointing to an "Omnivorous with meat prevalence" cluster. Finally, we named C4 as the "Omnivorous with plant-based foods prevalence" cluster: apart from the extreme coordinate on the "Vitamins and fiber" pattern, the remaining coordinates were both close to—or even lower than—the corresponding first quartile of the factor score.

The identified clusters were similar with respect to demographic, anthropometric, biochemical, and immunological characteristics (data not shown).

3.1.3. Gut Microbiota Profiling

The gut microbiota of the enrolled subjects was profiled at baseline and during the intervention at five time points (see Figure 1), for a total of 343 fecal samples subjected to 16S rRNA gene sequencing. Seventeen samples were missing or of low quality. A total of 6,682,079 high-quality reads (mean ± SD, 19,481 ± 9578) were obtained and analyzed.

The baseline profile was compared with that of 60 healthy Italians from previous studies, matched by age and gender [42–44], which are well-known microbiota-associated confounding factors [59]. According to the inverse Simpson index, alpha diversity was significantly lower in the enrolled subjects than in the healthy controls (p = 0.01, Wilcoxon test) (Figure 2A).

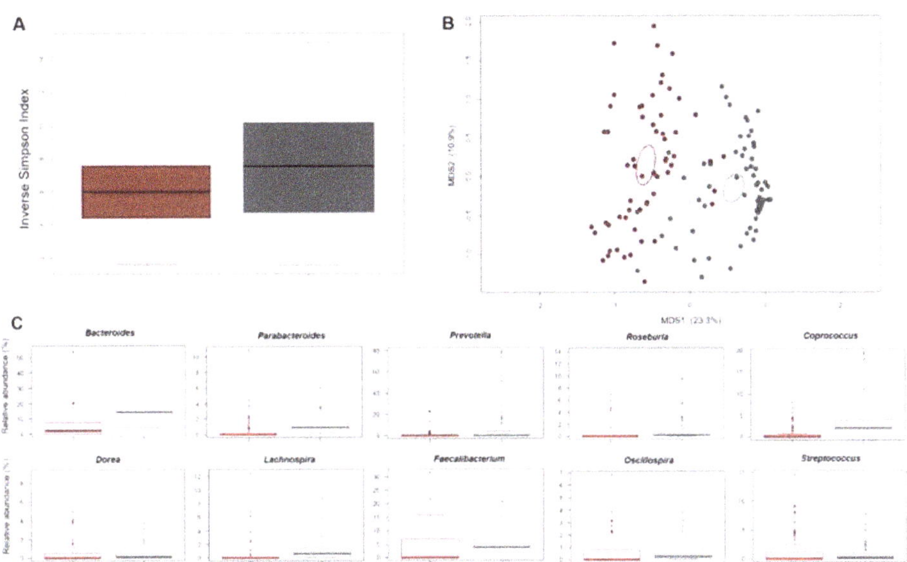

Figure 2. The gut microbiota of study subjects at risk for metabolic syndrome segregated from those of healthy Italian controls, matched by microbiota-associated confounding factors (i.e., age and gender). (**A**) Boxplots showing the distribution of alpha diversity, according to the inverse Simpson index, in study subjects (dark red) compared to healthy Italian controls (grey). A significantly reduced diversity was observed in the former group (p = 0.01, Wilcoxon test). (**B**) PCoA plot of beta diversity, based on Bray–Curtis dissimilarity between the genus-level microbial profiles. A significant separation between study subjects and healthy Italian controls was found (p = 0.001, permutation test with pseudo-F ratio). Samples are identified with colored dots as in (**A**). Ellipses include 95% confidence area based on the standard error of the weighted average of sample coordinates (dark red, subjects at risk for metabolic syndrome; grey, healthy controls). (**C**) Boxplots showing the relative abundance distribution of differentially represented genera between study subjects and healthy Italian controls ($p \leq 0.05$, Wilcoxon test).

Similarly, the PCoA of beta diversity, based on Bray–Curtis dissimilarity between the genus-level profiles, showed significant separation between the study samples and the healthy controls (p = 0.001, permutation test with pseudo-F ratio) (Figure 2B). In line with the available literature on gut microbiota in metabolic disorders [15,60–62], the study subjects showed a higher relative abundance of *Coriobacteriaceae* (p < 0.001, Wilcoxon test) and *Streptococcus* (p = 0.01), as well as reduced proportions of *Bacteroidaceae* members, including *Parabacteroides* (p < 0.001) (Figure 2C and Supplementary Figure S1). Interestingly, *Parabacteroides* has recently been suggested as a novel probiotic taxon for reducing obesity, inflammation levels and insulin resistance [63]. As expected [64], several health-associated SCFA-producing commensals belonging to the *Lachnospiraceae* and *Ruminococcaceae* families, including *Roseburia, Coprococcus, Lachnospira, Oscillospira* and *Faecalibacterium*, were also underrepresented in the gut microbiota of the study participants (p < 0.001).

Correlations between the relative abundances of bacterial taxa and anthropometric, biochemical and immunological parameters in the study subjects were next sought (Supplementary Figure S2). Despite the low correlation coefficients, it is worth noting that a *Coriobacteriaceae* member (i.e., *Adlercreutzia*) correlated positively with total cholesterol (tau = 0.239, p = 0.03, Kendall rank correlation test) and LDL cholesterol (tau = 0.237,

$p = 0.03$), while a negative correlation was found between *Akkermansia*, a mucus degrader associated with improved metabolic health [65] and insulin (tau = -0.226, $p = 0.03$). Furthermore, we found inverse correlations for *Bifidobacterium* (tau = -0.221, $p = 0.04$) and *Bacteroides* (tau = -0.216, $p = 0.03$) against IL-17A, as well as for *Bacteroides* (tau = -0.25, $p = 0.02$) and *Ruminococcus* (tau = -0.219, $p = 0.02$) against IFN-γ. *Ruminococcus* was also inversely correlated with IL-6 (tau = -0.215, $p = 0.02$).

As for associations with dietary habits (Supplementary Figure S3), the Bray–Curtis-based PCoA showed a significant separation between the microbiota structure of the "Omnivorous with plant-based foods prevalence" cluster and the others ($p = 0.05$, permutation test with pseudo-F ratio). When looking for a potential relationship with food groups, we found that consumption of milk ($p \leq 0.05$, "envfit" function) and white meat ($p \leq 0.1$) was associated, or tended to be, with the microbiota of individuals from the "High consumers", "Low consumers", and "Omnivorous with meat prevalence" clusters, where most animal products were represented to a greater or lesser extent. On the other hand, the microbiota of the "Omnivorous with plant-based foods prevalence" cluster subjects tended to be associated with the consumption of garlic and onion, and butter ($p \leq 0.1$). At the taxonomic level, the "High consumers"-related gut microbiota was characterized by greater proportions of *Enterobacteriaceae* members ($p = 0.02$, Kruskal-Wallis test) and a tendency to higher amounts of *Bifidobacterium* ($p = 0.1$), a well-known probiotic taxon associated with dairy consumption. The "Omnivorous with plant-based foods prevalence"-related microbiota tended to be discriminated by greater relative abundances of *Blautia* and *Butyricimonas* ($p = 0.1$). It is worth mentioning that both genera are SCFA producers, even if the former is acetogenic and the latter butyrogenic. However, conflicting data exist on the association between *Blautia* and metabolic health, with particular reference to abdominal fat [66,67], and its abundance was found to be positively associated with saturated and monounsaturated fatty acids [13], probably suggesting the existence of different oligotypes with various metabolic capacities.

3.1.4. Urine Nuclear Magnetic Resonance-Based Metabolomics

The urine metabolome was profiled at baseline and after intervention (see Figure 1), for a total of 120 urine samples. Six NMR spectra were excluded due to the presence of detectable ethanol as contaminant and high levels of glucose, thus obtaining a total of 114 urine samples suitable for successive multivariate analyses. Although very complex, the ^{1}H NMR spectra of urine contained thousands of sharp lines from predominantly low-molecular weight metabolites. Resonances were directly assigned on their chemical shifts, signal multiplicities (resolved by 2D NMR experiments, randomly performed on urine samples) and literature data [45]. Based on all subjects together, the main metabolites identified were creatinine, trimethylamine-N-oxide (TMAO), glycine, citrate, alanine, acetate, erythritol, trigonelline and hippurate. No substantial differences in the metabolomic profile between the two study groups and among the dietary clusters were observed. Moreover, samples were also homogeneous with respect to the information reported in Table 1 (Supplementary Figure S4).

As for associations with the gut microbiota (Supplementary Figure S2), again the correlation coefficients were very small, but it is interesting to mention that we found a negative correlation between the relative abundance of *Lactobacillus* and several metabolites, i.e., creatinine (tau = -0.234, $p = 0.04$), TMAO (tau = -0.226, $p = 0.05$) and phenylacetylglycine (tau = -0.239, $p = 0.03$). A positive correlation was found between *Blautia* and trigonelline (tau = 0.219, $p = 0.03$), an alkaloid with potential anti-diabetic activity [68].

3.2. Effects of the Dietary Intervention

3.2.1. Impact on Anthropometric, Biochemical and Immunological Parameters

Supplementary Table S2 shows the comparison of anthropometric, biochemical and immunological characteristics, as well as presence of MS, for all enrolled subjects and by SA- and PROB-groups, at the two time points T0 and T30. Compared to the baseline, some parameters changed after the dietary intervention. In particular, a statistically significant reduction in insulin values was observed ($p = 0.013$, Wilcoxon signed rank test). Other measures, such as cortisol, blood pressure (BP), and body mass index (BMI), showed an improvement, that is, a decrease, after the intervention, even if not statistically significant at a 5% level. Similarly, MS was detected in six of the nine subjects with the disease at baseline, thus representing a statistically significant improvement from the baseline to the end of the intervention ($p < 0.001$, McNemar test). When inspecting group-specific differences, statistically significant reductions were observed for systolic BP in the SA-group ($p = 0.032$, paired t-test), and for cortisol and insulin in PROB-group ($p = 0.020$ and $p = 0.006$, Wilcoxon signed rank test, respectively) (Supplementary Figure S5). With respect to insulin, one subject reported a very high value at baseline. However, after removal of this individual, the difference remained statistically significant ($p = 0.010$). In the SA-group, a slightly lower BMI was registered after the intervention (medians equal to 26.1 [IQR: 5.1] and 25.4 [IQR: 5.9] kg/m^2 for T0 and T30, respectively; $p = 0.057$, paired t-test), as well as reduced glucose levels (medians equal to 83.0 [IQR: 14.0] and 83.0 [IQR: 8.0] mg/dL for T0 and T30, respectively; $p = 0.067$, Wilcoxon signed rank test). In addition, in both groups, the number of subjects with MS decreased from the baseline to the end of the intervention, in a statistically significant way (SA-group: from six to four, $p < 0.001$, McNemar test; PROB-group: from three to two, $p = 0.002$, McNemar test). No other significant modifications were observed (data not shown).

3.2.2. Impact on the Gut Microbiota Composition

No differences in alpha and beta diversity were observed over time in the SA-group (Supplementary Figure S6). Similar results (i.e., no separation between fecal samples at different time points in the PCoA of beta diversity) were obtained in the PROB-group, for which, however, a temporal reduction in alpha diversity was found, with the lowest values after 30 days of intervention (Faith's Phylogenetic Diversity and number of observed ASVs: $p \leq 0.04$, Friedman test) (Supplementary Figure S6).

Interestingly, at the compositional level, some of the dysbiotic features identified at baseline were reversed after intervention with SA-derived foods, and others tended to be reversed (Figure 3A).

In particular, the relative abundance of *Coriobacteriaceae*, especially *Collinsella*, a potential pathobiont proposed as a target in future microbiome-based interventions for metabolic disorders [69], was significantly reduced in SA-group after 30 days of diet ($p \leq 0.007$, Wilcoxon test), with proportions of *Collinsella* tending to decrease already after 15 days ($p = 0.1$). Furthermore, we observed a rapid increase in the relative abundance of *Oscillospira* (T0 vs. T15, $p = 0.02$), a likely heritable taxon positively associated with leanness and health [70]. It is worth noting that such an increase persisted in the follow-up (T0 vs. TF15, $p = 0.04$), while other changes appeared only later on, namely the increase in *Clostridiaceae* (T0 vs. TF7, $p = 0.02$) and the tendency towards increased amounts of *Lachnospiraceae* (T0 vs. TF15, $p = 0.07$). For these taxa, the baseline variation in the 2 weeks prior to dietary intervention was not significant ($p > 0.05$, Friedman test) (Supplementary Figure S7), which supports that the aforementioned compositional changes were related to the change in diet and not to the typical oscillations of the gut microbiota in the absence of perturbation (see [71] for a recent discussion on the topic). When looking at the compositional variations within the four dietary clusters (Supplementary Figure S8), we found that "High consumers" cluster individuals showed a significant decrease in *Desulfovibrio* after 7 days of diet ($p = 0.04$, Wilcoxon test), which persisted over time, and a tendency towards increased proportions of *Roseburia* at the end of the intervention, until follow-up ($p = 0.1$, Friedman test). On the other hand, for the "Low consumers" and the "Omnivorous with meat prevalence" clusters, we found increasing trends in other SCFA producers, i.e., *Coprococcus* ($p = 0.06$) and *Oscillospira* ($p = 0.09$), respectively. Interestingly, *Coprococcus* correlated negatively with cortisol, whose levels decreased after intervention as reported above (Supplementary Figure S9). A negative correlation in the whole dataset was also found between *Phascolarctobacterium*, another SCFA (mainly propionate) producer, and blood pressure, in line with previous studies associating it with improved metabolic health [72]. Furthermore, the relative abundance of some taxa and precisely *Adlercreutzia*, *Prevotella*, *Butyricimonas* and *Blautia*, showed overall expected correlations with total cholesterol, LDL cholesterol and waist-to-hip ratio in all sample (Supplementary Figure S9).

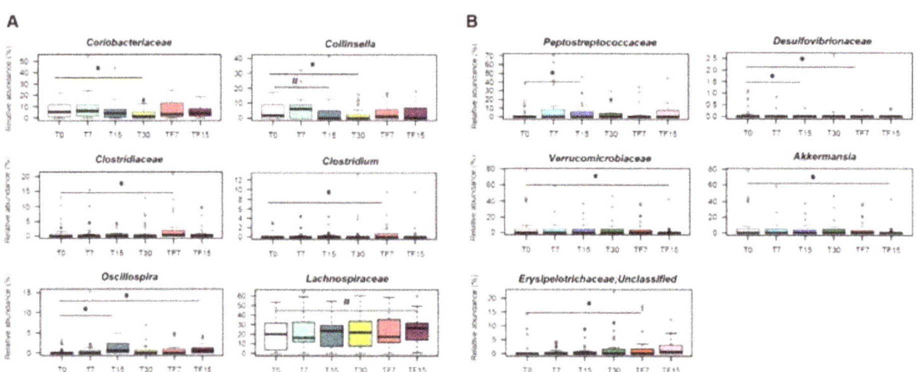

Figure 3. Impact on the gut microbiota of a diet with fresh foods from organic symbiotic agriculture versus probiotic supplementation. Boxplots showing the relative abundance distribution of differentially represented taxa over time, in subjects at risk for metabolic syndrome consuming fresh foods from organic symbiotic agriculture (SA-group) (**A**), or receiving probiotic supplementation (PROB-group) (B). The gut microbiota was profiled at baseline (T0), after 7 (T7), 15 (T15) and 30 (T30) days of intervention, and at follow-up, 7 (TF7) and 15 (TF15) days after the end of the intervention. *, $p \leq 0.05$; #, $0.05 < p \leq 0.1$; Wilcoxon test.

As for the PROB-group, apart from the reduction in the proportions of *Desulfovibrionaceae* (T0 vs. T15 and T0 vs. TF7, $p = 0.04$, Wilcoxon test), we observed some unfavorable changes, including the increase in pro-inflammatory taxa, such as *Peptostreptococcaceae* (T0 vs. T15, $p = 0.02$) and unclassified *Erysipelotrichaceae* members (T0 vs. TF7, $p = 0.02$), and the decrease in the metabolic health-associated genus, *Akkermansia* (T0 vs. TF15, $p = 0.04$) (Figure 3B). Again, these taxa showed no significant changes in the 2 weeks prior to intervention ($p > 0.05$, Friedman test) (data not shown).

When we inspected the relationship between changes in genus relative abundances and improvements in MS components over time within the SA-group, we identified two genera that showed peculiar trends: *Oscillospira* and *Akkermansia* (Supplementary Figure S10). The former showed an increasing trend over time ($p = 0.01$, ATS test for time effect), although similar in subjects who had an improvement in at least one MS factor and in those who didn't have it ($p = 0.14$, ATS test for group effect). The latter showed a difference between the two mentioned groups at T0, with lower values for those showing an improvement in the MS factors. However, this difference tended to disappear later on due to a subsequent increase of *Akkermansia* relative abundance over time for this group ($p = 0.08$, ATS test for the interaction effect). In line with the available literature [65], this further stresses the close association of *Akkermansia* with metabolic health.

The same approach was then applied to the single components of MS. Within the SA-group, *Oscillospira* showed a consistent increasing temporal trend for each MS component (all $p \leq 0.01$, ATS test for the time effect). An increase over time was also observed for *Roseburia* in relation to systolic BP ($p = 0.07$, ATS test for the time effect); however, this was similar for those who showed or not an improvement of at least 5% in this parameter ($p = 0.65$, ATS test for the interaction effect). Finally, *Lachnospira* showed a differential trend over time in relation to triglycerides and glycaemia, with an increase in those who improved only at the last time-point, TF15 ($p = 0.04$, ATS test for the interaction effect). When we analyzed PROB-group, other interesting trends were found. Among them,

Akkermansia behaved differently between subjects who then showed or not an improvement in triglyceride levels, with an increase in the former up to T30 ($p = 0.001$, ATS test for the interaction effect). Differential trends were also observed for *Oscillospira* in relation to both triglycerides and glycaemia, with an increase in follow-up for those who showed reduced triglyceride levels ($p = 0.01$, ATS test for the interaction effect) and at T15 for those with decreased blood glucose ($p = 0.03$, ATS test for the interaction effect) (data not shown).

3.2.3. Impact on the Urine Metabolome

Within the SA-group, after the dietary intervention, a good separation of the data points was observed as well as time-dependent discriminant metabolites. In particular, the corresponding S-line plot from OPLS-DA analysis showed increased levels of erythritol (3.78–3.68 ppm), glycine (3.57 ppm), citrate (2.68–2.54 ppm), acetate (1.93 ppm), and alanine (1.48 ppm), in samples after 30 days of treatment, with a concomitant reduction in the level of creatinine (4.06 ppm) and TMAO (3.27 ppm) (Supplementary Figure S11). Moreover, by comparing the predictive performances of the OPLS-DA models built for each dietary cluster, we observed that the "Omnivorous with meat prevalence" cluster showed a greater sample classification capacity (Q^2 of 0.18) as compared to the others. This indicates that the intervention with SA-foods had a greater overall impact on subjects with "Omnivorous with meat prevalence" behavior. Interestingly, a relatively higher level of hippurate and a particularly pronounced decrease in TMAO level were observed in these subjects after intervention. The decrease of TMAO is extremely favorable, as this molecule, which is formed in the liver from trimethylamine, a metabolite synthesized by the gut microbiota from dietary choline, is recognized as a cardiovascular risk factor and associated with various negative health outcomes [73]. Regarding glycine, its circulating levels have been reported to decrease in metabolic disorders associated with obesity [74]. Moreover, plasma glycine concentration is altered according to food choice, being higher in vegetarian and vegan groups than in meat eaters [75].

Interestingly, some of the metabolites whose levels varied following the intervention showed consistent correlations with the proportions of some of the aforementioned microorganisms (Supplementary Figure S9). In particular, in the SA-group, erythritol correlated negatively with *Collinsella* (tau = -0.204, $p = 0.04$, Kendall rank correlation test) while positively with *Coprococcus* (tau = 0.25, $p = 0.01$). Positive correlations were also observed for *Coprococcus* (tau = 0.249, $p = 0.01$) and *Faecalibacterium* (tau = 0.246, $p = 0.01$) against alanine, as well as between *Faecalibacterium* and glycine (tau = 0.263, $p = 0.009$). Finally, as expected based on its metabolic propensity, *Blautia* positively correlated with acetate (tau = 0.187, $p = 0.05$).

Within the PROB-group, a good sample discrimination after treatment was observed. In particular, the corresponding S-line plot of the OPLS-DA analysis (Supplementary Figure S11) allowed us to identify a decreased level of trigonelline (8.81–8.06 ppm, detectable only in the "High consumers" cluster). The analyses within dietary clusters did not reveal significant deviations, with the exception of relatively higher levels of hippurate in the "High consumers" and "Omnivorous with meat prevalence" clusters (data not shown).

4. Discussion

In this first pilot intervention study offering SA-derived products to subjects at risk for MS, we showed that mycorrhized farming products modulate certain components of the gut microbiota; this effect was accompanied by changes in some metabolic parameters and urinary metabolites and it was partly modulated by DPs at baseline. In addition, two out of six study participants suffering from MS at baseline no longer had MS after the intervention.

In line with the existing literature on gut microbiota in metabolic disorders [15,61,62], the study subjects, as compared to healthy age/sex-matched Italian adults, showed some dysbiotic features at baseline, namely: (i) reduced biodiversity; (ii) lower proportions of health-associated taxa, mainly SCFA producers from the *Lachnospiraceae* and *Ruminococ-*

caceae families, i.e., *Lachnospira*, *Coprococcus*, *Roseburia*, *Oscillospira* and *Faecalibacterium*, as well as *Parabacteroides*; and (iii) greater relative abundance of generally subdominant taxa with pathogenic potential, such as *Coriobacteriaceae* and *Streptococcus*. While the decrease in SCFA producers is frequently found in disparate diseases and pre-disease conditions, as a probably universal dysbiotic signature [64], the overrepresentation of *Coriobacteriaceae* could be regarded as a potentially specific alarm bell for metabolic disorders. Indeed, increased levels of *Coriobacteriaceae* members have been found in conditions of overweight and obesity, as well as in the context of type 2 diabetes and symptomatic atherosclerosis, and directly associated with metabolic risk factors, including those predisposing to MS, such as insulin, triglycerides and LDL cholesterol [15,17,60,76], as also observed in our sample. It has been hypothesized that a gut microbiota profile enriched in such microbes may influence intestinal absorption of cholesterol, hepatic glycogenesis and triglyceride synthesis, as well as interfere with the expression of tight junction proteins, resulting in loss of barrier integrity, metabolic endotoxemia and chronic low-grade inflammation [17,77]. On the other hand, *Parabacteroides*, which was found to be underrepresented in study participants, has recently been proposed as a probiotic candidate for its metabolic benefits, as observed in mouse models, probably through the production of succinate and secondary bile acids [63,78].

Dietary habits at baseline were well characterized by the use of DPs, which are combinations of dietary components meant to summarize total diet—or key aspects of the overall diet—in free-living individuals, as measured at one or more time points. As compared to analyzing single dietary components one at a time, the DP approach allows to capture well-known interactive effects among nutrients or foods, while solving statistical issues related to collinearity between food components and adjustment for multiple comparisons [79,80]. In the current application, we referred to both a priori (or index-based) and a posteriori (empirically derived) DPs [80]. Among available a priori DPs, we referred to the IMI to assess if study participants did follow a Mediterranean diet and to what extent it happened. We showed that our study sample adhered to the Mediterranean diet—as measured by the IMI—to the same extent as the more representative Italian sample of subjects belonging to the EPICOR study [47]. In addition, we applied a combination of EFA and CLU for identifying a posteriori DPs at baseline and relating them to microbiome or metabolome. The small number of subjects—as compared to nutrients—and the by-product of having the correlation structure of nutrients described in terms of EFA-based DPs suggested to perform an EFA before CLU. Cluster-based DPs provided an additional advantage within this project. As individual dietary habits were summarized with one belonging indicator—and not by multiple factors simultaneously—the assessment of the potential links between diet and microbiome or diet and metabolome was easier with CLU-based DPs rather than with the more common EFA based ones. In addition, the current paper explores the use of a more robust CLU partitioning algorithm, PAM, which is more suitable than the well-known k-means. The DPs derived from the application of EFA and CLU are very similar to those derived in an Italian network of case-control studies exploring the association between diet and cancer at several sites (e.g., [48,49,81]); the statistical approach was similar and the same Italian Food Composition Tables [34] were used to convert food items into nutrients, thus improving the possibility of finding similar patterns, as far as they are indeed present. In detail, our EFA-based "Animal products" DP—loading high on animal protein, fats, zinc, B-group vitamins, calcium, phosphorus, sodium, vitamin D, and retinol—was mostly overlapping with the corresponding "Animal products" identified in the previous network (e.g., [48,49,81]); minor differences between the two DPs likely included the major role of meat, including offal—represented by the additional presence of retinol—in our DP, as compared to the major role of dairy products—represented by the highest loadings on calcium and phosphorus—identified on similar DPs in the case-control studies. Similarly, our EFA-based "Vitamins and fiber" DP—loading high on vitamin C, beta-carotene, total fiber, total folate, vitamin E, potassium, monounsaturated fatty acids, and soluble carbohydrates—was similar to the corresponding "Vitamins and fiber" DP identified in the network, with both DPs pointing to consumption of fruit and vegetables; a minor difference between the 2 DPs dealt with the less dominant role of the "citrus fruits" food group in the current study, as compared to the case-control study network. Moreover, in our "Vitamins and fiber" DP, we have oils and vegetable consumption together (with high loadings on vitamin E and monounsaturated fatty acids), in the absence of an additional DP targeting vegetable fats. The "Regional" DP identified in the current study—loading high on vegetable protein, other polyunsaturated fatty acids, and starch—is in between the "Starch-rich" and the "Animal unsaturated fatty acids" DPs, as it combines vegetable protein and starch (but not sodium) from the formed DP with the other polyunsaturated fatty acids (but not vitamin D and niacin) from the latter DP. A paper of the same network applied CLU on the factor scores from a previous EFA [82]; like in the current application, each of the five selected clusters showed an extreme behavior in one of the center coordinates, except for one that is similar to our "Low consumers" DP. Specifically, two cluster centers were extreme on a "Vitamins and fiber" and on an "Animal products" DP that were comparable to our "Omnivorous with plant-based foods prevalence" and "Omnivorous with meat prevalence", respectively; the remaining two cluster centers were extremes on an "Unsaturated fats" and on a "Starch-rich" DP, which combine elements of our "High consumers" DP, although the fat profile is likely more oriented towards the vegetable source in the previous paper [82]. Within the Italian arm of the EPIC Elderly project, four EFA-based DPs (21% of explained variance) were identified on a comparable population interviewed with the same FFQ used in the current study [83,84]. Among them, the "prudent" (cooked vegetables, pulses, cabbage, seed oil and fish) and the "olive oil & salad" (raw vegetables, olive oil, soup and chicken) share similarities with our "Vitamins and fiber" DP, although we did not observe the simultaneous presence of vegetables and meat; their "pasta & meat" (pasta, tomato sauce, red meat, processed meat, bread and wine) is in between our "Animal products" and "Regional" DPs, but we were not able to identify any sort of "sweet & dairy" (sugar, cakes, ice cream, coffee and dairy) DP in our study sample. In conclusion, our study provided the possibility to confirm that, to some extent, Italian DPs derived with multivariate statistics show a good reproducibility across studies [85].

When looking for associations between CLU-based DPs and the gut microbiota, we found that the basal microbiota structure of the "Omnivorous with plant-based foods prevalence" cluster separated from all others, being discriminated by higher proportions of *Blautia* and *Butyricimonas*. Both genera are producers of SCFAs (mainly acetate and butyrate, respectively), which could play a multifactorial role in maintaining metabolic and immunological homeostasis [86]. With specific regard to *Blautia*, although the data have not been replicated in the elderly population [67], this taxon has recently been found to be inversely associated with visceral fat in a large population-based adult cohort [66], and hypothesized to have the potential to counteract MS risk factors. Unlike that study, in which no dietary factor correlated with *Blautia* amount, here we found that the *Blautia* and *Butyricimonas*-enriched "Omnivorous with plant-based foods prevalence" cluster was particularly associated with garlic, onion and butter consumption, suggesting a possible link between these foods and those microorganisms. However, as far as we know, no information is currently available on their impact on the microbiota, except for a correlation between *Blautia* and saturated and monounsaturated fatty acids [13], which are the major lipids of butter. As for the other clusters, it is worth mentioning that the "High consumers"-related microbiota was discriminated by higher proportions of enterobacteria and bifidobacteria, and associated with the intake of animal products, such as meat and milk. This was expected, as the link between *Bifidobacterium* and the intake of dairy products is well established, from early childhood [87,88]. After one month of dietary intervention, study participants experienced modest improvements in BMI, insulin and cortisol levels (especially in PROB-group), and BP (particularly in SA-group).

As for gut microbiota, some dysbiotic signatures were reversed and others tended to be reversed after intervention. In particular, the SA-based diet counteracted the increase in pathobionts, namely *Coriobacteriaceae*, as well as the decrease in SCFA producers, i.e., *Lachnospiraceae* and *Oscillospira*, a potentially heritable taxon capable of promoting leanness [70]. Interestingly, the increase in *Oscillospira* was already noticeable after only 15 days of diet, as well as a tendency towards reduced proportions of *Collinsella*, the dominant taxon of the *Coriobactaeriaceae* family, thus suggesting a modulatory effect even in the short term. Since these taxa did not show significant changes over the run-in period, we can reasonably argue that their variation is the result of introducing SA-derived food products into the diet. As expected, the extent of microbiota modulation was greater in participants not belonging to the "Omnivorous with plant-based foods prevalence" cluster, for whom the increase in *Lachnospiraceae* and *Ruminococcaceae* members, along with the decrease in *Desulfovibrio* (a sulphate-reducing pathobiont found to be increased in type 2 diabetic patients and those suffering from inflammatory bowel disease [89,90]) were more evident. Furthermore, it should be noted that an increase in *Akkermansia* specifically discriminated the participants who showed improvement in at least 1 risk factor for MS, further stressing the close association between this taxon and metabolic health [17,65]. *Akkermansia* is in fact a mucus degrader with promising metabolic benefits, as validated in a recent proof-of-concept exploratory study [65]. On the other hand, in the PROB-group, i.e., in subjects receiving probiotic supplementation, we observed a reduction of sulphate-reducing bacteria, but also several unfavorable microbiota changes, including reduced diversity and relative abundance of *Akkermansia*. In this group, we also found increased proportions of *Peptostreptococcaceae* and *Erysipelotrichaceae*, less characterized microorganisms, but generally associated with increased inflammatory tone [91,92].

Urine metabolomics confirmed a general beneficial effect of SA-derived products on the metabolic health of the participants, as exemplified by the decrease in TMAO levels and the increase in citrate levels in SA-group. As a result of the microbiota–host co-metabolism of dietary choline, TMAO has indeed been repeatedly associated with cardiovascular disease risk and atherosclerosis [73]. As for citrate, its urinary excretion rate mainly depends on the acid–base status of the body, and urinary citrate has long been recognized as an inhibitor of calcium salt crystallization [93]. Even small acid loads, such as meat-based or protein-rich meals in general, reduce urinary citrate excretion. It is, therefore, tempting to speculate that the intervention in SA-group may have a more protective role against kidney disease. Furthermore, subjects from the "Omnivorous with meat prevalence" cluster in both SA-group and PROB-group, and those from the "High consumers" cluster in PROB-group showed a relatively higher level of hippurate following the intervention. Hippurate has been strongly associated with increased gut microbiome diversity, consumption of polyphenol-rich foods, and reduced odds of MS [94]. On the other hand, subjects from the "High consumers" cluster of PROB-group experienced a lower amount of urine trigonelline, which could be unfavorable. Despite a possible dietary-related origin, trigonelline is mostly biosynthesized by the gut microbiota during the conversion of S-adenosylmethionine to S-adenosylhomocysteine (methionine cycle). Interestingly, this metabolite has been inversely correlated with obese and diabetic phenotypes [95].

This study has several strengths, including: (i) the collection of data over time, allowing for a strict control of potential changes over the study period; (ii) the collection of anthropometric, biochemical, and immunological information, as well as baseline dietary data, which allowed for a parallel exploration and interpretation of temporal effects in microbiome and metabolome; and (iii) the use of a validated instrument for the assessment of subjects' DPs. However, the present study has also some limitations: (i) its small sample size, even if legitimate for a hypothesis-generating study characterized by considerable organizational commitment; (ii) the lack of a control group who eat the same products of the SA-group but produced with conventional techniques, to disentangle the actual contribution of SA-based products; and (iii) the lack of an assessment of individual adherence to the dietary intervention to monitoring also the evolution of diet quality and potential shifting towards more balanced or controlled diets, including a higher adherence to the

Mediterranean diet, which is already known to provide some beneficial effects in obesity, type 2 diabetes, cardiometabolic disease risk and aging [15,96–98].

5. Conclusions

To our knowledge, this is the first study exploring the potential beneficial effects of SA-derived products on the gut microbiota and urinary metabolome in humans. Our preliminary evidence points to some improvements in the amounts of certain microorganisms and metabolites relevant to health, as well as in some risk factors for MS, in subjects receiving fresh food products. These benefits were greater in those who followed less healthy dietary habits. Participants receiving probiotics also showed some changes in microbial, metabolic and health parameters but the effects were marginal and not entirely favorable, in accordance with recent literature [99]. Diets based on foods from organic symbiotic crops may therefore be effective in modulating unbalanced microbiomes towards eubiotic configurations, and improving metabolomics profiles and metabolic health, with likely lower risk of developing MS and related disorders. Future studies in larger cohorts or randomized controlled trials, possibly including patients with MS or other metabolic diseases, and employing other omics techniques, such as shotgun metagenomics, are needed to validate these findings and provide further functional insights into the SA-based diet–microbiota–host axis.

Supplementary Materials: The following are available online at https://www.mdpi.com/article/10.3390/nu13062081/s1, Figure S1: Phylum- and family level composition of the gut microbiota of study subjects compared to healthy Italian controls, Figure S2: Scatter plots of correlations between taxon relative abundances and levels of biochemical parameters (A), cytokines (B) and urinary metabolites (C) in all study subjects, Figure S3: Relationship between the gut microbiota and dietary habits in subjects at risk for metabolic syndrome, Figure S4: PCA score plot from all study participants at baseline, Figure S5: Violin plots for statistically significant changes in systolic blood pressure (within SA-group), cortisol (within PROB-group), and insulin (within PROB-group) between T0 and T30, Figure S6: Impact on the gut microbiota diversity of a diet with fresh foods from organic symbiotic agriculture (SA-group), (A) versus probiotic supplementation (PROB-group), (B), Figure S7: Baseline variation in the relative abundances of genera that varied significantly during intervention within the SA-group, Figure S8: Microbiota compositional variations in SA-group subjects at risk for metabolic syndrome with different long-term dietary habits, Figure S9: Scatter plots and correlation coefficients between taxon relative abundances and levels of anthropometric/biochemical parameters (A) and urinary metabolites (B) in subjects at risk for metabolic syndrome, during the intervention in SA-group., Figure S10: Relationship between changes in genus relative abundances and improvements in metabolic syndrome components over time and within the SA-group, Figure S11: S-Line plot for the OPLS-DA model built for the SA-group (A) and for the OPLS-DA model built for PROB-group (B), Table S1: Factorability of the correlation matrix of the original nutrients: Bartlett's test of sphericity and measures of sampling adequacy, Table S2: Anthropometric, biochemical, and immunological characteristics for all study participants and separately for the dietary intervention groups at baseline (T0) and after intervention (T30).

Author Contributions: Conceptualization, S.T., E.P., O.N. and P.S.; Data curation, E.P., O.N. and P.S.; Formal analysis, S.T., E.P., V.E., F.D., L.D.C. and S.R.; Funding acquisition, P.S.; Investigation, D.G., C.R. and P.S.; Methodology, S.T., E.P. and O.N.; Project administration, P.S.; Resources, S.T., L.P., G.G. and P.S.; Software, S.T., E.P., V.E., F.D. and S.R.; Supervision, S.T., F.P.F., N.P., M.F. and P.S.; Visualization, S.T., E.P., V.E., F.D. and L.D.C.; Writing—original draft, S.T., E.P., V.E., A.M.G., F.D., L.P., L.D.C. and P.S.; Writing—review & editing, S.T., E.P., V.E., A.M.G., F.D., G.G., L.D.C., F.P.F., S.R., D.G., L.P., C.R., J.B., M.T., N.P., M.F., O.N. and P.S. All authors have read and agreed to the published version of the manuscript.

Funding: No competitive research grants were involved in the current project. Consorzio Eco-Simbiotico partially funded the study and PROBIOTICAL S.p.A provided the probiotics used during the study.

Institutional Review Board Statement: The study was conducted according to the guidelines of the Declaration of Helsinki and approved by the CEROM Ethics Committee (Study ID: IRST B086 L4P1755, Ethical approval ID: 6759/2018).

Informed Consent Statement: Informed consent was obtained from all subjects involved in the study.

Data Availability Statement: The data presented in this study are available on request from the corresponding author. Sequencing data are accessible at the National Center for Biotechnology Information Sequence Read Archive (NCBI SRA; Bioproject ID PRJNA726866).

Acknowledgments: The FFQ was developed with the contribution of the Italian Association for Research on Cancer (AIRC). We thank Claudia Agnoli for suggestions on the calculation of the Italian Mediterranean Index.

Conflicts of Interest: The authors declare no conflict of interest.

References

1. Lemieux, I.; Despres, J.P. Metabolic Syndrome: Past, Present and Future. *Nutrients* **2020**, *12*, 3501. [CrossRef] [PubMed]
2. Dabke, K.; Hendrick, G.; Devkota, S. The gut microbiome and metabolic syndrome. *J. Clin. Investig.* **2019**, *129*, 4050–4057. [CrossRef] [PubMed]
3. Vishram, J.K.; Borglykke, A.; Andreasen, A.H.; Jeppesen, J.; Ibsen, H.; Jorgensen, T.; Palmieri, L.; Giampaoli, S.; Donfrancesco, C.; Kee, F.; et al. Impact of age and gender on the prevalence and prognostic importance of the metabolic syndrome and its components in Europeans. The MORGAM Prospective Cohort Project. *PLoS ONE* **2014**, *9*, e107294. [CrossRef] [PubMed]
4. Mottillo, S.; Filion, K.B.; Genest, J.; Joseph, L.; Pilote, L.; Poirier, P.; Rinfret, S.; Schiffrin, E.L.; Eisenberg, M.J. The metabolic syndrome and cardiovascular risk a systematic review and meta-analysis. *J. Am. Coll. Cardiol.* **2010**, *56*, 1113–1132. [CrossRef]
5. Choi, I.Y.; Chun, S.; Shin, D.W.; Han, K.; Jeon, K.H.; Yu, J.; Chae, B.J.; Suh, M.; Park, Y.M. Changes in Metabolic Syndrome Status and Breast Cancer Risk: A Nationwide Cohort Study. *Cancers* **2021**, *13*, 1177. [CrossRef]
6. Esposito, K.; Chiodini, P.; Colao, A.; Lenzi, A.; Giugliano, D. Metabolic syndrome and risk of cancer: A systematic review and meta-analysis. *Diabetes Care* **2012**, *35*, 2402–2411. [CrossRef]
7. Han, F.; Wu, G.; Zhang, S.; Zhang, J.; Zhao, Y.; Xu, J. The association of Metabolic Syndrome and its Components with the Incidence and Survival of Colorectal Cancer: A Systematic Review and Meta-analysis. *Int. J. Biol. Sci.* **2021**, *17*, 487–497. [CrossRef]
8. Mendonca, F.M.; de Sousa, F.R.; Barbosa, A.L.; Martins, S.C.; Araujo, R.L.; Soares, R.; Abreu, C. Metabolic syndrome and risk of cancer: Which link? *Metabolism* **2015**, *64*, 182–189. [CrossRef]
9. Xia, B.; He, Q.; Pan, Y.; Gao, F.; Liu, A.; Tang, Y.; Chong, C.; Teoh, A.Y.B.; Li, F.; He, Y.; et al. Metabolic syndrome and risk of pancreatic cancer: A population-based prospective cohort study. *Int. J. Cancer* **2020**, *147*, 3384–3393. [CrossRef]
10. Gallardo-Becerra, L.; Cornejo-Granados, F.; Garcia-Lopez, R.; Valdez-Lara, A.; Bikel, S.; Canizales-Quinteros, S.; Lopez-Contreras, B.E.; Mendoza-Vargas, A.; Nielsen, H.; Ochoa-Leyva, A. Metatranscriptomic analysis to define the Secrebiome, and 16S rRNA profiling of the gut microbiome in obesity and metabolic syndrome of Mexican children. *Microb. Cell Fact.* **2020**, *19*, 61. [CrossRef]
11. He, Y.; Wu, W.; Wu, S.; Zheng, H.M.; Li, P.; Sheng, H.F.; Chen, M.X.; Chen, Z.H.; Ji, G.Y.; Zheng, Z.D.; et al. Linking gut microbiota, metabolic syndrome and economic status based on a population-level analysis. *Microbiome* **2018**, *6*, 172. [CrossRef]
12. Kootte, R.S.; Levin, E.; Salojarvi, J.; Smits, L.P.; Hartstra, A.V.; Udayappan, S.D.; Hermes, G.; Bouter, K.E.; Koopen, A.M.; Holst, J.J.; et al. Improvement of Insulin Sensitivity after Lean Donor Feces in Metabolic Syndrome Is Driven by Baseline Intestinal Microbiota Composition. *Cell Metab.* **2017**, *26*, 611–619. [CrossRef]
13. Org, E.; Blum, Y.; Kasela, S.; Mehrabian, M.; Kuusisto, J.; Kangas, A.J.; Soininen, P.; Wang, Z.; Ala-Korpela, M.; Hazen, S.L.; et al. Relationships between gut microbiota, plasma metabolites, and metabolic syndrome traits in the METSIM cohort. *Genome Biol.* **2017**, *18*, 70. [CrossRef]
14. Wutthi-In, M.; Cheevadhanarak, S.; Yasom, S.; Kerdphoo, S.; Thiennimitr, P.; Phrommintikul, A.; Chattipakorn, N.; Kittichotirat, W.; Chattipakorn, S. Gut Microbiota Profiles of Treated Metabolic Syndrome Patients and their Relationship with Metabolic Health. *Sci. Rep.* **2020**, *10*, 10085. [CrossRef]
15. Cancello, R.; Turroni, S.; Rampelli, S.; Cattaldo, S.; Candela, M.; Cattani, L.; Mai, S.; Vietti, R.; Scacchi, M.; Brigidi, P.; et al. Effect of Short-Term Dietary Intervention and Probiotic Mix Supplementation on the Gut Microbiota of Elderly Obese Women. *Nutrients* **2019**, *11*, 3011. [CrossRef]
16. Cani, P.D. Microbiota and metabolites in metabolic diseases. *Nat. Rev. Endocrinol.* **2019**, *15*, 69–70. [CrossRef]
17. Gomez-Arango, L.F.; Barrett, H.L.; Wilkinson, S.A.; Callaway, L.K.; McIntyre, H.D.; Morrison, M.; Dekker Nitert, M. Low dietary fiber intake increases Collinsella abundance in the gut microbiota of overweight and obese pregnant women. *Gut Microbes* **2018**, *9*, 189–201. [CrossRef]
18. Silveira-Nunes, G.; Durso, D.F.; de Oliveira, L.R.A., Jr.; Cunha, E.H.M.; Maioli, T.U.; Vieira, A.T.; Speziali, E.; Correa-Oliveira, R.; Martins-Filho, O.A.; Teixeira-Carvalho, A.; et al. Hypertension Is Associated With Intestinal Microbiota Dysbiosis and Inflammation in a Brazilian Population. *Front. Pharmacol.* **2020**, *11*, 258. [CrossRef]
19. Zhu, W.; Gregory, J.C.; Org, E.; Buffa, J.A.; Gupta, N.; Wang, Z.; Li, L.; Fu, X.; Wu, Y.; Mehrabian, M.; et al. Gut Microbial Metabolite TMAO Enhances Platelet Hyperreactivity and Thrombosis Risk. *Cell* **2016**, *165*, 111–124. [CrossRef]

20. Kolodziejczyk, A.A.; Zheng, D.; Elinav, E. Diet-microbiota interactions and personalized nutrition. *Nat. Rev. Micro Biol.* **2019**, *17*, 742–753. [CrossRef]
21. Zmora, N.; Suez, J.; Elinav, E. You are what you eat: Diet, health and the gut microbiota. *Nat. Rev. Gastroenterol. Hepatol.* **2019**, *16*, 35–56. [CrossRef]
22. Moszak, M.; Szulinska, M.; Bogdanski, P. You Are What You Eat-The Relationship between Diet, Microbiota, and Metabolic Disorders—A Review. *Nutrients* **2020**, *12*, 1096. [CrossRef]
23. Hirt, H. Healthy soils for healthy plants for healthy humans: How beneficial microbes in the soil, food and gut are interconnected and how agriculture can contribute to human health. *EMBO Rep.* **2020**, *21*, e51069. [CrossRef]
24. Raiola, A.; Tenore, G.C.; Petito, R.; Ciampaglia, R.; Ritieni, A. Improving of nutraceutical features of many important mediterranean vegetables by inoculation with a new commercial product. *Curr. Pharm. Biotechnol.* **2015**, *16*, 738–746. [CrossRef]
25. Longo, V.C.M.; Della Croce, C.M.; Giovannetti, G.; Masciandaro, G. Assessment of nutraceutical features of different foods from conventional and mycorrhized farming. *Bull. Sci. Inform.* **2018**, *36*, 40–46.
26. Giovannetti, G.; Polo, F.; Nutricato, S.; Masoero, G.; Nuti, M. Efficacy of a commercial symbiotic bio-fertilizer consortium for mitigating the olive quick decline syndrome (OQDS). *J. Agron. Res.* **2019**, *2*, 1. [CrossRef]
27. Alberti, K.G.; Eckel, R.H.; Grundy, S.M.; Zimmet, P.Z.; Cleeman, J.I.; Donato, K.A.; Fruchart, J.C.; James, W.P.; Loria, C.M.; Smith, S.C., Jr.; et al. Harmonizing the metabolic syndrome. *Circulation* **2009**, *120*, 1640–1645. [CrossRef]
28. Hsieh, P.S.; Ho, H.H.; Hsieh, S.H.; Kuo, Y.W.; Tseng, H.Y.; Kao, H.F.; Wang, J.Y. Lactobacillus salivarius AP-32 and Lactobacillus reuteri GL-104 decrease glycemic levels and attenuate diabetes-mediated liver and kidney injury in db/db mice. *BMJ Open Diabetes Res. Care* **2020**, *8*. [CrossRef]
29. Oh, Y.J.; Kim, H.J.; Kim, T.S.; Yeo, I.H.; Ji, G.E. Effects of Lactobacillus plantarum PMO 08 Alone and Combined with Chia Seeds on Metabolic Syndrome and Parameters Related to Gut Health in High-Fat Diet-Induced Obese Mice. *J. Med. Food* **2019**, *22*, 1199–1207. [CrossRef] [PubMed]
30. Russo, M.; Marquez, A.; Herrera, H.; Abeijon-Mukdsi, C.; Saavedra, L.; Hebert, E.; Gauffin-Cano, P.; Medina, R. Oral administration of Lactobacillus fermentum CRL1446 improves biomarkers of metabolic syndrome in mice fed a high-fat diet supplemented with wheat bran. *Food Funct.* **2020**, *11*, 3879–3894. [CrossRef] [PubMed]
31. Pala, V.; Sieri, S.; Palli, D.; Salvini, S.; Berrino, F.; Bellegotti, M.; Frasca, G.; Tumino, R.; Sacerdote, C.; Fiorini, L.; et al. Diet in the Italian EPIC cohorts: Presentation of data and methodological issues. *Tumori* **2003**, *89*, 594–607. [CrossRef]
32. Pisani, P.; Faggiano, F.; Krogh, V.; Palli, D.; Vineis, P.; Berrino, F. Relative validity and reproducibility of a food frequency dietary questionnaire for use in the Italian EPIC centres. *Int. J. Epidemiol.* **1997**, *26* (Suppl. S1), S152–S160. [CrossRef]
33. Riboli, E.; Hunt, K.J.; Slimani, N.; Ferrari, P.; Norat, T.; Fahey, M.; Charrondiere, U.R.; Hemon, B.; Casagrande, C.; Vignat, J.; et al. European Prospective Investigation into Cancer and Nutrition (EPIC): Study populations and data collection. *Public Health Nutr.* **2002**, *5*, 1113–1124. [CrossRef]
34. Salvini, S.; Parpinel, M.; Gnagnarella, P.; Maisonneuve, P.; Turrini, A. *Banca di Composizione degli Alimenti per Studi Epidemiologici in Italia*; Istituto Europeo di Oncologia: Milano, Italy, 1998.
35. Yu, Z.; Morrison, M. Improved extraction of PCR-quality community DNA from digesta and fecal samples. *Biotechniques* **2004**, *36*, 808–812. [CrossRef]
36. D'Amico, F.; Biagi, E.; Rampelli, S.; Fiori, J.; Zama, D.; Soverini, M.; Barone, M.; Leardini, D.; Muratore, E.; Prete, A.; et al. Enteral Nutrition in Pediatric Patients Undergoing Hematopoietic SCT Promotes the Recovery of Gut Microbiome Homeostasis. *Nutrients* **2019**, *11*, 2958. [CrossRef]
37. Klindworth, A.; Pruesse, E.; Schweer, T.; Peplies, J.; Quast, C.; Horn, M.; Glockner, F.O. Evaluation of general 16S ribosomal RNA gene PCR primers for classical and next-generation sequencing-based diversity studies. *Nucleic Acids Res.* **2013**, *41*, e1. [CrossRef]
38. Masella, A.P.; Bartram, A.K.; Truszkowski, J.M.; Brown, D.G.; Neufeld, J.D. PANDAseq: Paired-end assembler for illumina sequences. *BMC Bioinform.* **2012**, *13*, 31. [CrossRef]
39. Bolyen, E.; Rideout, J.R.; Dillon, M.R.; Bokulich, N.A.; Abnet, C.C.; Al-Ghalith, G.A.; Alexander, H.; Alm, E.J.; Arumugam, M.; Asnicar, F.; et al. Reproducible, interactive, scalable and extensible microbiome data science using QIIME 2. *Nat. Biotechnol.* **2019**, *37*, 852–857. [CrossRef]
40. Callahan, B.J.; McMurdie, P.J.; Rosen, M.J.; Han, A.W.; Johnson, A.J.; Holmes, S.P. DADA2: High-resolution sample inference from Illumina amplicon data. *Nat. Methods* **2016**, *13*, 581–583. [CrossRef]
41. Rognes, T.; Flouri, T.; Nichols, B.; Quince, C.; Mahe, F. VSEARCH: A versatile open source tool for metagenomics. *PeerJ* **2016**, *4*, e2584. [CrossRef]
42. De Filippis, F.; Pellegrini, N.; Vannini, L.; Jeffery, I.B.; La Storia, A.; Laghi, L.; Serrazanetti, D.I.; Di Cagno, R.; Ferrocino, I.; Lazzi, C.; et al. High-level adherence to a Mediterranean diet beneficially impacts the gut microbiota and associated metabolome. *Gut* **2016**, *65*, 1812–1821. [CrossRef] [PubMed]
43. Schnorr, S.L.; Candela, M.; Rampelli, S.; Centanni, M.; Consolandi, C.; Basaglia, G.; Turroni, S.; Biagi, E.; Peano, C.; Severgnini, M.; et al. Gut microbiome of the Hadza hunter-gatherers. *Nat. Commun.* **2014**, *5*, 3654. [CrossRef] [PubMed]
44. Biagi, E.; Franceschi, C.; Rampelli, S.; Severgnini, M.; Ostan, R.; Turroni, S.; Consolandi, C.; Quercia, S.; Scurti, M.; Monti, D.; et al. Gut Microbiota and Extreme Longevity. *Curr. Biol.* **2016**, *26*, 1480–1485. [CrossRef] [PubMed]
45. Bouatra, S.; Aziat, F.; Mandal, R.; Guo, A.C.; Wilson, M.R.; Knox, C.; Bjorndahl, T.C.; Krishnamurthy, R.; Saleem, F.; Liu, P.; et al. The human urine metabolome. *PLoS ONE* **2013**, *8*, e73076. [CrossRef] [PubMed]

46. Wishart, D.S.; Tzur, D.; Knox, C.; Eisner, R.; Guo, A.C.; Young, N.; Cheng, D.; Jewell, K.; Arndt, D.; Sawhney, S.; et al. HMDB: The Human Metabolome Database. *Nucleic Acids Res.* **2007**, *35*, D521–D526. [CrossRef]

47. Agnoli, C.; Krogh, V.; Grioni, S.; Sieri, S.; Palli, D.; Masala, G.; Sacerdote, C.; Vineis, P.; Tumino, R.; Frasca, G.; et al. A priori-defined dietary patterns are associated with reduced risk of stroke in a large Italian cohort. *J. Nutr.* **2011**, *141*, 1552–1558. [CrossRef]

48. Dalmartello, M.; Bravi, F.; Serraino, D.; Crispo, A.; Ferraroni, M.; La Vecchia, C.; Edefonti, V. Dietary Patterns in Italy and the Risk of Renal Cell Carcinoma. *Nutrients* **2020**, *12*, 134. [CrossRef]

49. Edefonti, V.; La Vecchia, C.; Di Maso, M.; Crispo, A.; Polesel, J.; Libra, M.; Parpinel, M.; Serraino, D.; Ferraroni, M.; Bravi, F. Association between Nutrient-Based Dietary Patterns and Bladder Cancer in Italy. *Nutrients* **2020**, *12*, 1584. [CrossRef]

50. Kaufman, L.; Rousseeuw, P.J. *Finding Groups in Data: An Introduction to Cluster Analysis*; John Wiley & Sons: Hoboken, NJ, USA, 2009; Volume 344.

51. Noguchi, K.; Gel, Y.R.; Brunner, E.; Konietschke, F. nparLD: An R software package for the nonparametric analysis of longitudinal data in factorial experiments. *J. Stat. Softw.* **2012**, *50*. [CrossRef]

52. Vignoli, A.; Rodio, D.M.; Bellizzi, A.; Sobolev, A.P.; Anzivino, E.; Mischitelli, M.; Tenori, L.; Marini, F.; Priori, R.; Scrivo, R.; et al. NMR-based metabolomic approach to study urine samples of chronic inflammatory rheumatic disease patients. *Anal. Bioanal. Chem.* **2017**, *409*, 1405–1413. [CrossRef] [PubMed]

53. R Development Core Team. *R: A Language and Environment for Statistical Computing 2021*; R Foundation for Statistical Computing: Vienna, Austria, 2021. Available online: http://www.R-project.org (accessed on 7 April 2021).

54. Revelle, W. Psych: Procedures for Psychological, Psychometric, and Personality Research; R Package Version 2.1.3. 2021. Available online: https://CRAN.R-project.org/package=psych (accessed on 7 April 2021).

55. Maechler, M.; Rousseeuw, P.; Struyf, A.; Hubert, M.; Hornik, K. Cluster: Cluster Analysis Basics and Extensions; R Package Version 2.1.1. 2021. Available online: https://CRAN.R-project.org/package=cluster (accessed on 7 April 2021).

56. Dimitriadou, E.; Hornik, K.; Hornik, M.K. Package 'cclust'. 2017. Available online: https://mran.microsoft.com/snapshot/2017-02-04/web/packages/cclust/index.html (accessed on 7 April 2021).

57. Scrucca, L.; Fop, M.; Murphy, T.B.; Raftery, A.E. mclust 5: Clustering, classification and density estimation using Gaussian finite mixture models. *R J.* **2016**, *8*, 289. [CrossRef] [PubMed]

58. Culhane, A.C.; Thioulouse, J.; Perriere, G.; Higgins, D.G. MADE4: An R package for multivariate analysis of gene expression data. *Bioinformatics* **2005**, *21*, 2789–2790. [CrossRef]

59. Vujkovic-Cvijin, I.; Sklar, J.; Jiang, L.; Natarajan, L.; Knight, R.; Belkaid, Y. Host variables confound gut microbiota studies of human disease. *Nature* **2020**, *587*, 448–454. [CrossRef]

60. Candela, M.; Biagi, E.; Soverini, M.; Consolandi, C.; Quercia, S.; Severgnini, M.; Peano, C.; Turroni, S.; Rampelli, S.; Pozzilli, P.; et al. Modulation of gut microbiota dysbioses in type 2 diabetic patients by macrobiotic Ma-Pi 2 diet. *Br. J. Nutr.* **2016**, *116*, 80–93. [CrossRef]

61. Smits, L.P.; Kootte, R.S.; Levin, E.; Prodan, A.; Fuentes, S.; Zoetendal, E.G.; Wang, Z.; Levison, B.S.; Cleophas, M.C.P.; Kemper, E.M.; et al. Effect of Vegan Fecal Microbiota Transplantation on Carnitine- and Choline-Derived Trimethylamine-N-Oxide Production and Vascular Inflammation in Patients With Metabolic Syndrome. *J. Am. Heart Assoc.* **2018**, *7*. [CrossRef]

62. Vrieze, A.; Van Nood, E.; Holleman, F.; Salojarvi, J.; Kootte, R.S.; Bartelsman, J.F.; Dallinga-Thie, G.M.; Ackermans, M.T.; Serlie, M.J.; Oozeer, R.; et al. Transfer of intestinal microbiota from lean donors increases insulin sensitivity in individuals with metabolic syndrome. *Gastroenterology* **2012**, *143*, 913–916. [CrossRef]

63. Wu, T.R.; Lin, C.S.; Chang, C.J.; Lin, T.L.; Martel, J.; Ko, Y.F.; Ojcius, D.M.; Lu, C.C.; Young, J.D.; Lai, H.C. Gut commensal Parabacteroides goldsteinii plays a predominant role in the anti-obesity effects of polysaccharides isolated from Hirsutella sinensis. *Gut* **2019**, *68*, 248–262. [CrossRef]

64. Duvallet, C.; Gibbons, S.M.; Gurry, T.; Irizarry, R.A.; Alm, E.J. Meta-analysis of gut microbiome studies identifies disease-specific and shared responses. *Nat. Commun.* **2017**, *8*, 1784. [CrossRef]

65. Depommier, C.; Everard, A.; Druart, C.; Plovier, H.; Van Hul, M.; Vieira-Silva, S.; Falony, G.; Raes, J.; Maiter, D.; Delzenne, N.M.; et al. Supplementation with Akkermansia muciniphila in overweight and obese human volunteers: A proof-of-concept exploratory study. *Nat. Med.* **2019**, *25*, 1096–1103. [CrossRef]

66. Ozato, N.; Saito, S.; Yamaguchi, T.; Katashima, M.; Tokuda, I.; Sawada, K.; Katsuragi, Y.; Kakuta, M.; Imoto, S.; Ihara, K.; et al. Blautia genus associated with visceral fat accumulation in adults 20–76 years of age. *NPJ Biofilms Microbiomes* **2019**, *5*, 28. [CrossRef]

67. Tavella, T.; Rampelli, S.; Guidarelli, G.; Bazzocchi, A.; Gasperini, C.; Pujos-Guillot, E.; Comte, B.; Barone, M.; Biagi, E.; Candela, M.; et al. Elevated gut microbiome abundance of Christensenellaceae, Porphyromonadaceae and Rikenellaceae is associated with reduced visceral adipose tissue and healthier metabolic profile in Italian elderly. *Gut Microbes* **2021**, *13*, 1–19. [CrossRef] [PubMed]

68. Zhou, J.; Chan, L.; Zhou, S. Trigonelline: A plant alkaloid with therapeutic potential for diabetes and central nervous system disease. *Curr. Med. Chem.* **2012**, *19*, 3523–3531. [CrossRef] [PubMed]

69. Frost, F.; Storck, L.J.; Kacprowski, T.; Gartner, S.; Ruhlemann, M.; Bang, C.; Franke, A.; Volker, U.; Aghdassi, A.A.; Steveling, A.; et al. A structured weight loss program increases gut microbiota phylogenetic diversity and reduces levels of Collinsella in obese type 2 diabetics: A pilot study. *PLoS ONE* **2019**, *14*, e0219489. [CrossRef] [PubMed]

70. Konikoff, T.; Gophna, U. Oscillospira: A Central, Enigmatic Component of the Human Gut Microbiota. *Trends Micro Biol.* **2016**, *24*, 523–524. [CrossRef]
71. Fassarella, M.; Blaak, E.E.; Penders, J.; Nauta, A.; Smidt, H.; Zoetendal, E.G. Gut microbiome stability and resilience: Elucidating the response to perturbations in order to modulate gut health. *Gut* **2021**, *70*, 595–605. [CrossRef] [PubMed]
72. Naderpoor, N.; Mousa, A.; Gomez-Arango, L.F.; Barrett, H.L.; Dekker Nitert, M.; de Courten, B. Faecal Microbiota Are Related to Insulin Sensitivity and Secretion in Overweight or Obese Adults. *J. Clin. Med.* **2019**, *8*, 452. [CrossRef]
73. Wang, Z.; Klipfell, E.; Bennett, B.J.; Koeth, R.; Levison, B.S.; Dugar, B.; Feldstein, A.E.; Britt, E.B.; Fu, X.; Chung, Y.M.; et al. Gut flora metabolism of phosphatidylcholine promotes cardiovascular disease. *Nature* **2011**, *472*, 57–63. [CrossRef]
74. Alves, A.; Bassot, A.; Bulteau, A.L.; Pirola, L.; Morio, B. Glycine Metabolism and Its Alterations in Obesity and Metabolic Diseases. *Nutrients* **2019**, *11*, 1356. [CrossRef]
75. Altorf-van der Kuil, W.; Brink, E.J.; Boetje, M.; Siebelink, E.; Bijlsma, S.; Engberink, M.F.; van't Veer, P.; Tome, D.; Bakker, S.J.; van Baak, M.A.; et al. Identification of biomarkers for intake of protein from meat, dairy products and grains: A controlled dietary intervention study. *Br. J. Nutr.* **2013**, *110*, 810–822. [CrossRef]
76. Karlsson, C.L.; Onnerfalt, J.; Xu, J.; Molin, G.; Ahrne, S.; Thorngren-Jerneck, K. The microbiota of the gut in preschool children with normal and excessive body weight. *Obesity* **2012**, *20*, 2257–2261. [CrossRef]
77. Chen, J.; Wright, K.; Davis, J.M.; Jeraldo, P.; Marietta, E.V.; Murray, J.; Nelson, H.; Matteson, E.L.; Taneja, V. An expansion of rare lineage intestinal microbes characterizes rheumatoid arthritis. *Genome Med.* **2016**, *8*, 43. [CrossRef]
78. Wang, K.; Liao, M.; Zhou, N.; Bao, L.; Ma, K.; Zheng, Z.; Wang, Y.; Liu, C.; Wang, W.; Wang, J.; et al. Parabacteroides distasonis Alleviates Obesity and Metabolic Dysfunctions via Production of Succinate and Secondary Bile Acids. *Cell Rep.* **2019**, *26*, 222–235. [CrossRef]
79. Hu, F.B. Dietary pattern analysis: A new direction in nutritional epidemiology. *Curr. Opin. Lipidol.* **2002**, *13*, 3–9. [CrossRef]
80. Moeller, S.M.; Reedy, J.; Millen, A.E.; Dixon, L.B.; Newby, P.K.; Tucker, K.L.; Krebs-Smith, S.M.; Guenther, P.M. Dietary patterns: Challenges and opportunities in dietary patterns research an Experimental Biology workshop, April 1, 2006. *J. Am. Diet. Assoc.* **2007**, *107*, 1233–1239. [CrossRef]
81. Edefonti, V.; Nicolussi, F.; Polesel, J.; Bravi, F.; Bosetti, C.; Garavello, W.; La Vecchia, C.; Bidoli, E.; Decarli, A.; Serraino, D.; et al. Nutrient-based dietary patterns and nasopharyngeal cancer: Evidence from an exploratory factor analysis. *Br. J. Cancer* **2015**, *112*, 446–454. [CrossRef]
82. Edefonti, V.; Randi, G.; Decarli, A.; La Vecchia, C.; Bosetti, C.; Franceschi, S.; Dal Maso, L.; Ferraroni, M. Clustering dietary habits and the risk of breast and ovarian cancers. *Ann. Oncol.* **2009**, *20*, 581–590. [CrossRef]
83. Masala, G.; Ceroti, M.; Pala, V.; Krogh, V.; Vineis, P.; Sacerdote, C.; Saieva, C.; Salvini, S.; Sieri, S.; Berrino, F.; et al. A dietary pattern rich in olive oil and raw vegetables is associated with lower mortality in Italian elderly subjects. *Br. J. Nutr.* **2007**, *98*, 406–415. [CrossRef]
84. Pala, V.; Sieri, S.; Masala, G.; Palli, D.; Panico, S.; Vineis, P.; Sacerdote, C.; Mattiello, A.; Galasso, R.; Salvini, S.; et al. Associations between dietary pattern and lifestyle, anthropometry and other health indicators in the elderly participants of the EPIC-Italy cohort. *Nutr. Metab. Cardiovasc. Dis.* **2006**, *16*, 186–201. [CrossRef]
85. Edefonti, V.; De Vito, R.; Dalmartello, M.; Patel, L.; Salvatori, A.; Ferraroni, M. Reproducibility and Validity of A Posteriori Dietary Patterns: A Systematic Review. *Adv. Nutr.* **2020**, *11*, 293–326. [CrossRef]
86. Koh, A.; De Vadder, F.; Kovatcheva-Datchary, P.; Backhed, F. From Dietary Fiber to Host Physiology: Short-Chain Fatty Acids as Key Bacterial Metabolites. *Cell* **2016**, *165*, 1332–1345. [CrossRef]
87. Goodrich, J.K.; Davenport, E.R.; Beaumont, M.; Jackson, M.A.; Knight, R.; Ober, C.; Spector, T.D.; Bell, J.T.; Clark, A.G.; Ley, R.E. Genetic Determinants of the Gut Microbiome in UK Twins. *Cell Host Microbe* **2016**, *19*, 731–743. [CrossRef] [PubMed]
88. Stewart, C.J.; Ajami, N.J.; O'Brien, J.L.; Hutchinson, D.S.; Smith, D.P.; Wong, M.C.; Ross, M.C.; Lloyd, R.E.; Doddapaneni, H.; Metcalf, G.A.; et al. Temporal development of the gut microbiome in early childhood from the TEDDY study. *Nature* **2018**, *562*, 583–588. [CrossRef] [PubMed]
89. Qin, J.; Li, Y.; Cai, Z.; Li, S.; Zhu, J.; Zhang, F.; Liang, S.; Zhang, W.; Guan, Y.; Shen, D.; et al. A metagenome-wide association study of gut microbiota in type 2 diabetes. *Nature* **2012**, *490*, 55–60. [CrossRef] [PubMed]
90. Rowan, F.; Docherty, N.G.; Murphy, M.; Murphy, B.; Calvin Coffey, J.; O'Connell, P.R. Desulfovibrio bacterial species are increased in ulcerative colitis. *Dis. Colon Rectum* **2010**, *53*, 1530–1536. [CrossRef]
91. Ahn, J.; Sinha, R.; Pei, Z.; Dominianni, C.; Wu, J.; Shi, J.; Goedert, J.J.; Hayes, R.B.; Yang, L. Human gut microbiome and risk for colorectal cancer. *J. Natl. Cancer Inst.* **2013**, *105*, 1907–1911. [CrossRef]
92. Kaakoush, N.O. Insights into the Role of Erysipelotrichaceae in the Human Host. *Front. Cell Infect. Micro Biol.* **2015**, *5*, 84. [CrossRef]
93. Granchi, D.; Baldini, N.; Ulivieri, F.M.; Caudarella, R. Role of Citrate in Pathophysiology and Medical Management of Bone Diseases. *Nutrients* **2019**, *11*, 2576. [CrossRef]
94. Pallister, T.; Jackson, M.A.; Martin, T.C.; Zierer, J.; Jennings, A.; Mohney, R.P.; MacGregor, A.; Steves, C.J.; Cassidy, A.; Spector, T.D.; et al. Hippurate as a metabolomic marker of gut microbiome diversity: Modulation by diet and relationship to metabolic syndrome. *Sci. Rep.* **2017**, *7*, 13670. [CrossRef]

95. Palau-Rodriguez, M.; Tulipani, S.; Isabel Queipo-Ortuno, M.; Urpi-Sarda, M.; Tinahones, F.J.; Andres-Lacueva, C. Metabolomic insights into the intricate gut microbial-host interaction in the development of obesity and type 2 diabetes. *Front. Micro Biol.* **2015**, *6*, 1151. [CrossRef]
96. Ghosh, T.S.; Rampelli, S.; Jeffery, I.B.; Santoro, A.; Neto, M.; Capri, M.; Giampieri, E.; Jennings, A.; Candela, M.; Turroni, S.; et al. Mediterranean diet intervention alters the gut microbiome in older people reducing frailty and improving health status: The NU-AGE 1-year dietary intervention across five European countries. *Gut* **2020**, *69*, 1218–1228. [CrossRef]
97. Haro, C.; Montes-Borrego, M.; Rangel-Zuniga, O.A.; Alcala-Diaz, J.F.; Gomez-Delgado, F.; Perez-Martinez, P.; Delgado-Lista, J.; Quintana-Navarro, G.M.; Tinahones, F.J.; Landa, B.B.; et al. Two Healthy Diets Modulate Gut Microbial Community Improving Insulin Sensitivity in a Human Obese Population. *J. Clin. Endocrinol. Metab.* **2016**, *101*, 233–242. [CrossRef]
98. Wang, D.D.; Nguyen, L.H.; Li, Y.; Yan, Y.; Ma, W.; Rinott, E.; Ivey, K.L.; Shai, I.; Willett, W.C.; Hu, F.B.; et al. The gut microbiome modulates the protective association between a Mediterranean diet and cardiometabolic disease risk. *Nat. Med.* **2021**, *27*, 333–343. [CrossRef]
99. Tenorio-Jimenez, C.; Martinez-Ramirez, M.J.; Gil, A.; Gomez-Llorente, C. Effects of Probiotics on Metabolic Syndrome: A Systematic Review of Randomized Clinical Trials. *Nutrients* **2020**, *12*, 124. [CrossRef]

Review

The Effects of Pro-, Pre-, and Synbiotics on Muscle Wasting, a Systematic Review—Gut Permeability as Potential Treatment Target

Sandra J. van Krimpen [1,†], Fleur A. C. Jansen [1,†], Veerle L. Ottenheim [1,†], Clara Belzer [2], Miranda van der Ende [1] and Klaske van Norren [1,*]

[1] Nutritional Biology, Division of Human Nutrition, Wageningen University,
    6708 WE Wageningen, The Netherlands; sandravk6@hotmail.com (S.J.v.K.); fleur.jansen@wur.nl (F.A.C.J.);
    veerle.ottenheim2@gmail.com (V.L.O.); miranda.vanderende@wur.nl (M.v.d.E.)
[2] Laboratory of Microbiology, Wageningen University, 6708 WE Wageningen, The Netherlands;
    clara.belzer@wur.nl
*   Correspondence: klaske.vannorren@wur.nl; Tel.: +31-317-485-093
†   These authors contributed equally to this work.

Abstract: Muscle wasting is a frequently observed, inflammation-driven condition in aging and disease, known as sarcopenia and cachexia. Current treatment strategies target the muscle directly and are often not able to reverse the process. Because a reduced gut function is related to systemic inflammation, this might be an indirect target to ameliorate muscle wasting, by administering pro-, pre-, and synbiotics. Therefore, this review aimed to study the potential of pro-, pre-, and synbiotics to treat muscle wasting and to elucidate which metabolites and mechanisms affect the organ crosstalk in cachexia. Overall, the literature shows that *Lactobacillus species pluralis* (spp.) and possibly other genera, such as *Bifidobacterium*, can ameliorate muscle wasting in mouse models. The beneficial effects of *Lactobacillus* spp. supplementation may be attributed to its potential to improve microbiome balance and to its reported capacity to reduce gut permeability. A subsequent literature search revealed that the reduction of a high gut permeability coincided with improved muscle mass or strength, which shows an association between gut permeability and muscle mass. A possible working mechanism is proposed, involving lactate, butyrate, and reduced inflammation in gut–brain–muscle crosstalk. Thus, reducing gut permeability via *Lactobacillus* spp. supplementation could be a potential treatment strategy for muscle wasting.

Keywords: muscle wasting; cachexia; sarcopenia; probiotics; prebiotics; *Lactobacillus*; intestinal permeability; gut–brain axis

Citation: van Krimpen, S.J.; Jansen, F.A.C.; Ottenheim, V.L.; Belzer, C.; van der Ende, M.; van Norren, K. The Effects of Pro-, Pre-, and Synbiotics on Muscle Wasting, a Systematic Review—Gut Permeability as Potential Treatment Target. *Nutrients* **2021**, *13*, 1115. https://doi.org/10.3390/nu13041115

Academic Editor: Panagiota Mitrou

Received: 19 February 2021
Accepted: 23 March 2021
Published: 29 March 2021

1. Introduction

Muscle wasting is a frequently observed condition that contributes to progressive functional impairment, psychologic distress, and overall reduced resilience [1,2]. Normally, the equilibrium between protein synthesis and breakdown is tightly regulated and influenced by external stimuli such as physical activity and protein intake. However, during muscle wasting, this equilibrium shifts toward muscle protein breakdown, which is often driven by inflammation, either disease- or age-induced. These inflammation-related muscle wasting syndromes are known as cachexia and sarcopenia, respectively [2,3]. Because chronic inflammatory diseases such as cancer primarily develop in the elderly, sarcopenia and cachexia can also co-occur [1,2]. Both syndromes negatively affect life expectancy, survival, and quality of life; however, especially for cachexia, current treatment strategies are limited, palliative, and often not able to reverse the muscle wasting process [1,3].

Current treatment strategies may not be effective yet, as they primarily focus on directly increasing muscle mass. However, not only muscle but also other organs are affected by cachexia such as the brain, kidneys, and gut [1,4]. Communication between these organs is mediated by inflammatory mediators and results in the disturbance of core processes such as appetite regulation, stress, and energy homeostasis [5]. These processes are all closely related to gut function because nutrient absorption, secretion of appetite-regulating hormones, and immune responses are involved. Thus, a disturbed gut function might play a central role in the organ crosstalk that contributes to cachexia [1,4,5]. In addition, it may also be an important factor in sarcopenia, as in both syndromes, gut barrier dysfunction and changed microbiota composition have been observed [4,6].

An intervention that has been described to support gut function is the administering of pro-, pre-, and synbiotics. Probiotics are defined as live microorganisms that confer health benefits when administered in adequate amounts, whereas prebiotics are substrates that are selectively utilized by host microorganisms conferring a health benefit [7,8]. Synbiotics refers to the mixture of pro- and prebiotics that positively affects the beneficial microorganisms in the gut [9]. The mechanisms for their positive effect on gut function still need to be elucidated; however, they may restore gut barrier dysfunction [7]. As more microbiota-related proinflammatory compounds can enter the body when gut barrier function is disrupted, gut permeability could be an important contributor to the inflammatory state during cachexia [5]. Because gut barrier dysfunction has indeed been observed in both cachectic mice and patients [10,11], decreasing gut permeability might be part of the mechanisms via which probiotics could ameliorate muscle wasting. Therefore, the aim of this review was to study the potential of pro-, pre-, and synbiotics to treat muscle wasting and to investigate the association between gut permeability and muscle mass or function.

2. Methods

A systematic search was conducted in the databases PubMed and Scopus for all studies published up to 28 October 2020 that used pro-, pre-, and synbiotics to treat muscle wasting and/or function loss in human or animal models of disease or aging. The following advanced search was applied in PubMed (All Fields) and Scopus (TITLE-ABS-KEY): (probiotic* OR prebiotic* OR symbiotic* OR synbiotic*) AND (mice OR male OR female OR men OR women OR patient OR human OR animal) AND (infect* OR tumor OR tumour OR cancer OR disease OR ag*ing) AND ("muscle mass" OR "muscle function" OR "muscle strength" OR "muscle wasting" OR "muscle weakness" OR sarcopen* OR cachexi* OR cachec*). By including the terms "ag*ing" and "sarcopen*," data on age-induced muscle wasting were also covered and could thus be compared to data on disease-induced muscle wasting. Study selection and data extraction were performed independently by three researchers.

Additionally, a systematic search was conducted for all studies published up to 28 October 2020 that measured gut permeability as well as muscle mass or muscle strength in human or animal models of disease or aging. The following advanced search was applied: ("gut permeability" OR "leaky gut" OR "intestinal permeability" OR "gut homeostasis" OR "intestinal homeostasis" OR "gut barrier") AND ("skeletal muscle" OR "muscle mass" OR "muscle atrophy" OR "muscle function" OR "muscle strength" OR "muscle weakness" OR cachexi* OR cachec*). This search and its study selection and data extraction were performed similarly to the first systematic search.

3. Results

Over the last five years, the application of pro-, pre-, and synbiotics as disease treatment has substantially gained attention. Due to their effect on gut function, which could influence the multiorgan crosstalk, the use of pro-, pre-, and synbiotics could also be effective to treat cachexia. However, an overview regarding the exact effect of both pro-, pre-, and synbiotics on muscle wasting was lacking. Therefore, a systematic search was conducted to determine whether pro-, pre-, and synbiotics can be used as a treatment for muscle wasting (Figure 1A). This search resulted in 48 articles in PubMed and 151 articles in Scopus, which, together, made a total of 199 articles (Figure 1A). Of those articles,

41 duplicates were removed. After checking their reference lists, we excluded reviews, book chapters, case reports, conference articles, and study protocols, which resulted in 35 research articles being left. These articles were screened fully to assess their eligibility, and eight articles were accepted for inclusion based on exposure, study population, and outcome measures [12–20]. After reviewing the reference lists of the included articles and broadening the term "disease" to ("heart failure" OR COPD OR "renal failure" OR HIV), one additional study was included in the review [21]. Moreover, additional research was performed to gain more insights into the relationship between gut permeability and muscle mass or function (Figure 1B). This study selection and data extraction was performed similarly to the first systematic search and resulted in the inclusion of eight studies [15,16,20,22–26].

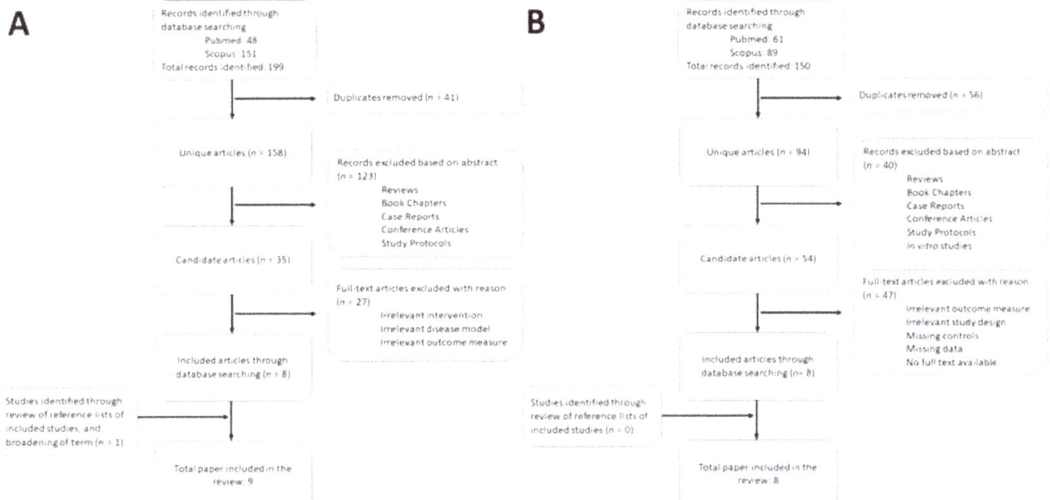

Figure 1. Flow diagram of (**A**) the identified, screened, and included studies on the effect of pro-, pre-, and synbiotics on muscle wasting [12–21] and (**B**) on the relationship between gut permeability and muscle wasting [15,16,20,22–26].

3.1. Effects of Pro-, Pre-, and Synbiotics on Muscle Wasting

All studies selected from the systematic search that included probiotic supplementation were performed in mice and used species from the genus of *Lactobacillus*, except for one that also used a strain of *Bifidobacterium* (Table 1). First of all, Varian et al. [12] found that supplementation with *Lactobacillus reuteri* significantly increased the muscle-to-body weight ratio and muscle fiber size in mice suffering from spontaneous intestinal adenoma. In line with these findings, Bindels et al. [21] showed that supplementation with a combination of *L. reuteri* and *L. gasseri* increased tibialis muscle weight by 8% ($p = 0.05$). However, gastrocnemius muscle weight was not significantly affected. This study was performed in mice injected with BaF cells, mimicking acute leukemia, which is a commonly used model for cancer cachexia. In addition to these studies in cachexia models, four studies with *Lactobacillus species pluralis* (spp.) supplementation in aging models were included. Firstly, Sugimura et al. [13] found that supplementation with *L. lactis* significantly increased the muscle-to-body weight ratio. Secondly, Chen et al. [14] showed that supplementation with *L. paracasei* significantly increased lean body mass. Furthermore, Varian et al. [12] reported that *L. reuteri* supplementation increased both the muscle-to-body weight ratio and muscle fiber size. Lastly, Ni et al. [15] found that *L. casei* significantly increased the muscle-to-body weight ratio as well as forelimb grip strength. Ni et al. [15] was the only study that also assessed the effects of another genus: *Bifidobacterium longum*. These effects were shown to be roughly comparable to the effects of *L. casei*. However, interestingly, the gut microbiota compositions of *Lactobacillus-* and *Bifidobacterium*-treated mice were changed differently upon treatment. For instance, *Lactobacillus* spp. significantly increased upon *L. reuteri* supplementation, but did not change upon B. longum supplementation. In summary, these studies collectively show that supplementation with *Lactobacillus* spp., and possibly also *Bifidobacterium* spp., has the potential to reduce cancer-induced and aging-induced muscle loss.

Furthermore, three studies on prebiotics were obtained from the systematic search, of which one was a randomized controlled clinical trial (Table 1). Firstly, supplementation with pectic oligosaccharides (POS) did not affect muscle mass in a mouse model of neuroblastoma. POS supplementation did also not change *Lactobacillus* spp. abundance in these diseased mice [16]. Secondly, Bindels et al. [17] also studied the effects of POS and, next to that, the effects of inulin in mice injected with BaF cells, inducing acute leukemia. In both groups, no treatment effect on muscle mass or *Lactobacillus* spp. abundance was found. Unfortunately, in the above-mentioned mice studies on prebiotics, cancer development failed to induce loss of muscle mass, which indicated there was no cachexia. Additionally, these prebiotic fibers have been tested in the elderly. A randomized controlled trial focusing on frailty in people over 65 years showed that supplementation with inulin and fructooligosaccharides (FOS) for 13 weeks resulted in reduced exhaustion and increased hand grip strength [18]. This indicates that prebiotics may increase muscle function. Unfortunately, muscle mass was not measured, so it remains unclear whether prebiotics can increase muscle mass in frail elderly.

Lastly, two studies were included focusing on synbiotics (Table 1). The effects of kimchi, a fermented product containing *Leuconostoc mesenteroides* and *Lactobacillus plantarum*, were tested in mice injected with adenocarcinoma [19]. Kimchi was found to significantly increase muscle mass in these mice. Interestingly, reduced expression and serum levels of interleukin (IL)-6 were also found, which is a proinflammatory cytokine involved in cachexia progression. Further, the effects of a combined treatment of *L. reuteri* with oligofructose were studied in mice suffering from acute leukemia [20]. This treatment significantly increased the percentage of lean body mass. So, prebiotics may produce a synergistic effect when combined with probiotics. This section may be divided by subheadings. It should provide a concise and precise description of the experimental results, their interpretation, and the experimental conclusions that can be drawn.

Table 1. Characteristics and results of the studies included in the systematic research on the effects of pre-, pro- and synbiotics on cachexia.

				Probiotics			
Family	Source	Model	Condition	Intervention	Muscle Outcome	Secondary Outcome	Reference
Lactobacillus	*reuteri*	C57BL/6 Apc^{mir}/+ mice	Spontaneous intestinal adenoma	3.5×10^5 CFU/day, 20 weeks	Muscle-to-BW ratio *, fiber size *	Intestinal polyps * and blood neutrophils *	[12]
Lactobacillus	*reuteri + gasseri*	BALB/c mice (female)	BaF acute leukemia	2×10^8 CFU/mL drinking water, from disease induction onwards	Muscle (mg)	*Lactobacillus* spp. *, food intake (-), body weight change (-), and IL-6 *	[21]
Lactobacillus	*lactis*	SAMP6 mice (female)	Aging (senescence-accelerated)	1 mg/day from 7 to 12 weeks of age	Muscle-to-BW ratio *	Survival *, senescence score *, and IL1beta *	[13]
Lactobacillus	*paracasei*	SAMP8 mice (female)	Aging (senescence-accelerated)	1×10^9 CFU/day from 16 to 28 weeks of age	Muscle (% of body) *, muscle strength *	Food intake (-), protein intake (-), TNFalfa *, and IL-6 *	[14]
Lactobacillus	*reuteri*	CD-1 mice	Aging	3.5×10^5 CFU/day from 2 to 12 months of age	Muscle-to-BW ratio *, fiber size *	Survival *, blood neutrophils *	[12]
Lactobacillus	*casei*	C57BL/6 mice (male)	Aging	2×10^9 CFU/day for 12 weeks from 10 months of age	Muscle-to-BW ratio *, forelimb grip strength *	Food intake (-), fatigue *, gut barrier proteins mRNA *, *Lactobacillus* spp. *, *Bifidobacterium* spp.	[15]
Bifidobacterium	*longum*	C57BL/6 mice (male)	Aging	2×10^9 CFU/day for 12 weeks from 10 months of age	Muscle-to-BW ratio *, forelimb grip strength	Food intake (-), fatigue *, gut barrier proteins mRNA *, *Bifidobacterium* spp.	[15]

Table 1. *Cont.*

				Prebiotics			
	Type	**Model**	**Condition**	**Intervention**	**Muscle Outcome**	**Secondary Outcome**	**Reference**
	POS	BALB/c Rj:ATHYM-Foxn1nu/numice (male)	Neuroblastoma	200 mg/day	Muscle/mm (-) (no cachexia developed)	*Lactobacillus* spp. (-), gut permeability (-), food consumption (-)	[16]
	POS	BALB/c mice (male)	BaF acute leukemia	5% POS for 2 weeks	Muscle (mg) (-) (no cachexia developed)	*Lactobacillus* spp. (-), anorexia * and propionate *	[17]
	Inulin	BALB/c mice (male)	BaF acute leukemia	5% inulin for 2 weeks	Muscle (mg) (-) (no cachexia developed)	*Lactobacillus* spp. (-), anorexia *, propionate and butyrate *	[17]
	Inulin + FOS	Elderly (aged 65 and over)	Frailty syndrome	3375 mg inulin + 3488 mg FOS/day for 13 weeks	Hand grip strength *	Energy intake (-), exhaustion *	[18]
				Synbiotics			
Probiotic	**Prebiotic**	**Model**	**Condition**	**Intervention**	**Muscle Outcome**	**Secondary Outcome**	**Reference**
Leuconostoc mesenteroides + Lactobacillus plantarum	Kimchi	BALB/c mice (male)	C26 colon carcinoma	Normal diet and cpKimchi diet for 3 weeks	Muscle mass *, ubiquitin *, AMPK *, PGC1-a *	Cachexia-induced lipolysis *, lipogenesis *, NF-κB *, AKT *, mTOR *, PI3K * and IL-6 *	[19]
Lactobacillus reuteri	OF	BALB/c mice (female)	BaF acute leukemia	2×10^8 CFU/mL probiotic + 0.2 g/day prebiotic from disease induction onwards	Muscle (% BW) *	Energy intake (-), survival (-), and gut barrier proteins mRNA *	[20]

BW: body weight; CFU: colony-forming unit; ↑: increased; (-): no change; ↓: decreased; * $p < 0.05$; POS: polyoligosaccharides; OF: oligofructans.

3.2. Gut Permeability and Muscle Wasting

As mentioned before, reducing gut permeability is hypothesized to be an effect of probiotic supplementation that may contribute toward the amelioration of muscle wasting. Therefore, a systematic search was performed to investigate the association between gut permeability and muscle mass or function. In total, eight studies were included, measuring both gut permeability and muscle mass or function in disease and aging models (Table 2) [15,16,20,22–26]. All studies were performed in mice, except for the studies of Cuoco et al. [22], Qi et al. [25], and van der Meij et al. [26], which concern human observational studies using matched controls to compare with patients and a young group to compare with elderly. Gut permeability was determined by measuring either the mRNA expression of one or multiple tight junction genes, levels of tight junction proteins, gut permeability markers, or by using an inert sugar permeability assay. Muscle mass was determined as the absolute weight of one or multiple muscles in the mice studies, while the human observational study of Cuoco et al. [22] measured muscle mass by Dual-Energy X-ray Absorptiometry (DEXA). Qi et al. [25] and van der Meij et al. [26] were the only studies that measured hand grip strength rather than muscle mass. In general, high gut permeability is associated with lower muscle mass in the majority of the studies [16,20,22–25], as only Obermüller et al. [17] showed no difference in muscle mass while gut permeability was increased in the diseased group. Van der Meij et al. [26] did not find an increase in gut permeability in cancer patients compared to matched controls but did find a significant negative correlation between small-intestinal barrier function and hand grip strength. Furthermore, four mice studies included an intervention group aiming to decrease gut permeability [15,16,20,24]. In these studies, decreased gut permeability was associated with decreased muscle mass loss. Thus, the results of the systematic search support the hypothesis of a correlation between gut permeability and muscle mass.

Table 2. Characteristics and results of the studies included in the systematic research on the relationship between gut permeability and muscle mass.

Model	Condition	Type of Intervention	Gut Permeability	Muscle Mass	Reference
BALB/c Rj:ATHYM-Foxn1nu/nu male mice	NB cells	Prebiotics: 200 mg/day oligosaccharides	Gut permeability in NB *, no difference after intervention (-)	Muscle mass in NB (-) (no cachexia developed), no difference after intervention (-)	[16]
Female Balb/c mice	Leukemia (BaF cells)	Synbiotic: inulin-type fructans (0.2 g/day) and *Lactobacillus reuteri* (average: 5.8×10^8 CFU/day)	mRNA expression tight junction genes after BaF injection * mRNA expression tight junction genes after intervention *	Muscle mass after BaF injection *, muscle mass after intervention	[20]
ICR-specific pathogen-free male mice	CKD	FMT	Expression tight junction protein in CKD *, expression tight junction protein after intervention	Muscle mass in CKD *, muscle mass in after intervention *	[24]
Male CD2F1 mice	C26 cells, cancer	N.A.	Gut permeability after C26 injection *	Muscle mass after C26 injection *	[23]
CD1 mice	Aging	Probiotics: *Lactobacillus casei* or *Bifidobacterium longum* (3.5×10^5 CFU/day) from 2 to 12 months of age	mRNA expression tight junction genes in old mice *, mRNA expression tight junction genes after intervention *	Muscle-to-BW ratio in old mice *, muscle-to-BW ratio after intervention * Forelimb grip strength in old mice *, forelimb strength after intervention	[15]
Patients with solid tumors undergoing chemotherapy (n = 16)	Cancer	N.A.	Small-intestinal membrane permeability (-)	Muscle strength *	[26]
Newly diagnosed patients (n = 13) 17–49 years	Crohn's disease	N.A.	Gut permeability *	Muscle mass *	[22]
Healthy elderly (n = 18) >70 years	Aging	N.A.	Gut permeability *	Muscle strength *	[25]

NB: neuroblastoma; CKD: chronic kidney disease; CFU: colony-forming unit; FMT: fecal microbial transplantation; ↑: increased; (-): no change; ↓: decreased; * $p < 0.05$.

4. Discussion

The aim of this review was to study the potential of pro-, pre-, and synbiotics to treat muscle wasting. Our systematic literature analysis showed that *Lactobacillus* spp. and possibly other genera, such as *Bifidobacterium*, can ameliorate muscle wasting in mouse models; however, it has not been studied yet in humans. The effect of prebiotics on muscle wasting could not be established, as the disease models in which this was studied did not develop cachexia. Our second literature search showed that reduced gut permeability often coincides with improved muscle mass or strength in both mouse and human models. Therefore, the described effect of *Lactobacillus* spp. on muscle wasting could be mediated via reduced gut permeability. Unfortunately, the included studies on gut permeability did not elaborate on *Lactobacillus* spp. or the possible working mechanism. Therefore, the following paragraphs will discuss the relation between *Lactobacillus* spp. and gut permeability in the context of the multiorgan nature of cachexia.

4.1. Lactobacillus spp., Microbiome Composition, and Gut Permeability

As mentioned earlier, gut permeability appears to be an important factor during muscle wasting. Increased gut permeability in cachectic mice can be linked to microbiome imbalance characterized by an increase in bacteria such as *Klebsiella oxytoca* and *Enterobacteriaceae* spp. [20,27,28]. Interestingly, an increase of these bacteria is associated with a decrease in *Lactobacillus* spp. [20,23,28]. Thus, this suggests that *Lactobacillus* spp. plays an important role in microbiome balance and gut permeability. Yet, in cachexia models, little research has been conducted on the effect of *Lactobacillus* spp. on gut permeability. This effect has, however, been studied in in vitro models [29–31] and other diseases/conditions such as inflammation and aging [10,11]. Several in vitro studies showed that *Lactobacillus* spp. positively affects gut permeability, as incubation of CaCo2 cells with different strains of *Lactobacillus* spp. restored the transepithelial resistance after induced intestinal barrier impairment [29–31]. In line with these findings, Cui et al. [10] showed that mice injected with lipopolysaccharide (LPS) and supplemented with *L. rhamnosus* had a reduced gut permeability compared to the mice supplemented with the placebo. Moreover, they showed that *L. rhamnosus* affected gut permeability via the regulation of tight junction proteins. These findings were further supported by van Beek et al. [11], as they found that supplementation with *L. plantarum* in aging mice led to a thicker mucus layer compared to control, which is a marker of a balanced microbiome. All three of these studies suggest that supplementation with *Lactobacillus* spp. could improve microbiome balance and decrease gut permeability.

The above-mentioned studies were performed either in in vitro or in mice models. In human studies, there is some indirect evidence supporting the findings mentioned above. The in vitro and mice studies directly showed the effect of *Lactobacillus* spp. on gut permeability, while the human studies only showed gut microbiome composition changes. For example, during chronic kidney disease, systemic inflammation can result in an increase in *Enterobacteriaceae* and *Pseudomonadaceae* genera and a decrease in *Lactobacillus* spp., *Prevotella* spp., and *Bifidobacterium* spp. [32]. In these patients, the changes in the microbiome composition are similar to those found in cachectic mice [20,27,28]. Furthermore, Haran et al. [33] found that with increasing frailty, during aging, there was an increase of *Ruminococcus gnavus* and a decrease of the *Lachnospira* spp. and *Ruminococcus* spp. families. Both bacteria are associated with the maintenance of microbiome balance and production of short-chain fatty acids (SCFAs), specifically butyrate [34,35]. Furthermore, the abundance of other butyrate-producing bacteria decreased, while the elderly with better fitness had a high abundance of *Lactobacillus* spp. [33]. All in all, these human studies show unfavorable changes in the microbiota composition, specifically the reduction of *Lactobacillus* spp., upon inflammation. These findings are in line with the microbiota changes in cachectic mice models.

4.2. Lactobacillus spp. and Metabolites

As a result of the transition toward a more imbalanced microbiome during cachexia, the gut permeability can increase. A balanced microbiome breaks down indigestible food components into favorable metabolites that can influence body function. However, when the abundance of certain unfavorable species increases, the metabolites produced by the microbiota can differ from those of a balanced microbiome. Several of these metabolites, such as ammonia and hydrogen sulfide, may increase the risk of high gut permeability and consequently inflammation [36]. The gut microbiome balance can be restored using pro-, pre-, and synbiotics because they stimulate the growth of specific, more favorable groups of bacteria. Consequently, the metabolite profile will be influenced, leading to a more beneficial metabolite profile and lower gut permeability.

Both *Lactobacillus* spp. and *Bifidobacterium* spp. are known to have beneficial effects on the metabolite profile. They belong to the group of lactic acid bacteria and thus produce lactate via fermentative processes. This metabolite is often associated with a balanced microbiome because of its immunomodulating properties via suppression of the LPS/Toll-like receptor 4 signaling pathway [37]. Next to lactate, other metabolites that are considered to have a wide range of beneficial effects are the SCFAs. SCFAs, such as acetate, propionate, and butyrate, are produced by the direct bacterial fermentation of complex carbohydrates. In addition to direct fermentation, the products of one bacterium can be further converted by another bacterium, resulting in the indirect production of metabolites. This mechanism is called cross-feeding and especially contributes to butyrate production. For instance, *Lactobacillus* spp. and *Bifidobacterium* spp. produce lactate, as mentioned before, and acetate, which can both be converted into butyrate by other gut bacteria [38,39]. Butyrate is a frequently studied metabolite in microbiome research because of its "butyrogenic effects." These effects include providing energy for the colon epithelial cells, maintaining the gut barrier functions, and modulating the immune system in an anti-inflammatory manner [40]. In vitro studies on the relation between butyrate and other SCFAs and gut permeability showed that supplementation with these metabolites normalized the transepithelial resistance after induced intestinal barrier impairment [41,42]. Moreover, these studies showed that SCFAs stimulated the formation of tight junctions. This might be a possible mechanism via which probiotics positively affect gut permeability.

Butyrate production is not only increased by probiotics such as *Lactobacillus* spp. and *Bifidobacterium* spp. but probably also by prebiotics, in particular oligosaccharides and inulin-type fructans (ITFs). Both prebiotics were found to increase *Bifidobacterium* spp., which resulted in increased butyrate production via cross-feeding [18,43]. Research has shown that oligofructose and (pectic) oligosaccharides indeed promote the production of butyrate from lactate and acetate [44–46]. In addition, ITFs and (pectic) oligosaccharides can also be consumed by certain butyrate-producing bacteria [47,48]. In conclusion, it is hypothesized that the effect of pro-, pre-, and synbiotics on gut and immune function is partly attributed to increased production of lactate and SCFAs, specifically butyrate, both directly as well as via cross-feeding.

4.3. Lactobacillus spp., Inflammation, and Organ Crosstalk

Lactobacillus spp. supplementation was hypothesized to ameliorate muscle wasting by modulating gut permeability and immune function through SCFA production. Because both increased gut permeability and muscle wasting are related to inflammation, this could be a central mediator in the underlying mechanism (Figure 2). IL-6 has especially been described as an important cytokine with respect to these conditions [49]. Several studies on pro- and/or prebiotic supplementation included in Table 1 also measured the levels of IL-6 and other proinflammatory cytokines [13,14,19,21]. In these studies, ameliorated muscle wasting after *Lactobacillus* spp. supplementation was accompanied by decreased levels of these cytokines. This association between inflammation, gut permeability, and muscle wasting is further supported by the observation that injection of anti-IL-6 antibody in cachectic mice leads to a decrease of both gut permeability as well as muscle wasting [23].

So, in general, inflammation and particularly IL-6 can be suggested to play a mediating role in the mechanism via which *Lactobacillus* spp. supplementation ameliorates muscle wasting.

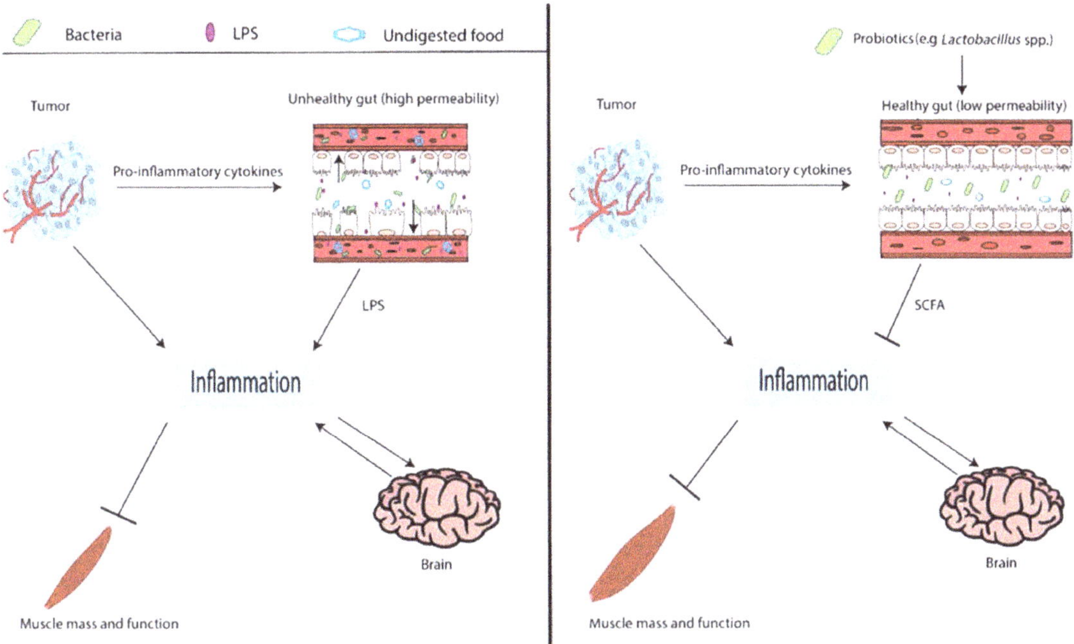

Figure 2. Hypothesis on the mechanism behind the reported effects of probiotics on muscle wasting, involving the organ crosstalk during cancer cachexia. On the left-hand side, the situation when gut permeability is high is illustrated. On the right-hand side, the effect of probiotics is shown. The probiotics inhibit gut permeability and thus improve gut function, reduce inflammation, and consequently ameliorate muscle wasting. LPS: lipopolysaccharide; SCFA: short-chain fatty acid

In cachexia, inflammation can stimulate muscle wasting directly, as well as via organ crosstalk. The latter could include the gut–brain axis because this crosstalk is affected by the inflammation resulting from increased gut permeability. Moreover, the gut–brain axis influences core processes involved in muscle wasting such as stress and appetite regulation [5]. Braun et al. [50] found that in cancer cachexia models, hypothalamic inflammation plays a key role in muscle mass loss via the hypothalamus–pituitary–adrenal axis induced production of cortisol. To discover whether influencing these processes could be part of the underlying mechanism, we explored the effects of *Lactobacillus* spp. supplementation on cortisol production and food intake in relation to gut function and inflammation. With respect to cortisol production, Gareau et al. [51] and Ait-Belganoui et al. [52] showed that *Lactobacillus* spp. supplementation decreased corticosterone release as well as gut permeability in stressed rats. In addition, Ait-Belganoui et al. [52] showed in a model with antibiotic-induced disruption of the gut microbiota that increased corticosterone release was a result of more bacterial compounds crossing a disturbed gut barrier. With respect to appetite, decreased levels of IL-6 have been reported to increase food intake. This effect has been associated with the altered expression of inflammation-related genes in the hypothalamus [53]. Contradictorily, supplementation of *Lactobacillus* spp. was found to increase the release of appetite-suppressing hormone glucagon-like peptide (GLP)-1. This effect is thought to be induced via an increased butyrate production [54,55]. When directly assessing the effect of *Lactobacillus* spp. supplementation on food intake in disease and aging, no significant changes were reported (Table 1) [14,15,20,21]. This could be the result of *Lactobacillus* spp. supplementation stimulating appetite via lowering IL-6 on the one hand, while suppressing appetite via stimulating GLP-1 release on the other hand. The effect of *Lactobacillus* spp. stimulating GLP-1 release via butyrate could, however, also potentially improve the delivery and uptake of insulin in the muscle, thereby promoting muscle protein synthesis [56]. Furthermore, GLP-1 agonists have been shown to ameliorate muscle wasting via suppressing myostatin, stimulating myogenic factors, and thus supporting muscle regeneration which has been linked to improved muscle function [57]. Altogether, supplementation of *Lactobacillus* spp. is suggested to ameliorate muscle wasting via increasing butyrate production and decreasing gut permeability, which alters gut–brain interactions and exerts multiorgan effects.

4.4. Translatability of Mouse Models

Current findings are foremostly based on mouse studies, and even though the mouse models provide intriguing results, it is important to note that these findings are not directly translatable to humans. Although the gastrointestinal tracts of mice and humans are anatomically comparable, the colon of humans consists of different sections, while in mice it is rather smooth, and there is no division [58]. Moreover, mice have a larger colon than humans when compared to their body weight [59]. These small differences in anatomy can make a big difference in the translatability of the results. Furthermore, many variables can influence the gastrointestinal tract, such as diet, exercise, environment, and stress [59,60]. As these variables are more difficult to control in humans compared to mice, it also makes it harder to translate the results directly. In addition, cancer patients often receive chemotherapy as an anticancer treatment. This has been associated with microbiome imbalance and increased gut permeability, wherefore the chemotherapy may worsen cachexia development. Additionally, it may interfere with interventions that aim to improve microbiome balance such as probiotic supplementation [61]. When the interaction between chemotherapy and probiotic supplementation is further elucidated, an intervention can be developed in which probiotics are combined with other treatment strategies to achieve optimal efficacy. Such multitarget treatment is hypothesized to be more effective because cachexia is a multiorgan syndrome [5].

4.5. Future Research

All in all, only a limited number of studies specifically measured the effect of pro-, pre-, and synbiotics on muscle mass or function in cachexia-inducing disease models and could therefore be included in this review. Notably, all included studies on pro- and synbiotics in cachexia used *Lactobacillus* strains. Interestingly, our review showed that although various strains of *Lactobacillus* were used in different cachectic mice models, all studies consistently showed increased muscle mass and function after its supplementation. In addition, the potential of several strains of *Lactobacillus* to reduce muscle wasting was also established in models of age-induced muscle mass loss. This suggests that a generic property of *Lactobacillus* spp. influences muscle wasting regardless of cause. Based on the literature, we hypothesize that the genus of *Lactobacillus*, solely or combined with prebiotics, decreases gut permeability in cachexia by improving gut function via increased lactate and butyrate production. Other lactate-producing bacteria such as *Bifidobacterium* spp. may have similar effects [37], as shown by Ni et al. [15] However, they only investigated age-induced muscle wasting, and the effects of *Bifidobacterium* spp. on disease-induced muscle wasting have not yet been studied. Next to that, it might be of interest to take electrolyte changes into account. In rodent hypertension models, sodium has been reported to influence inflammatory factors both directly as indirectly via the microbiome [62].

To determine whether the effects of pro- and synbiotics are genera- and/or strain-specific and to investigate the underlying mechanism(s), more research is necessary. Future research should focus on the effect of different bacterial genera and strains on microbiome balance, metabolite profiles, gut function, and muscle mass in cachexia and sarcopenia. Based on deviations observed in the microbiota composition of cachectic mice and expected metabolite profile changes [20,27,28], multiple strains of both *Lactobacillus* and *Bifidobacterium* should be investigated in future studies. If these studies show that amelioration of muscle mass loss is mainly driven by improved microbiome balance and gut function rather than by genera-specific effects, other nutritional interventions, such as prebiotics, may also be effective. To improve the validity and comparability of these studies, lean body mass measurements by DEXA are recommended [63]. In the included mice studies, only one or two isolated muscles were measured. Differences occur between studies regarding which muscles are measured and in which units these measurements are expressed. In addition, a few studies define muscle wasting solely based on muscle function, while muscle mass was not taken into account. By measuring total lean body mass by DEXA, validity, as well as comparability, will be increased [64].

5. Conclusions

To summarize, *Lactobacillus* spp. has the potential to ameliorate muscle wasting via influencing organ crosstalk, presumably by inducing the production of lactate and butyrate, decreasing the gut permeability, and consequently reducing inflammation. Other genera of bacteria might have similar effects but are not studied yet in relation to muscle mass loss in disease models, although *Bifidobacterium* spp. was found to reduce muscle mass loss in an aging model. To investigate whether the described effects are genus-specific or related to improved gut function in general, more research is needed. Altogether, *Lactobacillus* spp. and possibly other pro-, pre-, and synbiotics have the potential to contribute to effective multitarget cachexia treatment.

Author Contributions: Drafted the manuscript, S.J.v.K., F.A.C.J. and V.L.O.; supervised and revised the draft manuscript, K.v.N.; revised the draft manuscript as independent critical experts, C.B. and M.v.d.E. All authors have read and agreed to the published version of the manuscript.

Funding: This research received no external funding.

Institutional Review Board Statement: Not applicable.

Informed Consent Statement: Not applicable.

Data Availability Statement: Not applicable.

Conflicts of Interest: The authors declare no conflict of interest.

References

1. Siddiqui, J.A.; Pothuraju, R.; Jain, M.; Batra, S.K.; Nasser, M.W. Advances in cancer cachexia: Intersection between affected organs, mediators, and pharmacological interventions. *Biochim. Biophys. Acta Rev. Cancer* **2020**, *1873*, 188359. [CrossRef] [PubMed]
2. Lenk, K.; Schuler, G.; Adams, V. Skeletal muscle wasting in cachexia and sarcopenia: Molecular pathophysiology and impact of exercise training. *J. Cachexia Sarcopenia Muscle* **2010**, *1*, 9–21. [CrossRef]
3. Cole, C.L.; Kleckner, I.R.; Jatoi, A.; Schwarz, E.M.; Dunne, R.F. The role of systemic inflammation in cancer-associated muscle wasting and rationale for exercise as a therapeutic intervention. *JCSM Clin. Rep.* **2018**, *3*, e00065. [CrossRef]
4. Argilés, J.M.; Stemmler, B.; López-Soriano, F.J.; Busquets, S. Inter-tissue communication in cancer cachexia. *Nat. Rev. Endocrinol.* **2018**, *15*, 9–20. [CrossRef]
5. Witkamp, R.F.; van Norren, K. Let thy food be thy medicine … when possible. *Eur. J. Pharmacol.* **2018**, *836*, 102–114. [CrossRef] [PubMed]
6. Ferrucci, L. Inflammageing: Chronic inflammation in ageing, cardiovascular disease, and frailty. *Nat. Rev. Cardiol.* **2018**, *15*, 505–522. [CrossRef]
7. Hill, C.; Guarner, F.; Reid, G.; Gibson, G.R.; Merenstein, D.J.; Pot, B.; Morelli, L.; Canani, R.B.; Flint, H.J.; Salminen, S.; et al. Expert consensus document: The international scientific association for probiotics and prebiotics consensus statement on the scope and appropriate use of the term probiotic. *Nat. Rev. Gastroenterol. Hepatol.* **2014**, *11*, 506–514. [CrossRef] [PubMed]
8. Gibson, G.R.; Hutkins, R.; Sanders, M.E.; Prescott, S.L.; Reimer, R.A.; Salminen, S.J.; Scott, K.; Stanton, C.; Swanson, K.S.; Cani, P.D.; et al. Expert consensus document: The International Scientific Association for Probiotics and Prebiotics (ISAPP) consensus statement on the definition and scope of prebiotics. *Nat. Rev. Gastroenterol. Hepatol.* **2017**, *14*, 491–502. [CrossRef] [PubMed]
9. Pandey, K.R.; Naik, S.R.; Vakil, B.V. Probiotics, prebiotics and synbiotics—A review. *J. Food Sci. Technol.* **2015**, *52*, 7577–7587. [CrossRef]

10. Cui, Y.; Liu, L.; Dou, X.; Wang, C.; Zhang, W.; Gao, K.; Liu, J.; Wang, H. Lactobacillus reuteri ZJ617 maintains intestinal integrity via regulating tight junction, autophagy and apoptosis in mice challenged with lipopolysaccharide. *Oncotarget* **2017**, *8*, 77489–77499. [CrossRef] [PubMed]
11. Van Beek, A.; Sovran, B.; Hugenholtz, F.; Meijer, B.; Hoogerland, J.; Mihailova, V.; van der Ploeg, C.; Belzer, C.; Boekschoten, M.; Hoeijmakers, J.; et al. Supplementation with Lactobacillus plantarum WCFS1 Prevents Decline of Mucus Barrier in Colon of Accelerated Aging Ercc1 − /Δ7 Mice. *Front. Immunol.* **2016**, *7*, 408. [CrossRef] [PubMed]
12. Varian, B.J.; Goureshetti, S.; Poutahidis, T.; Lakrits, J.R.; Levkovich, T.; Kwok, C.; Teliousis, K.; Ibrahim, Y.M.; Mirabal, S.; Erdman, S.E. Beneficial bacteria inhibit cachexia. *Oncotarget* **2016**, *7*, 11803–11816. [CrossRef] [PubMed]
13. Sugimura, T.; Jounai, K.; Ohshio, K.; Suzuki, H.; Kirisako, T.; Sugihara, Y.; Fujiwara, D. Long-term administration of pDC-Stimulative Lactococcus lactis strain decelerates senescence and prolongs the lifespan of mice. *Int. Immunopharmacol.* **2018**, *58*, 166–172. [CrossRef] [PubMed]
14. Chen, L.; Huang, S.; Huan, K.; Hsu, C.; Yang, K.; Li, L.; Chan, C.; Huang, H. Lactobacillus paracasei PS23 decelerated age-related muscle loss by ensuring mitochondrial function in SAMP8 mice. *Aging* **2019**, *11*, 756–770. [CrossRef] [PubMed]
15. Ni, Y.; Yang, X.; Zheng, L.; Wang, Z.; Wu, L.; Jiang, J.; Yang, T.; Ma, L.; Fu, Z. Lactobacillus and Bifidobacterium Improves Physiological Function and Cognitive Ability in Aged Mice by the Regulation of Gut Microbiota. *Mol. Nutr. Food Res.* **2019**, *63*, e1900603. [CrossRef]
16. Obermüller, B.; Singer, G.; Kienesberger, B.; Klymiuk, I.; Sperl, D.; Stadlbauer, V.; Horvath, A.; Miekisch, W.; Gierschner, P.; Grabherr, R.; et al. The Effects of Prebiotic Supplementation with OMNi-LOGiC® FIBRE on Fecal Microbiome, Fecal Volatile Organic Compounds, and Gut Permeability in Murine Neuroblastoma-Induced Tumor-Associated Cachexia. *Nutrients* **2020**, *12*, 2029. [CrossRef]
17. Bindels, L.B.; Nayrinck, A.; Salazar, N.; Taminiau, B.; Druart, C.; Muccioli, G.; Francois, E.; Blecker, C.; Richel, A.; Daube, G.; et al. Non Digestible Oligosaccharides Modulate the Gut Microbiota to Control the Development of Leukemia and Associated Cachexia in Mice. *PLoS ONE* **2015**, *10*, e0131009. [CrossRef]
18. Buigues, C.; Gernandez-Garrido, J.; Pruimboom, L.; Hoogland, A.J.; Navarro-Martinez, R.; Martinez-Martinez, M.; Verdejo, Y.; Mascaros, M.C.; Peris, C.; Cauli, O. Effect of a prebiotic formulation on frailty syndrome: A randomized, double-blind clinical trial. *Int. J. Mol. Sci.* **2016**, *17*, 932. [CrossRef]
19. An, J.M.; Kang, E.A.; Han, Y.; Oh, J.Y.; Lee, D.Y.; Choi, S.H.; Kim, D.H.; Hahm, K.B. Dietary intake of probiotic kimchi ameliorated IL-6-driven cancer cachexia. *J. Clin. Biochem. Nutr.* **2019**, *65*, 109–117. [CrossRef] [PubMed]
20. Bindels, L.B.; Neyrinck, A.M.; Claus, S.P.; Le Roy, C.I.; Grangette, C.; Pot, B.; Martinez, I.; Walter, J.; Cani, P.D.; Delzenne, N.M. Synbiotic approach restores intestinal homeostasis and prolongs survival in leukaemic mice with cachexia. *ISME J.* **2016**, *10*, 1456–1470. [CrossRef]
21. Bindels, L.B.; Beck, R.; Schakman, O.; Martin, J.C.; De Backer, F.; Sohet, F.M.; Dewulf, E.M.; Pachiklan, B.D.; Neyrinck, A.M.; Thissen, J.; et al. Restoring specific lactobacilli levels decreases inflammation and muscle atrophy markers in an acute leukemia mouse model. *PLoS ONE* **2012**, *7*, 37971. [CrossRef]
22. Cuoco, L.; Vescovo, G.; Castaman, R.; Ravara, B.; Cammarota, G.; Angelini, A.; Salvaginini, M.; Dalla Libera, L. Skeletal muscle wastage in Crohn's disease: A pathway shared with heart failure? *Int. J. Cardiol.* **2008**, *127*, 219–227. [CrossRef] [PubMed]
23. Bindels, L.B.; Neyrinck, A.; Loumaye, A.; Catry, E.; Walgrave, H.; Cherbuy, C.; Leclercq, S.; Van Hul, M.; Plovier, H.; Pachikian, B.; et al. Increased gut permeability in cancer cachexia: Mechanisms and clinical relevance. *Oncotarget* **2018**, *9*, 18224–18238. [CrossRef] [PubMed]
24. Uchiyama, K.; Wakino, S.; Irie, J.; Miyamoto, J.; Matsui, A.; Tajima, T.; Itoh, T.; Oshima, Y.; Ysohifuji, A.; Kimura, I.; et al. Contribution of uremic dysbiosis to insulin resistance and sarcopenia. *Nephrol. Dial. Transplant.* **2020**, *35*, 501–1517. [CrossRef] [PubMed]
25. Qi, Y.; Goel, R.; Kim, S.; Richards, E.; Carter, C.; Pepine, C.; Raizada, M.; Buford, T. Intestinal Permeability Biomarker Zonulin is Elevated in Healthy Aging. *J. Am. Med. Dir. Assoc.* **2017**, *18*, e1–e810. [CrossRef] [PubMed]
26. Van der Meij, B.S.; Deutz, N.E.P.; Rodriguez, R.E.; Engelen, M.K.P.J. Early Signs of Impaired Gut Function Affect Daily Functioning in Patients With Advanced Cancer Undergoing Chemotherapy. *J. Parenter. Enter. Nutr.* **2020**. [CrossRef]
27. Pötgens, S.A.; Brossel, H.; Sboarina, M.; Catry, E.; Cani, P.; Neyrinck, A.; Delzenne, N.; Bindels, L. Klebsiella oxytoca expands in cancer cachexia and acts as a gut pathobiont contributing to intestinal dysfunction. *Sci. Rep.* **2018**, *8*, 1–12. [CrossRef] [PubMed]
28. Bindels, L.B.; Delzenne, N.M. Muscle wasting: The gut microbiota as a new therapeutic target? *Int. J. Biochem. Cell Biol.* **2013**, *45*, 2186–2190. [CrossRef]
29. Ko, J.S.; Yang, H.R.; Chang, J.Y.; Seo, J.K. Lactobacillus plantarum inhibits epithelial barrier dysfunction and interleukin-8 secretion induced by tumor necrosis factor-α. *World J.Gastroenterol.* **2007**, *13*, 1962–1965. [CrossRef]
30. Rao, R.K.; Polk, D.B.; Seth, A.; Yan, F. Probiotics the good neighbor: Guarding the gut mucosal barrier. *Am. J. Infect. Dis.* **2009**, *5*, 195–199. [CrossRef]
31. Parassol, N.; Freita, M.; Thoreaux, K.; Dalmasso, G.; Bourdet-Sicard, R.; Rampal, P. Lactobacillus casei DN-114001 inhibits the increase in paracellular permeability of enteropathogenic Escherichia coli-infected T84 cells. *Res. Microbiol.* **2005**, *156*, 256–262. [CrossRef] [PubMed]
32. Vaziri, N.; Wong, J.; Pahl, M.; Piceno, Y.; Yuan, J.; Desantis, T.; Ni, Z.; Nguyen, T.; Andersen, G. Chronic kidney disease alters intestinal microbial flora. *Kidney Int.* **2013**, *83*, 308–315. [CrossRef]

33. Haran, J.P.; Bucci, V.; Dutta, P.; Ward, D.; McCormick, B. The nursing home elder microbiome stability and associations with age, frailty, nutrition and physical location. *J. Med. Microbiol.* **2018**, *67*, 40–51. [CrossRef] [PubMed]

34. Venegas, D.; De La Fuente, M.; Landskron, G.; Gonzalez, M.; Quera, R.; Dijkstra, G.; Harmsen, H.; Faber, K.; Hermoso, M. Short chain fatty acids (SCFAs)mediated gut epithelial and immune regulation and its relevance for inflammatory bowel diseases. *Front. Immunol.* **2019**, *10*, 277. [CrossRef] [PubMed]

35. Biddle, A.; Stewart, L.; Blanchard, J.; Leschine, S. Untangling the genetic basis of fibrolytic specialization by lachnospiraceae and ruminococcaceae in diverse gut communities. *Diversity* **2013**, *5*, 627–640. [CrossRef]

36. Tarashi, S.; Siadat, S.; Badi, S.; Zali, M.; Biassoni, R.; Ponzoni, M.; Moshiri, A. Gut Bacteria and their Metabolites: Which One Is the Defendant for Colorectal Cancer? *Microorganisms* **2019**, *7*, 561. [CrossRef] [PubMed]

37. Watanabe, T.; Nishio, H.; Tanigawa, T.; Yamagami, H.; Okazaki, H.; Watanabe, K.; Tominaga, K.; Fujiwara, Y.; Oshitani, N.; Asahara, T.; et al. Probiotic Lactobacillus casei strain Shirota prevents indomethacin-induced small intestinal injury: Involvement of lactic acid. *Am. J. Physiol. Liver Physiol.* **2009**, *297*, G506–G513. [CrossRef] [PubMed]

38. De Vuyst, L.; Leroy, F. Cross-feeding between bifidobacteria and butyrate-producing colon bacteria explains bifidobacterial competitiveness, butyrate production, and gas production. *Int. J. Food Microbiol.* **2011**, *149*, 73–80. [CrossRef] [PubMed]

39. Duncan, S.H.; Louis, P.; Flint, H.J. Lactate-utilizing bacteria, isolated from human feces, that produce butyrate as a major fermentation product. *Appl. Environ. Microbiol.* **2004**, *70*, 5810–5817. [CrossRef] [PubMed]

40. Rivière, A.; Selak, M.; Lantin, D.; Leroy, F.; De Vuyst, L. Bifidobacteria and Butyrate-Producing Colon Bacteria: Importance and Strategies for Their Stimulation in the Human Gut. *Front. Microbiol.* **2016**, *7*, 979. [CrossRef]

41. Peng, L.; Li, Z.R.; Green, R.S.; Holzman, I.R.; Lin, J. Butyrate enhances the intestinal barrier by facilitating tight junction assembly via activation of AMP-activated protein kinase in Caco-2 cell monolayers. *J. Nutr.* **2009**, *139*, 1619–1625. [CrossRef]

42. Feng, Y.; Wang, Y.; Wang, P.; Huang, Y.; Wang, F. Short-chain fatty acids manifest stimulative and protective effects on intestinal barrier function through the inhibition of NLRP3 inflammasome and autophagy. *Cell Physiol. Biochem.* **2018**, *49*, 190–205. [CrossRef]

43. Roberfroid, M.; Gibson, G.; Hoyles, L.; McCartney, A.; Rastall, R.; Rowland, I.; Wolvers, D.; Watzl, B.; Szajewska, H.; Stahl, B.; et al. Prebiotic effects: Metabolic and health benefits. *Br. J. Nutr.* **2010**, *104*, S1–S63. [CrossRef] [PubMed]

44. Louis, P.; Scott, K.P.; Duncan, S.H.; Flint, H.J. Understanding the effects of diet on bacterial metabolism in the large intestine. *J. Appl. Microbiol.* **2007**, *102*, 1197–1208. [CrossRef] [PubMed]

45. Gómez, B.; Gullón, B.; Remoroza, C.; Schols, H.A.; Parajó, J.C.; Alonso, J.L. Purification, Characterization, and Prebiotic Properties of Pectic Oligosaccharides from Orange Peel Wastes. *J. Agric. Food Chem.* **2014**, *62*, 9769–9782. [CrossRef]

46. Chen, J.; Liang, R.; Liu, W.; Li, T.; Liu, C.; Wu, S.; Wang, Z. Pectic-oligosaccharides prepared by dynamic high-pressure microfluidization and their in vitro fermentation properties. *Carbohydr. Polym.* **2013**, *91*, 175–182. [CrossRef] [PubMed]

47. Falony, G.; Verschaeren, A.; De Bruycker, F.; De Preter, V.; Verbeke, K.; Leroy, F.; De Vuyst, L. In vitro kinetics of prebiotic inulin-type fructan fermentation by butyrate-producing colon bacteria: Implementation of online gas chromatography for quantitative analysis of carbon dioxide and hydrogen gas production. *Appl. Environ. Microbiol.* **2009**, *75*, 5884–5892. [CrossRef]

48. Chung, W.S.F.; Meijerink, M.; Zeuner, B.; Holck, J.; Louis, P.; Meyer, A.; Wells, J.; Flint, H.; Duncan, S. Prebiotic potential of pectin and pectic oligosaccharides to promote anti-inflammatory commensal bacteria in the human colon. *FEMS Microbiol. Ecol.* **2017**, *93*, fix127. [CrossRef] [PubMed]

49. Narsale, A.A.; Carson, J.A. Role of interleukin-6 in cachexia: Therapeutic Implications. *Curr. Opin. Support Palliat. Care* **2014**, *8*, 321–327. [CrossRef] [PubMed]

50. Braun, T.; Szumowski, M.; Levasseur, P.; Grossberg, A.; Zhu, X.; Agarwal, A.; Marks, D. Muscle Atrophy in Response to Cytotoxic Chemotherapy Is Dependent on Intact Glucocorticoid Signaling in Skeletal Muscle. *PLoS ONE* **2014**, *9*, e106489. [CrossRef]

51. Gareau, M.; Jury, J.; MacQueen, G.; Sherman, P.; Perdue, M. Probiotic treatment of rat pups normalises corticosterone release and ameliorates colonic dysfunction induced by maternal separation. *Gut* **2007**, *56*, 1522–1528. [CrossRef]

52. Ait-Belgnaoui, A.; Durand, H.; Cartier, C.; Chaumaz, G.; Eutamene, H.; Ferrier, L.; Houdeau, E.; Fioramonti, J.; Bueno, L.; Theodorou, V. Prevention of gut leakiness by a probiotic treatment leads to attenuated HPA response to an acute psychological stress in rats. *Psychoneuroendocrinology* **2012**, *37*, 1885–1895. [CrossRef] [PubMed]

53. Dwarkasing, T.; Witkamp, R.; Boekschoten, M.; Ter Laak, M.; Heins, M.; van Norren, K. Increased hypothalamic serotonin turnover in inflammation-induced anorexia. *BMC Neurosci.* **2016**, *17*, 26. [CrossRef] [PubMed]

54. Simon, M.; Strassburger, K.; Nowotny, B.; Kolb, H.; Nowotny, P.; Burkart, V.; Zivehe, F.; Hwang, J.; Stehle, P.; Pacini, G.; et al. Intake of lactobacillus reuteri improves incretin and insulin secretion in Glucose-Tolerant humans: A proof of concept. *Diabetes Care* **2015**, *38*, 1827–1834. [CrossRef]

55. Yadav, H.; Lee, J.; Lloyd, J.; Walter, P.; Rane, S. Beneficial metabolic effects of a probiotic via butyrate-induced GLP-1 hormone secretion. *J. Biol. Chem.* **2013**, *288*, 25088–25097. [CrossRef]

56. Ayala, J.E.; Bracy, D.P.; James, F.D.; Julien, B.M.; Wasserman, D.H.; Drucker, D.J. The glucagon-like peptide-1 receptor regulates endogenous glucose production and muscle glucose uptake independent of its incretin action. *Endocrinology* **2009**, *150*, 1155–1164. [CrossRef]

57. Hong, Y.; Lee, J.H.; Jeong, K.W.; Choi, C.S.; Jun, H. Amelioration of muscle wasting by glucagon-like peptide-1 receptor agonist in muscle atrophy. *J. Cachexia Sarcopenia Muscle* **2019**, *10*, 903–918. [CrossRef] [PubMed]

58. Nguyen, T.L.; Vieira-Silva, S.; Liston, A.; Raes, J. How informative is the mouse for human gut microbiota research? *Dis. Model. Mech.* **2015**, *8*, 1–16. [CrossRef]
59. Hugenholtz, F.; de Vos, W.M. Mouse models for human intestinal microbiota research: A critical evaluation. *Cell. Mol. Life Sci.* **2018**, *75*, 149–160. [CrossRef]
60. JanssenDuijghuijsen, L.M.; van Norren, K.; Grefte, S.; Koppelman, S.J.; Lenaerts, K.; Keijer, J.; Witkamp, R.F.; Wichers, H.J. Endurance Exercise Increases Intestinal Uptake of the Peanut Allergen Ara h 6 after Peanut Consumption in Humans. *Nutrients* **2017**, *9*, 84. [CrossRef]
61. Deleemans, J.M.; Chleilat, F.; Reimer, R.A.; Henning, J.-W.; Baydoun, M.; Piedalue, K.-A.; McLennan, A.; Carlson, L.E. The chemo-gut study: Investigating the long-term effects of chemotherapy on gut microbiota, metabolic, immune, psychological and cognitive parameters in young adult Cancer survivors; Study protocol. *BMC Cancer* **2019**, *19*, 1243. [CrossRef]
62. Smiljanec, K.; Lennon, S.L. Sodium, hypertension, and the gut: Does the gut microbiota go salty? *Am. J. Physiol. Heart Circ. Physiol.* **2019**, *317*, H1173–h1182. [CrossRef] [PubMed]
63. Buckinx, F.; Landi, F.; Cesari, M.; Fielding, R.A.; Visser, M.; Engelke, K.; Maggi, M.; Dennison, E.; Al-Daghri, N.M.; Allepaerts, S.; et al. Pitfalls in the measurement of muscle mass: A need for a reference standard. *J. Cachexia Sarcopenia Muscle* **2018**, *9*, 269–278. [CrossRef] [PubMed]
64. Cole, C.L.; Beck, C.A.; Robinson, D.; Ye, J.; Mills, B.; Gerber, S.A.; Schwarz, E.M.; Linehan, D. Dual Energy X-ray Absorptiometry (DEXA) as a longitudinal outcome measure of cancer-related muscle wasting in mice. *PLoS ONE* **2020**, *15*, e0230695. [CrossRef] [PubMed]

 *nutrients*

MDPI

Review

Does Folic Acid Protect Patients with Inflammatory Bowel Disease from Complications?

Alicja Ewa Ratajczak *,†, Aleksandra Szymczak-Tomczak †, Anna Maria Rychter, Agnieszka Zawada, Agnieszka Dobrowolska and Iwona Krela-Kaźmierczak *

Department of Gastroenterology, Dietetics and Internal Diseases, Poznan University of Medical Sciences, 61-701 Poznan, Poland; aleksandra.szymczak@o2.pl (A.S.-T.); a.m.rychter@gmail.com (A.M.R.); a.zawada@ump.edu.pl (A.Z.); agdob@ump.edu.pl (A.D.)
* Correspondence: alicjaewaratajczak@gmail.com (A.E.R.); krela@op.pl (I.K.-K.);
  Tel.: +48-667-385-996 (A.E.R.); +48-8691-343 (I.K.-K.); Fax: +48-8691-686 (A.E.R.)
† These authors contributed equally to this work: Alicja Ewa Ratajczak and Aleksandra Szymczak-Tomczak.

Abstract: Folic acid, referred to as vitamin B9, is a water-soluble substance, which participates in the synthesis of nucleic acids, amino acids, and proteins. Similarly to B12 and B6, vitamin B9 is involved in the metabolism of homocysteine, which is associated with the *MTHFR* gene. The human body is not able to synthesize folic acid; thus, it must be supplemented with diet. The most common consequence of folic acid deficiency is anemia; however, some studies have also demonstrated the correlation between low bone mineral density, hyperhomocysteinemia, and folic acid deficiency. Patients with inflammatory bowel disease (IBD) frequently suffer from malabsorption and avoid certain products, such as fresh fruits and vegetables, which constitute the main sources of vitamin B9. Additionally, the use of sulfasalazine by patients may result in folic acid deficiency. Therefore, IBD patients present a higher risk of folic acid deficiency and require particular supervision with regard to anemia and osteoporosis prevention, which are common consequences of IBD.

Keywords: folic acid; homocysteine; inflammatory bowel disease; microbiota

Citation: Ratajczak, A.E.; Szymczak-Tomczak, A.; Rychter, A.M.; Zawada, A.; Dobrowolska, A.; Krela-Kaźmierczak, I. Does Folic Acid Protect Patients with Inflammatory Bowel Disease from Complications? *Nutrients* **2021**, *13*, 4036. https://doi.org/10.3390/nu13114036

Academic Editor: Panagiota Mitrou

Received: 29 September 2021
Accepted: 9 November 2021
Published: 12 November 2021

Publisher's Note: MDPI stays neutral with regard to jurisdictional claims in published maps and institutional affiliations.

1. Introduction

Folic acid (FA), also known as vitamin B9, is a fully oxidated synthetic form of pteroyl-glutamic acid monoglutamate and a water-soluble vitamin, whose name originates from the Latin "folium" meaning "leaf." Naturally occurring folic acid has a reduced form, referred to as folate [1,2]. Since the human body is unable to produce vitamin B9 by itself, it must either be derived from a traditional or fortified diet, as the human intestinal microbiome is capable of synthesizing it [3], or by means of potential supplementation. This vitamin is essential for the growth of new cells and remethylation of homocysteine (Hcy), which is vital for the process of nucleotide synthesis [4], while appropriate vitamin B9 intake during pregnancy is a preventive factor of neural tube defects (NTDs) in gastrulation [5]. Conversely, folic acid deficiency is associated with several types of adverse health conditions, and although the most discussed conditions comprise anemia and cardiovascular disease, it is common among patients suffering from inflammatory bowel disease (IBD) [6]. Some studies have also reported an increased risk of osteoporosis as a consequence of low folic acid concentration, suggesting low bone mineral density may be due to low folic acid levels or hyperhomocysteinemia. Additionally, genetic factors are also known to affect the FA concentration [7].

2. The Role of Folic Acid in the Human Body

Folate performs many functions in the human body—it acts as a coenzyme in the synthesis of purines and pyrimidines, but it is also involved both in the transformation of one-carbon units and in the methylation cycle. Folate deficiency may be the result of insufficient intake, higher demand, malabsorption, or administration of certain drugs. Hence, it may lead to an increased risk of numerous diseases, such as cardiovascular disease, neoplasms, and cognitive impairment. In fact, B9 deficiency may result in hyperhomocysteinemia and disorders of protein and DNA synthesis [8]. Folic acid is absorbed by the high-affinity folate transporter in the active process in the duodenum and jejunum. In food, folates occur as polyglutamates, which may be absorbed in the intestine following enzymatic conversion into folate monoglutamates by the jejunal mucosal [9,10].

The molecular structure of folic acid comprises three different moieties—glutamic acid residue, pteroyl group, and para-aminobenzoic acid [11]. However, it must be subjected to a two-step enzymatic reaction with the dihydrofolic acid intermediate and the dihydrofolate reductase (DHFR) enzyme before it becomes an active coenzyme, i.e., the THF (tetrahydrofolian) form [5]. Subsequently, THF converts to 5-10-MTHF (methylenetetrahydrofolate), and it is reduced by MTHFR to 5-MTHF, which is involved in homocysteine conversion to methionine, donating the remaining methyl group in the process via methionine synthase. Figure 1 shows a diagram of folic acid metabolic pathways. Apart from the abovementioned elements, vitamin B12 and B6 also play a major role in the metabolism of homocysteine [12]. In fact, folic acid takes part in one-carbon metabolism, because 1-carbon units are transferred to THF for reduction or oxidation and are essential for DNA synthesis. S-adenosylmethionine is a methyl donor for biological methylation, including DNA and protein methylation. Low concentrations of folic acid in cell division inhibit the conversion of dUMP (deoxyuridine monophosphate) to dTMP (thymidylate) and uracil may be substituted in the DNA sequence [13]. Therefore, folic acid deficiency may lead to genomic hypomethylation [14]. It has been proven that cytosine methylation in the DNA sequence plays an important role in gene expression [15]. The control of gene transcription and particularly its suppression can stem from changes in methylation levels within the promoter regions of genes. Hypomethylation caused by folic acid deficiency can lead to the induction of protooncogenes, which promote tumorigenesis [16]. Hypermethylation, in turn, causes inactivation of the promoter regions of suppressor genes.

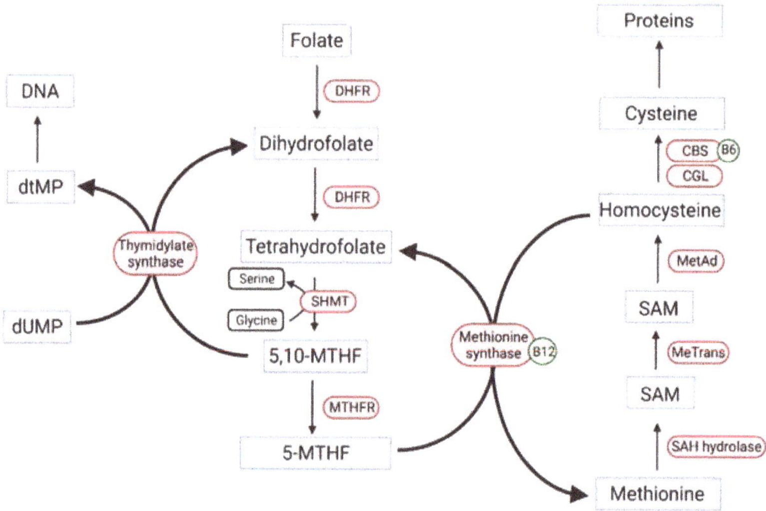

Figure 1. Folic acid cycle and homocysteine metabolism. DNA, deoxyribonucleic acid; dTMP, deoxythymidine monophosphate; dUMP, deoxyuridine monophosphate; DHFR, dihydrofolate reductase; SHMT, serine hydroxymethyltransferase; 5,10-MTHF, 5,10-methylenetetrahydrofolate; 5-MTHF, 5-methylenetetrahydrofolate; CBS, cystathionine-β-synthase; CGL, cystathionine gamma-lyase; MetAd, methionine adenosyltransferase; MeTrans, methyltransferase; SAM, S-adenosylomethionine; SAH, S-adenosylhomocysteine hydrolase.

Folic acid is fundamental in the proper development of pregnancy, as its deficiency exacerbates the risk of neural tube defects in children, whereas its supplementation in the course of pregnancy mitigates the risk of heart disorders [1]. In fact, an elevated demand for vitamin B9 during gestation may lead to a decrease in its concentration. Crucially, folic acid participates in the synthesis of protein and amino acids, as well as in the multiplication of cells, which is particularly important in the first weeks of gestation. Current research indicates that the supplementation of multivitamin formulations, including folic acid, may lower the risk of pre-eclampsia in pregnant women [17]. Finally, vitamin B9, as well as B12, participate in erythropoiesis; thus, the deficiency of the aforementioned substances may lead to macrocytosis, erythroblasts apoptosis, and anemia [18].

As a result, folic acid affects numerous human metabolic pathways; therefore, inadequate folic acid intake may cause many diseases, including cancers, cardiovascular diseases, cognitive disorders, birth defects, and anemia [19].

2.1. Dietary Sources of Folic Acid

Green leafy vegetables, nuts, beans, and vitamin B9-supplemented products (e.g., rice, breakfast cereals, and pasta) are rich sources of folic acid [20]; Table 1 shows the folate content of the selected products [21]. It should be noted that the bioavailability of folate in food (estimated to equal approximately 50%) is approximately half of that of synthetic folic acid found in supplements [1] (amounting to 85–100%), although it is worth bearing in mind that taking supplements with meals reduces bioavailability [22]. Nevertheless, it is difficult to determine the bioavailability of folic acid in different groups of products. In fact, the supplementation of folic acid (400 µg/day) for five weeks significantly increases the serum folate levels in pregnant women [23]. An enhancement of serum folate levels upon folic acid supplementation in adults has also been observed [24].

Table 1. Folate content in the selected products [21].

Product	Folate Content in 100 g of a Product (µg)
Milk	5
Quark	27
Egg yolk	152
Chicken liver	590
Beef liver	330
Rice	29
Broccoli	119
Parsley	170
Spinach	193
Avocado	62
Apple	6

2.2. Recommendations Regarding Folic Acid Intake

The daily requirements for folic acid depend on the patient's medical condition. Folic acid, similarly to other water-soluble vitamins, does not accumulate in the human body and is rarely known to cause toxic effects [1,2]. Since the human body is unable to synthesize folate, supplementation or a dietary intake is essential. It is noteworthy that the daily intake of folic acid in food amounts to 150–250 µg, which is significantly below the recommended dietary allowance (RDA). Since folic acid plays a role in embryonic and fetal development, pregnant women, or women attempting pregnancy, its supplementation should be initiated 12 weeks prior to pregnancy and should be continued throughout the entire pregnancy, as well as during the post-partum period and breastfeeding. The recommended supplementation dosage of folic acid depends on the risk of neural birth tube defects, i.e., 0.4 mg/day for women at low risk and 5 mg/day for women in a high risk group [25]. Additionally, folic acid supplements are recommended for smokers, individuals treated with aspirin, patients suffering from kidney disease, in whom serum homocysteine levels often increase albuminuria, individuals taking certain medications following bariatric surgery, as well as in patients with diseases related to malabsorption (e.g., inflammatory bowel disease) [20].

3. The Role of Folic Acid in Inflammatory Bowel Disease

Folic acid deficiency constitutes a serious health factor, particularly for patients suffering from IBD. Yun et al. demonstrated in their meta-analysis that the level of folate in IBD patients was significantly lower when compared to that of healthy groups. Additionally, the folate concentration was lower in UC patients than in healthy individuals, although not in CD patients [6]. On the contrary, according to Ehrlich et al., folic acid deficiency occurred in approximately 92% of CD patients, as well as in more than 94% of UC patients and patients with an unclassified form of IBD. Moreover, 10–13% and 3.8–9.7% of children suffering from CD and UC, respectively, displayed vitamin B9 deficiency [26].

One of the significant risk factors in IBD is a poor folic acid diet, as patients often avoid vitamin B9-rich products, mostly for fear of the exacerbation of symptoms following consumption. Another one is total parenteral nutrition, as folic acid is absorbed in the duodenum and proximal jejunum. Therefore, severe inflammation, fibrosis, resection, and the occurrence of fistulas in this region may impair absorption due to a reduced active absorption area. Finally, active inflammation leads to a higher demand for folic acid, which is associated with the increased production of granulocytes and inflammatory cells [27].

Folic acid plays an important role in IBD patients—it affects the growth and regulation of cell functions because it is involved in the production of nucleic acids, protein synthesis, and amino acid transformation. IBD treatment involves the administration of methotrexate, a chemical compound included in the antimetabolite group, which has immunomodulatory and anti-inflammatory activity. The structure of methotrexate is similar to that of folic acid, and it inhibits the activity of dihydrofolate reductase, which catalyzes the transformation of dihydrofolate to tetrahydrofolate [28]. The most common adverse reactions comprise nausea, vomiting, diarrhea, bloating, liver damage, bone marrow suppression, pneumonitis, teratogenic effects, and folic acid deficiency. Studies have shown that the supplementation of folic acid may decrease the occurrence of side effects, mainly gastrointestinal disease, inflammation of the mucosa, and myelotoxicity [29]. The causes and consequences of folic acid deficiency in IBD are presented on Figure 2.

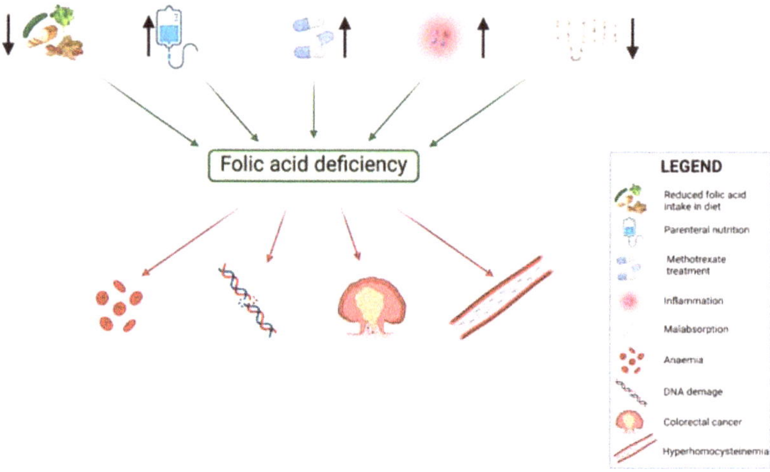

Figure 2. Causes and consequences of folic acid deficiency.

According to the guidelines of the European Crohn's and Colitis Organization (ECCO), folic acid supplementation of 5 mg is a recommended dose within two to three days of methotrexate administration [30], while the British Society of Gastroenterology recommends discontinuation of methotrexate in women who are planning pregnancy during therapy and a high dose (15 mg daily) of folic acid supplementation for a minimum period of six months [31]. Moreover, since sulfasalazine impairs folic acid absorption, selected patients treated with sulfasalazine should be supplemented with folic acid [32]. Additionally, pregnant women suffering from IBD treated with sulfasalazine should increase their dose of folic acid to 2 mg/day [33]. In the case of IBD, folic acid deficiency is one of the most common causal agents of non-iron deficiency anemia (NIDA) [34], which may decrease the quality of life in patients. Simultaneously, patients suffering from IBD are at a higher risk of folic deficiency (e.g., with inflammation in the small intestine or following resection) and require adequate supervision.

In folic acid deficiency, the production of red blood cells is disturbed, their volume increases, the survival time is shortened, and the bone marrow is damaged prematurely. The main reason for this phenomenon is the disruption of the synthesis of nucleic acids, mainly purine precursors. In normal circumstances, the level of folic acid should be controlled at least once a year, whereas in patients following extensive resection of the small intestine, extensive involvement of the ileum, with intestinal reservoir or with symptoms of deficiency its levels, should be monitored more frequently [35,36].

According to the guidelines of ECCO-ESGAR (the European Society of Gastrointestinal and Abdominal Radiology), the folic acid concentration should be tested every three to six months in patients suffering from IBD with a damaged small intestine, or following small intestine resection [37]. Samblas et al. suggested that folic acid may be effective in controlling the chronic inflammation in inflammatory diseases, mainly by means of DNA methylation. Folic acid is a methyl donor, and this activity is associated with the synthesis of S-adenosyl methionine (SAM), which is also a methyl donor. As researchers have observed, folic acid and other methyl donors decreases the expression of interleukin 1β (IL-1β), as well as tumor necrosis factor (TNF). Moreover, they also reduce the level of C-C motif chemokine ligand 2 (*CCL2*) mRNA, and increases the methylation in CpGs located in the genes of IL-1β, SERPINE1, and IL-18 [38].

Patients with long-term UC or CD present an increased risk of colorectal cancer (CRC) associated with inflammation and dysplasia. Additionally, CRC related to IBD is associated with chromosomal instability, microsatellite instability, and hypermethylation. In sporadic CRC, a low intake of folic acid is associated with a higher risk of adenomas and CRC, and the mechanisms are possibly related to the maintenance of normal methylation of DNA [39]. DNA methylation may affect gene expression. Therefore, folic acid deficiency might cause hypomethylation of DNA, leading to disorders of protooncogene participating in cancerogenesis. According to another hypothesis, deficiency of folic acid may induce uracil misincorporation during the synthesis of DNA, which leads to the breakage of DNA strands and chromosome damage [16].

According to Lashner et al., the supplementation of folic acid decreases the risk of neoplasms by approximately 62%, although these changes are not significant. Moreover, researchers have reported no considerable correlation between the folic acid dose and the development of neoplasms in a six-month supplementation of folate [40,41]. In contrast, as pointed out in the experimental study by Biasco et al., a three-month long supplementation of folic acid resulted in decreased cell proliferation [39]. A meta-analysis indicated that the supplementation of folic acid plays a protective role against the development of colorectal cancer [42]. It is vital to note that the role of folic acid in the prevention of CRC in patients suffering from IBD has not been substantiated in a large, randomized study yet. Furthermore, ECCO does not recommend routine supplementation of folic acid for the prevention of CRC in patients suffering from IBD [43].

Hyperhomocysteinemia, associated with group B vitamin deficiency, may increase the activity of Th17 cells in the bowel mucosa [44]. It is suggested that the supplementation of folic acid—due to decreased homocysteine levels—may reduce the activity of IBD and the occurrence of other autoimmune diseases [45]. The supplementation of folic acid, vitamin B12, and vitamin D did not change the lumbar spine and femoral neck BMD values more than in the placebo group (supplemented with vitamin D only). However, the concentration of Hcy was significantly reduced in the study group when compared to patients administered the placebo [46]. As Salari et al. reported, a six-month folic acid supplementation in postmenopausal women resulted in a significantly lower vitamin B12 level and a higher concentration of osteocalcin in the study group as compared to the control group [47].

3.1. Methylenotetrahydrofolate Reductase Gene

The *MTHFR* gene is located in the shorter arm of chromosome 1 (loci: 1p36.22) [48]. The metabolism of Hcy depends on MTHFR, which is coded by the *MTHFR* gene. A single nucleotide polymorphism (SNP) in the position 677 C $\rightarrow$ T causes a decrease in enzyme activity by 60% [49]. As pointed by Hanks et al., the 677CC genotype occurred in 55% of subjects, 677CT in 35%, and 677TT in 10% [50]. It is interesting to observe that the 677TT genotype occurred in 17.5% of patients presenting with ulcerative colitis (UC), in 16.8% of subjects with Crohn's disease (CD), and in 7.3% of healthy individuals. Moreover, in patients suffering from inflammatory bowel disease (IBD), the Hcy levels were significantly higher in persons with the 677TT genotype when compared to those with 677CT [51]. Nevertheless, no significant differences were observed in the frequency of the 677TT genotype in IBD patients and healthy persons [52,53]. As a meta-analysis showed, the mutation of the *MTHFR* gene in the position 677C/T did not increase the risk of IBD development [54], although in the Chinese population, 677TT polymorphism occurs more frequently in pancolitis than in other CD patients [55]. The mutation has been associated with a higher serum level of homocysteine; in fact, subjects with the 677TT and 677CT genotypes present higher homocysteine levels than subjects with 677CC. Moreover, serum folate levels decreased as the number of T alleles increased [56]. Therefore, using 677CC and 677CT as reference values, the odds ratio for folic acid deficiency amounted to 2.34 for individuals with the 677TT genotype. [57].

Another polymorphism of the *MTHFR* gene is 1298A $\rightarrow$ C. It is vital to notice that individuals with 1298AA and 1298AC demonstrate higher folic acid concentrations than patients with 1298CC, which is considered a wild-type [57].

Additionally, the differences in the previously discussed polymorphisms vary in different populations; for instance, the frequency of allele 1298C and genotype 1298CC is lower in western Africa and Mexico than in European countries (Italy—Sicily and France). Moreover, the 677TT genotype considerably increases the Hcy levels in individuals from western Africa, whereas only moderately in subjects from France and Italy. However, the abovementioned genotype does not elevate the concentration of Hcy in Mexicans [58].

Patients suffering from IBD are thought to be at a risk of developing osteoporosis, although no research has been conducted in regard to the risk of low bone mineral density (BMD) in IBD patients and the specific polymorphisms of the *MTHFR* gene. Gjesdal et al. presented various genotypes of the MTHFR gene in position 677C/T, and claimed that 1298A/C did not affect the risk of hip fracture [59]. On the contrary, as another meta-analysis showed, postmenopausal women with 677TT presented a higher femoral neck BMD than women with 677CC/CT, in spite of the fact that no such observation was made concerning lumbar spine BMD [60]. Furthermore, the TT genotype occurred more frequently in women with vertebral fractures than in female patients without such fractures [61]. Finally, studies have shown that the 677C/T polymorphism also affects spine BMD in nine-year-old children, which may further affect peak bone mass [62].

3.2. Homocysteine and Bone Mineral Density in Inflammatory Bowel Disease Patients

Inflammatory bowel disease patients present an increased risk of low BMD, leading to osteoporosis. Among others, risk factors of osteoporosis in IBD include malnutrition, low body mass, malabsorption, or use of corticosteroids [63]. According to Adriani et al., osteopenia and osteoporosis affect 46% and 11% of patients suffering from IBD, respectively [64].

Research has indicated that the mean homocysteine concentrations are, in general, higher in IBD patients than in control groups, with no significant difference between UC and CD; however, it has been pointed out that Hcy levels are higher in men than in women and also correlate with age in the study group, although not in the control group [65,66]. Similarly, Akbult et al. noted that the Hcy level among patients suffering from UC was higher than in healthy individuals [67], while Zezes et al. demonstrated a greater incidence of hyperhomocysteinemia in UC patients than in healthy subjects [68].

Homocysteine (Hcy) is an amino acid containing sulfur and does not occur in proteins, but it is formed in the course of methionine transmethylation [69], and the average Hcy level is below 15 µmol/L. Vitamin B6, folic acid, vitamin B12, and methylene tetrahydrofolate reductase are essential for homocysteine remethylation. If a deficiency of the abovementioned substances occurs, homocysteine is accumulated in serum, leading to hyperhomocysteinemia; moreover, it may be metabolized into cysteine and excreted in urine [70]. Homocysteine is a well-known risk factor of atherosclerosis, which is a chronic inflammation of the endothelium with an increased plasma permeability and lipid deposition in the atherosclerotic plaque. The accumulated atherosclerotic plaque, in turn, undergoes calcification and fibrosis. Additionally, homocysteine affects bone tissue, since the substance increases the activity of osteoclasts and inhibits apoptosis. The increased activity of osteoclasts may lead to increased bone resorption, a reduction of BMD, and an elevated risk of fracture [69]. What is more, Hcy exacerbates apoptosis in bone marrow and osteoclastogenesis, simultaneously decreasing blood flow in bone tissue, which may lead to reduced BMD [71], as well as increases intracellular reactive oxygen spices, which causes an increased differentiation of osteoclasts [72]. Finally, the concentration of Hcy is associated with bone resorption markers, e.g., β-CTX (C-terminal cross-linked telopeptide of type I collagen) [73].

No association between folic acid and vitamin B12 treatment to reduce the Hcy level and risk of hip fracture was shown in a previous meta-analysis [74]. On the contrary, Hcy levels have been shown to be correlated negatively with BMD among women [75]. In a study by Bailey et al., the researchers observed a lower serum level of vitamin B12 and folic acid in erythrocytes, as well as a higher level of bone turnover markers (serum of alkaline phosphatase and urine excreted of *N*-terminal cross-linked telopeptide of type I collagen) in postmenopausal women with a higher concentration of Hcy in comparison to individuals presenting normal Hcy levels. Additionally, Hcy levels correlated negatively with total BMD and BMD of the lumbar spine [76]. Nevertheless, Tariq et al. reported no association between Hcy levels and T- and Z-scores in postmenopausal women [77], and, according to Mittal et al., the concentration of Hcy is correlated positively with the level of parathormone (PTH) and phosphate. In contrast, Hcy levels were not related to BMD of the hip, lumbar spine, and forearm [78]. Postmenopausal women with osteoporosis presented a higher level of homocysteine than the control group. Moreover, the concentration of folic acid was lower in the study group than in the controls, although not significantly. In fact, Hcy levels were associated with the concentration of PTH, CTX (type I collagen C-telopeptides), and bone-specific alkaline phosphatase [79,80]. The adjusted hazard ratio for fracture was 2.42 (women) and 1.37 (men) for a concentration of homocysteine above 15 µM when compared to Hcy levels below 9 µM. The Hcy level correlated positively with the risk of fracture and, in fact, there was no association between the MTHFR genotype, vitamin B level, and risk of fracture [59]. As a meta-analysis showed, an increase in the homocysteine level by 1 µM elevates the risk of fracture by 4% [80]. According to Stone et al., the supplementation of folic acid and vitamins B6 and B12 did not alter the risk of fracture in women with higher levels of Hcy, or lower concentrations of folic acid, vitamin B6, or vitamin B12. Supplementation did not affect bone turnover markers such as CTX and P1NP (type I procollagen N-propeptide) [81]. Additionally, the supplementation of folic acid and vitamins B12 and B6 (individually or in combination) did not change the concentration of bone turnover markers and osteoporotic fracture in individuals suffering from the vascular disease with a normal level of Hcy [82].

Although the studies describing the association between IBD and homocysteine focused on non-IBD individuals, they suggest that hyperhomocysteinemia may constitute a potential additional factor of low BMD among IBD patients.

4. Microbiota and Folate Metabolism in IBD Patients

There are many factors that can affect the gut microbiota in patients with inflammatory bowel disease [83], such as age, inhabitable environment, culture habits, medical history, and the applied treatment. Nevertheless, the most critical factor affecting the diversity of species in the gut microbiota is diet, where saturated fatty acids and sugar have demonstrated a negative impact [84]. On the contrary, a vegetarian diet, rich in fiber, fruits, and vegetables, stimulates the growth of eubiotic bacteria and improves the function of enterocytes. The type of diet is particularly important for patients suffering from IBD, as a well-balanced one provides all macro- and micronutrients and eliminates products exacerbating symptoms of the disease. Conversely, fiber-rich products, vegetables, and fruits, which influence the microbiota content and the metabolism of folate, often exacerbate the symptoms. Additionally, using bacteriostatic and immunosuppressive drugs inhibiting the synthesis of PGE2 and leukotrienes may modify the gut microbiota and affect species diversity. In contrast, dietary carbon, nitrogen, water, and other nutrients provide a healthy development of intestinal bacteria, which are a source of vitamins and macro- and micronutrients. The most significant amount of folic acid is found in plant products, but folate may also be provided through animal products and various supplements [85], though the absorption of folic acid from plant products is decreased due to the presence of a conjugate inhibitor. Moreover, although the bioavailability of folic acid is higher in animal products, the content of vitamin B9 is significantly lower. As a result, the gut microbiota constitutes an essential element of folate absorption, which is twofold and occurs via folate receptors, as well as by means of specific receptors [86].

Most bacterial strains of *Bifidobacteria*, except for *B. gallicum* and *biavatii*, possess genes responsible for folate synthesis [87], while most lactic acid bacteria cannot synthesize folate. In fact, *Lactobacillus plantarum* is the only strain capable of producing folate, although the presence of 4-aminobenzoic acid is necessary [86]. According to Strozzi et al., the supplementation of *Bifidobacterium adolscentis* and *pseudocatenulatum* results in an increased level of fecal folate [88]. Additionally, as pointed out by LeBlanc et al., lactic acid bacteria and *Bifidobacteria*—found in fermentable dairy products—are able to synthesize vitamins de novo, particularly group B vitamins such as folic acid and vitamin B12 [89]. Furthermore, the supplementation of *L. plantarum* decreases the level of Il-8 and TNF-α and the expression of the occludin gene [90]. MacFarlane, on the other hand, evaluated how diet and microflora affect the course of IBD. The effect of folate on colonic microflora and the development of colonic tumors was determined in chemically induced ulcerative colitis in mice. According to the study results, the concentration of folic acid depended on the diet, and mice with colitis did present lower levels of circulating folic acid. However, folic acid had a minimal effect on tumor initiation and no effect on intestinal microflora. These data suggest that folic acid intake has little or no effect on the alleviation of IBD symptoms, or the risk of developing colon cancer in patients with IBD [91].

In another study, folate synthesized in the colon was absorbed and utilized by the host, and the local production of folic acid in the colon could help patients with IBD and reduce the risk of carcinogenesis [92]. In a study by Laiño, *Lactobacillus delbrueckii* subsp. *bulgaricus* CRL 863 and *S. thermophilus* CRL 415 and CRL 803 produced folic acid in fat-free milk, and increased the initial concentration of folic acid by approximately 190% [93]. Therefore, probiotic supplementation containing *Bifidobacterium* and *Lactobacillus plantarum* strains may increase folic acid production in patients with IBD and have a protective function for colonocytes in the course of this disease. Additionally, *Bifidobacteria* are involved in the regulation of intestinal homeostasis and have the ability to modulate the immune response [94]. Finally, it has been demonstrated that the use of *B. adolescentis* and *B. pseudocatenulatum* strains in humans increases the fecal folic acid concentration [95].

Microbial folate synthesis may also be affected by certain drugs, e.g., metformin, used in type 2 diabetes mellitus and insulin resistance, decreasing folate synthesis due to an increase of *Coenorhabditis elegans*, which also leads to decreasing serum folate levels [96]. Similarly, sulfonamides (structural analogs of p-aminobenzoic acid (PABA)) inhibit the synthesis of dihydrofolate (DHF) [97].

In summary, a proper supply of folic acid has a beneficial impact on the development of the gut microbiota, and certain bacteria strains provide an optimal level of folate in patients suffering from IBD.

5. Summary and Conclusions

Folic acid is a water-soluble group B vitamin, and its deficiency may lead to clinical complications, especially among patients suffering from IBD. Folic acid participates in the metabolism of homocysteine, high levels of which are associated with an increased risk of cardiovascular diseases and osteoporosis. Additionally, vitamin B9 is essential for the synthesis of nucleic acids and proteins. Therefore, providing adequate amounts of folic acid may prevent complications in this specific group of patients. Nevertheless, research regarding the association between folic acid, IBD, and bone mineral density is scarce. Therefore, future studies are necessary in order to investigate the potential benefits of folic acid use among IBD patients, which may improve both the course of the disease and the quality of life.

Summary:

1. In patients suffering from IBD, the concentration of folic acid should be evaluated more frequently than once per year, as it will help to diagnose a potential deficiency and macrocytic anemia.
2. Following the recommendations of ECCO, IBD patients treated with methotrexate should be supplied with 5 mg of folic acid at two- to three-day intervals during the administration of methotrexate.
3. Pregnant women, or women attempting pregnancy, should supplement folic acid. The recommended dosage is 0.4–5 mg/day (depending on the risk of neural birth tube defects).
4. The supplementation of folic acid may be a protective factor against the development of CRC. However, this hypothesis requires further research.

Author Contributions: Conceptualization, A.E.R., A.S.-T. and I.K.-K.; writing—original draft preparation, A.E.R., A.S.-T., A.M.R. and A.Z.; critical revision of the manuscript, I.K.-K. and A.D.; supervision, I.K.-K.; acceptance of the final version, all authors. All authors have read and agreed to the published version of the manuscript.

Funding: This research received no external funding.

Institutional Review Board Statement: Not applicable.

Informed Consent Statement: Not applicable.

Data Availability Statement: Not applicable.

Acknowledgments: We would like to thank Dagmara Mahadea for linguistic improvement.

Conflicts of Interest: The authors declare no conflict of interest.

References

1. Shelke, N.; Keith, L. Folic Acid Supplementation for Women of Childbearing Age versus Supplementation for the General Population: A Review of the Known Advantages and Risks. *Int. J. Fam. Med.* **2011**, *2011*, 173705. [CrossRef] [PubMed]
2. Merrell, B.J.; McMurry, J.P. Folic Acid. In *StatPearls*; StatPearls Publishing: Treasure Island, FL, USA, 2020.
3. Sijilmassi, O. Folic Acid Deficiency and Vision: A Review. *Graefes Arch. Clin. Exp. Ophthalmol.* **2019**, *257*, 1573–1580. [CrossRef] [PubMed]
4. Enderami, A.; Zarghami, M.; Darvishi-Khezri, H. The Effects and Potential Mechanisms of Folic Acid on Cognitive Function: A Comprehensive Review. *Neurol. Sci.* **2018**, *39*, 1667–1675. [CrossRef] [PubMed]

5. Scaglione, F.; Panzavolta, G. Folate, Folic Acid and 5-Methyltetrahydrofolate Are Not the Same Thing. *Xenobiotica* **2014**, *44*, 480–488. [CrossRef]
6. Pan, Y.; Liu, Y.; Guo, H.; Jabir, M.S.; Liu, X.; Cui, W.; Li, D. Associations between Folate and Vitamin B12 Levels and Inflammatory Bowel Disease: A Meta-Analysis. *Nutrients* **2017**, *9*, 382. [CrossRef]
7. De Martinis, M.; Sirufo, M.M.; Nocelli, C.; Fontanella, L.; Ginaldi, L. Hyperhomocysteinemia Is Associated with Inflammation, Bone Resorption, Vitamin B12 and Folate Deficiency and MTHFR C677T Polymorphism in Postmenopausal Women with Decreased Bone Mineral Density. *Int. J. Environ. Res. Public Health* **2020**, *17*, 4260. [CrossRef]
8. Karakoyun, I.; Duman, C.; Demet Arslan, F.; Baysoy, A.; Isbilen Basok, B. Vitamin B12 and Folic Acid Associated Megaloblastic Anemia: Could It Mislead the Diagnosis of Breast Cancer? *Int. J. Vitam. Nutr. Res.* **2019**, *89*, 255–260. [CrossRef]
9. Milman, N. Intestinal Absorption of Folic Acid—New Physiologic & Molecular Aspects. *Indian J. Med. Res.* **2012**, *136*, 725–728.
10. Bernstein, L.H.; Gutstein, S.; Weiner, S.; Efron, G. The Absorption and Malabsorption of Folic Acid and Its Polyglutamates. *Am. J. Med.* **1970**, *48*, 570–579. [CrossRef]
11. Gazzali, A.M.; Lobry, M.; Colombeau, L.; Acherar, S.; Azaïs, H.; Mordon, S.; Arnoux, P.; Baros, F.; Vanderesse, R.; Frochot, C. Stability of Folic Acid under Several Parameters. *Eur. J. Pharm. Sci.* **2016**, *93*, 419–430. [CrossRef]
12. Nuru, M.; Muradashvili, N.; Kalani, A.; Lominadze, D.; Tyagi, N. High Methionine, Low Folate and Low Vitamin B6/B12 (HM-LF-LV) Diet Causes Neurodegeneration and Subsequent Short-Term Memory Loss. *Metab. Brain Dis.* **2018**, *33*, 1923–1934. [CrossRef]
13. Crider, K.S.; Yang, T.P.; Berry, R.J.; Bailey, L.B. Folate and DNA Methylation: A Review of Molecular Mechanisms and the Evidence for Folate's Role. *Adv. Nutr.* **2012**, *3*, 21–38. [CrossRef]
14. Duthie, S.J.; Narayanan, S.; Brand, G.M.; Pirie, L.; Grant, G. Impact of Folate Deficiency on DNA Stability. *J. Nutr.* **2002**, *132*, 2444S–2449S. [CrossRef] [PubMed]
15. Otani, T.; Iwasaki, M.; Hanaoka, T.; Kobayashi, M.; Ishihara, J.; Natsukawa, S.; Shaura, K.; Koizumi, Y.; Kasuga, Y.; Yoshimura, K.; et al. Folate, Vitamin B6, Vitamin B12, and Vitamin B 2 Intake, Genetic Polymorphisms of Related Enzymes, and Risk of Colorectal Cancer in a Hospital-Based Case-Control Study in Japan. *Nutr. Cancer* **2005**, *53*, 42–50. [CrossRef]
16. Duthie, S.J. Folic Acid Deficiency and Cancer: Mechanisms of DNA Instability. *Br. Med. Bull.* **1999**, *55*, 578–592. [CrossRef] [PubMed]
17. Liu, C.; Liu, C.; Wang, Q.; Zhang, Z. Supplementation of Folic Acid in Pregnancy and the Risk of Preeclampsia and Gestational Hypertension: A Meta-Analysis. *Arch. Gynecol. Obstet.* **2018**, *298*, 697–704. [CrossRef] [PubMed]
18. Ghishan, F.K.; Kiela, P.R. Vitamins and Minerals in IBD. *Gastroenterol. Clin.* **2017**, *46*, 797–808. [CrossRef]
19. Field, M.S.; Stover, P.J. Safety of Folic Acid. *Ann. N. Y. Acad. Sci.* **2018**, *1414*, 59–71. [CrossRef] [PubMed]
20. Likis, F. Folic Acid. *J. Midwifery Women's Health* **2016**, *61*, 797–798. [CrossRef]
21. Kunachowicz, H.; Nadolna, I.; Iwanow, K.; Przygoda, B. *Wartość Odżywcza Wybranych Produktów Spożywczych i Typowych Potraw*; VI uaktualnione i rozszerzone; Wydawnictwo Lekarskie PZWL: Warszawa, Poland, 2010.
22. Caudill, M.A. Folate Bioavailability: Implications for Establishing Dietary Recommendations and Optimizing Status1234. *Am. J. Clin. Nutr.* **2010**, *91*, 1455S–1460S. [CrossRef]
23. Kondo, A.; Asada, Y.; Shibata, K.; Kihira, M.; Ninomiya, K.; Suzuki, M.; Oguchi, H.; Hayashi, Y.; Narita, O.; Watanabe, J.; et al. Dietary Folate Intakes and Effects of Folic Acid Supplementation on Folate Concentrations among Japanese Pregnant Women. *J. Obstet. Gynaecol. Res.* **2011**, *37*, 331–336. [CrossRef] [PubMed]
24. Anderson, C.A.M.; Jee, S.H.; Charleston, J.; Narrett, M.; Appel, L.J. Effects of Folic Acid Supplementation on Serum Folate and Plasma Homocysteine Concentrations in Older Adults: A Dose-Response Trial. *Am. J. Epidemiol.* **2010**, *172*, 932–941. [CrossRef] [PubMed]
25. Bomba-Opoń, D.; Hirnle, L.; Kalinka, J.; Seremak-Mrozikiewicz, A. Folate Supplementation during the Preconception Period, Pregnancy and Puerperium. Polish Society of Gynecologists and Obstetricians Guidelines. *Ginekol. Pol.* **2017**, *88*, 633–636. [CrossRef] [PubMed]
26. Ehrlich, S.; Mark, A.G.; Rinawi, F.; Shamir, R.; Assa, A. Micronutrient Deficiencies in Children with Inflammatory Bowel Diseases. *Nutr. Clin. Pract.* **2020**, *35*, 315–322. [CrossRef] [PubMed]
27. Hoffbrand, A.V.; Stewart, J.S.; Booth, C.C.; Mollin, D.L. Folate Deficiency in Crohn's Disease: Incidence, Pathogenesis, and Treatment. *Br. Med. J.* **1968**, *2*, 71–75. [CrossRef] [PubMed]
28. Klimczak, K.; Łykowska-Szuber, L.; Krela-Kazmierczak, I.; Eder, P.; Szymczak, A.; Stawczyk-Eder, K.; Linke, K. Zastosowanie Metotreksatu w Nieswoistych Chorobach Zapalnych Jelit Na Podstawie Przeglądu Aktualnego Piśmiennictwa. *Wiad. Lek.* **2016**, *69*, 262–266.
29. Schröder, O.; Stein, J. Low Dose Methotrexate in Inflammatory Bowel Disease: Current Status and Future Directions. *Am. J. Gastroenterol.* **2003**, *98*, 530–537. [CrossRef]
30. Gomollón, F.; Dignass, A.; Annese, V.; Tilg, H.; Van Assche, G.; Lindsay, J.O.; Peyrin-Biroulet, L.; Cullen, G.J.; Daperno, M.; Kucharzik, T.; et al. 3rd European Evidence-Based Consensus on the Diagnosis and Management of Crohn's Disease 2016: Part 1: Diagnosis and Medical Management. *J. Crohns Colitis* **2017**, *11*, 3–25. [CrossRef]
31. Lamb, C.A.; Kennedy, N.A.; Raine, T.; Hendy, P.A.; Smith, P.J.; Limdi, J.K.; Hayee, B.; Lomer, M.C.E.; Parkes, G.C.; Selinger, C.; et al. British Society of Gastroenterology Consensus Guidelines on the Management of Inflammatory Bowel Disease in Adults. *Gut* **2019**, *68*, s1–s106. [CrossRef]

32. Bischoff, S.C.; Escher, J.; Hébuterne, X.; Kłęk, S.; Krznaric, Z.; Schneider, S.; Shamir, R.; Stardelova, K.; Wierdsma, N.; Wiskin, A.E.; et al. ESPEN Practical Guideline: Clinical Nutrition in Inflammatory Bowel Disease. *Clin. Nutr.* **2020**, *39*, 632–653. [CrossRef]

33. Restellini, S.; Biedermann, L.; Hruz, P.; Mottet, C.; Moens, A.; Ferrante, M.; Schoepfer, A.M.; on behalf of Swiss IBDnet, an Official Working Group of the S.S. of Gastroenterology. Update on the Management of Inflammatory Bowel Disease during Pregnancy and Breastfeeding. *Digestion* **2020**, *101*, 27–42. [CrossRef]

34. Gasche, C.; Lomer, M.C.E.; Cavill, I.; Weiss, G. Iron, Anaemia, and Inflammatory Bowel Diseases. *Gut* **2004**, *53*, 1190–1197. [CrossRef]

35. Dignass, A.U.; Gasche, C.; Bettenworth, D.; Birgegård, G.; Danese, S.; Gisbert, J.P.; Gomollon, F.; Iqbal, T.; Katsanos, K.; Koutroubakis, I.; et al. European Consensus on the Diagnosis and Management of Iron Deficiency and Anaemia in Inflammatory Bowel Diseases. *J. Crohns Colitis* **2015**, *9*, 211–222. [CrossRef]

36. Ebara, S. Nutritional Role of Folate. *Congenit. Anom.* **2017**, *57*, 138–141. [CrossRef]

37. Maaser, C.; Sturm, A.; Vavricka, S.R.; Kucharzik, T.; Fiorino, G.; Annese, V.; Calabrese, E.; Baumgart, D.C.; Bettenworth, D.; Borralho Nunes, P.; et al. ECCO-ESGAR Guideline for Diagnostic Assessment in IBD Part 1: Initial Diagnosis, Monitoring of Known IBD, Detection of Complications. *J. Crohns Colitis* **2019**, *13*, 144–164. [CrossRef]

38. Samblas, M.; Martínez, J.A.; Milagro, F. Folic Acid Improves the Inflammatory Response in LPS-Activated THP-1 Macrophages. *Mediat. Inflamm.* **2018**, *2018*, 1312626. [CrossRef]

39. Biasco, G.; Zannoni, U.; Paganelli, G.M.; Santucci, R.; Gionchetti, P.; Rivolta, G.; Miniero, R.; Pironi, L.; Calabrese, C.; Di Febo, G.; et al. Folic Acid Supplementation and Cell Kinetics of Rectal Mucosa in Patients with Ulcerative Colitis. *Cancer Epidemiol. Biomark. Prev.* **1997**, *6*, 469–471.

40. Lashner, B.A.; Heidenreich, P.A.; Su, G.L.; Kane, S.V.; Hanauer, S.B. Effect of Folate Supplementation on the Incidence of Dysplasia and Cancer in Chronic Ulcerative Colitis. A Case-Control Study. *Gastroenterology* **1989**, *97*, 255–259. [CrossRef]

41. Lashner, B.A.; Provencher, K.S.; Seidner, D.L.; Knesebeck, A.; Brzezinski, A. The Effect of Folic Acid Supplementation on the Risk for Cancer or Dysplasia in Ulcerative Colitis. *Gastroenterology* **1997**, *112*, 29–32. [CrossRef]

42. Burr, N.E.; Hull, M.A.; Subramanian, V. Folic Acid Supplementation May Reduce Colorectal Cancer Risk in Patients with Inflammatory Bowel Disease: A Systematic Review and Meta-Analysis. *J. Clin. Gastroenterol.* **2017**, *51*, 247–253. [CrossRef] [PubMed]

43. Sebastian, S.; Hernández, V.; Myrelid, P.; Kariv, R.; Tsianos, E.; Toruner, M.; Marti-Gallostra, M.; Spinelli, A.; van der Meulen-de Jong, A.E.; Yuksel, E.S.; et al. Colorectal Cancer in Inflammatory Bowel Disease: Results of the 3rd ECCO Pathogenesis Scientific Workshop (I). *J. Crohns Colitis* **2014**, *8*, 5–18. [CrossRef]

44. Wang, N.; Tang, H.; Wang, X.; Wang, W.; Feng, J. Homocysteine Upregulates Interleukin-17A Expression via NSun2-Mediated RNA Methylation in T Lymphocytes. *Biochem. Biophys. Res. Commun.* **2017**, *493*, 94–99. [CrossRef]

45. Lin, X.; Meng, X.; Song, Z. Homocysteine and Psoriasis. *Biosci. Rep.* **2019**, *39*, BSR20190867. [CrossRef]

46. Enneman, A.W.; Swart, K.M.A.; van Wijngaarden, J.P.; van Dijk, S.C.; Ham, A.C.; Brouwer-Brolsma, E.M.; van der Zwaluw, N.L.; Dhonukshe-Rutten, R.A.M.; van der Cammen, T.J.M.; de Groot, L.C.P.G.M.; et al. Effect of Vitamin B12 and Folic Acid Supplementation on Bone Mineral Density and Quantitative Ultrasound Parameters in Older People with an Elevated Plasma Homocysteine Level: B-PROOF, a Randomized Controlled Trial. *Calcif. Tissue Int.* **2015**, *96*, 401–409. [CrossRef]

47. Salari, P.; Abdollahi, M.; Heshmat, R.; Meybodi, H.A.; Razi, F. Effect of Folic Acid on Bone Metabolism: A Randomized Double Blind Clinical Trial in Postmenopausal Osteoporotic Women. *DARU J. Pharm. Sci.* **2014**, *22*, 62. [CrossRef] [PubMed]

48. Genetics Home Reference. MTHFR Gene. Available online: https://ghr.nlm.nih.gov/gene/MTHFR (accessed on 11 May 2020).

49. Xu, B.; Kong, X.; Xu, R.; Song, Y.; Liu, L.; Zhou, Z.; Gu, R.; Shi, X.; Zhao, M.; Huang, X.; et al. Homocysteine and All-Cause Mortality in Hypertensive Adults without Pre-Existing Cardiovascular Conditions: Effect Modification by MTHFR C677T Polymorphism. *Medicine* **2017**, *96*, e5862. [CrossRef] [PubMed]

50. Hanks, J.; Ayed, I.; Kukreja, N.; Rogers, C.; Harris, J.; Gheorghiu, A.; Liu, C.L.; Emery, P.; Pufulete, M. The Association between MTHFR 677C>T Genotype and Folate Status and Genomic and Gene-Specific DNA Methylation in the Colon of Individuals without Colorectal Neoplasia. *Am. J. Clin. Nutr.* **2013**, *98*, 1564–1574. [CrossRef] [PubMed]

51. Mahmud, N.; Molloy, A.; McPartlin, J.; Corbally, R.; Whitehead, A.S.; Scott, J.M.; Weir, D.G. Increased Prevalence of Methylenetetrahydrofolate Reductase C677T Variant in Patients with Inflammatory Bowel Disease, and Its Clinical Implications. *Gut* **1999**, *45*, 389–394. [CrossRef] [PubMed]

52. Stocco, G.; Martelossi, S.; Sartor, F.; Toffoli, G.; Lionetti, P.; Barabino, A.; Fontana, M.; Decorti, G.; Bartoli, F.; Giraldi, T.; et al. Prevalence of Methylenetetrahydrofolate Reductase Polymorphisms in Young Patients with Inflammatory Bowel Disease. *Dig. Dis. Sci.* **2006**, *51*, 474–479. [CrossRef]

53. Senhaji, N.; Serbati, N.; Diakité, B.; Arazzakou, S.; Hamzi, K.; Badre, W.; Nadifi, S. Methylenetetrahydrofolate Reductase C677T Variant in Moroccan Patients with Inflammatory Bowel Disease. *Gene* **2013**, *521*, 45–49. [CrossRef]

54. Lu, L.; Wang, W.; Li, L.; Pang, X.; Fei, S. Risk of Inflammatory Bowel Disease Associated with MTHFRC677T and Prothrombin G20210A Mutation: A Meta-Analysis. *Int. J. Clin. Exp. Med.* **2017**, *10*, 743–7442.

55. Chen, M.; Peyrin-Biroulet, L.; Xia, B.; Guéant-Rodriguez, R.-M.; Bronowicki, J.-P.; Bigard, M.-A.; Guéant, J.-L. Methionine Synthase A2756G Polymorphism May Predict Ulcerative Colitis and Methylenetetrahydrofolate Reductase C677T Pancolitis, in Central China. *BMC Med. Genet.* **2008**, *9*, 78. [CrossRef] [PubMed]

56. Hustad, S.; Midttun, Ø.; Schneede, J.; Vollset, S.E.; Grotmol, T.; Ueland, P.M. The Methylenetetrahydrofolate Reductase 677C–>T Polymorphism as a Modulator of a B Vitamin Network with Major Effects on Homocysteine Metabolism. *Am. J. Hum. Genet.* **2007**, *80*, 846–855. [CrossRef] [PubMed]

57. Li, W.-X.; Dai, S.-X.; Zheng, J.-J.; Liu, J.-Q.; Huang, J.-F. Homocysteine Metabolism Gene Polymorphisms (MTHFR C677T, MTHFR A1298C, MTR A2756G and MTRR A66G) Jointly Elevate the Risk of Folate Deficiency. *Nutrients* **2015**, *7*, 6670–6687. [CrossRef] [PubMed]

58. Guéant-Rodriguez, R.-M.; Guéant, J.-L.; Debard, R.; Thirion, S.; Hong, L.X.; Bronowicki, J.-P.; Namour, F.; Chabi, N.W.; Sanni, A.; Anello, G.; et al. Prevalence of Methylenetetrahydrofolate Reductase 677T and 1298C Alleles and Folate Status: A Comparative Study in Mexican, West African, and European Populations. *Am. J. Clin. Nutr.* **2006**, *83*, 701–707. [CrossRef]

59. Gjesdal, C.G.; Vollset, S.E.; Ueland, P.M.; Refsum, H.; Meyer, H.E.; Tell, G.S. Plasma Homocysteine, Folate, and Vitamin B 12 and the Risk of Hip Fracture: The Hordaland Homocysteine Study. *J. Bone Miner. Res.* **2007**, *22*, 747–756. [CrossRef]

60. Li, D.; Wu, J. Association of the MTHFR C677T Polymorphism and Bone Mineral Density in Postmenopausal Women: A Meta-Analysis. *J. Biomed. Res.* **2010**, *24*, 417–423. [CrossRef]

61. Villadsen, M.M.; Bünger, M.H.; Carstens, M.; Stenkjaer, L.; Langdahl, B.L. Methylenetetrahydrofolate Reductase (MTHFR) C677T Polymorphism Is Associated with Osteoporotic Vertebral Fractures, but Is a Weak Predictor of BMD. *Osteoporos. Int.* **2005**, *16*, 411–416. [CrossRef]

62. Cd, S.; Pm, E.; Sj, L.; Gd, S.; Jh, T. Methylenetetrahydrofolate Reductase (MTHFR) C677T Polymorphism Is Associated with Spinal BMD in 9-Year-Old Children. *J. Bone Miner. Res.* **2009**, *24*, 117–124. [CrossRef]

63. Krela-Kaźmierczak, I.; Michalak, M.; Szymczak-Tomczak, A.; Łykowska-Szuber, L.; Stawczyk-Eder, K.; Waszak, K.; Kucharski, M.A.; Dobrowolska, A.; Eder, P. Prevalence of Osteoporosis and Osteopenia in a Population of Patients with Inflammatory Bowel Diseases from the Wielkopolska Region. *Pol. Arch. Intern. Med.* **2018**, *128*, 447–454. [CrossRef]

64. Adriani, A.; Pantaleoni, S.; Luchino, M.; Ribaldone, D.G.; Reggiani, S.; Sapone, N.; Sguazzini, C.; Isaia, G.; Pellicano, R.; Astegiano, M. Osteopenia and Osteoporosis in Patients with New Diagnosis of Inflammatory Bowel Disease. *Panminerva Med.* **2014**, *56*, 145–149.

65. Oussalah, A.; Guéant, J.-L.; Peyrin-Biroulet, L. Meta-Analysis: Hyperhomocysteinaemia in Inflammatory Bowel Diseases. *Aliment. Pharm. Ther.* **2011**, *34*, 1173–1184. [CrossRef]

66. Owczarek, D.; Cibor, D.; Sałapa, K.; Jurczyszyn, A.; Mach, T. Homocysteine in Patients with Inflammatory Bowel Diseases. *Przegląd Lek.* **2014**, *71*, 189–192.

67. Akbulut, S.; Altiparmak, E.; Topal, F.; Ozaslan, E.; Kucukazman, M.; Yonem, O. Increased Levels of Homocysteine in Patients with Ulcerative Colitis. *World J. Gastroenterol.* **2010**, *16*, 2411–2416. [CrossRef] [PubMed]

68. Zezos, P.; Papaioannou, G.; Nikolaidis, N.; Vasiliadis, T.; Giouleme, O.; Evgenidis, N. Hyperhomocysteinemia in Ulcerative Colitis Is Related to Folate Levels. *World J. Gastroenterol.* **2005**, *11*, 6038–6042. [CrossRef] [PubMed]

69. Škovierová, H.; Vidomanová, E.; Mahmood, S.; Sopková, J.; Drgová, A.; Červeňová, T.; Halašová, E.; Lehotský, J. The Molecular and Cellular Effect of Homocysteine Metabolism Imbalance on Human Health. *Int. J. Mol. Sci.* **2016**, *17*, 1733. [CrossRef] [PubMed]

70. Moll, S.; Varga, E.A. Homocysteine and MTHFR Mutations. *Circulation* **2015**, *132*, e6–e9. [CrossRef]

71. Bahtiri, E.; Islami, H.; Rexhepi, S.; Qorraj-Bytyqi, H.; Thaçi, K.; Thaçi, S.; Karakulak, C.; Hoxha, R. Relationship of Homocysteine Levels with Lumbar Spine and Femur Neck BMD in Postmenopausal Women. *Acta Reum. Port.* **2015**, *40*, 355–362.

72. Behera, J.; Bala, J.; Nuru, M.; Tyagi, S.C.; Tyagi, N. Homocysteine as a Pathological Biomarker for Bone Disease. *J. Cell. Physiol.* **2017**, *232*, 2704–2709. [CrossRef]

73. Álvarez-Sánchez, N.; Álvarez-Ríos, A.I.; Guerrero, J.M.; García-García, F.J.; Rodríguez-Mañas, L.; Cruz-Chamorro, I.; Lardone, P.J.; Carrillo-Vico, A. Homocysteine Levels Are Associated with Bone Resorption in Pre-Frail and Frail Spanish Women: The Toledo Study for Healthy Aging. *Exp. Gerontol.* **2018**, *108*, 201–208. [CrossRef]

74. Garcia Lopez, M.; Baron, J.A.; Omsland, T.K.; Søgaard, A.J.; Meyer, H.E. Homocysteine-Lowering Treatment and the Risk of Fracture: Secondary Analysis of a Randomized Controlled Trial and an Updated Meta-Analysis. *JBMR Plus* **2018**, *2*, 295–303. [CrossRef]

75. Saoji, R.; Das, R.S.; Desai, M.; Pasi, A.; Sachdeva, G.; Das, T.K.; Khatkhatay, M.I. Association of High-Density Lipoprotein, Triglycerides, and Homocysteine with Bone Mineral Density in Young Indian Tribal Women. *Arch. Osteoporos.* **2018**, *13*, 108. [CrossRef] [PubMed]

76. Bailey, R.L.; Looker, A.C.; Lu, Z.; Fan, R.; Eicher-Miller, H.A.; Fakhouri, T.H.; Gahche, J.J.; Weaver, C.M.; Mills, J.L. B-Vitamin Status and Bone Mineral Density and Risk of Lumbar Osteoporosis in Older Females in the United States. *Am. J. Clin. Nutr.* **2015**, *102*, 687–694. [CrossRef]

77. Tariq, S.; Lone, K.P.; Tariq, S. Comparison of Parameters of Bone Profile and Homocysteine in Physically Active and Non-Active Postmenopausal Females. *Pak. J. Med. Sci.* **2016**, *32*, 1263–1267. [CrossRef] [PubMed]

78. Mittal, M.; Verma, R.; Mishra, A.; Singh, A.; Kumar, V.; Sawlani, K.K.; Ahmad, M.K.; Mishra, P.; Gaur, R. Relation of Bone Mineral Density with Homocysteine and Cathepsin K Levels in Postmenopausal Women. *Indian J. Endocrinol. Metab.* **2018**, *22*, 261. [CrossRef] [PubMed]

79. Haliloglu, B.; Aksungar, F.B.; Ilter, E.; Peker, H.; Akin, F.T.; Mutlu, N.; Ozekici, U. Relationship between Bone Mineral Density, Bone Turnover Markers and Homocysteine, Folate and Vitamin B12 Levels in Postmenopausal Women. *Arch. Gynecol. Obstet.* **2010**, *281*, 663–668. [CrossRef]

80. Van Wijngaarden, J.P.; Doets, E.L.; Szczecińska, A.; Souverein, O.W.; Duffy, M.E.; Dullemeijer, C.; Cavelaars, A.E.J.M.; Pietruszka, B.; Van't Veer, P.; Brzozowska, A.; et al. Vitamin B12, Folate, Homocysteine, and Bone Health in Adults and Elderly People: A Systematic Review with Meta-Analyses. *J. Nutr. Metab.* **2013**, *2013*, 486186. [CrossRef]

81. Stone, K.L.; Lui, L.-Y.; Christen, W.G.; Troen, A.M.; Bauer, D.C.; Kado, D.; Schambach, C.; Cummings, S.R.; Manson, J.E. Effect of Combination Folic Acid, Vitamin B6, and Vitamin B12 Supplementation on Fracture Risk in Women: A Randomized, Controlled Trial. *J. Bone Miner. Res.* **2017**, *32*, 2331–2338. [CrossRef]

82. Ruan, J.; Gong, X.; Kong, J.; Wang, H.; Zheng, X.; Chen, T. Effect of B Vitamin (Folate, B6, and B12) Supplementation on Osteoporotic Fracture and Bone Turnover Markers: A Meta-Analysis. *Med. Sci. Monit.* **2015**, *21*, 875–881. [CrossRef]

83. Glassner, K.L.; Abraham, B.P.; Quigley, E.M.M. The Microbiome and Inflammatory Bowel Disease. *J. Allergy Clin. Immunol.* **2020**, *145*, 16–27. [CrossRef]

84. Ley, R.E.; Turnbaugh, P.J.; Klein, S.; Gordon, J.I. Human Gut Microbes Associated with Obesity. *Nature* **2006**, *444*, 1022–1023. [CrossRef] [PubMed]

85. Czeczot, H. Kwas foliowy w fizjologii i patologii. *Postepy Hig. Med. Dosw.* **2008**, *62*, 405–419.

86. Rossi, M.; Amaretti, A.; Raimondi, S. Folate Production by Probiotic Bacteria. *Nutrients* **2011**, *3*, 118–134. [CrossRef]

87. Sugahara, H.; Odamaki, T.; Hashikura, N.; Abe, F.; Xiao, J. Differences in Folate Production by Bifidobacteria of Different Origins. *Biosci. Microbiota Food Health* **2015**, *34*, 87–93. [CrossRef] [PubMed]

88. Pompei, A.; Cordisco, L.; Amaretti, A.; Zanoni, S.; Matteuzzi, D.; Rossi, M. Folate Production by Bifidobacteria as a Potential Probiotic Property. *Appl. Environ. Microbiol.* **2007**, *73*, 179–185. [CrossRef]

89. LeBlanc, J.G.; Milani, C.; de Giori, G.S.; Sesma, F.; van Sinderen, D.; Ventura, M. Bacteria as Vitamin Suppliers to Their Host: A Gut Microbiota Perspective. *Curr. Opin. Biotechnol.* **2013**, *24*, 160–168. [CrossRef]

90. Wu, Y.; Zhu, C.; Chen, Z.; Chen, Z.; Zhang, W.; Ma, X.; Wang, L.; Yang, X.; Jiang, Z. Protective Effects of Lactobacillus Plantarum on Epithelial Barrier Disruption Caused by Enterotoxigenic Escherichia Coli in Intestinal Porcine Epithelial Cells. *Vet. Immunol. Immunopathol.* **2016**, *172*, 55–63. [CrossRef]

91. MacFarlane, A.J.; Behan, N.A.; Matias, F.M.G.; Green, J.; Caldwell, D.; Brooks, S.P.J. Dietary Folate Does Not Significantly Affect the Intestinal Microbiome, Inflammation or Tumorigenesis in Azoxymethane-Dextran Sodium Sulphate-Treated Mice. *Br. J. Nutr.* **2013**, *109*, 630–638. [CrossRef]

92. Choi, S.-W.; Mason, J.B. Folate Status: Effects on Pathways of Colorectal Carcinogenesis. *J. Nutr.* **2002**, *132*, 2413S–2418S. [CrossRef]

93. Laiño, J.E.; Leblanc, J.G.; Savoy de Giori, G. Production of Natural Folates by Lactic Acid Bacteria Starter Cultures Isolated from Artisanal Argentinean Yogurts. *Can. J. Microbiol.* **2012**, *58*, 581–588. [CrossRef]

94. Rossi, M.; Amaretti, A. Probiotic Properties of Bifidobacteria. In *Bifidobacteria: Genomics and Molecular Aspects*; van Synderen, D., Mayo, B., Eds.; Caister Academic Press: Norfolk, UK, 2010.

95. Strozzi, G.P.; Mogna, L. Quantification of Folic Acid in Human Feces after Administration of Bifidobacterium Probiotic Strains. *J. Clin. Gastroenterol.* **2008**, *42*, S179–S184. [CrossRef] [PubMed]

96. Storelli, G.; Téfit, M.; Leulier, F. Metformin, Microbes, and Aging. *Cell Metab.* **2013**, *17*, 809–811. [CrossRef] [PubMed]

97. Brooks, G.F. *Jawetz, Melnick & Adelberg's Medical Microbiology*; McGraw-Hill Medical: New York, NY, USA, 2007; ISBN 978-0-07-147666-9.

 *nutrients*

Systematic Review

Does Sodium Intake Induce Systemic Inflammatory Response? A Systematic Review and Meta-Analysis of Randomized Studies in Humans

Eirini D. Basdeki [1,2], Anastasios Kollias [3], Panagiota Mitrou [4], Christiana Tsirimiagkou [1,2], Marios K. Georgakis [5], Antonios Chatzigeorgiou [6], Antonios Argyris [1], Kalliopi Karatzi [7], Yannis Manios [2], Petros P. Sfikakis [8] and Athanase D. Protogerou [1,*]

[1] Cardiovascular Prevention & Research Unit, Clinic & Laboratory of Pathophysiology, Department of Medicine, National and Kapodistrian University of Athens, 11527 Athens, Greece; eirinibasdeki@gmail.com (E.D.B.); tsirimiagou.ch@gmail.com (C.T.); and1dr@gmail.com (A.A.)

[2] Department of Nutrition and Dietetics, School of Health Science and Education, Harokopio University of Athens, 17671 Kallithea, Greece; manios@hua.gr

[3] Hypertension Center STRIDE-7, Third Department of Medicine, School of Medicine, National and Kapodistrian University of Athens, Sotiria Hospital, 11527 Athens, Greece; taskollias@gmail.com

[4] Hellenic Republic Ministry of Health, 10433 Athens, Greece; mitroup@moh.gov.gr

[5] Institute for Stroke and Dementia Research, University Hospital, Ludwig Maximilians University, 81377 Munich, Germany; mgeorgakis91@gmail.com

[6] Department of Physiology, Medical School, National & Kapodistrian University of Athens, 11527 Athens, Greece; achatzig@med.uoa.gr

[7] Department of Food Science and Human Nutrition, Agricultural University of Athens, Iera Odos 75, 11855 Athens, Greece; pkaratzi@aua.gr

[8] First Department of Propaedeutic and Internal Medicine, Athens University Medical School, Laiko Hospital, 11527 Athens, Greece; psfikakis@med.uoa.gr

* Correspondence: aprotog@med.uoa.gr; Tel.: +30-210-746-2566

Citation: Basdeki, E.D.; Kollias, A.; Mitrou, P.; Tsirimiagkou, C.; Georgakis, M.K.; Chatzigeorgiou, A.; Argyris, A.; Karatzi, K.; Manios, Y.; Sfikakis, P.P.; et al. Does Sodium Intake Induce Systemic Inflammatory Response? A Systematic Review and Meta-Analysis of Randomized Studies in Humans. *Nutrients* **2021**, *13*, 2632. https://doi.org/10.3390/nu13082632

Academic Editors: Francesca Di Sole and Roberto Iacone

Received: 29 May 2021
Accepted: 27 July 2021
Published: 30 July 2021

Publisher's Note: MDPI stays neutral with regard to jurisdictional claims in published maps and institutional affiliations.

Abstract: Experimental studies suggest that sodium induced inflammation might be another missing link leading to atherosclerosis. To test the hypothesis that high daily sodium intake induces systemic inflammatory response in humans, we performed a systematic review according to PRISMA guidelines of randomized controlled trials (RCTs) that examined the effect of high versus low sodium dose (HSD vs. LSD), as defined per study, on plasma circulating inflammatory biomarkers. Eight RCTs that examined CRP, TNF-a and IL-6 were found. Meta-analysis testing the change of each biomarker in HSD versus LSD was possible for CRP ($n = 5$ studies), TNF-a ($n = 4$ studies) and IL-6 ($n = 4$ studies). The pooled difference (95% confidence intervals) per biomarker was for: CRP values of 0.1($-$0.3, 0.4) mg/L; TNF-a $-0.7(-5.0, 3.6)$ pg/mL; IL-6 $-1.1(-3.3$ to 1.1) pg/mL. Importantly, there was inconsistency between RCTs regarding major population characteristics and the applied methodology, including a very wide range of LSD (460 to 6740 mg/day) and HSD (2800 to 7452 mg/day). Although our results suggest that the different levels of daily sodium intake are not associated with significant changes in the level of systemic inflammation in humans, this outcome may result from methodological issues. Based on these identified methodological issues we propose that future RCTs should focus on young healthy participants to avoid confounding effects of comorbidities, should have three instead of two arms (very low, "normal" and high) of daily sodium intake with more than 100 participants per arm, whereas an intervention duration of 14 days is adequate.

Keywords: sodium; sodium intake; inflammation; systemic inflammation; sodium induced inflammation; CRP; TNF-a; IL-6

1. Introduction

The first observational evidence showing that the higher the sodium intake the higher the systemic inflammation—as measured by plasma c-reactive protein (CRP)—was published twelve years ago [1]. However, the magnitude of this association was only marginal (increase in CRP of 1.06 mg/L per 2.3 g/L of urinary sodium excretion) with the authors suggesting that dietary sodium consumption is unlikely to be an important modifiable risk factor for increased systemic inflammation [1]. However, since robust evidence suggest that higher sodium intake is associated with higher incidence of cardiovascular (CV) disease [2], it seems plausible that sodium induced inflammation might be another missing link, beyond blood pressure (BP) increase, leading to atherosclerosis, a per se inflammatory process [3].

Since then, several studies have been conducted in humans but mostly in animal models to test the hypothesis of sodium induced systemic inflammation [4]. In animal models, the majority of the available data, but not all [5,6], indicate that high sodium intake is associated with increased levels of circulating inflammatory biomarkers [7–10]. However, the evidence regarding the association of sodium intake and systemic inflammation in humans remain limited and controversial, in part due to the high heterogeneity and differences in the methodology applied e.g., regarding type of study (observation or intervention (acute or chronic) [11–14], the level of daily sodium intake tested [13,15], and the type of inflammatory biomarker that was evaluated [16,17]).

Of note, over the years this hypothesis has become even more intriguing because: (i) of the presence of a J-shape association between daily sodium intake and mortality in epidemiological studies [18–20], (ii) both very high and very low sodium intake are implicated in the pathogenesis of arterial damage [5,21–24], (iii) data derived from in vitro and in vivo experimental animal studies, as well as preliminary human studies, suggest an association between high sodium intake and autoimmune disease [25], and finally (iv) not only high, but also very low sodium intake seems to be proinflammatory [26,27]. However, given the fact that other factors (including certain types of foods, sedentary life, sleep apnea) [28,29] may have proinflammatory effects, the task to delineate and quantify the potential effects of sodium intake on systemic inflammation is quite complex.

In the present study we aimed to investigate the hypothesis that sodium intake induces systemic inflammatory response in humans in a dose response manner. To this end, we performed a systematic review and meta-analysis of all randomized studies comparing the effect of at least two different levels of dietary sodium intake on the magnitude of systematic inflammatory response, as described by predefined circulating inflammatory biomarkers. The primary endpoint of the meta-analysis was the magnitude of inflammatory response (i.e., the difference in the levels of each inflammatory biomarker) after the intervention with a high sodium diet (HSD) versus low sodium diet (LSD); the magnitude of BP response was defined as a secondary endpoint (i.e., the difference in BP).

2. Materials and Methods

2.1. Search Strategy

This systematic review and meta-analysis was performed according to PRISMA guidelines (Supplement Table S1) [30]. A systematic search of potentially relevant studies was performed throughout April 2020 by two separate investigators (E.D. Basdeki & C. Tsirimiagkou) in PUBMED and SCOPUS databases. Search terms applied included: ("dietary sodium" OR "sodium intake" OR "sodium excretion" OR "urinary sodium" OR salt) AND (inflammation OR "inflammatory biomarkers" OR "inflammatory indices" OR "inflammatory cells" OR "white blood cells" OR "C-reactive protein" OR crp OR interleukin OR lymphocyte OR leukocytes OR il-5 OR il-6 OR il-10 OR il-12 OR il-23 OR il-17 OR cytokines OR "tumor necrosis factor" OR tnf OR tnf-a OR tnf-b OR cd4 OR cd8)). Articles were also identified from reference lists of relevant papers and hand search. Studies were limited to English language, human, and randomized controlled studies (RCTs). Disagreements were resolved by consensus with a senior author (A.D. Protogerou).

2.2. Inclusion & Exclusion Criteria

Eligible studies were full-text peer-reviewed articles in English that: 1. were RCTs with parallel-arm (different patients) or crossover (same patients) design, 2. conducted in males and/or females regardless of diseases (chronic or acute), 3. examined the effect of at least two different daily doses of sodium intake on circulating inflammatory biomarkers. The following exclusion criteria were applied: epidemiological studies, non-RCTs, animal studies, reviews, systematic reviews, meta-analyses, comments/letters.

2.3. Selection of Studies & Data Extraction

Two reviewers screened the available titles, abstracts and keywords from all of the available articles. Discrepancies were resolved after discussion. After agreement, full text screening was carried out. Both reviewers extracted independently qualitative and quantitative data from all included articles, concerning study design, population characteristics and data regarding primary endpoints from included studies where available. Authors of the included studies were contacted by e-mail to obtain additional details not reported in the published paper (i.e., mean and SD of difference regarding the variable of interest). The risk of bias was assessed using a Cochrane Collaboration's tool for assessing risk of bias in randomized trials [31].

2.4. Statistical Analysis

Meta-analysis was performed using the Stata/SE 11 (Texas) software. Sensitivity analyses were performed to compensate for the observed methodological heterogeneity among the included studies. Meta-regression analysis was performed for assessing associations between the difference in (a) CRP or systolic/diastolic BP (SBP/DBP) and (b) sex, age, duration of intervention, and difference in sodium intake between the examined diet arms across the included studies. Sensitivity analyses were performed according to the design of the study (crossover or parallel), the mean age of the studied population, the average sodium intake in each arm, and by excluding studies with patients on hemodialysis. Mean values of subgroups were combined where feasible [32]. Median values were converted to mean values using appropriate formulas [33]. In the case of missing values regarding the mean (SD) of difference in the outcome of interest between the examined groups, these were calculated from the groups' mean values using appropriate calculators [34]. The latter procedure was also implemented for paired comparisons in crossover studies as a rough approximation. Heterogeneity was tested using an I^2 statistic. A value of I^2 statistic >50% was considered to indicate significant heterogeneity between studies. When significant heterogeneity was present, a random-effects model of analysis was used; otherwise, a fixed-effects model of analysis was used. Publication bias was assessed by inspecting funnel plots, as well as Egger's test (linear regression method) and Begg's test (rank correlation method) [35,36]. Two-sided p values of <0.05 were considered significant.

3. Results

3.1. Number of Studies Screened and Selected—General Description

The PRISMA Checklist for the present systematic review and meta-analysis is presented in Table S1. Three thousand six hundred and twenty-three (3623) studies were identified through a systematic search. The flow chart for study selection is shown in Figure 1. Eight studies [13,27,37–42] met the inclusion criteria for examining the effect of sodium intake on circulating inflammatory biomarkers and were included in the systematic review. Detailed descriptive data, as well as results for all the included studies are provided in Tables 1 and 2.

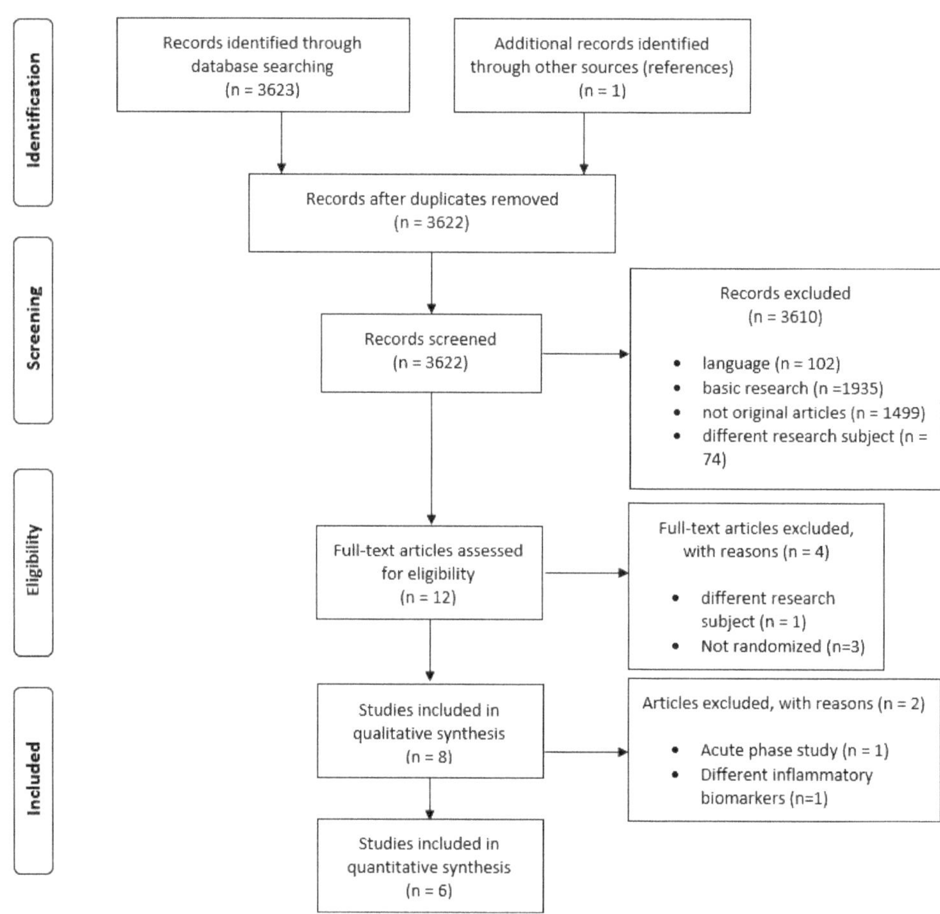

Figure 1. Identification and selection of the eligible studies according to the PRISMA criteria.

Table 1. Descriptive characteristics of all of the selected studies ($n = 8$).

Author (Year)	Study Design	Population Description	N	Sex (% Male)	Age (Mean ± SD or Range)	Run-in Period (Days) *	Duration (Days)	Intervention (mg Sodium/d)	Sodium Intake Method	Sodium Assessment Method	Inflammatory Biomarkers at Baseline (Mean ± SD)	Salt Sensitivity Assessment
Mickleborough T. (2005) [ ]	randomized, db, cross-over, placebo controlled	Treated mild asthma	24	62.5	24 ± 1.8	no	14	LSD: 1500 HSD: 5500	diet + sodium capsules or placebo	24 hU	IL-1β: n/a ** IL-8: n/a **	no
Parrinello G. (2009) [ ]	randomized, db, 2 parallel arms	HF patients	173	60.7	72.5 ± 7	no	180	LSD: 1800 HSD: 2800	diet	dietary methods	TNF-a: 19.1 ± 8.6 (LSD) 17.8 ± 9 (HSD) IL-6: 20.8 ± 6.9 (LSD) 21.3 ± 12.5 (HSD) IL-10: 68.7 ± 5.6 (LSD) 62.8 ± 5.4 (HSD)	no
Forrester G. (2010) [ ]	randomized, db, placebo controlled, 2 parallel arms	Asthma & measurable bronchial reactivity to metha-choline	171	37.5	44.2 ± 12.2	7	42	LSD: 1840 HSD: 3680	diet + sodium capsules or placebo	24 hU	hs-CRP: n/a	no
Mallamaci F. (2013) [ ]	randomized, sb, cross-over, placebo controlled	mild-to-moderate HTN, CVD free, no anti-HTN drugs	32	72	48 ± 9	no	14	LSD: 460 HSD: 4600	diet + sodium capsules or placebo	24 hU	hs-CRP: n/a TNF-a: n/a IL-6: n/a hs-PCT: n/a	yes
Dickinson K. (2014) [ ]	randomized, cross-over	NT, BMI: 18–27 Kg/m^2	16	43.75	18–70	no	0.1	LSD: 115 HSD: 1495	diet	N/A	CRP: n/a	no
Campbell K. (2014) [ ]	randomized, db, cross-over, placebo controlled	(P)HT, Stage III & IV CKD	20	75	68.5 ± 11	7	14	LSD: 1380 to 1840 HSD: 1380 to 1840 plus 2760	diet + sodium capsules or placebo	24 hU	CRP: 3.6 ± 3.4 TNF-a: n/a IL-6: n/a Interferon-γ: n/a	no
Telini L. (2014) [ ]	randomized controlled study, 2 parallel arms	CKD—hemodialysis for at least 90 days	39	38.5	57.9 ± 12.8	no	112	LSD: habitual diet minus 2000 HSD: habitual diet	diet	dietary methods	CRP: 11.3± 3.9 (LSD) 11.8± 4.8 (HSD) TNF-a: 694.7 ± 101 (LSD) 651± 96.5 (HSD) IL-6: 5.4± 0.7 (LSD) 5.7± 0.6 (HSD)	no

Table 1. *Cont.*

Author (Year)	Study Design	Population Description	N	Sex (% Male)	Age (Mean ± SD or Range)	Run-in Period (Days) *	Duration (Days)	Intervention (mg Sodium/d)	Sodium Intake Method	Sodium Assessment Method	Inflammatory Biomarkers at Baseline (Mean ± SD)	Salt Sensitivity Assessment
Wenstedt E. (2019) [41]	randomized, cross-over	healthy, non-smoking	11	100	28 ± 5	no	14	LSD: <1200 HSD: >4800	diet	24 hU	CRP: n/a TNF-a: n/a ** IL-6: n/a ** IL-8: n/a ** IL-12: n/a **	no

ABBREVATIONS: sb: single blind; db: double—blind; HT: hypertensives; NT: normotensives; PHT: pro-hypertensives; prosp: prospective; 24 hU: 24-h urine collection; HF: heart failure; CKD: chronic kidney disease; CVD: cardiovascular disease; LSD: low sodium diet; HSD: high sodium diet; BMI: body mass index; HS: high sodium; LS: low sodium; N/A: Not available; PCT: procalcitonin; SS: salt sensitivity; * During the run in period all participants received standard daily Na intake diet; n/a **:only diagrams were provided, accurate data not available.

Table 2. Brief qualitative description of results, primary and secondary end points, per study (*n* = 8).

Author (Year)	Duration (Days)	Intervention (mg Sodium/d)	Primary End Point									Secondary End Point BP Change
			CRP	TNF-a	IL-6	IL-8	IL-10	IL-12	IL-1β	Interferon-γ	hs-PCT	
Mickleborough T. (2005) [39]	14	Baseline: n/a LSD: 1500 HSD: 5500	-	-	-	LSD vs. HSD: ↓	-	-	LSD vs. HSD: ↓	-	-	ns
Parrinello G. (2009) [40]	180 **	Baseline: 2600 LSD: 1800 HSD: 2800	-	LSD vs. HSD: ↑	LSD vs. HSD: ↑	-	LSD vs. HSD: ↓	-	-	-	-	ns
Forrester G. (2010) [38]	42	Baseline: n/a LSD: 1840 HSD: 3680	LSD vs. HSD: ↓	-	-	-	-	-	-	-	-	n/a
Mallamaci F. (2013) [27]	14	Baseline: n/a LSD: 460 HSD: 4600	LSD vs. HSD: ns	LSD vs. HSD: ↑	LSD vs. HSD: ns	-	-	-	-	-	LSD vs. HSD: ↑	After HSD: ↑24 h & night-time & daytime BP
Dickinson K. (2014) [13]	0.1	Baseline: n/a LSD: 115 HSD: 1495	LSD vs. HSD: ns	-	-	-	-	-	-	-	-	ns
Campbell K. (2014) [42]	14	Baseline: 3200 LSD: 1725 HSD: 3864	LSD vs. HSD: ns	LSD vs. HSD: ns	LSD vs. HSD: ns	-	-	-	-	LSD vs. HSD: ns	-	Peripheral SBP & DBP & central SPB: ↓ in LSD vs. HSD

Table 2. *Cont.*

Author (Year)	Duration (Days)	Intervention (mg Sodium/d)	Primary End Point									Secondary End Point BP Change
			CRP	TNF-a	IL-6	IL-8	IL-10	IL-12	IL-1β	Interferon-γ	hs-PCT	
Telini L. (2014) [ ]	112	Baseline: n/a LSD: 6740 HSD: 9240	* LSD: ↓ * HSD: ns	* LSD: ↓ * HSD: ns	* LSD: ↓ * HSD: ns	-	-	-	-	-	-	ns
Wenstedt E. (2019) [ ]	14	Baseline: 4000 LSD: 736 HSD: 7452	LSD vs. HSD: ns	LSD vs. HSD: ns	LSD vs. HSD: ↓	LSD vs. HSD: ns	-	LSD vs. HSD: ns	-	-	-	SBP: ↑ in HSD

ABBREVIATIONS: BP: blood pressure; SBP: systolic blood pressure; DBP: diastolic blood pressure; MAP: mean arterial pressure; LSD: low sodium diet; HSD: high sodium diet; n/a: not available; ns: non-significant; vs: versus; ns: no statistically significant change; HS: high sodium; LS: low sodium; N/A: not available; PCT: procalcitonin; ↑ or ↓: statistically significant difference (higher or lower) between LSD & HSD. * Indicates statistically significant differences between LSD or HSD and baseline. ** The study evaluated outcome both at 180 and 365 days; the presented data correspond to the 180 days intervention in order to minimize the duration gap with the rest of the studies which presented maximum duration 112 days. Level of statistical significance $p < 0.05$ for differences presented either between LSD and HSD or between baseline and modified sodium diet.

All 8 eligible studies identified (published from 2005 to 2019) were predefined RCTs, 5 of which had a cross-over study design [13,27,39,41,42] and 3 had a parallel-arm study design [37,38,40] (Table 1). Each study had 2 sodium intervention periods (for cross-over design studies) or 2 sodium intervention groups (for parallel-arm design studies); no study with more than 2 sodium periods or groups was identified (Table 1). Half of the studies [27,38,39,42] used sodium capsules versus placebo as an add-on intervention to the diet; only 5 out of the 8 studies used 24-h urine collection as a sodium intake assessment method (Table 1).

Throughout this present text, the different levels/groups of daily sodium consumption in each study are quoted for simplicity as "HSD" and "LSD". However, there was high heterogeneity between the identified studies regarding the level of daily sodium intake. LSD was highly variable, ranging from 115 to 6740 mg/day, and likewise HSD ranged from 1380 to 9240 mg/day (Table 1).

The examined inflammatory biomarkers were CRP in 6 out of 8 studies [13,27,37,38,41,42], TNF-a and IL-6 in 5 out of 8 studies [27,37,40–42], IL-8 in 2 out of 8 studies [39,41] and IL-10 [40], IL-12 [41], IL-1β [39], interferon-γ [42], and procalcitonin (PCT) [27], in 1 out of 8 studies for each biomarker.

All studies included small to moderate size populations (from 11 to 173 participants; male sex from 37% to 100%) with high heterogeneity (Table 1) regarding: (a) the type of populations investigated, (b) their age level (from 18 to 72 years old), and (c) duration of intervention (from 14 to 365 days; as well as one acute effect study) [13]. Finally, only 2 studies had a run-in period [38,42], and baseline sodium levels were described only in 3 studies [40–42] (Table 2); in only 1 out of 8 studies, salt sensitivity assessment was conducted [27]. No differences in the changes of the inflammation biomarkers (CRP, TNF-a, IL-6, PCT) were detected between salt sensitive and salt resistant individuals.

3.2. Systematic Review Results per Inflammatory Biomarker: Qualitative Description per Inflammatory Biomarker

Results on CRP (Tables 1 and 2 and Figure 2): out of the 6 studies [13,27,37,38,41,42] investigating the effect of different levels of sodium intake on CRP, 1 was an acute effect study (evaluating the effect of sodium for 2 h, every 30 min), in which participants were asked to consume a meal that was low or high in sodium [13]. This acute effect study indicated no statistically significant changes in CRP results. Only 2 out of the 5 remaining studies showed statistically significant results [37,38].

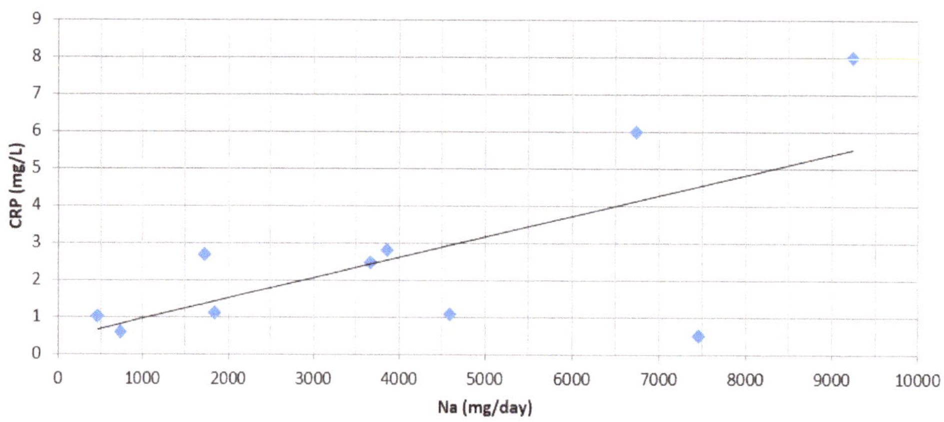

Figure 2. Scatter dot plot of average daily sodium intake (mg/day) and CRP plasma levels (mg/L) per study group. Data from 5 available randomized studies [27,37,38,41,42], after excluding 1 study [13], which evaluated the acute effect (after one meal, every 30 min, for 2 h) of sodium intake. Range of duration of sodium intervention was 14 to 112 days. r = 0.663; *p* = 0.037.

Results of TNF-a (Tables 1 and 2): 3 of the 5 studies [27,37,40–42] examining sodium intake and TNF-a showed statistically significant results [27,37,40]. In 2 of the studies [27,40], TNF-a levels after the LSD (460 mg/day [27] and 1800 mg/day [40]) intervention were significantly higher than that after the HSD (4600 mg/day [27] and 2800 mg/day [40]). In the third study [37], TNF-a decreased significantly only after LSD (6740 mg/day) compared to the baseline. No statistically significant results were found after HSD intervention (9240 mg/day) compared to baseline.

Results on IL-6 (Tables 1 and 2): 3 out of the 5 studies [27,37,40–42] investigating sodium intake and IL-6 showed statistically significant results. In 1 study [41], IL-6 levels after the LSD (736 mg/day) intervention were significantly lower than that after the HSD (7452 mg/day). In 1 study [40], IL-6 levels after the LSD (1800 mg/day) intervention were significantly higher than those after the HSD (2800 mg/day). In the third study [37], IL-6 decreased significantly after LSD (6740 mg/day) compared to the baseline. No statistically significant results were found after HSD intervention (9240 mg/day) compared to baseline.

Results on IL-8 (Tables 1 and 2): 1 of the 2 studies [39,41] investigating IL-8 found statistically significant results [39]; IL-8 levels after the intervention with LSD were significantly lower than that after the HSD intervention period.

Results on IL-10 (Tables 1 and 2): only 1 study [40] examined sodium intake and IL-10, indicating that IL-10 levels after the intervention with LSD were significantly lower than that after the HSD intervention period.

Results on IL-12 (Tables 1 and 2): only 1 study [41] investigated sodium intake and IL-12, but no statistically significant results were found.

Results on IL-1β (Tables 1 and 2): only 1 study [39] investigated sodium intake and L-1β, indicating statistically significant results; IL-1β levels after the intervention with LSD were significantly lower than that after the HSD intervention period.

Results on Interferon-γ (Tables 1 and 2): only 1 study [42] investigated sodium intake and interferon-γ, but no statistically significant results were found.

Results on PCT (Tables 1 and 2): only 1 study [27] investigated sodium intake and PCT levels. PCT levels after the intervention with LSD were significantly higher than that after the HSD intervention period.

3.3. Meta-Analysis Results: Primary Endpoints

Out of the 8 studies, 6 [27,37,38,40–42] were included in the meta-analysis (Figure 1). One study [39] was not included since it investigated inflammatory biomarkers not measured in any of the rest of the studies. A second study [13] was excluded because it was the only acute effect study (2 h duration of intervention period); therefore, the results were not comparable to the rest of the long-duration studies (14 to 180 days). The assessment of the risk of bias is presented in Table S2.

Five studies provided data on the CRP difference between HSD versus LSD and were included in the meta-analysis [27,37,38,41,42] (n = 273, weighted age 47.7 ± 8.4 years, men 47%, hypertension 23.4%). HSD versus LSD resulted in a pooled difference in CRP values (HSD–LSD) of 0.1 (95% confidence intervals [CI] −0.3, 0.4) mg/L (Figure 3). No publication bias was identified (all p = NS, Begg's funnel plot is presented in Figure S1a). Meta-regression analysis did not reveal any significant associations between the difference in CRP and male percentage, mean age, duration of intervention and difference in sodium intake between the examined diet arms across the included studies (all p = NS). In a sensitivity analysis excluding the study of Telini et al. [37] (which included patients on hemodialysis and had parallel arm design), the pooled difference in CRP values was similar at 0.1 (−0.3, 0.4) mg/L. In another sensitivity analysis including only the 3 studies [27,41,42] with crossover design, the pooled estimate was −0.1 (−0.4, 0.3) mg/L. By selecting the 2 studies with average sodium intake at the lowest (<1000 mg/day) and highest (>4500 mg/day) range for each arm [27,41], the pooled difference was −0.1 (−0.4, 0.3) mg/L. Three studies had populations with an average age <50 years [27,38,41] and the pooled difference calculated from these was 0.2 (−0.4, 0.7) mg/L.

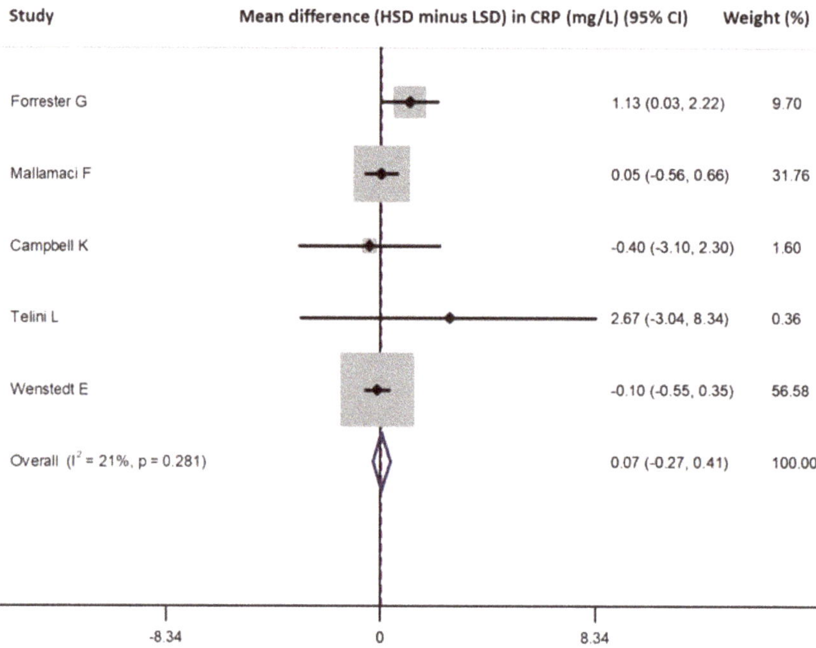

Figure 3. Forest plot of the mean difference in CRP levels for high sodium dose (HSD) versus low sodium dose (LSD), as defined per study [27,37,38,41,42].

Four studies provided data on the TNF-difference between HSD versus LSD and were included in the meta-analysis [27,37,40,42] (*n* = 264, weighted age 67.1 ± 8.7 years, men 60%, hypertension 45.8%). They showed that the HSD versus the LSD resulted in a pooled difference in TNF-a values (HSD-LSD) of −0.7 (−5.0, 3.6) pg/mL (Figure 4). No publication bias was identified (all *p* = NS, Begg's funnel plot is presented in Figure S1b). Meta-regression analysis did not reveal any significant associations between the difference in TNF-a and male percentage, mean age, duration of intervention and difference in sodium intake between the examined diet arms across the included studies (all *p* = NS). In a sensitivity analysis excluding the study of Telini et al. [37] (which included patients on hemodialysis and had a parallel arm design), the pooled difference in TNF-a values showed a pooled difference of −1.3 (−3.5, 0.8) pg/mL. Three studies had populations with an average age >55 years [37,40,42] and the pooled difference calculated from these was 2.8 (−7.8, 13.5) pg/mL.

Four studies provided data on the IL-6 difference between HSD versus LSD and were included in the meta-analysis [27,37,40,42] (*n* = 264, weighted age 67.1 ± 8.7 years, men 60%, hypertension 45.8%). They showed that the HSD versus the LSD resulted in a pooled difference in IL-6 values (HSD-LSD) of −1.1 (−3.3, 1.1) pg/mL (Figure 5). No publication bias was identified (all *p* = NS, Begg's funnel plot is presented in Figure S1c). Meta-regression analysis did not reveal any significant associations between the difference in IL-6 and male percentage, mean age, duration of intervention and difference in sodium intake between the examined diet arms across the included studies (all *p* = NS). In a sensitivity analysis excluding the study of Telini et al. [37], the pooled difference in IL-6 was −2.4 (−4.9, 0.2) pg/mL. Three studies had populations with an average age >55 years [37,40,42] and the pooled difference calculated from these was −1.8 (−5.3, 1.7) pg/mL.

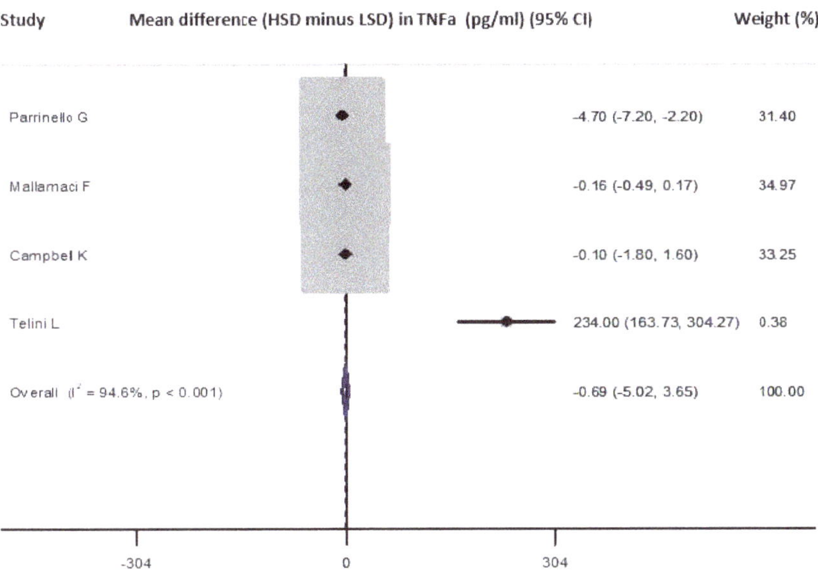

Figure 4. Forest plot of the mean difference in TNF-a levels for high sodium dose (HSD) versus low sodium dose (LSD) as defined per study [27,37,40,42].

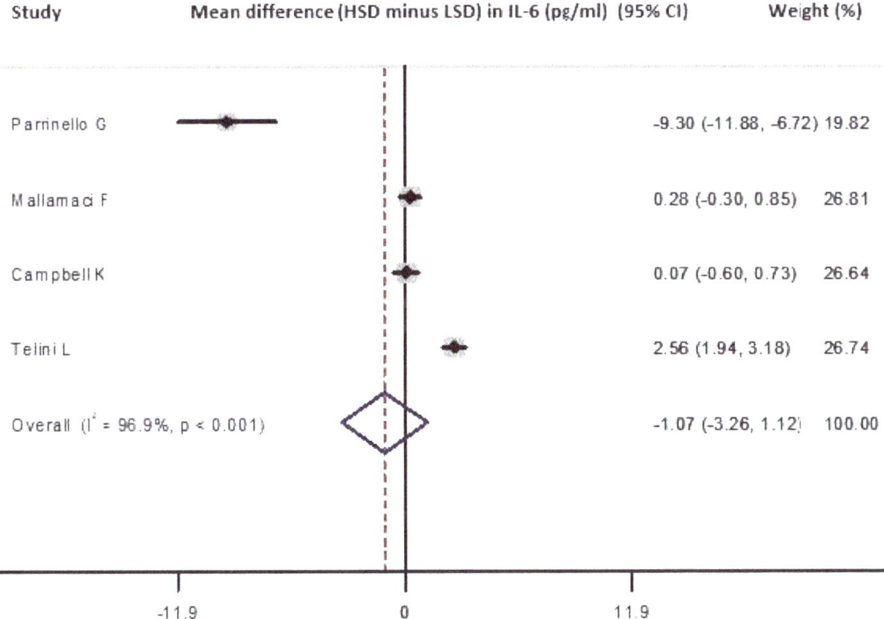

Figure 5. Forest plot of the mean difference in IL-6 levels for high sodium dose (HSD) versus low sodium dose (LSD) as defined per study [27,37,40,42].

3.4. Meta-Analysis Results: Secondary End-Points—Blood Pressure

Five studies [27,37,40–42] (*n* = 275, weighted age 65.5 ± 11.5 years, men 61%, hypertension 44%) reported the effect of HSD vs. LSD on SBP/DBP difference with a pooled estimate of 4.5 (−1.4, 10.4)/2.2 (−0.1, 4.4) mmHg (Figure 6). No publication bias was identified (all *p* = ns, Begg's funnel plots are presented in Figure S2). Meta-regression analysis showed a higher difference in SBP with a shorter duration of the intervention (in days) across the included studies, whereas this was not evident for DBP (*p* = 0.03/ns respectively; Figure S3). No significant associations were observed between the difference in SBP/DBP and male percentage, mean age and difference in sodium intake between the examined diet arms across the included studies (all *p* = ns). The following sensitivity analyses were conducted: the first analysis excluded the study of Telini et al. [37] and showed a pooled SBP/DBP difference of 5.0 (−1.8 to 11.8)/2.4 (0.1 to 4.7) mm Hg. The second analysis included only the 3 studies [27,41,42] with crossover design (same patients) and showed a pooled SBP/DBP difference of 8.1 (4.1,12.2)/2.6 (−0.3, 5.5) mm Hg. The third analysis included studies with an average age of the population >55 years [37,40,42] and showed a pooled SBP/DBP difference of 2.2 (−5.6, 10.1)/2.2 (−1.0, 5.4) mm Hg. The fourth analysis included the 2 studies with average sodium intake at the lowest (<1000 mg/day) and highest (>4500 mg/day) range for each arm [27,41], and showed a pooled SBP/DBP difference of 7.7 (3.4, 12.0)/2.1 (−1.0, 5.3) mm Hg.

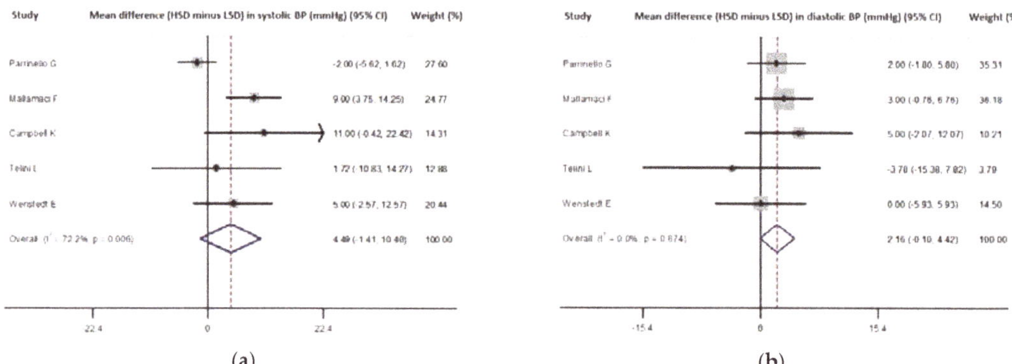

(a) (b)

Figure 6. (**a**) Forest plot of the mean difference in systolic blood pressure (SBP) levels for high sodium dose (HSD) versus low sodium dose (LSD) as defined per study [27,37,40–42]. (**b**) Forest plot of the mean difference in diastolic blood pressure (DBP) levels for high sodium dose (HSD) versus low sodium dose (LSD) as defined per study [27,37,40–42].

4. Discussion

In this present systematic review, we identified eight RCT studies that compared the effect of two different levels of dietary sodium intake on the magnitude of systematic inflammatory response, by assessing overall nine different circulating biomarkers (CRP, TNF-a, IL-1β, IL-6, IL-8, IL-10, IL-12, interferon-γ and hs-PCT). The meta-analysis of studies was feasible only for three inflammatory biomarkers (CRP, TNF-a, IL-6); all three of them showed non-significant differences in the primary endpoint i.e., the difference in circulating inflammatory biomarkers after HSD versus LSD.

The qualitative description of the included studies revealed inconsistent results for all biomarkers, as well as major methodological limitations (e.g., small sample size, poor sodium intake assessment methods) and high heterogeneity regarding major methodological traits (e.g., sodium doses, age, underlying diseases). Of note, there was extremely high heterogeneity and variability regarding the level of daily sodium intake that renders the used terms "LSD" and "HSD" relative and valid mostly for within each study comparison. This major limitation should be taken into account for the interpretation of meta-analysis results.

CRP was the inflammatory biomarker most commonly studied in the RCTs studies (overall six studies [13,27,37,38,41,42]; five included in the meta-analysis), however the overall negative result of the meta-analysis is limited by numerous methodological limitations, as previously discussed. Only two of the studies examined high sensitivity CRP (hs-CRP) [27,38], and only one study investigated more than 40 participants (the higher sample size was 171 [38]), whereas the overall sample size included in the metanalysis was 273. Actually, the most convincing data regarding a positive inflammatory response in HSD were derived from a single study [38], which is the only one that satisfied all of the following necessary methodological characteristics, i.e., evaluated population without low- or high-grade inflammation, had adequate sample size and methodology for sodium intake assessment and, most importantly, compared two "reasonable" levels of daily LSD and HSD (i.e., LSD of 1840 mg/day versus HSD of 3680 mg/day). On the contrary, other studies examined healthy populations either with a very small sample size ($n = 11$) [41], or examined a small number of diseased populations that exhibited at least low (if not high) grade inflammation (e.g., chronic kidney disease) [37,42]. Moreover, two studies [27,41] evaluated the effect of very low doses of daily sodium intake (460 to 736 mg/day) versus high doses of daily sodium intake; given the fact that both very low and high doses of sodium may be proinflammatory, the conclusions from these studies are not easy to interpret [26,27]; therefore, this may have limited the detection of any significant differences in inflammatory response. The only acute phase study that evaluated inflammatory response after a few hours of sodium intake used extremely low doses versus the usual dose of sodium intake [13].

TNF-a and IL-6 were examined in five studies; the same four of which were included in the respective meta-analysis with no overall significant effects for both biomarkers. All the previous discussed limitations are also present in these four studies. One study [27], that evaluated the effect of a very low sodium intake (i.e., LSD: 460 mg sodium/day versus HSD: 4600 mg sodium/day) showed that LSD significantly increased TNF-a levels versus HSD; however, as already discussed, it has been suggested that a very low sodium diet may also be proinflammatory [26,27]. On the other hand, two of the studies [40,42] that evaluated close to the internationally recommended LSD level (i.e., LSD: 1800 mg sodium/day and LSD: 1700 mg sodium/day) versus reasonable HSD, showed conflicting results.

All of the other identified inflammatory biomarkers (IL-1β, IL-8, IL-10, IL-12, interferon-γ and hs-PCT) have been barely evaluated since they were all examined just once [27,39–42], except for IL-8, which was examined in two different studies [39,41]; however, only one found significant results [39]. Overall, all of the above studies but one [41] were conducted in non-healthy subjects (i.e., asthmatic patients, chronic kidney disease, hypertensives, heart failure patients), with a limited number of participants: less than 33 in each included study, with the exception of one study that was conducted in 173 heart failure patients and evaluated IL-10 levels [40]. Although three of the studies indicated statistically significant results [27,39,40] regarding IL-8, IL-1β, IL-10 and hs-PCT, safe conclusions cannot be drawn given the described limitations above.

Given the fact that the association between daily sodium intake and BP levels is very well established [2,43], we included in the present meta-analysis—as a secondary end point—the change of BP levels, between HSD versus LSD, using the extracted data from the very same studies that were used to evaluate inflammatory biomarkers. The fact that the principal BP meta-analysis showed marginally no significant associations between sodium and SBP/DBP does verify that the previously described methodological limitations of the included studies are important and may have indeed limited our ability to identify an association between sodium intake and systematic inflammation biomarkers.

The present systematic review and meta-analysis has some major limitations, mainly due to the methodological heterogeneity of the included studies. First, the available studies have been conducted in populations with different characteristics and types of chronic diseases. The existence of chronic diseases might have also influenced participants' levels of inflammation. Second, there was heterogeneity in the duration of intervention and the intake of sodium among the included studies. Third, the number of the studies included was small, as this topic is understudied. Four, several other undetermined factors (diet, quality of sleep, exercise) might have confounded the final results. However, the abovementioned limitations were, at least partly, compensated by sensitivity and stratified analyses. In addition, the significant effect of the increased sodium intake on BP levels is reassuring for the use of validated methodology in the included studies.

Beside the negative results, the present systematic review helped us to critically revise all the literature and to identify major issues that must be addressed in future RCT efforts to address this hypothesis. Based on previously discussed data, we recommend that such RCT should have three arms (very low, "normal" and high) of daily sodium intake, with sample sizes of more than 100 participants per arm, and should focus on healthy and young participants to avoid the confounding effect of aging and comorbidities. A short study duration of 14 days with a seven days run-in period has been proven to be sufficient, and seems to be more adequate than longer studies, in order to accomplish maximal adherence to the study diets which, however, should be very closely monitored and verified with repeated 24-h urine selections. Most importantly, the above studies were all focused on circulatory biomarkers of inflammation and cannot provide evidence of tissue response or other types of inflammatory (e.g., cellular) response. Therefore, future studies should not only use state-of-the-art methodology to assess circulating biomarkers of inflammation, but should also provide evidence at a cellular level.

5. Conclusions

To conclude, this present study failed to verify the hypothesis that sodium intake induces systemic inflammatory response in humans in a dose response manner. However, due to the aforementioned major limitations, the negative results of the meta-analysis are neither convincing nor provide a definite response to the hypothesis. Future better designed RCTs addressing all of the issues that were discussed above are needed, since a potentially weak association between dietary sodium intake and systemic inflammation, as suggested by Fogarty et al. in the very first publication [1] on the topic, needs a very carefully designed trial to be properly quantified.

Supplementary Materials: The following are available online at https://www.mdpi.com/article/10.3390/nu13082632/s1, Figure S1: (a), (b), (c): Begg's Funnel Plot for CRP, TNF-a and IL-6, Figure S2: Begg's Funnel Plot for SPB and DBP, Figure S3: Meta-regression analysis for SBB and DBP and duration of the intervention, Table S1: PRISMA Checklist, Table S2: Assessment of risk of bias.

Author Contributions: Conceptualization, A.D.P., K.K. and P.P.S.; methodology, E.D.B. and C.T.; software, A.K.; validation, A.A. and A.C.; formal analysis, A.K.; investigation, E.D.B.; data curation, A.D.P. and A.K.; writing—original draft preparation, E.D.B.; writing—review and editing, E.D.B., A.D.P., A.K. and M.K.G.; supervision, A.D.P. and Y.M.; project administration, A.D.P. and P.M. All authors have read and agreed to the published version of the manuscript.

Funding: This research received no external funding.

Conflicts of Interest: The authors declare no conflict of interest.

References

1. Fogarty, A.W.; Lewis, S.A.; McKeever, T.M.; Britton, J.R. Is higher sodium intake associated with elevated systemic inflammation? A population-based study. *Am. J. Clin. Nutr.* **2009**, *89*, 1901–1904. [CrossRef] [PubMed]
2. Mozaffarian, D.; Fahimi, S.; Singh, G.M.; Micha, R.; Khatibzadeh, S.; Engell, R.E.; Lim, S.; Danaei, G.; Ezzati, M.; Powles, J.; et al. Global sodium consumption and death from cardiovascular causes. *N. Engl. J. Med.* **2014**, *371*, 624–634. [CrossRef]
3. Ross, R. Atherosclerosis–An inflammatory disease. *N. Engl. J. Med.* **1999**, *340*, 115–126. [CrossRef]
4. Wilck, N.; Balogh, A.; Marko, L.; Bartolomaeus, H.; Muller, D.N. The role of sodium in modulating immune cell function. *Nat. Rev. Nephrol.* **2019**, *15*, 546–558. [CrossRef] [PubMed]
5. Tikellis, C.; Pickering, R.J.; Tsorotes, D.; Huet, O.; Chin-Dusting, J.; Cooper, M.E.; Thomas, M.C. Activation of the Renin-Angiotensin system mediates the effects of dietary salt intake on atherogenesis in the apolipoprotein E knockout mouse. *Hypertension* **2012**, *60*, 98–105. [CrossRef] [PubMed]

6. Heras-Garvin, A.; Refolo, V.; Reindl, M.; Wenning, G.K.; Stefanova, N. High-salt diet does not boost neuroinflammation and neurodegeneration in a model of alpha-synucleinopathy. *J. Neuroinflamm.* **2020**, *17*, 35. [CrossRef] [PubMed]
7. Kamari, Y.; Shimoni, N.; Koren, F.; Peleg, E.; Sharabi, Y.; Grossman, E. High-salt diet increases plasma adiponectin levels independent of blood pressure in hypertensive rats: The role of the renin-angiotensin-aldosterone system. *J. Hypertens.* **2010**, *28*, 95–101. [CrossRef] [PubMed]
8. Hernandez, A.L.; Kitz, A.; Wu, C.; Lowther, D.E.; Rodriguez, D.M.; Vudattu, N.; Deng, S.; Herold, K.C.; Kuchroo, V.K.; Kleinewietfeld, M.; et al. Sodium chloride inhibits the suppressive function of FOXP3+ regulatory T cells. *J. Clin. Investig.* **2015**, *125*, 4212–4222. [CrossRef] [PubMed]
9. Kleinewietfeld, M.; Manzel, A.; Titze, J.; Kvakan, H.; Yosef, N.; Linker, R.A.; Muller, D.N.; Hafler, D.A. Sodium chloride drives autoimmune disease by the induction of pathogenic TH17 cells. *Nature* **2013**, *496*, 518–522. [CrossRef]
10. Willebrand, R.; Kleinewietfeld, M. The role of salt for immune cell function and disease. *Immunology* **2018**, *154*, 346–353. [CrossRef] [PubMed]
11. Azak, A.; Huddam, B.; Gonen, N.; Yilmaz, S.R.; Kocak, G.; Duranay, M. Salt intake is associated with inflammation in chronic heart failure. *Int. Cardiovasc. Res. J.* **2014**, *8*, 89–93.
12. Scrivo, R.; Massaro, L.; Barbati, C.; Vomero, M.; Ceccarelli, F.; Spinelli, F.R.; Riccieri, V.; Spagnoli, A.; Alessandri, C.; Desideri, G.; et al. The role of dietary sodium intake on the modulation of T helper 17 cells and regulatory T cells in patients with rheumatoid arthritis and systemic lupus erythematosus. *PLoS ONE* **2017**, *12*, e0184449. [CrossRef]
13. Dickinson, K.M.; Clifton, P.M.; Burrell, L.M.; Barrett, P.H.; Keogh, J.B. Postprandial effects of a high salt meal on serum sodium, arterial stiffness, markers of nitric oxide production and markers of endothelial function. *Atherosclerosis* **2014**, *232*, 211–216. [CrossRef]
14. Yi, B.; Titze, J.; Rykova, M.; Feuerecker, M.; Vassilieva, G.; Nichiporuk, I.; Schelling, G.; Morukov, B.; Choukèr, A. Effects of dietary salt levels on monocytic cells and immune responses in healthy human subjects: A longitudinal study. *Transl. Res. J. Lab. Clin. Med.* **2015**, *166*, 103–110. [CrossRef] [PubMed]
15. Wang, K.; Chu, C.; Hu, J.; Wang, Y.; Zheng, W.; Lv, Y.; Yan, Y.; Ma, Q.; Mu, J. Effect of Salt Intake on the Serum Cardiotrophin-1 Levels in Chinese Adults. *Ann. Nutr. Metab.* **2018**, *73*, 302–309. [CrossRef] [PubMed]
16. Arutyunov, G.P.; Dragunov, D.O.; Sokolova, A.V.; Mitrokhin, V.M.; Kamkin, A.G.; Latyshev, T.V. Correlations of IL-18 and IL-6 with sodium consumption in patients with arterial hypertension and diabetes mellitus. *Kardiologiia* **2017**, *57* (Suppl. S1), 355–359. [PubMed]
17. Luo, T.; Ji, W.J.; Yuan, F.; Guo, Z.Z.; Li, Y.X.; Dong, Y.; Ma, Y.; Zhou, X.; Li, Y. Th17/Treg Imbalance Induced by Dietary Salt Variation Indicates Inflammation of Target Organs in Humans. *Sci. Rep.* **2016**, *6*, 26767. [CrossRef] [PubMed]
18. Alderman, M.H.; Cohen, H.; Madhavan, S. Dietary sodium intake and mortality: The National Health and Nutrition Examination Survey (NHANES I). *Lancet* **1998**, *351*, 781–785. [CrossRef]
19. O'Donnell, M.; Mente, A.; Rangarajan, S.; McQueen, M.J.; Wang, X.; Liu, L.; Yan, H.; Lee, S.F.; Mony, P.; Devanath, A.; et al. Urinary sodium and potassium excretion, mortality, and cardiovascular events. *N. Engl. J. Med.* **2014**, *371*, 612–623. [CrossRef]
20. Graudal, N.; Jurgens, G.; Baslund, B.; Alderman, M.H. Compared with usual sodium intake, low- and excessive-sodium diets are associated with increased mortality: A meta-analysis. *Am. J. Hypertens.* **2014**, *27*, 1129–1137. [CrossRef]
21. Tsirimiagkou, C.; Basdeki, E.D.; Argyris, A.; Manios, Y.; Yannakoulia, M.; Protogerou, A.D.; Karatzi, K. Current Data on Dietary Sodium, Arterial Structure and Function in Humans: A Systematic Review. *Nutrients* **2019**, *12*, 5. [CrossRef] [PubMed]
22. Ivanovski, O.; Szumilak, D.; Nguyen-Khoa, T.; Dechaux, M.; Massy, Z.A.; Phan, O.; Mothu, N.; Lacour, B.; Drueke, T.B.; Muntzel, M. Dietary salt restriction accelerates atherosclerosis in apolipoprotein E-deficient mice. *Atherosclerosis* **2005**, *180*, 271–276. [CrossRef]
23. Mercier, N.; Labat, C.; Louis, H.; Cattan, V.; Benetos, A.; Safar, M.E.; Lacolley, P. Sodium, arterial stiffness, and cardiovascular mortality in hypertensive rats. *Am. J. Hypertens.* **2007**, *20*, 319–325. [CrossRef]
24. Tsirimiagkou, C.; Karatzi, K.; Argyris, A.; Chalkidou, F.; Tzelefa, V.; Sfikakis, P.P.; Yannakoulia, M.; Protogerou, A.D. Levels of dietary sodium intake: Diverging associations with arterial stiffness and atheromatosis. *Hell. J. Cardiol. HJC* **2021**, in press.
25. Mavropoulos, A. On the Role of Salt in Immunoregulation and Autoimmunity. *Mediterr. J. Rheumatol.* **2020**, in press.
26. Afsar, B.; Kuwabara, M.; Ortiz, A.; Yerlikaya, A.; Siriopol, D.; Covic, A.; Rodriguez-Iturbe, B.; Johnson, R.J.; Kanbay, M. Salt Intake and Immunity. *Hypertension* **2018**, *72*, 19–23. [CrossRef] [PubMed]
27. Mallamaci, F.; Leonardis, D.; Pizzini, P.; Cutrupi, S.; Tripepi, G.; Zoccali, C. Procalcitonin and the inflammatory response to salt in essential hypertension: A randomized cross-over clinical trial. *J. Hypertens.* **2013**, *31*, 1424–1430, discussion 30. [CrossRef]
28. Christ, A.; Lauterbach, M.; Latz, E. Western Diet and the Immune System: An Inflammatory Connection. *Immunity* **2019**, *51*, 794–811. [CrossRef]
29. Kheirandish-Gozal, L.; Gozal, D. Obstructive Sleep Apnea and Inflammation: Proof of Concept Based on Two Illustrative Cytokines. *Int. J. Mol. Sci.* **2019**, *20*, 459. [CrossRef] [PubMed]
30. PRISMA. PRISMA Guidelines. 2015. Available online: http://www.prisma-statement.org/ (accessed on 1 July 2020).
31. Higgins, J.P.; Altman, D.G.; Gotzsche, P.C.; Juni, P.; Moher, D.; Oxman, A.D.; Savović, J.; Schulz, K.F.; Weeks, L.; Sterne, J.A.C.; et al. The Cochrane Collaboration's tool for assessing risk of bias in randomised trials. *BMJ* **2011**, *343*, d5928. [CrossRef] [PubMed]
32. StatsToDo: Combine Means and SDs into One Group Program. 2020. Available online: https://www.statstodo.com/CombineMeansSDs.php (accessed on 1 November 2020).

33. MEDCALC. Comparison of Means Calculator. 2020. Available online: https://www.medcalc.org/calc/comparison_of_means.php (accessed on 1 November 2020).

34. Wan, X.; Wang, W.; Liu, J.; Tong, T. Estimating the sample mean and standard deviation from the sample size, median, range and/or interquartile range. *BMC Med. Res. Methodol.* **2014**, *14*, 135. [CrossRef]

35. Begg, C.B.; Mazumdar, M. Operating characteristics of a rank correlation test for publication bias. *Biometrics* **1994**, *50*, 1088–1101. [CrossRef]

36. Egger, M.; Davey Smith, G.; Schneider, M.; Minder, C. Bias in meta-analysis detected by a simple, graphical test. *BMJ* **1997**, *315*, 629–634. [CrossRef]

37. Rodrigues Telini, L.S.; de Carvalho Beduschi, G.; Caramori, J.C.; Castro, J.H.; Martin, L.C.; Barretti, P. Effect of dietary sodium restriction on body water, blood pressure, and inflammation in hemodialysis patients: A prospective randomized controlled study. *Int. Urol. Nephrol.* **2014**, *46*, 91–97. [CrossRef]

38. Forrester, D.L.; Britton, J.; Lewis, S.A.; Pogson, Z.; Antoniak, M.; Pacey, S.J.; Purcell, G.; Fogarty, A.W. Impact of adopting low sodium diet on biomarkers of inflammation and coagulation: A randomised controlled trial. *J. Nephrol.* **2010**, *23*, 49–54.

39. Mickleborough, T.D.; Lindley, M.R.; Ray, S. Dietary salt, airway inflammation, and diffusion capacity in exercise-induced asthma. *Med. Sci. Sports Exerc.* **2005**, *37*, 904–914.

40. Parrinello, G.; di Pasquale, P.; Licata, G.; Torres, D.; Giammanco, M.; Fasullo, S.; Mezzero, M.; Paterna, S. Long-term effects of dietary sodium intake on cytokines and neurohormonal activation in patients with recently compensated congestive heart failure. *J. Card. Fail.* **2009**, *15*, 864–873. [CrossRef]

41. Wenstedt, E.F.E.; Verberk, S.G.S.; Kroon, J.; Neele, A.E.; Baardman, J.; Claessen, N.; Pasaoglu, Ö.T.; Rademaker, E.; Schrooten, E.M.; Wouda, R.D.; et al. Salt increases monocyte CCR2 expression and inflammatory responses in humans. *JCI Insight* **2019**, *4*, e130508. [CrossRef]

42. Campbell, K.L.; Johnson, D.W.; Bauer, J.D.; Hawley, C.M.; Isbel, N.M.; Stowasser, M.; Whitehead, J.P.; Dimeski, G.; McMahon, E. A randomized trial of sodium-restriction on kidney function, fluid volume and adipokines in CKD patients. *BMC Nephrol.* **2014**, *15*, 57. [CrossRef] [PubMed]

43. Huang, L.; Trieu, K.; Yoshimura, S.; Neal, B.; Woodward, M.; Campbell, N.R.C.; Li, Q.; Lackland, D.T.; Leung, A.A.; Anderson, C.A.M. Effect of dose and duration of reduction in dietary sodium on blood pressure levels: Systematic review and meta-analysis of randomised trials. *BMJ* **2020**, *368*, m315. [CrossRef]

Systematic Review

Effect of *Crocus sativus* (Saffron) Intake on Top of Standard Treatment, on Disease Outcomes and Comorbidities in Patients with Rheumatic Diseases: Synthesis without Meta-Analysis (SWiM) and Level of Adherence to the CONSORT Statement for Randomized Controlled Trials Delivering Herbal Medicine Interventions

Sotirios G. Tsiogkas [1,2,†], Maria G. Grammatikopoulou [1,3,†], Konstantinos Gkiouras [1,2,†], Efterpi Zafiriou [4], Iordanis Papadopoulos [3], Christos Liaskos [1], Efthimios Dardiotis [5], Lazaros I. Sakkas [1] and Dimitrios P. Bogdanos [1,*]

[1] Department of Rheumatology and Clinical Immunology, Faculty of Medicine, School of Health Sciences, University of Thessaly, Biopolis, GR-41334 Larissa, Greece; stsiogkas@gmail.com (S.G.T.); mariagram@auth.gr (M.G.G.); kostasgkiouras@hotmail.com (K.G.); liaskosch@uth.gr (C.L.); lsakkas@med.uth.gr (L.I.S.)

[2] Faculty of Medicine, School of Health Sciences, Aristotle University of Thessaloniki, GR-54124 Thessaloniki, Greece

[3] Department of Nutritional Sciences & Dietetics, Faculty of Health Sciences, Alexander Campus, International Hellenic University, GR-57400 Thessaloniki, Greece; driordanis@yahoo.gr

[4] Department of Dermatology, Faculty of Medicine, School of Health Sciences, University of Thessaly, Biopolis, GR-41334 Larissa, Greece; zafevi@med.uth.gr

[5] Department of Neurology, Faculty of Medicine, School of Health Sciences, University of Thessaly, Biopolis, GR-41334 Larissa, Greece; edar@med.uth.gr

[*] Correspondence: bogdanos@med.uth.gr

[†] These authors contributed equally to this work.

Citation: Tsiogkas, S.G.; Grammatikopoulou, M.G.; Gkiouras, K.; Zafiriou, E.; Papadopoulos, I.; Liaskos, C.; Dardiotis, E.; Sakkas, L.I.; Bogdanos, D.P. Effect of *Crocus sativus* (Saffron) Intake on Top of Standard Treatment, on Disease Outcomes and Comorbidities in Patients with Rheumatic Diseases: Synthesis without Meta-Analysis (SWiM) and Level of Adherence to the CONSORT Statement for Randomized Controlled Trials Delivering Herbal Medicine Interventions. *Nutrients* 2021, 13, 4274. https://doi.org/10.3390/nu13124274

Academic Editor: Panagiota Mitrou

Received: 30 October 2021
Accepted: 25 November 2021
Published: 27 November 2021

Publisher's Note: MDPI stays neutral with regard to jurisdictional claims in published maps and institutional affiliations.

Abstract: Rheumatic diseases (RDs) are often complicated by chronic symptoms and frequent side-effects associated with their treatment. Saffron, a spice derived from the *Crocus sativus L.* flower, is a popular complementary and alternative medicine among patients with RDs. The present systematic review aimed to summarize the available evidence regarding the efficacy of supplementation with saffron on disease outcomes and comorbidities in patients with RD diagnoses. PubMed, CENTRAL, clinicaltrials.gov and the grey literature were searched until October 2021, and relevant randomized controlled trials (RCTs) were screened for eligibility using Rayyan. Risk of bias was assessed using the Cochrane's Risk of Bias-2.0 (RoB) tool. A synthesis without meta-analysis (SWiM) was performed by vote counting and an effect direction plot was created. Out of 125 reports, seven fulfilled the eligibility criteria belonging to five RCTs and were included in the SWiM. The RCTs involved patients with rheumatoid arthritis, osteoarthritis and fibromyalgia, and evaluated outcomes related to pain, disease activity, depression, immune response, inflammation, oxidative stress, health, fatigue and functional ability. The majority of trials demonstrated some concerns regarding overall bias. Moreover, the majority of trialists failed to adhere to the formula elaborations suggested by the CONSORT statement for RCTs incorporating herbal medicine interventions. Standardization of herbal medicine confirms its identity, purity and quality; however, the majority of trials failed to adhere to these guidelines. Due to the great heterogeneity and the lack of important information regarding the standardization and content of herbal interventions, it appears that the evidence is not enough to secure a direction of effect for any of the examined outcomes.

Keywords: crocin; crocetin; safranal; effect direction plot; complementary and alternative medicine; medicinal plant; CAM; herbal medicine; TNF-α; qualitative synthesis; medicinal plant; dietary supplements

1. Introduction

Rheumatic and musculoskeletal diseases have the highest population impact across all adverse health outcomes, including greater disability-adjusted life years (DALY) [1,2]. Due to the chronic nature of these conditions and the frequent side-effects associated with their treatment, patients often resort to complementary and alternative medicines (CAMs), in search of "less toxic" therapies [3,4].

Garlic, ginger, curcumin, cinnamon, or saffron are a few of the most popular CAMs used in rheumatic diseases (RDs) [5–7]. Saffron, in particular, is the dried stigma of the flowers of *Crocus sativus L.* (family *Iridaceae*), cultivated mainly in Southern Europe, India and Iran, and is considered as one of the most expensive culinary spices globally [8]. The medicinal properties of saffron and its constituents (safranal, crocin, and crocetin) include anti-inflammatory, antioxidant, analgesic, antihypertensive, hypolipidemic, antitussive, anticonvulsant, antidepressant, anxiolytic, anticancer, and antinociceptive characteristics [9–15]. Nevertheless, although saffron supplementation has been tested in patients with various RDs employing a randomized controlled trial (RCT) design, we have insufficient evidence regarding its efficacy, as no systematic reviews have attempted to synthetize these data in order to aid in the formulation of recommendations.

A common issue in CAM research, however, is the lack of standardization of the administered products, often resulting in an inability to reproduce the findings and understand which active ingredients may in fact propel the observed outcomes. The standardization of herbal medicine confirms its identity, purity and quality, and for this, relevant trials ought to disclose information regarding formula elaborations [16]. This information is required to judge the internal validity, external validity, and reproducibility of the administered interventions [17,18].

The aim of the present systematic review was to evaluate the efficacy of saffron oral nutrient supplementation (ONS) on top of standard treatment, on disease outcomes and comorbidities in patients with RDs and evaluate the quality of these trials.

2. Materials and Methods

2.1. Systematic Review Protocol and PIO

The Preferred Reporting Items for Systematic reviews and Meta-Analyses (PRISMA) [19] and the Synthesis Without Meta-analysis (SWiM) extension [20] were used for the presentation of the present review. The study's protocol was published at the center for open science framework (OSF) (https://bit.ly/3pHeSa7, accessed on 26 November 2021).

The PICO of the study's research question is detailed in Table 1.

Table 1. PICO components of the study's research question.

Population	Patients with any rheumatic disease diagnosis
Intervention	Saffron (tabs, sachets, pills, tea, etc.)
Comparison	Placebo, or any other intervention
Outcomes	Any disease-specific (immediate/intermediate) or comorbidity-related outcome

2.2. Search Strategy and Algorithm

Studies related to the research question were identified through PubMed, the Cochrane Central Register of Controlled Trials (CENTRAL), clinicaltrials.gov and grey literature searches from inception until October 2021 by three independent reviewers (S.G.T., M.G.G. and K.G.). Any disagreement between reviewers was resolved by a senior researcher (D.P.B.). The search syntax used in each database is presented in Figure 1.

Rayyan [21], a web and mobile application for conducting systematic reviews, was used to scan and identify all studies fulfilling the study's criteria. All identified references were imported into Rayyan using reference manager software, and duplicate entries were excluded.

Combinations of relevant keywords were used to identify relevant RCTs in the literature. The keywords used included (Crocus sativus), (saffron), (crocin), (crocetin), (safranal), (rheumatoid arthritis), (scleroderma), (fibromyalgia), (Behçet's syndrome), (osteoarthritis), (hyperuricemia), (gout), (ankylosing spondylitis), (psoriatic arthritis), (psoriasis), (psoriatic plaque), (spondylarthritis), (systemic lupus erythematosus), (lupus), (SLE), (Sjogren's syndrome), (systemic sclerosis), and (rheumatic disease*).

Database	Search syntax used	
PubMed	((SSc OR Systemic Sclerosis[MeSH Terms] OR "Systemic Sclerosis" OR Scleroderma OR "Systemic Scleroderma" OR "Scleroderma Diffuse" OR "Diffuse Cutaneous Systemic Sclerosis" OR "Crest Syndrome" OR "Limited Cutaneous Systemic Sclerosis" OR "Scleroderma Limited" OR "Rheumatoid Arthritis" OR "Arthritis, Rheumatoid"[MESH] OR RA OR Osteoarthritis[MESH] OR Osteoarthritis OR Arthritis OR Arthritis[MESH] OR Fibromyalgia OR Fibromyalgia[MeSH Terms] OR Fibromyositis OR Fibrositis OR "Myofascial pain" OR "Chronic Generalized Pain" OR "Chronic Widespread Pain" OR Psoriasis[MeSH Terms] OR Psoriasis OR Psoriases OR Psoriasi OR Psoriasiform OR "Lupus Erythematosus, Systemic"[MESH Terms] OR "Systemic Lupus Erythematosus" OR SLE OR Lupus OR Spondylitis Ankylosing[MeSH Terms] OR "Spondylitis Ankylosing" OR "Spondyloarthritis Ankylopoietica" OR "Ankylosing Spondylarthritis" OR "Ankylosing Spondylarthritides" OR "Spondylarthritides Ankylosing" OR "Ankylosing Spondylitis" OR "Behcet Syndrome"[MeSH Terms] OR "Behcet's Disease" OR "Behcet Syndrome" OR "Adamantiades-Behcet's Syndrome" OR "Touraine's Aphthosis" OR "Sjogren's Syndrome"[MeSH Terms] OR "Sjogren Syndrome" OR "Sjogren's Syndrome" OR "Sjogrens Syndrome" OR "Sicca Syndrome" OR Sjogren's OR Psoriatic Arthritis[MeSH Terms] OR "Psoriatic Arthritis" OR "Psoriasis, Arthritic" OR "Arthritic Psoriasis" OR "Psoriasis Arthropathica" OR "Psoriatic Arthropathy" OR "Arthropathies, Psoriatic" OR "Arthropathy, Psoriatic" OR "Psoriatic Arthropathies" OR "Rheumatic Disease*" OR Pain OR Gout OR Hyperuricemia) AND (Crocus Sativus OR Croci Stigma OR Iridaceae OR Zafran OR Crocus Sativus Linn OR Safranal OR Saffron OR Crocus OR Crocin OR Crocetin)) AND ((randomized controlled trial[pt] OR controlled clinical trial[pt] OR randomized[tiab] OR placebo[tiab] OR drug therapy[sh] OR randomly[tiab] OR trial[tiab] OR groups[tiab] NOT (animals [mh] NOT humans [mh])))	
Cochrane Library	**ID** **Search**	**Hits**
	#1 Systemic Sclerosis or SSc Scleroderma or "Systemic Scleroderma" or "Scleroderma Diffuse" or "Diffuse Cutaneous Systemic Sclerosis" or "Crest Syndrome" or "Limited Cutaneous Systemic Sclerosis" or "Scleroderma Limited"	1843
	#2 MeSH descriptor: [Scleroderma, Systemic] explode all trees	599
	#3 "Rheumatoid Arthritis" or RA	40,907
	#4 MeSH descriptor: [Arthritis, Rheumatoid] explode all trees	6338
	#5 Osteoarthritis or Arthritis	44,080
	#6 MeSH descriptor: [Osteoarthritis] explode all trees	8119
	#7 MeSH descriptor: [Arthritis] explode all trees	16,661
	#8 MeSH descriptor: [Fibromyalgia] explode all trees	1500
	#9 Fibromyalgia	3411
	#10 Fibromyositis or Fibrositis or "Myofascial Pain" or "Chronic Generalized Pain" or "Chronic Widespread Pain"	2078
	#11 MeSH descriptor: [Psoriasis] explode all trees	3547
	#12 Psoriasis or Psoriases or Psoriasi or Psoriasiform	8701
	#13 MeSH descriptor: [Lupus Erythematosus, Systemic] explode all trees	1103
	#14 "Lupus Erythematosus, Systemic" or "Systemic Lupus Erythematosus" or SLE or Lupus	4104
	#15 MeSH descriptor: [Spondylitis, Ankylosing] explode all trees	719
	#16 "Spondylitis Ankylosing" or "Spondyloarthritis Ankylopoietica" or "Ankylosing Spondylarthritis" or "Ankylosing Spondylarthritides" or "Spondylarthritides Ankylosing" or "Ankylosing Spondylitis"	2402
	#17 MeSH descriptor: [Behcet Syndrome] explode all trees	134
	#18 "Behcet Syndrome" or "Behcet's Disease" or "Behcet Syndrome" or "Adamantiades-Behcet's Syndrome" or "Touraine's Aphthosis"	325
	#19 MeSH descriptor: [Sjogren's Syndrome] explode all trees	312
	#20 "Sjogren's Syndrome" or "Sjogren Syndrome" or "Sjogrens Syndrome" or "Sicca Syndrome" or Sjogren's	1220
	#21 MeSH descriptor: [Arthritis, Psoriatic] explode all trees	503
	#22 "Psoriatic Arthritis" or "Psoriasis, Arthritic" or "Arthritic Psoriasis" or "Psoriasis Arthropathica" or "Psoriatic Arthropathy" or "Arthropathies, Psoriatic" or "Arthropathy, Psoriatic" or "Psoriatic Arthropathies"	2504
	#23 "Rheumatic Disease*" or Pain or Gout or Hyperuricemia	208,232
	#24 #1 or #2 or #3 or #4 or #5 or #6 or #7 or #8 or #9 or #10 or #11 or #12 or #13 or #14 or #15 or #16 or #17 or #18 or #19 or #20 or #21 or #22 or #23	266,353
	#25 "Crocus Sativus" or "Croci Stigma" or Iridaceae or Zafran or "Crocus sativus Linn" or Safranal or Saffron or Crocus or Crocin or Crocetin	514
	#26 #24 and #25	86
Scopus	Condition: Rheumatic Disease OR Scleroderma OR osteoarthritis OR Rheumatoid Arthritis OR Systemic Lupus Erythematosus OR Ankylosing Spondylarthritis OR Behcet's disease OR psoriasis OR sjogren's syndrome OR Psoriatic Arthritis Intervention: Crocus sativus OR Croci Stigma OR iridaceae OR Zafran Crocus sativus Linn OR Safranal OR saffron OR crocus OR crocin OR crocetin OR crocus sativus linn OR safranal OR saffron	

Figure 1. Search syntax used in the databases.

Although not belonging to the RDs, osteoarthritis (OA) was also included in the search strategy, since many patients with RA are often misdiagnosed with OA and vice versa [22].

2.3. Inclusion and Exclusion Criteria

Studies were included in the synthesis when they (1) had an RCT design, (2) were parallel or cross-over, (3) used an active *per os* intervention with saffron in any form (tabs, caps, powder, syrup, sachets, tea), (4) were conducted in patients with a RD diagnosis (or osteoarthritis), (5) examined any age group, and (6) used a placebo or any other intervention as a comparator (comparative effectiveness studies).

Exclusion criteria involved (1) all other study designs (non-interventional) including those lacking a comparator arm, (2) studies not including patients with RDs, (3) or using interventions lacking saffron, (4) interventions with curcumin, and (5) published protocols without published results, as well as (6) studies on animals or preclinical studies.

Special caution was taken not to include RCTs investigating the effects of curcumin, which is also known as "Indian saffron" [23].

2.4. Outcomes of Interest

Outcomes of interest involved any specific index/score for RDs, including disease activity scores, pain, inflammation markers, antioxidant and oxidative stress status, anxiety, depression, quality of life (QoL), health assessment, immune response indicators, etc.

2.5. Risk of Bias

Eligible studies were independently assessed for bias using the Cochrane's revised Risk of Bias (RoB) tool 2.0 [24] by two authors (K.G. and M.G.G.). Judgments were made if there was a low risk, some concerns or high risk of bias in terms of the randomization process, deviations from intended interventions, missing outcome data, measurement of the outcomes, selection of the reported results and the final assessment regarding the overall bias.

2.6. Data Extraction

Two independent researchers (M.G.G. and K.G.) extracted data in Excel spreadsheets. Information regarding the sample (size, RD diagnosis, age, % female), recruitment, country of origin, funding, design and methodology (randomization particularities, masking, etc.), intervention (standardization particularities and dosage) and comparator arms, outcomes of interest, drop-outs, adverse events, presented analysis, and general results was extracted for all studies.

2.7. Data Synthesis

At least three RCTs investigating the same outcome for each RD were required for an effective data synthesis. Since a meta-analysis was not feasible, vote counting was applied, based on the direction of effect (mean differences) for each outcome [25] in order to accompany the narrative synthesis [26].

The methodological characteristics of each study (RD diagnosis, overall risk of bias, etc.) were used to assess heterogeneity, according to the Cochrane Handbook [26] and the SWiM guidelines [20].

3. Results

3.1. Search Results

Out of 139 studies screened in total, five RCTs and seven publications (two studies with duplicate publications) [27–33] fulfilled the protocol's criteria and were included in the systematic review. Figure 2 details the PRISMA 2020 flow diagram of the study selection process [19].

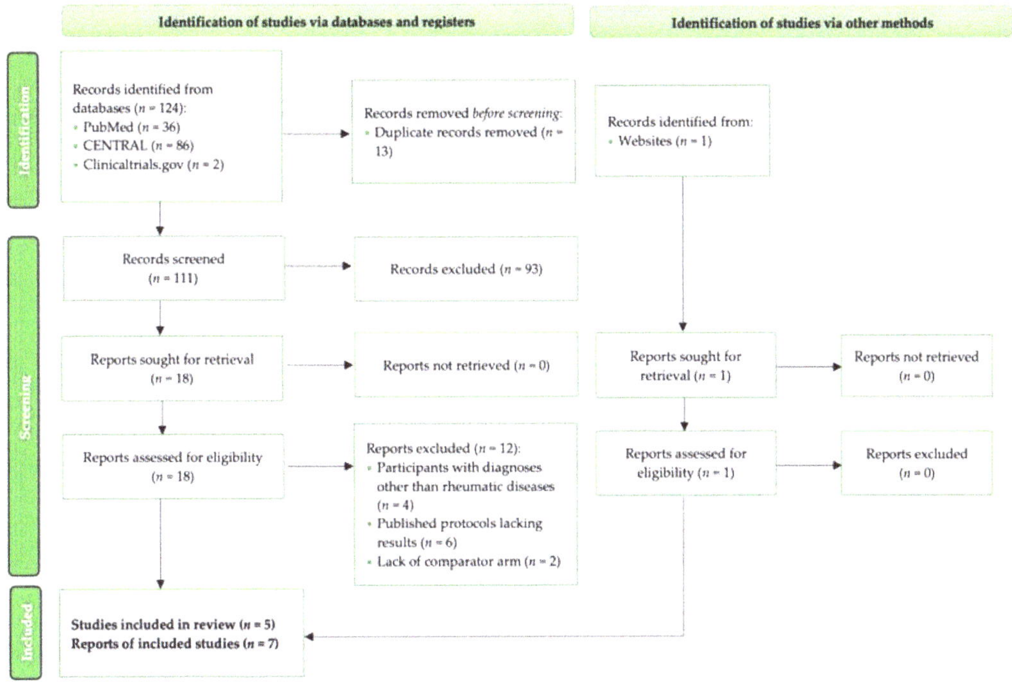

Figure 2. PRISMA [19] flow diagram of the studies' selection process.

3.2. Characteristics of RCTs with Saffron Interventions in Patients with Rheumatic Diseases

3.2.1. RD Diagnoses

Details of the RCTs fulfilling the study's criteria, evaluating saffron interventions in patients with rheumatoid arthritis (RA), osteoarthritis (OA), or fibromyalgia (FM), the respective trials are detailed in Figure 3. The effect of saffron supplementation was evaluated in two trials using participants with RA [27,31,33], an additional two RCTs with patients with a knee OA diagnosis [28,29,32] and finally, on one RCT performed in patients with FM [30] (Figure 3). In RA, two different diagnostic criteria were employed, including the American College of Rheumatology/European league against Rheumatism (ACR/EULAR) 2010 [34] and the revised ACR 2017 [35]. For patients with OA and FB, the ACR [36] and ACR [37] criteria were employed, respectively.

No relevant completed trials were retrieved for spondylarthritis, ankylosing spondylitis, Sjogren's syndrome, hyperuricemia, systemic lupus erythematosus (SLE), scleroderma, psoriatic arthritis, psoriasis, or Behçet's syndrome (BS).

3.2.2. Trial Design and Origin

All trials were conducted in Iran and were published between the years 2018 and 2021. The RCTs employed a parallel intervention design. No cross-over trials were retrieved, fulfilling the PICO question of the study. All included RCTs were double blinded [27–30,32].

3.2.3. Intervention and Comparator Particularities

The administered doses of *Crocus sativus L.* ranged between 15 mg/day [28,30] and 100 mg daily [27,29,31–33]. All studies used pills, tablets or capsules for the delivery of saffron supplements. Sahebari and associates [27] used pure saffron powder made of saffron flowers (Saharkhiz Saffron Factory, Mashhad, Iran), Hamidi et al. [31,33] administered saffron Sargol (Saharkhiz Saffron Factory, Mashhad, Iran), and Poursamimi and associates [28] applied interventions with Krocina™ (Samisaz Pharmaceutical Company, Mashhad, Iran). In the Shakiba trial [30], dried and milled *Crocus sativus L.* stigma (IMPI-RAN, Tehran, Iran) was used for the preparation of tablets, and Firoozabadi et al. [29,32] administered saffron pills (not-other defined). Extraction information and methods were only provided in two trials [27,30]. Additional compounds in the administered tabs were reported in two trials [27,30], but the exact composition of the final products was not declared in any RCT. Standardization of the final product was only reported by Shakiba and associates [30], based on the crocin and safranal content of the capsules via spectrometry. Although Poursamimi et al. [28] administered ready-to-buy supplements, no information is currently provided on the manufacturer's website [38]. Intervention duration spanned between 8 weeks [30] and 4 months [28].

Four RCTs used placebos as comparators [27–29,31–33] and one used duloxetine [30], but the aim in the latter was to assess the comparative effectiveness of saffron versus duloxetine for depression in patients with FM.

3.2.4. Sample Size

The sample size was rather small in all RCTs, spanning from 40 [28] to a maximum of 82 [27] patients per trial, prior to randomization. The included RCTs involved a total of 148 patients with RA, 106 patients with OA, and 54 patients with FM. In the pooled sample, 104 patients received a saffron intervention and 104 were allocated to the control arms. One trial which was only published in abstract format [29,32] did not report the number of patients allocated in the intervention/comparator arms.

3.3. Outcomes Assessed in the Included Interventions
3.3.1. Sensation of Pain

One important outcome of interest among the included trials involved pain, which was evaluated using the visual analogue scale (VAS) [39], the pain scale (not defined), the brief pain inventory (BPI) [40], or the Western Ontario and McMaster universities (WOMAC) OA index pain subscale [41].

3.3.2. Immune Response

Immune response post-saffron supplementation was evaluated in one RCT [28] assessing CD8+ and CD4+ T helper (Th) cells, Th17 cell percentage (%), T-regulatory (Treg) cells percentage (%), and the geometric mean fluorescence intensity (gMFI) of forkhead box protein P3 (FOXP3) of Treg cells, as well as the Treg/Th17 ratio.

3.3.3. Inflammation

Assessed inflammation markers included the erythrocyte sedimentation rate (ESR), C-reactive protein (CRP) and hs-CRP (high sensitivity CRP), tumor necrosis factor-α (TNF-α), interferon-γ (IFN-γ), interleukine-17 (IL-17), and interleukine-1β (IL-1β) levels.

3.3.4. Health Assessment, Depression and Fatigue

Health was self-assessed by the patients themselves using the health assessment questionnaire-disability index (HAQ-DI) [42], or by their physicians using the physician global assessment (PGA) [43]. Fatigue was evaluated using the global fatigue index (GFI) [44] in one trial.

Depression was assessed using Beck's depression inventory (BDI), the Hamilton depression rating scale [45], or the Hospital Anxiety and Depression Scale (HADS) [46].

First author	Sahebari [27]	Hamidi [31,33]	Poursamimi [28]	Firoozabadi [29,32]	Shakiba [30]
Publication	Full-text	Full-text and abstract	Full-text	2 x Abstract	Full-text
Journal	*Avicenna J Phytomed*, 2021	*Phytotherapy Res*, 2020; *Curr Develop Nutr*, 2021	*Iran J Allergy Asthma Immunol*, 2020	*Clin Exp Rheumatol*, 2018; *Osteoporos Int*, 2020	*Avicenna J Phytomed*, 2018
Origin	Iran	Iran	Iran	Iran	Iran
Registry	IRCT201/1407 1218453N1	IRCT201707 139472N14	IRCT201502 1910507N2	IRCT201160N 1029777N1	IRCT201604 2615563N91
Design	Parallel	Parallel	Parallel	Parallel	Parallel
Funding	NR	Iran University of Medical Sciences	Mashhad University of Medical Science	NR	Tehran University of Medical Sciences
Masking	Double blind	Double blind	Double blind	Double blind	Double blind
Randomization	Using a random number table	Using permuted block randomization with 2 size blocks and a random number table	Using a PC-generated code	NR	On a 1:1 ratio through PC-random number generation
Condition	RA	RA	Knee OA	Knee OA	FM
Recruitment	Rheumatic Diseases Research Center, Mashhad University	Shariati Hospital and Rasoul e-Akram Hospital	Imamreza clinic	NR	Rheumatology clinic, Imam Khomeini Hospital
Study duration	NR	2017-8	2016-8	NR	2016-7
Participants	N=82 patients with newly diagnosed RA	N=66 patients with active RA	N=40 patients with OA (WOMAC 2-3)	N=66 patients with symptomatic unilateral knee OA	N=54 patients with FM with a pain score >40 (VAS/100)
Participant age (years)	49.3 ± 12.4[M]	10.7 ± 5.7[M] (intervention); 9.6 ± 5.1[M] (placebo)	40-75[R]	57.3 ± 6.0[M]	18-60[R]
Men/Women (n)	21/61	0/66	26/58	7/59	16/34
Diagnostic criteria	ACR/EULAR 2010 [34]	ACR 2017 [35]	ACR [36]	NR	ACR [37]
Anti-CCP + (active/control)	89.2%/81.6%	NR	N/A	N/A	N/A
Intervention	ONS with saffron (100 mg pills/day) (n=41)	ONS with saffron (1 x 100 mg tabs/day) (n=33)	ONS with Krocina™ (15 mg tabs/day) (n=20)	ONS with saffron (1 x 100 mg tab/day) (n=NR)	ONS with saffron (1 x 15 mg caps/day), with an additional 15 mg each ascending week (n=27)
Comparator	Placebo (n=41)	Placebo (HPMC) (n=33)	Placebo tabs (n=20)	Placebo (n=NR)	1 x duloxetine caps (30 mg)/day, with an additional 30 mg each ascending week (n=27)
Intervention duration	3 months	12 weeks	4 months	12 weeks	8 weeks
Standard therapy	Prednisolone, methotrexate, folic acid, vitamin D, Calcium, alendronate	Methotrexate, hydroxychloroquine, sulfasalazine, prednisolone	Sodium diclofenac (NSAID)	Apart from the NSAIDs, no other therapy was reported	NR
Treatment adherence	NR	By returning the tabs boxes	NR	NR	By caps counts and reports
Ban of other antioxidants	yes	yes	Long history of AID intake was an exclusion criterion	NR	History of saffron treatment was an exclusion criterion
Main hypothesis	Δ in proxy markers of disease severity	Δ in clinical outcomes and metabolic profiles	Δ in clinical and paraclinical parameters	Δ in pain relief and inflammatory markers, saffron efficacy-tolerability	Comparative effectiveness of saffron versus duloxetine for depression
Outcomes of interest	DAS-28-ESR, HAQ-DI, VAS, PS, physical function	DAS-28-ESR, morning stiffness, hs-CRP, ESR, TNF-α, IFN-γ, TAC, pain (VAS), SJC, TJC, PGA, MDA	ESR, VAS, CRP, gMFI-FOXP3 Treg-cells, IL-17, CD8+ and CD4+ T cells and Th 17 cells	WOMAC, WOMAC pain subscale, n of NSAIDs/day, IL-1β, TNF-α	HRSD, VAS, GFI, BPI, FIQ, HADS
Assays	N/A	Westergren method (ESR), immunoturbidometry (hs-CRP), ELISA, chemical colorimetry (TAC, MDA)	Westergren method (ESR), flow cytometry (Th17, CD8+ T cells, and CD3+, CD4+ T-cells)	N/A	N/A
Dietary assessment	NR	At baseline and at 12-weeks, collection of 3 x 24h diet recalls (2 typical days, 1 holiday)	NR	NR	NR
PE assessment	NR	Via the IPAQ	NR	NR	NR
Control for intra-articular injection	The need for intra-articular injection was assessed at the end of treatment	‚	NR	NR	N/A
Adverse events in the saffron receiving arm	Xerostomia (12.2%), constipation (2.4%), palpitation (9.8%), restlessness (4.9%), anxiety (9.8%), nausea (4.9%), reflux (2.4%), abdominal pain (4.9%), headache (4.8%), dizziness (2.4%), vomiting (2.5%), paresthesia (4.9%), AUB (2.4%)	Stomach pain (n=1)	NR	NR	Abdominal pain (n=1), nausea (n=2), diarrhea (n=1)
Discontinued (n)	n=27 for non-compliance, slight side effects (NOD) or loss to follow-up (n=14 active; n=13 placebo)	n=1 non-compliance (active arm), n=1 reluctance to attend final session, n=1 lost to follow-up (placebo arm)	n=1 risk of breast cancer, n=1 personal reasons (active), n=2 personal reasons, n=1 GI stress (placebo)	NR	n=7 withdrew consent (n=4 in saffron and n=3 in duloxetine arms), n=1 was ineligible to continue (duloxetine)
N included in final analysis	n=27 active arm n=28 placebo	n=32 active arm[‡] n=31 placebo[‡]	n=18 active arm n=17 placebo	NR	n=27 saffron n=28 duloxetine
Analysis	ITT	PP	PP	NR	ITT (using LOCF)
Results	No difference between arms regarding to the DAS-28 at the end of the study. The trend of reduction in DAS-28, VAS, poor physical function and PS was significant in each arm, but not different between arms	Saffron ONS decreased the number of TJC and SJC, pain intensity (VAS), DAS-28, and the levels of hs-CRP, TNF-α, IFN-γ and MDA. Moreover, PGA and ESR were both improved in the saffron arm	CRP, gMFI-IL-17 levels, % Th17 cell were reduced in the saffron arm. In parallel, the % of Treg-cells, Treg/Th17 ratio and gMFI-FOXP3 were increased	WOMAC and WOMAC PS were improved in each arm at the end of the study without difference between arms. NSAIDs consumption in the intervention group was reduced after 6 weeks of saffron intake in comparison with the control arm	No difference were detected for any of the scales, neither in terms of score changes from baseline to endpoint between the two treatment arms, nor in terms of time-treatment interactions

Figure 3. Characteristics of the parallel RCTs evaluating interventions with saffron in patients with RA, OA, or FM included in the qualitative synthesis. ACR, American College Of Rheumatology; AID, anti-inflammatory drug; anti-CCP, anti-cyclic citrullinated peptide; AUB, abnormal uterine bleeding; BPI, Brief Pain Inventory [40]; DAS-28, disease activity score -28 [47]; ELISA, enzyme-linked immunosorbent assay; ESR, erythrocyte sedimentation rate; EULAR, European League Against Rheumatism; FIQ, Fibromyalgia Impact Questionnaire [48,49]; FM, fibromyalgia; FOXP3, forkhead box protein P3; GFI, global fatigue index [44]; GI, gastrointestinal; gMFI, geometric mean fluorescence intensity; HADS, Hospital Anxiety and Depression Scale [46]; HAQ-DI, health assessment questionnaire-disability index [42]; HPMC, hydroxy-propyl methyl-cellulose; hs-CRP, high sensitivity C-reactive protein; HRSD, Hamilton Rating Scale for Depression [45]; IFN-γ, interferon-γ; IPAQ, International Physical Activity Questionnaire [50]; LOCF, last-observation carry forward; MDA, malondialdehyde; N/A, not applicable; NR, not reported; NSAIDs, non-steroid anti-inflammatory drugs; ITT, intention-to-treat; OA, osteoarthritis; ONS, oral nutrient supplementation; PE, physical exercise; PGA, Physician Global Assessment [43]; PP, per protocol; PS, Pain score; RA, Rheumatoid arthritis; SJC, swollen joint count; Th, T helper; TJC, tender joint count; TNF-a, tumor necrosis factor α; Treg-cells, regulatory T cells; VAS, visual analogue scale [39]; WOMAC, Western Ontario and McMaster Universities Osteoarthritis Index [41]. * Within the manuscript text, 31 participants were reported to have completed the active arm intervention and 30 controls, in the CONSORT flow chart it appears that 32 women from the intervention and 31 from the control arm were analyzed, but in the tables, the respective number of reported participants in active and comparator arms was 33 and 32. ‡ Reported data refer to the PP analysis; [R] range; [M] mean ± standard deviation.

3.3.5. Antioxidant Status

Antioxidant activity and oxidative stress were assessed according to the malondialdehyde (MDA) levels, and total antioxidant capacity (TAC) via the ferric reducing ability of plasma (FRAP) method.

3.3.6. Disease-Specific Scores

Disease-specific scores were also evaluated, depending on the diagnosis of the participants in each trial. For RA, the disease-specific scores involved the disease activity score-28 (DAS-28) including the ESR assay [47] (DAS-28-ESR) and the swollen and tender joint count (SJC, TJC). For the RCT performed in patients with FM [30], the fibromyalgia impact questionnaire (FIQ) [48,49] was applied. In the case of OA, one trial [29,32] reported using the WOMAC [41].

3.4. Risk of Bias Summary

The summary of risk of bias for the included RCTs is presented in Figure 4. The majority of RCTs (60%) exhibited some concerns for overall risk of bias, with the remaining 40% having a high risk for overall bias. The greatest proportion of trials with unclear bias involved the randomization process and the deviations from intended interventions domains.

Figure 4. Summary of risk of bias [24] assessment for the included RCTs. HADS, Hospital Anxiety and Depression Scale; HDRS, Hamilton Depression Rating Scale; RCT, randomized controlled trial; VAS, visual analogue scale.

3.5. Other Bias

3.5.1. Treatment Adherence

Treatment adherence was assessed only in two trials [30,31,33], with the remaining studies failing to control for this issue. Furthermore, the ban of antioxidant supplements at the beginning of the trials was not reported by any trialist, whereas in Shakiba's trial [30], only a history of treatment with saffron was an exclusion criterion, without controlling for other antioxidants.

3.5.2. Dietary Intake and Exercise Patterns

Diet was only recorded and assessed by Hamidi et al. [31,33], despite the fact that it can alter antioxidant intake. Similarly, physical activity a known mediator of disease activity and stress was only assessed by Hamidi [31,33].

3.6. Adherence to the CONSORT Statement for RCTs with Herbal Medicine

Among the included trials, the majority failed to adhere to the formula elaborations suggested by the Consolidated Standards of Reporting Trials (CONSORT) statement for RCTs including herbal medicine interventions [17] (Figure 5). Thus, it appears that the exact composition and dosage of active saffron ingredients, including crocin, crocetin or safranal, cannot be calculated, with the exception of one trial [30]. Shakiba's RCT [30] adhered to the majority of CONSORT components involving the standardization and procedures required for RCTs with herbal medicine interventions. On the other hand, Sahebari [27] also reported all added constituents, but failed to define the exact dosage per administered unit. Firoozabadi and associates [29,32] demonstrated the least adherence; however, their results were only published in abstract format, and thus limited space was available.

Domains	Components	Sahebari [27]	Hamidi [31,33]	Poursamimi [28]	Firoozabadi [29,32]	Shakiba [30]
Herbal medicinal product name	The Latin binomial name for each herbal constituent.	✓	✓	✗	✓	✓
	The botanical authority and family name for each herbal constituent.	✓	✓	✗	✗	✗
	Common name(s) for each herbal constituent.	✓	✓	✓	✓	✓
	The proprietary product (brand) or the extract name (e.g., EGb-761).	✓	✗	✓	✗	✓
	The name of the product manufacturer.	✓	✓	✓	✗	✓
	If the product is authorized (licensed, registered) in the country where the study was conducted.	✗	✗	✓	✗	✓
Characteristics of the herbal product	The part(s) of plant used to produce the product or extract.	✓	✗	✗	✗	✓
	The type of product used [raw (fresh or dry), extract].	✓	✗	✗	✗	✓
	The type and concentration of extraction solvent used (e.g. 80% H$_2$O or 40% ethanol, etc.) and the herbal drug/extract ratio (drug:extract; e.g., 2:1).	✗	✗	✗	✗	✓
	The method of authentication of raw material (i.e., how performed and by whom) and the lot number.	✗	✗	✗	✗	✗
	State if a voucher specimen was retained and, if so, where it is kept or deposited, and the reference number.	✗	✗	✗	✗	✗
Dosage regimen, quantitative data	All quantified herbal product constituents (native or added), including fillers, binders, and other excipients must be reported per unit of dosage form (e.g., 20% maltodextrin, 3% silicon dioxide/capsule).	✗	✗	✗	✗	✗
	For standardized products, the quantity of active/marker constituents must be provided per unit of dosage.	✗	✗	✗	✗	✓
Qualitative testing	The product's chemical fingerprint and methodology applied (equipment, chemical reference standards) and which laboratory performed it.	✗	✗	✗	✗	✓
	If samples of the product (retention samples) were retained, and if so, where these are deposited.	✗	✗	✗	✗	✗
	Description of any special testing/purity testing (heavy metal or contaminant testing) was undertaken and if unwanted components were removed and how (related methodology).	✗	✗	✗	✗	✗
	Standardization: what to (which component(s) of the product) and how (chemical processes or biological/functional measures of activity).	✗	✗	✗	✗	✓

Figure 5. Summary of adherence to the formula elaborations suggested by the CONSORT statement for RCTs implementing herbal medicine interventions [17]. CONSORT, Consolidated Standards of Reporting Trials; RCT, randomized controlled trial. ✗ not reported; ✓ reported.

3.7. Results

Regarding the sensation of pain, ONS with *Crocus sativus L.* either reduced [27,28,33], or did not appear to have an effect [27,29,30,32] when administered to patients with RA, OA or FM. The use of NSAIDs was reduced in one trial using a sample of patients with OA [29,32]. On the other hand, no change was recorded regarding the sensation of fatigue in FM (one trial) [30].

Markers of inflammation were examined in RA and OA and were either reduced or remained unchanged post-intervention, with trials indicating conflicting results. Indicators of antioxidant activity and oxidative damage remained unchanged (MDA [31,33] and TAC [31,33] in one RCT each), raising concerns regarding the efficacy of saffron.

Immune response was evaluated in one OA RCT [28], which reported an increase in the percentage of Treg-cells, the Treg/Th17 ratio and a decrease in the Th17 cell percentage.

In RA, disease-specific indexes such as the DAS-28-ESR, morning stiffness, TJC and SJC were reduced in one RCT [31,33] and were not affected in another [27]. In OA, results concerning the total WOMAC score were not reported by the trialists [29,32]. In FM, saffron ONS did not affect the FIQ among participants [30].

Depression scores remained unchanged in individual trials in patients with RA [27] and FM [30]. ONS with saffron failed to induce improvement in physical function among patients with RA (one RCT) [27]. Last, self-rated health assessment remained unchanged in RA post-intervention [27], but was improved when assessed by the physicians using the PGA in patients with RA [31,33].

3.8. Adverse Events

In the present systematic review, two [28,29,32] out of five RCTs failed to report adverse events. The most frequently reported issues following saffron supplementation involved xerostomia, abdominal pain, vomiting, anxiety, palpitations, etc.

3.9. Synthesis without Meta-Analysis (SWiM)

Figure 6 details the effect direction plot of the outcomes assessed in the included RCTs. For the majority of outcomes, conflicting results are apparent. Moreover, for most outcomes, less than three RCTs have provided results regarding similar outcomes.

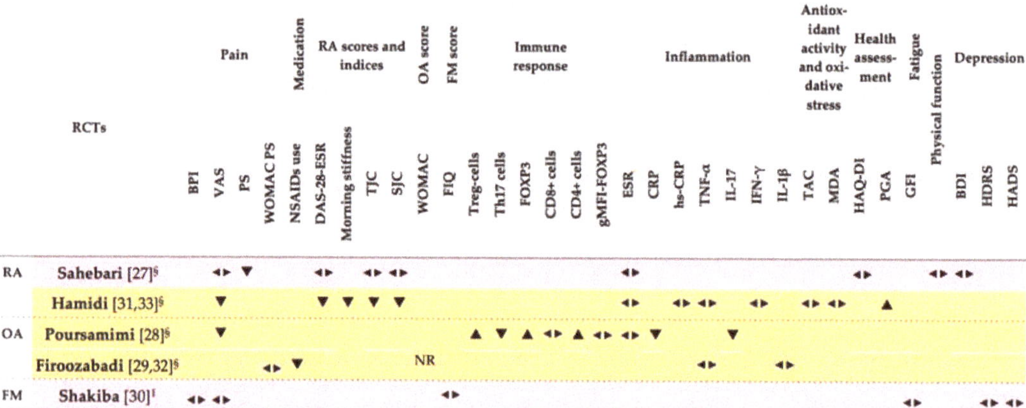

Figure 6. Qualitative synthesis without meta-analysis of the outcomes in each RCT, favoring the saffron arms post-intervention. All RCTs had less than 50 participants in each arm. Row background colors denote study quality according to the RoB assessment. BDI, Beck Depression Inventory [51]; BPI, Brief Pain Inventory [40]; DAS-28, disease activity score 28 [47]; ESR, erythrocyte sedimentation rate; FIQ, Fibromyalgia Impact Questionnaire [48,49]; FM, fibromyalgia; FOXP3, forkhead box protein P3; GFI, global fatigue index [44]; gMFI, geometric mean fluorescence intensity; HADS, Hospital Anxiety and Depression Scale [46]; HAQ-DI, health assessment questionnaire-disability index [42]; HDRS, Hamilton Depression Rating Scale [45]; hs-CRP, high sensitivity C-reactive protein; IFN-γ, interferon-γ; MDA, malondialdehyde; NR, not reported; NSAIDs, non-steroid anti-inflammatory drugs; OA, osteoarthritis; PGA, Physician Global Assessment [43]; RA, rheumatoid arthritis; RCT, randomized controlled trial; RoB, risk of bias [24]; SJC, swollen joint count; TAC, total antioxidant capacity; Th, T helper; TJC, tender joint count; Treg, T regulatory; ‖ intervention duration <3 months; § intervention duration between 3–4 months; ▲ increased; ▼ reduced; ◄► no change.

For the outcome of pain (VAS), four RCTs provided results, with two indicating a reduction in pain and two failing to reveal an effect. Regarding the levels of ESR, and TNF-α, three and two of the included RCTs, respectively, provided results. However, for both ESR and TNF-α, the direction plot was similar in all, indicating a lack of effect following saffron supplementation in patients with RA and OA.

Due to the heterogeneity of the RCTs and the lack of data regarding the standardization of the herbal medicine interventions, a meta-analysis was not deemed as a safe option.

4. Discussion

The present SWiM assessed the effects of supplementary *Crocus sativus L.* intake on disease-related outcomes among patients with a RD diagnosis. It appears that limited RCTs have been performed on this issue, thus demonstrating1 that the evidence is not enough to secure a positive direction of effect for any of the examined outcomes. Moreover, serious pitfalls regarding the reporting of the intervention formulas are apparent, further reducing the quality of the trials.

Consumption of saffron can reduce inflammation through inhibition of the cyclooxygenase enzyme activity [52]. According to a recent meta-analysis [9], saffron is effective in improving the levels of inflammatory markers such as TNF-α, IL-6 and CRP when administered at specific doses ($\leq$30 mg/day) in young adults (<50 years old) lacking a diabetes diagnosis. In the present review, only four trials administered a dose not exceeding 30 mg/day [28,30], with only one [28] evaluating inflammatory markers among participants. Interestingly, CRP and IL-17 were improved in this trial post-intervention. Thus, it is possible that the higher doses administered in the rest of the trials [29,31–33] might have produced a negative or null effect. Nevertheless, another meta-analysis [9] failed to detect any differences regarding CRP, TNF-α, and IL-6 between the saffron and placebo arms. These discrepancies, however, may lay in the underline pathologies of the participants, the duration of interventions, or differences in the standardization of the administered supplements.

Research indicates that saffron can reduce the concentrations of endogenously generated reactive oxygen species, inhibiting oxidative damage, while reducing the production of pro-inflammatory biomarkers [9,53]. According to a recent meta-analysis [54], supplementation can induce improvements in the MDA and TAC levels. However, no improvements were revealed in the present SWiM, due to the small number of studies evaluating these outcomes, most of which indicated a null effect.

Depression and anxiety are common problems in patients with chronic disease and rheumatic disease in particular. Moreover, recent meta-analyses indicate that ONS with *Crocus sativus L.* may improve depressive symptoms and anxiety [12,55]. This effect is persistent even when used as an adjunct to antidepressants, as in the present RCTs. Moreover, specific depression batteries such as the BDI appear to be more sensitive to saffron ONS, whereas the HDRS has been reported to be less flexible [56]. Saffron has been suggested to entail relaxant, inhibitory effects on both histamine (H1) and the muscarinic receptors [57]. By inducing relaxation and reducing anxiety, supplementation with *Crocus sativus* can also improve sleep quality [58]. On the other hand, improved sleep is associated with less fatigue. Overall, previous evidence synthesis indicates that saffron is more efficient compared to placebo and additionally equally effective with synthetic antidepressants [59,60]. These findings, however, were not akin to the present SWiM due to the probable methodological pitfalls of the included trials, heterogeneity and lack of information regarding the standardization of the intervention formulations.

Regarding pain, no meta-analyses have evaluated the effect of saffron ONS, although individual RCTs performed on patients with distinct diagnoses indicate possible improvement in the sensation of pain [61].

According to research, the dried stigmas and tops of the plant styles have the majority of medicinal properties, including immunomodulating responses. Saffron contains a variety of mineral agents, glycosides, anthocyanins, alkaloids, carotenoids and flavonoids including quercetin and kaempferol, which further increase its immunoregulatory properties [62,63]. Studies using animal models have revealed that saffron acts on selective Th2 upregulation, naming it a "nutraceutical" spice [64]. Other preclinical and animal studies showed that saffron can increase the expression levels of FOXP3, a transcriptional factor, in Treg cells, and suppress IL-10 and IFN-γ secretion [65–67]. In the present SWiM, only one trial [28] evaluated immune response post-saffron supplementation, indicating improved immunomodulation. However, further studies are required, assessing similar outcomes.

4.1. Methodological Limitations of the Included Trials

4.1.1. Assessment of Treatment Adherence Rate

According to research, treatment adherence in clinical trials is suboptimal, affecting the economic costs, while impacting the methodological quality of the trials [68]. Nearly half of the RCTs involving oral pharmacological interventions failed to report adherence rates [69], indicating that proper adherence consideration is the exception instead of the rule [68]. In the present review 40% ($n = 2$) of the included RCTs reported assessing treatment adherence, although the exact rates were not presented. Moreover, none of the trials adhered to the ESPACOMP Medication Adherence Reporting Guideline (EMERGE) reporting guidelines regarding treatment adherence assessment [68]. A high non-adherence rate can reduce a trial's ability to detect a true treatment effect [70]. If adherence was considered and reported, the results regarding *Crocus sativus L.* supplementation in RDs might have been different.

4.1.2. Possible Cross-Treatment Effect

The standard treatment of participants was not reported in all trials. In the Sahebari et al. RCT [27], vitamin D ONS was among the standard therapy received by the participants and changes the sensation of pain was one of the outcomes of interest. Although pain was improved post-saffron administration [27], the scientific literature indicates that vitamin D might influence immune cells and pain sensitization through a variety of hormonal and neurological pathways [71,72]. Thus, the improved pain sensation noted in the trial might well be the synergistic result of vitamin D and *Crocus sativus L.*

Similarly, in the trial conducted by Poursamimi and associates [28], as the authors promptly noted, the improved pain relief observed may be the result of sodium diclofenac, which was administered to all participants during the trial. For this, significant improvements regarding pain were noted in both arms [28].

4.1.3. Differentiation between OA and RA

In the present SWiM, RCTs performed in patients with an OA diagnosis were also included, as often, patients with RA are misdiagnosed with OA, and vice versa [22]. Thus, it is possible that some of the patients included in the trials might have belonged in the opposite diagnosis, despite recruitment intentions.

Among the included RCTs, the one conducted by Sahebari and associates [27] was the only one where the anti-cyclic citrullinated peptide (anti-CCP)-positive patients were assessed within the sample, reporting that 89.2% of those allocated in the intervention and 81.6% of the controls were positive. The remaining RA/OA trials [28,29,31–33] failed to address this issue. Since this is a common problem in arthritis research, including both diagnosis without merging them was deemed as the safest option for the SWiM.

4.1.4. Effect of Lifestyle on RD Outcomes

Lifestyle has an impact on disease activity and outcomes in patients with RDs. In further detail, exercise can reduce disease activity and diet can either improve or amplify symptoms related to the diseases [73–77]. For this, the diet of participants in each RCT with ONS interventions must be recorded, and in parallel, physical activity should also be monitored. Among the included RCTs, however, only one [31,33] evaluated the diet of participants and their physical activity levels. The remaining failed to control for this important factor, introducing bias to their results.

4.1.5. Standardization of the Herbal Intervention and Reporting Quality of Formula Elaborations

As Ali [78] noted, herbal medicines tend to suffer from lack of standardization parameters. In more detail, there appears to be a lack of standardization regarding the raw materials used, the harvesting, drying, storage and processing methods, as well as the final products and dosage formulation [16,79]. Moreover, quality control procedures are inexistent in most of the trials [79]. According to the World Health Organization (WHO), all medicines, whether they are of plant origin or synthetic, must fulfill the basic requirements of safety and effectiveness [16,80]. Nevertheless, it appears that trials implementing herbal medicine interventions often fail to report information required to judge internal validity, external validity, and reproducibility [17,18]. From the bush to the content of a pill, herbal substances undergo a variety of procedures that define the final product's active ingredients and may greatly affect efficacy. As a result, most frequently, batch-to-batch uniformity of the active constituents and quality control using various analytical techniques are inexistent [81], leading to substantial variations in the formulation and bioactivity of herbal medicine supplements from lot to lot [82], and it is unclear if single and consistent batches are used for the formulations applied in the trials. Moreover, the need to quantify the test substance using high-performance liquid chromatography, gas chromatography, or other techniques is required to understand the exact dose of active ingredient that produces a significant effect [81].

According to Guo [83], the often non-standardized nature of the prepared interventions increases the probability for adverse events, indicating that in all cases of RCTs with herbal medicine, standards of safety and efficacy must be implemented. Today, poor reporting of adverse events consists of a frequent criticism regarding CAM research [84,85] and in the present systematic review 2/3 of the RCTs failed to report any adverse reactions. Moreover, serious adverse events have been reported by the FDA; however, as they are rare, they often fail to be manifested in small or underpowered RCTs [82].

Apart from the CONSORT for herbal medicine interventions [17], a variety of additional guidelines have been published with regard to quality standards and good clinical practice in herbal medicine trials, including WHO recommendations and International Union of Pure and Applied Chemistry (IUPAC) protocols [86–89]. Furthermore, information regarding fingerprinting analyses for the quality assessment of herbal medicine have also been proposed for interested stakeholders [90].

In the present systematic review, it was shown that regarding RCTs with saffron interventions in patients with RDs, the majority failed to adhere to the CONSORT-specific requirements for herbal medicine interventions. Similar issues have also been reported to exist in Cochrane systematic reviews evaluating herbal medicine [91]. For this important limitation, despite the plethora of meta-research evaluating herbal medicine interventions that have been published in high-end academic journals without considering this limitation [84,92], we considered that any quantitative synthesis would be misleading for the authors and clinical practice, and was avoided.

4.1.6. Intervention Duration

The duration of the intervention varied greatly in the included RCTs, spanning from as low as 8 weeks [30] to 4 months [28]. It is possible that a longer intervention duration might have changed the results in several trials, as other trials administering saffron for other conditions have, in their majority, applied the interventions for 3–4 months [9,15,93], with a respective follow-up session. Moreover, according to a recent meta-analysis, longer saffron supplementation durations have been shown to improve outcomes with regard to blood pressure [15]. Suffice to say, the exact intervention duration required to produce beneficial effects for each outcome has not yet been delineated.

4.1.7. Country of Origin

All trials included in the present SWiM were conducted in Iran. Today, 80% of the global saffron production is harvested from Iran, and this is why Iranian researchers are keen on investigating the plant's properties [57]. Nevertheless, according to an umbrella review [94], when studying the available literature, the need to conduct higher-quality trials outside of Iran becomes apparent, in order to reduce bias.

4.2. Ongoing Trials

Figure 7 details the ongoing trials investigating the effect of saffron in patients with RDs. A total of four RCTs were identified in the Iranian registry of clinical trials (IRCT) and none in the clinicaltrials.gov database. These trials are recruiting patients with BS, RA or FM, investigating similar outcomes as in the present review. Their results are expected to aid in understanding the possible results of saffron supplementation among patients with rheumatic diseases.

CTI	Sample	Affiliation	Intervention/comparator duration	Arms		Outcomes	
				Intervention	Comparator(s)	Primary	Secondary
IRCT201304 18013058N12[DB]	Patients with active BS and ocular complications	Mashhad University of Medical Sciences	3 mo	ONS with 1 crocin tab (15 mg)	1 tab of placebo	Ocular inflammation (Fluorescein Angiography)	BCVA (Snellen Chart)
IRCT201911 25045496N2[OL]	Women diagnosed with FM	Mazandaran University of Medical Sciences	6 wks	1) ONS with safrotin caps (30 mg standardized dry saffron extract) once/day 2) Aerobic exercise (3 sessions/wk, 60–70% of HR_{max}, 24–39') + ONS with safrotin caps (30 mg standardized dry saffron extract) once/day	1) Placebo caps (30 mg of starch) once/day 2) Aerobic exercise (3 sessions/wk, 60–70% of HR_{max}, 24–39') + placebo caps (30 mg of starch) once/day	QoL (SF-36) CRP (ELISA)	NR
IRCT201707 309472N15[DB]	Patients with active RA	Iran University of Medical Sciences	3 mo	ONS with powdered saffron (1 x 100 mg tabs/day)	ONS with maltodextrin (1 x 100 mg tabs)	FOXP3, GATA3, NF-kB, t-bet, PPAR-γ and ROR-γt gene expression (RT-PCR) IL-17, IL-4 (ELISA)	BMI
IRCT201711 06037265N1[DB]	Patients with RA with pain intensity > 40 (VAS)	Zahedan University of Medical Sciences	6 wks	ONS with saffron (15 mg) caps/day in the first wk and then twice/day	1) duloxetine (30 mg) caps/day in the first wk and then 60 mg in 2 divided doses/day	Pain (BPI) Depression (HDRS)	NR

Figure 7. Parallel RCTs investigating ONS with saffron in patients with rheumatic diseases. BCVA, best corrected visual acuity; BMI, body mass index; BPI, brief pain inventory; BS, Behcet's syndrome; CRP, C-reactive-protein; CTI, clinical trial identifier; ELISA, enzyme-linked immunosorbent assay; FM, fibromyalgia; FOXP3, forkhead box P3; HDRS, Hamilton Depression Survey Questionnaire; HR_{max}, maximum heart rate; IL-4, interleukin-4; IL-17, interleukin-17; mo, months; NF-kB, nuclear factor kappa-light-chain-enhancer of activated B cells; NR, not reported; ONS, oral nutrient supplementation; PPAR-γ, peroxisome proliferator-activated receptors γ; QoL, quality of life; RA, rheumatoid arthritis; RCT, randomized controlled trial; ROR-γt, RAR-related orphan receptors γt (thymus-specific isoform); RT-PCR, real-time polymerase chain reaction; SF-36, Short Form 36; T-bet, T-Box protein expressed in T cells; VAS, visual analogue scale; wks, weeks; [DB] double blind; [OL] open label.

4.3. Limitations of the Present Qualitative Synthesis

The limitations of the present qualitative synthesis primarily involve the lack of an adequate number of trials investigating similar outcomes in distinct RDs. Furthermore, a gap in the literature is apparent, with null saffron RCTs conducted for specific RDs (psoriasis, SLE, ankylosing spondylitis, Sjogren's syndrome, etc.)

As in every meta-research, the present review also carries the limitations of the included trials, indicating that there is room for the methodological improvement of RCTs investigating saffron in RDs. Interestingly, most of the included trials failed to assess and report changes in disease-activity specific scores (e.g., WOMAC), an issue that should be accounted for when designing future trials. Moreover, the high clinical and methodological heterogeneity among the included trials did not allow for a meta-analysis to be performed. According to a recent umbrella systematic review [94], RCTs evaluating saffron interventions entail a variety of biases, and their methodology should be improved.

The need for evaluating herbal medicine interventions is indisputable. Today, it is estimated that 2/3 of the global population uses herbal medicines, with some countries having incorporated them into the public health system [88]. Nevertheless, serious doubts regarding their safety and effectiveness remain [95]. According to Ernst and Pittler [96], the majority of studies published in CAM journals report positive findings and the concerns regarding the variation in formulation and bioactivity of some supplements remain a challenge [82]. As suggested by the European research network for CAM [97], CAM constitutes a neglected research area requiring more activities; however, specific standards of reporting must be met in advance. Although the assessment of the adherence to the CONSORT guidelines for the conduction and reporting of herbal medicine RCTs was not included in the initial aims of the present systematic review or the protocol, during the peer review process, it became clear that this issue constitutes an important factor affecting trial quality and intervention efficacy. This additional analysis added value to the present review, highlighting an area in need of improvement regarding the reporting of these trials.

5. Conclusions

Pedanio Dioscorides, an ancient Greek medical practitioner, was the first to report the medicinal properties of saffron [52,57]. In an extensive review of the history and the literature, Christodoulou [57] underlined the value of saffron over the centuries, with the "Saffron war" taking place in the Middle Ages and the execution of those who dared to tamper with saffron's composition due to its medicinal properties. Today, in the era of evidence-based medicine, whether this value can also be evidence-based greatly depends on the appraisal of the existing primary studies.

Research has suggested that *Crocus sativus* can form an effective adjuvant therapy for many conditions, and a promising one for RDs. RCTs performed in patients with RDs indicate that saffron may target many different outcomes, including inflammation, antioxidant status, depression and anxiety, pain, immune response and many others. If its efficacy is demonstrated, then it will undoubtedly be the "golden spice" for RDs. Nevertheless, at the moment, more primary studies are required to help us find the appropriate dose and conclude with certainty on the efficacy of saffron ONS in rheumatic diseases-related outcomes. Furthermore, the present systematic review raised concerns regarding the importance of reporting standards in herbal medicine research, with chemical fingerprinting being a required prerequisite for the standardization, safety and efficacy evaluation of the active ingredients.

Author Contributions: Conceptualization, D.P.B., M.G.G. and L.I.S.; methodology, M.G.G. and K.G.; formal analysis, M.G.G. and K.G.; investigation, S.G.T., M.G.G. and K.G.; data curation, M.G.G., S.G.T. and K.G.; writing—original draft preparation, M.G.G., K.G., S.G.T. and D.P.B.; writing—review and editing, D.P.B., E.D., E.Z., C.L., I.P., M.G.G., K.G., S.G.T., L.I.S.; visualization, M.G.G., S.G.T., K.G.; supervision, D.P.B. and L.I.S.; project administration, D.P.B.; All authors have read and agreed to the published version of the manuscript.

Funding: This research received no external funding.

Institutional Review Board Statement: Ethical review and approval were waived for this study, as this is a meta-research.

Informed Consent Statement: Not applicable.

Data Availability Statement: All data are presented within the manuscript text.

Conflicts of Interest: The authors declare no conflict of interest.

References

1. Safiri, S.; Kolahi, A.-A.; Cross, M.; Hill, C.; Smith, E.; Carson-Chahhoud, K.; Mansournia, M.A.; Almasi-Hashiani, A.; Ashrafi-Asgarabad, A.; Kaufman, J.; et al. Prevalence, Deaths, and Disability-Adjusted Life Years Due to Musculoskeletal Disorders for 195 Countries and Territories 1990–2017. *Arthritis Rheumatol.* **2021**, *73*, 702–714. [CrossRef]
2. Simões, D.; Araújo, F.A.; Monjardino, T.; Severo, M.; Cruz, I.; Carmona, L.; Lucas, R. The population impact of rheumatic and musculoskeletal diseases in relation to other non-communicable disorders: Comparing two estimation approaches. *Rheumatol. Int.* **2018**, *38*, 905–915. [CrossRef] [PubMed]
3. Ernst, E. Herbal medicine in the treatment of rheumatic diseases. *Rheum. Dis. Clin. North Am.* **2011**, *37*, 95–102. [CrossRef]
4. Kolasinski, S.L. Herbal medicine for rheumatic diseases: Promises kept? *Curr. Rheumatol. Rep.* **2012**, *14*, 617–623. [CrossRef]

5. Mauricio Marquez, A.; Evans, C.; Boltson, K.; Kesselman, M. Systematic Review Nutritional interventions and supplementation for rheumatoid arthritis patients: A systematic review for clinical application, Part 3: Fruits and Herbs. *Curr. Rheumatol. Res.* **2020**, *1*, 39–48.

6. Letarouilly, J.-G.; Sanchez, P.; Nguyen, Y.; Sigaux, J.; Czernichow, S.; Flipo, R.-M.; Sellam, J.; Daïen, C. Efficacy of Spice Supplementation in Rheumatoid Arthritis: A Systematic Literature Review. *Nutrients* **2020**, *12*, 3800. [CrossRef]

7. Grammatikopoulou, M.; Gkiouras, K.; Theodoridis, X.; Asteriou, E.; Forbes, A.; Bogdanos, D. Oral Adjuvant Curcumin Therapy for Attaining Clinical Remission in Ulcerative Colitis: A Systematic Review and Meta-Analysis of Randomized Controlled Trials. *Nutrients* **2018**, *10*, 1737. [CrossRef] [PubMed]

8. Singletary, K. Saffron: Potential Health Benefits. *Nutr. Today* **2020**, *55*, 294–303. [CrossRef]

9. Asbaghi, O.; Sadeghian, M.; Sadeghi, O.; Rigi, S.; Tan, S.C.; Shokri, A.; Mousavi, S.M. Effects of saffron (*Crocus sativus* L.) supplementation on inflammatory biomarkers: A systematic review and meta-analysis. *Phyther. Res.* **2021**, *35*, 20–32. [CrossRef] [PubMed]

10. Amin, B.; Hosseinzadeh, H. Analgesic and Anti-Inflammatory Effects of *Crocus sativus* L. (Saffron). In *Bioactive Nutraceuticals and Dietary Supplements in Neurological and Brain Disease: Prevention and Therapy*; Watson, R.R., Preedy, V.R., Eds.; Academic Press: Cambridge, MA, USA, 2015; pp. 319–324.

11. Sohaei, S.; Hadi, A.; Karimi, E.; Arab, A. Saffron supplementation effects on glycemic indices: A systematic review and meta-analysis of randomized controlled clinical trials. *Int. J. Food Prop.* **2020**, *23*, 1386–1401. [CrossRef]

12. Marx, W.; Lane, M.; Rocks, T.; Ruusunen, A.; Loughman, A.; Lopresti, A.; Marshall, S.; Berk, M.; Jacka, F.; Dean, O.M. Effect of saffron supplementation on symptoms of depression and anxiety: A systematic review and meta-analysis. *Nutr. Rev.* **2019**, *77*, 557–571. [CrossRef] [PubMed]

13. Ayati, Z.; Yang, G.; Ayati, M.H.; Emami, S.A.; Chang, D. Saffron for mild cognitive impairment and dementia: A systematic review and meta-analysis of randomised clinical trials. *BMC Complement. Med. Ther.* **2020**, *20*, 1–10. [CrossRef]

14. Zare, M.; Bazrafshan, A.; Afshar, R.M.; Mazloomi, S.M. Saffron (adjunct) for people with schizophrenia who have antipsychotic-induced metabolic syndrome. *Cochrane Database Syst. Rev.* **2018**, *2018*. [CrossRef]

15. Setayesh, L.; Ashtary-Larky, D.; Clark, C.C.T.; Kelishadi, M.R.; Khalili, P.; Bagheri, R.; Asbaghi, O.; Suzuki, K. The Effect of Saffron Supplementation on Blood Pressure in Adults: A Systematic Review and Dose-Response Meta-Analysis of Randomized Controlled Trials. *Nutrients* **2021**, *13*, 2736. [CrossRef]

16. Kunle, O.F.; Egharevba, H.O.; Ahmadu, P.O. Standardization of herbal medicines—A review. *Int. J. Biodivers. Conserv.* **2012**, *4*, 101–112. [CrossRef]

17. Gagnier, J.J.; Boon, H.; Rochon, P.; Moher, D.; Barnes, J.; Bombardier, C. Reporting randomized, controlled trials of herbal interventions: An elaborated CONSORT statement. *Ann. Intern. Med.* **2006**, *144*, 364–367. [CrossRef] [PubMed]

18. Gagnier, J.J.; Oltean, H.; Van Tulder, M.W.; Berman, B.M.; Bombardier, C.; Robbins, C.B. Herbal Medicine for Low Back Pain: A Cochrane Review. *Spine* **2016**, *41*, 116–133. [CrossRef]

19. Page, M.J.; McKenzie, J.E.; Bossuyt, P.M.; Boutron, I.; Hoffmann, T.C.; Mulrow, C.D.; Shamseer, L.; Tetzlaff, J.M.; Akl, E.A.; Brennan, S.E.; et al. The PRISMA 2020 statement: An updated guideline for reporting systematic reviews. *BMJ* **2021**, *372*. [CrossRef]

20. Campbell, M.; McKenzie, J.E.; Sowden, A.; Katikireddi, S.V.; Brennan, S.E.; Ellis, S.; Hartmann-Boyce, J.; Ryan, R.; Shepperd, S.; Thomas, J.; et al. Synthesis without meta-analysis (SWiM) in systematic reviews: Reporting guideline. *BMJ* **2020**, *368*, l6890. [CrossRef]

21. Ouzzani, M.; Hammady, H.; Fedorowicz, Z.; Elmagarmid, A. Rayyan-a web and mobile app for systematic reviews. *Syst. Rev.* **2016**, *5*, 210. [CrossRef]

22. Andrade, O.; Gomez, D.; Santos-Moreno, P.; Saavedra-Martinez, G. AB0378 Osteoarthritis is the Most Frequent Cause of Rheumathoid Arthritis Misdiagnosis in a Colombian Specialized Center. *Ann. Rheum. Dis.* **2015**, *74*. [CrossRef]

23. Goel, A.; Aggarwal, B.B. Curcumin, the golden spice from Indian saffron, is a chemosensitizer and radiosensitizer for tumors and chemoprotector and radioprotector for normal organs. *Nutr. Cancer* **2010**, *62*, 919–930. [CrossRef]

24. Sterne, J.A.C.; Savović, J.; Page, M.; Elbers, R.; Blencowe, N.; Boutron, I.; Cates, C.; Cheng, H.-Y.; Corbett, M.; Eldridge, S.; et al. RoB 2: A revised tool for assessing risk of bias in randomised trials. *Br. Med. J.* **2019**, *366*, l4898. [CrossRef] [PubMed]

25. Thomson, H.J.; Thomas, S. The effect direction plot: Visual display of non-standardised effects across multiple outcome domains. *Res. Synth. Methods* **2013**, *4*, 95. [CrossRef] [PubMed]

26. McKenzie, J.E.; Brennan, S.E. Synthesizing and presenting findings using other methods. In *Cochrane Handbook for Systematic Reviews of Interventions*; Higgins, J.P.T., Thomas, J., Chandler, J., Cumpston, M., Li, T., Page, M.J., Welch, V.A., Eds.; John Wiley & Sons, Ltd.: Hoboken, NJ, USA, 2019; pp. 321–347.

27. Sahebari, M.; Heidari, H.; Nabavi, S.; Khodashahi, M.; Rezaieyazdi, Z.; Dadgarmoghaddam, M.; Hosseinzaheh, H.; Abbasi, S.; Hashemzadeh, K. A double-blind placebo-controlled randomized trial of oral saffron in the treatment of rheumatoid arthritis. *Avicenna J. Phytomed.* **2021**, *11*, 332. [CrossRef] [PubMed]

28. Poursamimi, J.; Shariati-Sarabi, Z.; Tavakkol-Afshari, J.; Mohajeri, S.A.; Ghoryani, M.; Mohammadi, M. Immunoregulatory Effects of Krocina™, a Herbal Medicine Made of Crocin, on Osteoarthritis Patients: A Successful Clinical Trial in Iran. *Iran. J. Allergy Asthma Immunol.* **2020**, *19*, 253–263. [CrossRef]

29. Firoozabadi, M.; Abdolahi, N.; Mirfeyzi, S. P976. Anti-inflammatory effect of saffron in knee osteoarthritis: A double blind clinical trial. *Osteoporos. Int.* **2020**, *31*, S488.
30. Shakiba, M.; Moazen-Zadeh, E.; Noorbala, A.A.; Jafarinia, M.; Divsalar, P.; Kashani, L.; Shahmansouri, N.; Tafakhori, A.; Bayat, H.; Akhondzadeh, S. Saffron (*Crocus sativus*) versus duloxetine for treatment of patients with fibromyalgia: A randomized double-blind clinical trial. *Avicenna J. Phytomed.* **2018**, *8*, 513–523. [CrossRef]
31. Aryaeian, N.; Hamidi, Z.; Shirani, F.; Hadidi, M.; Abolghasemi, J.; Moradi, N.; Fallah, S. The Effect of Saffron Supplement on Clinical Outcomes, Inflammatory and Oxidative Markers in Patients With Active Rheumatoid Arthritis. *Curr. Dev. Nutr.* **2021**, *5*, 1118. [CrossRef]
32. Firoozabadi, M.; Aghei, M.; Mirfeizi, S.Z.; Roshandel, G.R.; Abdol Ahi, N. PT3:026 The effectiveness of saffron pill in treatment of osteoarthritis: A randomized, double-blind controlled trial. *Clin. Exp. Rheumatol.* **2018**, *36*, S62.
33. Hamidi, Z.; Aryaeian, N.; Abolghasemi, J.; Shirani, F.; Hadidi, M.; Fallah, S.; Moradi, N. The effect of saffron supplement on clinical outcomes and metabolic profiles in patients with active rheumatoid arthritis: A randomized, double-blind, placebo-controlled clinical trial. *Phyther. Res.* **2020**, *34*, 1650–1658. [CrossRef] [PubMed]
34. Aletaha, D.; Neogi, T.; Silman, A.J.; Funovits, J.; Felson, D.T.; Bingham, C.O.; Birnbaum, N.S.; Burmester, G.R.; Bykerk, V.P.; Cohen, M.D.; et al. Rheumatoid arthritis classification criteria: An American College of Rheumatology/European League Against Rheumatism collaborative initiative. *ARTHRITIS Rheum.* **2010**, *62*, 2569–2581. [CrossRef] [PubMed]
35. Mil, A.V.D.H.-V.; Zink, A. What is rheumatoid arthritis? Considering consequences of changed classification criteria. *Ann. Rheum. Dis.* **2017**, *76*, 315–317. [CrossRef]
36. Altman, R.; Asch, E.; Bloch, D.; Bole, G.; Borenstein, D.; Brandt, K.; Christy, W.; Cooke, T.D.; Greenwald, R.; Hochberg, M.; et al. Development of criteria for the classification and reporting of osteoarthritis: Classification of osteoarthritis of the knee. *Arthritis Rheum.* **1986**, *29*, 1039–1049. [CrossRef]
37. Wolfe, F.; Clauw, D.J.; Fitzcharles, M.-A.; Goldenberg, D.L.; Katz, R.S.; Mease, P.; Russell, A.S.; Russell, I.J.; Winfield, J.B.; Yunus, M.B. The American College of Rheumatology Preliminary Diagnostic Criteria for Fibromyalgia and Measurement of Symptom Severity. *Arthritis Care Res.* **2010**, *62*, 600–610. [CrossRef]
38. Samisaz Pharmaceutical Company KROCINA | Sami Saz. Available online: http://samisaz.com (accessed on 22 November 2021).
39. Anderson, J.; Caplan, L.; Yazdany, J.; Robbins, M.L.; Neogi, T.; Michaud, K.; Saag, K.G.; O'dell, J.R.; Kazi, S. Rheumatoid arthritis disease activity measures: American College of Rheumatology recommendations for use in clinical practice. *Arthritis Care Res.* **2012**, *64*, 640–647. [CrossRef]
40. Cleeland, C.S.; Ryan, K.M. Pain assessment: Global use of the Brief Pain Inventory—PubMed. *Ann. Acad. Med. Singap.* **1994**, *23*, 129–138. [PubMed]
41. Theiler, R.; Spielberger, J.; Bischoff, H.A.; Bellamy, N.; Huber, J.; Kroesen, S. Clinical evaluation of the WOMAC 3.0 OA Index in numeric rating scale format using a computerized touch screen version. *Osteoarthr. Cartil.* **2002**, *10*, 479–481. [CrossRef]
42. Wolfe, F.; Michaud, K.; Pincus, T. Development and validation of the health assessment questionnaire II: A revised version of the health assessment questionnaire. *Arthritis Rheum.* **2004**, *50*, 3296–3305. [CrossRef]
43. Pascoe, V.L.; Enamandram, M.; Corey, K.C.; Cheng, C.E.; Javorsky, E.J.; Sung, S.M.; Donahue, K.R.; Kimball, A.B. Using the Physician Global Assessment in a Clinical Setting to Measure and Track Patient Outcomes. *JAMA Dermatol.* **2015**, *151*, 375–381. [CrossRef]
44. Belza, B.L. Comparison of self-reported fatigue in rheumatoid arthritis and controls—PubMed. *J. Rheumatol.* **1995**, *22*, 639–643.
45. Hamilton, M. A rating scale for depression. *J. Neurol. Neurosurg. Psychiatry* **1960**, *23*, 62. [CrossRef]
46. Zigmond, A.S.; Snaith, R.P. The Hospital Anxiety and Depression Scale. *Acta Psychiatr. Scand.* **1983**, *67*, 361–370. [CrossRef] [PubMed]
47. Wells, G.; Becker, J.-C.; Teng, J.; Dougados, M.; Schiff, M.; Smolen, J.; Aletaha, D.; van Riel, P.L.C.M. Validation of the 28-joint Disease Activity Score (DAS28) and European League Against Rheumatism response criteria based on C-reactive protein against disease progression in patients with rheumatoid arthritis, and comparison with the DAS28 based on erythrocyte sedimentation rate. *Ann. Rheum. Dis.* **2009**, *68*, 954–960. [CrossRef]
48. Burckhardt, C.S.; Clark, S.R.; Bennet, R.M. The fibromyalgia impact questionnaire: Development and validation—PubMed. *J. Rheumatol.* **1991**, *18*, 728–733. [PubMed]
49. Ghavidel Parsa, B.; Amir Maafi, A.; Haghdoost, A.; Arabi, Y.; Khojamli, M.; Chatrnour, G.; Bidari, A. The validity and reliability of the Persian version of the Revised Fibromyalgia Impact Questionnaire. *Rheumatol. Int.* **2014**, *34*, 175–180. [CrossRef] [PubMed]
50. Craig, C.L.; Marshall, A.L.; Sjöström, M.; Bauman, A.E.; Booth, M.L.; Ainsworth, B.E.; Pratt, M.; Ekelund, U.; Yngve, A.; Sallis, J.F.; et al. International Physical Activity Questionnaire: 12-Country Reliability and Validity. *Med. Sci. Sport. Exerc.* **2003**, *35*, 1381–1395. [CrossRef] [PubMed]
51. Beck, A.; Steer, R.; Brown, G. *Beck Depression Inventory*, 2nd ed.; Psychological Corporation: San Antonio, TX, USA, 1996.
52. Husaini, A.M.; Jan, N.; Wani, G.A. Saffron: A potential drug-supplement for severe acute respiratory syndrome coronavirus (COVID) management. *Heliyon* **2021**, *7*, e07068. [CrossRef]
53. Thushara, R.M.; Hemshekhar, M.; Paul, M.; Shanmuga Sundaram, M.; Shankar, R.L.; Kemparaju, K.; Girish, K.S. Crocin prevents sesamol-induced oxidative stress and apoptosis in human platelets. *J. Thromb. Thrombolysis* **2014**, *38*, 321–330. [CrossRef]

54. Morvaridzadeh, M.; Agah, S.; Estêvão, M.D.; Hosseini, A.S.; Heydari, H.; Toupchian, O.; Abdollahi, S.; Persad, E.; Abu-Zaid, A.; Rezamand, G.; et al. Effect of saffron supplementation on oxidative stress parameters: A systematic review and meta-analysis of randomized placebo-controlled trials. *Food Sci. Nutr.* **2021**, *9*, 5819. [CrossRef]

55. Tóth, B.; Hegyi, P.; Lantos, T.; Szakács, Z.; Kerémi, B.; Varga, G.; Tenk, J.; Pétervári, E.; Balaskó, M.; Rumbus, Z.; et al. The Efficacy of Saffron in the Treatment of Mild to Moderate Depression: A Meta-analysis. *Planta Med.* **2018**, *85*, 24–31. [CrossRef]

56. Ghaderi, A.; Asbaghi, O.; Reiner, Ž.; Kolahdooz, F.; Amirani, E.; Mirzaei, H.; Banafshe, H.R.; Maleki Dana, P.; Asemi, Z. The effects of saffron (*Crocus sativus* L.) on mental health parameters and C-reactive protein: A meta-analysis of randomized clinical trials. *Complement. Ther. Med.* **2020**, *48*, 102250. [CrossRef]

57. Christodoulou, E.; Kadoglou, N.P.; Kostomitsopoulos, N.; Valsami, G. Saffron: A natural product with potential pharmaceutical applications. *J. Pharm. Pharmacol.* **2015**, *67*, 1634–1649. [CrossRef]

58. Pachikian, B.D.; Copine, S.; Suchareau, M.; Deldicque, L. Effects of Saffron Extract on Sleep Quality: A Randomized Double-Blind Controlled Clinical Trial. *Nutrients* **2021**, *13*, 1473. [CrossRef] [PubMed]

59. Dai, L.; Chen, L.; Wang, W. Safety and Efficacy of Saffron (*Crocus sativus* L.) for Treating Mild to Moderate Depression: A Systematic Review and Meta-analysis. *J. Nerv. Ment. Dis.* **2020**, *208*, 269–276. [CrossRef]

60. Khaksarian, M.; Behzadifar, M.; Behzadifar, M.; Alipour, M.; Jahanpanah, F.; Re, T.S.; Firenzuoli, F.; Zerbetto, R.; Bragazzi, N.L. The efficacy of Crocus sativus (Saffron) versus placebo and Fluoxetine in treating depression: A systematic review and meta-analysis. *Psychol. Res. Behav. Manag.* **2019**, *12*, 305. [CrossRef] [PubMed]

61. Azimi, P.; Abrishami, R. Comparison of the Effects of Crocus Sativus and Mefenamic Acid on Primary Dysmenorrhea. *J. Pharm. Care* **2016**, *4*, 75–78.

62. Kubo, I.; Kinst-Hori, I. Flavonols from saffron flower: Tyrosinase inhibitory activity and inhibition mechanism. *J. Agric. Food Chem.* **1999**, *47*, 4121–4125. [CrossRef]

63. Hosseini, A.; Razavi, B.M.; Hosseinzadeh, H. Saffron (*Crocus sativus*) petal as a new pharmacological target: A review. *Iran. J. Basic Med. Sci.* **2018**, *21*, 1091–1099. [CrossRef] [PubMed]

64. Bani, S.; Pandey, A.; Agnihotri, V.K.; Pathania, V.; Singh, B. Selective Th2 upregulation by crocus sativus: A neutraceutical spice. *Evid. Based Complement. Altern. Med.* **2010**, *2011*, 639862. [CrossRef]

65. Chen, B.; Hou, Z.H.; Dong, Z.; Li, C.D. Crocetin downregulates the proinflammatory cytokines in methylcholanthrene-induced rodent tumor model and inhibits COX-2 expression in cervical cancer Cells. *Biomed Res. Int.* **2015**, *2015*. [CrossRef] [PubMed]

66. Ding, J.; Su, J.; Zhang, L.; Ma, J. Crocetin Activates Foxp3 Through TIPE2 in Asthma-Associated Treg Cells. *Cell. Physiol. Biochem.* **2015**, *37*, 2425–2433. [CrossRef]

67. Khazdair, M.R.; Gholamnezhad, Z.; Rezaee, R.; Boskabady, M.H. A qualitative and quantitative comparison of *Crocus sativus* and Nigella sativa immunomodulatory effects. *Biomed. Pharmacother.* **2021**, *140*, 111774. [CrossRef] [PubMed]

68. Eliasson, L.; Clifford, S.; Mulick, A.; Jackson, C.; Vrijens, B. How the EMERGE guideline on medication adherence can improve the quality of clinical trials. *Br. J. Clin. Pharmacol.* **2020**, *86*, 687–697. [CrossRef]

69. Zhang, Z.; Peluso, M.J.; Gross, C.P.; Viscoli, C.M.; Kernan, W.N. Adherence reporting in randomized controlled trials. *Clin. Trials* **2014**, *11*, 195–204. [CrossRef]

70. Murali, K.M.; Mullan, J.; Chen, J.H.C.; Roodenrys, S.; Lonergan, M. Medication adherence in randomized controlled trials evaluating cardiovascular or mortality outcomes in dialysis patients: A systematic review. *BMC Nephrol.* **2017**, *18*, 1–11. [CrossRef]

71. Shipton, E.A.; Shipton, E.E. Vitamin D and pain: Vitamin D and its role in the aetiology and maintenance of chronic pain states and associated comorbidities. *Pain Res. Treat.* **2015**, *2015*, 904967. [CrossRef] [PubMed]

72. Habib, A.M.; Nagi, K.; Thillaiappan, N.B.; Sukumaran, V.; Akhtar, S. Vitamin D and Its Potential Interplay With Pain Signaling Pathways. *Front. Immunol.* **2020**, 820. [CrossRef]

73. Vranou, P.; Gkoutzourelas, A.; Athanatou, D.; Zafiriou, E.; Grammatikopoulou, M.G.; Bogdanos, D.P. Let food be thy medicine: The case of the Mediterranean diet in rheumatoid arthritis. *Mediterr. J. Rheumatol.* **2020**, in press. [CrossRef]

74. Westerlind, H.; Palmqvist, I.; Saevarsdottir, S.; Alfredsson, L.; Klareskog, L.; Di Giuseppe, D. Is tea consumption associated with reduction of risk of rheumatoid arthritis? A Swedish case-control study. *Arthritis Res. Ther.* **2021**, *23*, 209. [CrossRef]

75. Guagnano, M.T.; D'Angelo, C.; Caniglia, D.; Di Giovanni, P.; Celletti, E.; Sabatini, E.; Speranza, L.; Bucci, M.; Cipollone, F.; Paganelli, R. Improvement of Inflammation and Pain after Three Months' Exclusion Diet in Rheumatoid Arthritis Patients. *Nutrients* **2021**, *13*, 3535. [CrossRef]

76. Skyvalidas, D.N.; Mavropoulos, A.; Tsiogkas, S.; Dardiotis, E.; Liaskos, C.; Mamuris, Z.; Roussaki-Schulze, A.; Sakkas, L.A.; Zafiriou, E.; Bogdanos, D.P. Curcumin mediates attenuation of pro-inflammatory interferon γ and interleukin 17 cytokine responses in psoriatic disease, strengthening its role as a dietary immunosuppressant. *Nutr. Res.* **2020**, *75*, 95–108. [CrossRef]

77. Nelson, J.; Sjöblom, H.; Gjertsson, I.; Ulven, S.M.; Lindqvist, H.M.; Bärebring, L. Do Interventions with Diet or Dietary Supplements Reduce the Disease Activity Score in Rheumatoid Arthritis? A Systematic Review of Randomized Controlled Trials. *Nutrients* **2020**, *12*, 2991. [CrossRef] [PubMed]

78. Ali, R.; Rooman, M.; Mussarat, S.; Norin, S.; Ali, S.; Adnan, M.; Khan, S.N. A Systematic Review on Comparative Analysis, Toxicology, and Pharmacology of Medicinal Plants Against Haemonchus contortus. *Front. Pharmacol.* **2021**, *12*, 807. [CrossRef] [PubMed]

79. Sachan, A.K.; Vishnoi, G.; Kumar, R. Need of standardization of herbal medicines in Modern era. *Int. J. Phytomed.* **2016**, *8*, 300–307. [CrossRef]
80. World Health Organization. *WHO Guidelines on Safety Monitoring of Herbal Medicines in Pharmacovigilance Systems*; World Health Organization: Geneva, Switzerland, 2004.
81. Parveen, A.; Parveen, B.; Parveen, R.; Ahmad, S. Challenges and guidelines for clinical trial of herbal drugs. *J. Pharm. Bioallied Sci.* **2015**, *7*, 333. [CrossRef]
82. Shekelle, P.G.; Morton, S.C.; Suttorp, M.J.; Buscemi, N.; Friesen, C. Challenges in systematic reviews of complementary and alternative medicine topics. *Ann. Intern. Med.* **2005**, *142*, 1042–1047. [CrossRef]
83. Guo, R.; Canter, P.H.; Ernst, E. A systematic review of randomised clinical trials of individualised herbal medicine in any indication. *Postgrad. Med. J.* **2007**, *83*, 637. [CrossRef]
84. Perry, R.; Leach, V.; Davies, P.; Penfold, C.; Ness, A.; Churchill, R. An overview of systematic reviews of complementary and alternative therapies for fibromyalgia using both AMSTAR and ROBIS as quality assessment tools. *Syst. Rev.* **2017**, *6*, 1–23. [CrossRef]
85. Ernst, E. "First, do no harm" with complementary and alternative medicine. *Trends Pharmacol. Sci.* **2007**, *28*, 48–50. [CrossRef]
86. UNICEF/UNDP/World Bank/WHO Special Programme for Research and Training in Tropical Diseases (TDR). *Operational Guidance: Information Needed to Support Clinical Trials of Herbal products*; World Health Organization: Geneva, Switzerland, 2005.
87. Cheng, C.W.; Wu, T.X.; Shang, H.C.; Li, Y.P.; Altman, D.G.; Moher, D.; Bian, Z.X. CONSORT Extension for Chinese Herbal Medicine Formulas 2017: Recommendations, Explanation, and Elaboration (Traditional Chinese Version). *Ann. Intern. Med.* **2017**, *167*, W7–W20. [CrossRef] [PubMed]
88. World Health Organization. *General Guidelines for Methodologies on Research and Evaluation of Traditional Medicine*; World Health Organization: Geneva, Switzerland, 2000.
89. World Health Organization. *Annex 1 WHO Guidelines for Selecting Marker Substances of Herbal Origin for Quality Control of Herbal Medicines*; World Health Organization: Geneva, Switzerland, 2017.
90. Li, Y.; Shen, Y.; Yao, C.-L.; Guo, D.-A. Quality assessment of herbal medicines based on chemical fingerprints combined with chemometrics approach: A review. *J. Pharm. Biomed. Anal.* **2020**, *185*, 113215. [CrossRef]
91. Zhang, X.; Aixinjueluo, Q.Y.; Li, S.Y.; Song, L.L.; Lau, C.T.; Tan, R.; Bian, Z.X. Reporting quality of Cochrane systematic reviews with Chinese herbal medicines. *Syst. Rev.* **2019**, *8*, 1–11. [CrossRef]
92. Gamret, A.C.; Price, A.; Fertig, R.M.; Lev-Tov, H.; Nichols, A.J. Complementary and Alternative Medicine Therapies for Psoriasis: A Systematic Review. *JAMA Dermatol.* **2018**, *154*, 1330–1337. [CrossRef]
93. Heitmar, R.; Brown, J.; Kyrou, I. Saffron (*Crocus sativus* L.) in Ocular Diseases: A Narrative Review of the Existing Evidence from Clinical Studies. *Nutrients* **2019**, *11*, 649. [CrossRef] [PubMed]
94. Lu, C.; Ke, L.; Li, J.; Zhao, H.; Lu, T.; Mentis, A.F.A.; Wang, Y.; Wang, Z.; Polissiou, M.G.; Tang, L.; et al. Saffron (*Crocus sativus* L.) and health outcomes: A meta-research review of meta-analyses and an evidence mapping study. *Phytomedicine* **2021**, *91*, 153699. [CrossRef] [PubMed]
95. Mosihuzzaman, M. Herbal Medicine in Healthcare-An Overview. *Nat. Prod. Commun.* **2012**, *7*, 807–812. [CrossRef]
96. Ernst, E.; Pittler, M.H. Alternative therapy bias. *Nature* **1997**, *385*, 480. [CrossRef]
97. CAMbrella—A Pan-European Research Network for Complementary and Alternative Medicine (CAM) Final Report Summary. Available online: https://cordis.europa.eu/project/id/241951/reporting (accessed on 22 November 2021).

 *nutrients*

Article

Diabetes Mellitus and Associated Factors in Slovakia: Results from the European Health Interview Survey 2009, 2014, and 2019

Nour Mahrouseh [1], Carlos Alexandre Soares Andrade [1], Nóra Kovács [1], Diana Wangeshi Njuguna [1] and Orsolya Varga [1,2,*]

[1] Department of Public Health and Epidemiology, Faculty of Medicine, University of Debrecen, 4032 Debrecen, Hungary; nour.mahrouseh@med.unideb.hu (N.M.); soares.andrade@med.unideb.hu (C.A.S.A.); kovacs.nora@med.unideb.hu (N.K.); diana.njuguna@med.unideb.hu (D.W.N.)

[2] Office for Supported Research Groups, Eötvös Loránd Research Network, 1052 Budapest, Hungary

* Correspondence: varga.orsolya@med.unideb.hu

Abstract: Diabetes mellitus (DM) is a high-risk non-communicable disease with an emerging burden for the European Union (EU) member states in the past decades. The unfavorable trend of the burden is striking compared to the declining disease burden due to cardiovascular diseases or stagnation of neoplasms. The goal of this study is to describe the temporal changes of diabetes in the adult population of Slovakia through the three European Health Interview Survey (EHIS) waves and to assess the association between DM and socioeconomic and/or lifestyle characteristics. These cross-sectional studies were carried out using microdata derived from Slovakia's EHISs conducted in the years 2009 (n = 4972), 2014 (n = 5490), and 2019 (n = 5527). The DM variable was compared to the independent variables such as sociodemographic and lifestyle characteristics including dietary patterns and physical activity. DM prevalence for the EHIS in 2009, 2014, and 2019 were 6.1%, 8.2%, and 9.8%, respectively. In bivariate analysis, the relationship between DM and age, education level, job status, BMI, walking for at least 10 min, and physical activity was significant in the three EHISs. In 2014 and 2019, there was an inverse association between the risk of DM and walking regularly. There was no association between the frequency of eating fruits or vegetables and DM, with the exception of 2009, where a negative association between eating vegetables one to six times a week and DM was observed. Present health policies and activities in Slovakia were unable to reverse the increasing DM burden, indicating that a more systematic approach is needed. Complex policy strategies and legislative measures must be developed and implemented at both the national and EU levels.

Keywords: EHIS; diabetes; disease burden; European Union; policies

Citation: Mahrouseh, N.; Andrade, C.A.S.; Kovács, N.; Njuguna, D.W.; Varga, O. Diabetes Mellitus and Associated Factors in Slovakia: Results from the European Health Interview Survey 2009, 2014, and 2019. *Nutrients* **2021**, *13*, 2156. https://doi.org/10.3390/nu13072156

Academic Editor: Panagiota Mitrou

Received: 15 May 2021
Accepted: 21 June 2021
Published: 23 June 2021

1. Introduction

Diabetes mellitus (DM) and its complications rank high in the global list of diseases with high burden values [1]. DM is one of the four most common chronic non-communicable diseases (NCDs), which occurs due to inadequate insulin production in the pancreas, or when the body does not use its insulin production properly, which impact blood sugar regulation. Uncontrolled DM can lead to several systemic complications in the body. According to the World Health Organization (WHO), the three major types of DM are type 1 diabetes mellitus, type 2 diabetes mellitus (T2DM), and gestational diabetes. Among these, the T2DM is by far the most common, accounting for 90% of total DM cases. With regard to its frequency and preventability, policies and strategies for the prevention of DM are mainly tackling diabetes T2DM risk factors [2,3]. The European Union (EU) member states are especially facing an increasing burden of DM; projections for 2030 and 2045 show that the prevalence of DM among adults will increase to 50.48% and 50.51%, respectively [4]. Such a trend is very unfavorable compared to the two other major NCDs—cardiovascular diseases and neoplasms. According to the Global Burden of Disease (GBD) database, the trend of DALYs (age-standardized rate per 100,000) due to cardiovascular diseases showed a considerable decline from 8477.91 in 1990 to 5779.75 in 2019. Neoplasms had a plateau in trend with modest change on DALYs from 5752.73 in 1990 to 6022.41 in 2019. In the same time period, DM was reported to show a significant increase of burden, from 741.63 to 1098.57 [1].

In European countries, efforts have been made to tackle DM since the establishment of the St. Vincent Declaration in 1989. Although some political initiatives by the European Parliament were taken [5], in the last 20 years, DM has been mostly addressed as one of the NCDs—for example, the Action Plan for the Prevention and Control of Non-communicable Diseases in the WHO European Region [6].

However, according to Article 168 of the Treaty on the Functioning of the European Union, the member states have the primary role of organizing the healthcare services; the EU provides support and funds for prevention and research to reduce the burden due to DM in many ways [7]—for instance, by facilitating the production of comparable datasets for the identification and adoption of effective health policy measures. The European Health Interview Survey (EHIS) is the most significant health data collection instrument. It provides information on the health status, health determinants and healthcare services facilitating data comparability between the member states [8]. These comparable data allow stakeholders to develop and select tailored responses targeting the causal factors of NCDs.

According to the WHO, the onset of type 2 diabetes can be delayed by maintaining a healthy body weight, being physically active, eating a healthy diet and avoiding tobacco use [9]. Poor diet is one of the major issues implicated in the incidence of DM. High intake of foods with a high quantity of sugar, such sweetened beverages, can lead to worsened DALYs in countries of the EU. Moreover, the suboptimal consumption of whole grains, nuts, fruit, fish and legumes have a negative impact on the burden of DM [10]. Increased physical activity, exercise or training, and reduced sedentary lifestyle are also behaviors needed to avoid the occurrence of DM [11].

In addition to lifestyle related factors, socioeconomic factors also influence the development of DM. Population-based studies have reported an association with older age, lower socioeconomic status, and a lower level of education [12]. The burden due to DM varies significantly among member states, and the diversity is present at a subnational and regional level [13,14]. Thus, individual-based strategies addressing socioeconomic disadvantages also seem necessary for DM prevention [15].

This study aims to describe the temporal changes of DM in the adult population of Slovakia through the three EHIS waves, and to assess the association between DM and socioeconomic and/or lifestyle characteristics. This study shows important results from an ongoing project analyzing data from the three EHIS waves with a focus on DM in the EU.

2. Methods

Cross-sectional studies were carried out using microdata derived from Slovakia's EHISs conducted in 2009 ($n = 4972$), 2014 ($n = 5490$), and 2019 ($n = 5527$). The microdata was obtained from Eurostat for European Health Interview survey 2009 and 2014, and The Statistical Office of the Slovak Republic for European Health Interview Survey 2019. These samples are representative of the Slovakian adult population (aged 15 years or over) residing in private households. The three surveys included different participants. The EHIS data collection method differs between countries, which may include face-to-face interviews, telephone interviews, postal, web interviews or a combination of these methods. Self-administered questionnaires were also applied for some questions; through papers, the Internet, or both. According to the quality report of wave 2 in Slovakia, face-to-face interviews and self-administered paper-based questionnaires were used. The sampling frame varied by countries as some used population registers, dwelling registers, population censuses, and others. Slovakia's sampling frame was drawn from the dwelling register.

Some countries used proxy interviews; in Slovakia, proxy interviews were not applied. Details of the methodology and sampling are reported by the European Commission [ ]. Our study variables were based on questions consequently asked in the 2009, 2014, and 2019 EHISs. Respondents who answered "yes" to the question: "During the past 12 months, have you had diabetes?" were considered in the group with DM, including any types of diabetes mellitus. Self-reporting sociodemographic and lifestyle characteristics were analyzed as independent variables; a definition of each variable as derived from the survey is available in Supplementary.

Demographic and socioeconomic characteristics included sex, age (15 to 44, 45 to 64, and 65 and above), regions based on level 2 of nomenclature of territorial units for statistics (NUTS2) of Slovakia (Bratislavský kraj, Západné Slovensko, Stredné Slovensko, and Východné Slovensko), degree of urbanization (cities, towns and suburbs and rural areas), education level (less than primary/primary education, secondary education and higher education), labor status (employed, unemployed and others).

The included lifestyle variables were body mass index (BMI, kg/m^2) (<18.5, 18.5–24.9, 25–29.9 and $\geq$30), frequency of walking for transportation purposes (to get to and from places) at least 10 min per day (everyday, one to six days, and never), physical activity per week (two days and more, one day per week and never), the frequency of eating fruits and frequency of eating vegetables per week (one or more per day, one to six times a week, and less than once a week and never).

The distribution of the variables was described and compared within surveys and for diabetic and nondiabetic respondents. Proportions were used as descriptive statistics. The estimated prevalence of DM for each year was based on the study sample. To perform bivariate comparisons, Pearson chi-square test was used to analyze the association between the study variables and DM. A multivariable unconditional logistic regression model was conducted, including variables which were statistically significant in the bivariate analysis, and variables that were not statistically significant, but were of interest from an epidemiological perspective. The regression results are shown as odds ratios (ORs) with 95% confidence interval (CIs). We used the three datasets separately for analysis. Sampling weights were available in the database; svy function in Stata was used to preserve the EHIS survey weighting only in the multivariable analysis. Statistical analyses were performed by using STATA IC version 13.0 software.

3. Results

DM prevalence for the 2009 EHIS was 6.1%, 8.2% for 2014, and 9.8% for 2019. The distribution by numbers of the study population by DM occurrence according to demographic, socioeconomic variables, and lifestyle are shown in Table 1. The distribution by numbers and relative frequencies of the study population divided by the presence of DM according to demographic, socioeconomic variables, and lifestyle are shown in Table S1 of the Supplementary File.

The results of the bivariate analysis showed that the percentage of individuals who had DM differed by gender (*p*-value < 0.05) in 2009; female respondents had higher percentages in the diabetes group than males. Frequencies were significantly different by age groups in 2009, 2014, and 2019: diabetic respondents belonged to older age categories of 65 and older—3.59%, 4.81%, and 6.06%, respectively. It is observed that respondents with secondary education were more affected by DM than any other education levels and the relation between DM and education level is significant. The degree of urbanization presented a significant relationship with DM only in 2009 and 2014; the respondents who had DM and lived in the rural areas were higher in 2009. In 2014, the most individuals with DM lived in towns and suburbs. The occurrence of DM differed by labor status; respondents who were diabetic were higher in other groups (e.g., students, pensioners) of labor status. As seen by the frequencies in the cross-tabulated Table 1, there is a significant relationship between the presence of DM and BMI in 2009, 2014, and 2019. The majority of diabetic individuals belonged to the overweight or obesity group. Individuals with

DM were higher in the BMI group between 25 to 29.9 in 2009 and 2014, but in 2019, the frequency of diabetes was higher in the BMI group $\geq$30. The frequency of having DM differed by the number of days walking for at least 10 min per week. Of the total number of respondents, most of the diabetic individuals in 2009 and 2014 walked to get to and from places one to six times per week for 10 min. Walking every day for transportation purposes was higher and less diabetic individuals never walked in 2009 and 2014. In 2019, respondents with DM had a higher frequency for walking for transportation for 10 min every day than walking one to six times per week for 10 min or never groups. A significant relationship was shown between the presence of DM and physical activity categories in 2009, 2014, and 2019. Higher frequencies of diabetic individuals were found in the group of individuals who never performed any kind of physical activity per week in 2009, 2014, and 2019. The relationship between the consumption of fruits and vegetables and the presence of DM was significant only in 2009. Diabetic respondents had the highest frequency of eating one or more fruits and vegetables per day in 2009 than eating one to six times a week or less than once a week or never.

The results of the multivariable model are shown in Table 2. The results indicated that females in 2019 were 27% less likely to have DM compared to males (OR 0.77; 95% CI 0.62 to 0.95). Age groups of 15 to 44 and 45 to 64 had a negative association to DM compared to the reference age group (65 and above) in all three surveys. The degree of urbanization was not associated with the presence of DM. Primary or less than primary education was positively associated with having DM in 2019 (OR 3.25; 95% CI 1.12 to 9.46) compared to respondents who were in the higher education category. Employment as a labor status had significant lower likelihood of DM, compared to the reference group of other labor status (OR 0.40; 95% CI 0.27–0.60) (OR 0.35; 95% CI 0.25–0.49) (OR 0.36; 95% CI 0.25 to 0.50) in the years of 2009, 2014 and 2019. Unemployment (OR 0.56; 95% CI 0.33 to 0.95) presented lower probability of having DM compared to reference group of other labor status, only in 2014.

People with BMI of 30 or higher had a greater probability of developing DM in 2009 and 2019 as compared to the overweight group (BMI 25 to 29.9), and accordingly, lower BMI (<18.5 to 24.9) was associated with a low probability of having DM in 2009 (OR 0.59; 95% CI 0.41 to 0.84), 2014 (OR 0.57; 95% CI 0.43 to 0.76), 2019 (OR 0.65; 95% CI 0.49 to 0.88), respectively. Regarding physical activity and movement, a negative association between the risk of DM and walking to get to and from places for at least 10 min every day (OR 0.67;95% CI 0.49 to 0.92) (OR 0.57; 95% CI 0.42 to 0.77) or at least one to six times per week (OR 0.70; 95% CI 0.52 to 0.96) (OR 0.69; 95% CI 0.50 to 0.94) compared to our reference category of "never" was found in the years of 2014 and 2019, respectively. In 2014, a lack of physical activity, "never", increased the probability of DM (OR 2.55; 95% 1.02 to 6.37). There was no association between the frequency of eating fruits and the presence of DM. The frequency of eating vegetables one to six times per week compared to our reference category of one and more a day decreased the risk of DM presence in the year 2009 (OR 0.66; 95% CI 0.48 to 0.92). There was no significant association regarding regions of Slovakia (data from 2019, exclusively).

Table 1. Distribution of the study population.

Variable	Category	EHIS 2009			EHIS 2014			EHIS 2019		
		With Diabetes n	Without Diabetes n	p-Value	With Diabetes n	Without Diabetes n	p-Value	With Diabetes n	Without Diabetes n	p-Value
Sex	Male	132	2257	0.022	184	2270	0.098	233	2087	0.799
	Female	183	2392		265	2771		317	2898	
Age	15 to 44	Below 20	2689	<0.001	Between 20 and 49	2467	<0.001	Between 20 and 49	1933	<0.001
	45 to 64	118	1427		149	1673		180	1843	
	65 and older	178	533		264	901		335	1209	
Region *	Bratislavský kraj							Between 20 and 49	621	0.104
	Západné Slovensko							194	1674	
	Stredné Slovensko							138	1220	
	Východné Slovensko							160	1470	
Degree of urbanization	Cities	54	1109	0.006	92	1415	<0.001	113	1127	0.342
	Towns and suburbs	120	1445		196	2045		178	1721	
	Rural areas	141	2095		161	1581		249	2137	
Education level	Primary/less than primary education	Below 20	68	0.049	Below 20	Between 20 and 49	<0.001	Below 20	Below 20	<0.001
	Secondary education	270	3724		405	4009		482	3930	
	Higher education	Between 20 and 49	857		Between 20 and 49	991		51	1027	
Labor activity status	Employed	66	2730	<0.001	59	2393	<0.001	86	2503	<0.001
	Unemployed	Below 20	299		Below 20	476		Below 20	263	
	Others	241	1620		372	2172		437	2219	

Table 1. *Cont.*

Variable	Category	EHIS 2009			EHIS 2014			EHIS 2019		
		With Diabetes n	Without Diabetes n	*p*-Value	With Diabetes n	Without Diabetes n	*p*-Value	With Diabetes n	Without Diabetes n	*p*-Value
BMI (kg/m^2)	<18.5	Below 20	154	<0.001	Below 20	131	<0.001	Below 20	102	<0.001
	18.5 to 24.9	60	2212		88	2206		81	1883	
	25 to 29.9	130	1529		222	1888		205	1949	
	$\geq$30	109	598		138	816		243	971	
Frequency of walking for transportation purposes for at least 10 min continuously per week	Everyday	113	2049	<0.001	176	2381	<0.001	245	2864	<0.001
	One to six days	116	1989		188	2114		195	1690	
	Never	65	429		85	546		99	422	
Physical activity	2 Days and more	138	2996	<0.001	Between 20 and 29	1503	<0.001	53	1406	<0.001
	One day per week	Between 20 and 49	259		below 20	230		Below 20	200	
	Never	143	1132		399	3308		476	3378	
Frequency of eating fruits	One or more per day	230	2953	0.001	210	2402	0.185	281	2682	0.701
	One to six times a week	69	1492		204	2353		226	1993	
	Less than once a week and never	Below 20	193		Between 20 and 49	286		Between 20 and 49	308	
Frequency of eating vegetables or salad	Once and more a day	191	2366	0.001	192	2220	0.843	248	2363	0.693
	One to six times a week	104	2023		230	2539		255	2333	
	Less than a week and never	Below 20	247		Between 20 and 49	282		Between 20 and 49	287	

Legend: "below 20" represents below 20 observations in the cell, "between 20 and 45" represents observations between 20 and 45 in the cell. "Below 20", "between 20 and 45" are used according to the database guideline of statistical disclosure control. * Sorted by gross domestic product (GDP) at current market prices by NUTS 2 regions of Slovakia in Euros (€) per inhabitant, Bratislavský kraj 39,700 € per inhabitant, Západné Slovensko 15,800 € per inhabitant, Stredné Slovensko 14,100 € per inhabitant, Východné Slovensko 12,200 € per inhabitant. BMI body mass index (kg/m^2).

Table 2. Factors associated with diabetes Variable.

Variable	Category	EHIS 2009	EHIS 2014	EHIS 2019
		OR (95% CI)	OR (95% CI)	OR (95% CI)
Sex (ref: males)	Female	0.85 (0.64–1.14)	0.91 (0.73–1.13)	0.77 (0.62–0.95) *
Age (ref: 65 and older)	15 to 44	0.06 (0.04–0.10) *	0.14 (0.09–0.21) *	0.12 (0.07–0.19) *
	45 to 64	0.46 (0.32–0.65) *	0.54 (0.41–0.71) *	0.71 (0.54–0.95) *
Region (ref: Bratislavský kraj)	Západné Slovensko			0.83 (0.52–1.31)
	Stredné Slovensko			0.90 (0.57–1.44)
	Východné Slovensko			0.92 (0.58–1.46)
Degree of urbanization (ref: rural areas)	Cities	0.82 (0.55–1.22)	0.82 (0.60–1.11)	1.19 (0.87–1.63)
	Towns and suburbs	1.19 (0.87–1.63)	1.14 (0.89–1.45)	1.03 (0.82–1.30)
Education level (ref: higher education)	Primary/less than primary education	0.38 (0.10–1.43)	1.23 (0.47–3.22)	3.25 (1.12–9.46) *
	Secondary education	0.71 (0.48–1.06)	1.29 (0.89–1.88)	1.36 (0.97–1.92)
Labor status (ref: others)	Employed	0.40 (0.27–0.60) *	0.35 (0.25–0.49) *	0.36 (0.25–0.50) *
	Unemployed	0.72 (0.33–1.56)	0.56 (0.33–0.95) *	0.65 (0.36–1.18)
BMI (kg/m^2) (ref: 25 to 29.9)	<18.5	0.64 (0.15–2.69)	0.08 (0.01–0.56)	1.04 (0.34–3.13)
	18.5 to 24.9	0.59 (0.41–0.84) *	0.57 (0.43–0.76) *	0.65 (0.49–0.88) *
	≥30	1.81 (1.30–2.52) *	1.11 (0.86–1.44)	2.04 (1.62–2.57) *
Frequency of walking for transportation purposes for at least 10 min continuously per week (ref: never)	Everyday	0.76 (0.51–1.14)	0.67 (0.49–0.92) *	0.57 (0.42–0.77) *
	One to six days	0.76 (0.51–1.13)	0.70 (0.52–0.96) *	0.69 (0.50–0.94) *
Physical activity (ref: one day per week)	Two days and more	0.65 (0.38–1.11)	1.53 (0.59–3.96)	0.63 (0.31–1.31)
	Neve	0.95 (0.55–1.64)	2.55 (1.02–6.37) *	1.43 (0.73–2.81)
Frequency of eating fruits (ref: one or more per day)	One to six times a week	0.96 (0.67–1.39)	0.98 (0.73–1.32)	1.10 (0.81–1.51)
	Less than once a week and never	1.18 (0.55–2.51)	1.54 (0.87–2.71)	1.07 (0.59–1.92)
Frequency of eating vegetables or salad (ref: one or more per day)	One to six times a week	0.66 (0.48–0.92) *	1.07 (0.80–1.44)	0.82 (0.60–1.11)
	Less than a week and never	0.67 (0.34–1.30)	0.79 (0.41–1.52)	0.80(0.46–1.39)

Legend: * significant association ($p < 0.05$) between with diabetes and without diabetes in regression model. BMI body mass index (kg/m^2).

4. Discussion

To the best of our knowledge, this is the first study analyzing DM burden throughout the three waves of EHIS. Our study presents nationwide, representative data on an adult population covering basic health monitoring indicators, establishing the required baseline data for future evaluation. According to our analysis, the prevalence of DM patients has shifted upwards from 2009 to 2019. This increase is similar to the increase of prevalence in most of the EU member states and the EU average [16].

The results showed that the degree of urbanization was not associated with the risk of DM, contrary to results from a meta-analysis, which considered living in cities would elevate the risk of DM [17]. Socioeconomic factors such as lower education attainment level and labor status may function as factors due to various disparities in the population [18,19]. Results from Denmark and other European countries demonstrate that individuals who are less educated are prone to develop DM than those without. Employment is inversely linked to DM in the three waves; studies have found that individuals with DM exit labor earlier, in addition to reducing their quality of work [20]. Studies suggest that the work environment, type of work and working hours increase the risk of DM [21–23].

Dietary habits have shifted in all EU countries to a "Westernized" diet based on the globalization of food production and distribution, which is most striking in the Mediterranean countries [24]. The vegetable intake in Europe increased by approximately 20% from the middle of the last century until 2006. Historically, countries of southern Europe reported the highest vegetable intake (double other European regions), but in the beginning of this century, it has started to decrease. Fruit consumption has increased across Europe over the past 60 years, in line with vegetable consumption. Market sales of fruit increased until the beginning of the 21st century, followed by a slow decline. At a regional level, the most significant growth was observed in Northern European countries, where it has been slowly declining in recent years [25].

Our study has supported the already established evidence of the association between obesity and overweight with higher DM risk, which was revealed significantly through the three waves. Fruit and vegetable intake is often at the center of health policies tackling DM. In some prospective studies, fruit and vegetable consumption was found to reduce the risk of DM [26,27]. However, in our cross-sectional analyses, their consumption did not show a consistent association with DM occurrence. People with DM may consume similar amounts of fruits and vegetables as individuals without DM following healthcare recommendations [6,7]. This contradiction is not surprising; the lack of a clear link between total vegetable and fruit consumption and the incidence of T2DM was already reported by a meta-analysis [28].

In high-income countries such as Slovakia, DM occurrence may be more associated with obesity and physical inactivity than other socioeconomic factors and urbanization. This might be caused by diet transitioning towards Westernized diet in recent years. The dietary quality of a country may depend on interrelated factors including traditional food patterns, local food availability, the food supply chain, and food policies.

Sedentary lifestyle, decreased physical activity, is more so than an unhealthy diet considered to be associated directly or indirectly with DM. A study using data from Sport and Physical Activity EU Special Eurobarometers reported that sedentary behavior became more prevalent from 2005 to 2017 in the EU member states (except for Finland) and this increasing prevalence occurred in the total population, and men and women separately. The higher prevalence of sedentary lifestyle was observed among men than women, except for Bulgaria, Estonia, Hungary, Latvia and Lithuania [29].

Based on our results, walking as part of active transportation was a protective factor for DM, which aligns with results from a dose–response meta-analysis; walking up to two to three hours per week reduces the risk of DM, but above these levels, there is no reduction in the risk [30]. Individuals who had a sedentary lifestyle were associated with DM. Clinical trials and cohorts found that both aerobic and resistance exercise have an inverse association with DM risk [31]. Meta-analysis suggests a greater risk of DM associated with a large duration of sedentary behavior [32].

Human studies have supported that DM can be delayed or managed integrating a multi-component approach including the main affecting socioeconomic and lifestyle factors through regulating food intake, behavioral changes, and physical activity. A recent randomized controlled trial found that intensive lifestyle interventions resulted in significant weight loss over 12 months, and that more than 60% of participants experienced the remission of diabetes and 30% achieved normoglycemia [33].

Although Slovakia has already established legislative efforts in these domains [34,35]—which include a policy on the organization of sport in educational settings and promoting sport in younger ages, as well as a national action plan to promote sport and physical activity, several nutritional and labeling policies targeting obesity and improving health, and a national DM plan—the burden of DM is still high [36]. However, the experience with policy interventions at the population level is controversial. While population-based interventions are often followed by some successes, these do not necessarily translate into long-term reductions in disease burden. A systematic review, for example, about regulatory interventions targeting population nutrition found that some "isolated regulatory interventions" may have a positive impact on intermediate outcomes, but this change has not reached clinically significant levels—e.g., having such an impact on food intake that can result in reduced incidence of obesity or NCDs [37]. Similarly, another systematic review has not found evidence of the impact of any of the studied interventions on the prevalence of overweight, obesity, or T2DM [38]. Simulation studies project that a network of interventions is needed to achieve the targets in disease burden reduction. In a model with all potential interventions incorporated, the population risk ratios could be reduced both for obesity and T2DM [39].

The major limitation of this study is that, due to the cross-sectional design, causal relationship between DM and risk factors cannot be established. In our analysis, the employment position was not considered. Regions of level 2 of nomenclature of territorial units for statistics (NUTS2) were only available in 2019; thus, the association between DM and living in different regions of Slovkia was studied in 2019 exclusively. EHIS is a self-reported survey; all answers were subjective, which may affect the accuracy and reliability of the reported data and estimated associations. Additionally, due to the homogenous category of DM that was used in the EHIS waves, the different types of DM were not distinguished in our analysis, although the background pathologies of each type are different.

5. Conclusions

Lifestyle characteristics such as dietary habits of eating fruit and vegetables were not associated with DM, but results from 2009 demonstrated that eating vegetables several times a week may reduce the risk of DM and movement for at least 10 min, as walking may prevent DM. However, more lifestyle characteristics and socioeconomic conditions should be studied to evaluate their role in the increased prevalence in Slovakia and in comparison to other EU countries. In the EU, the member states have a leading role in combating DM and its risk factors, to make legislation and provide healthcare services. In Slovakia, existing health policies and actions could not reverse the gradually growing DM burden, indicating that a more systematic approach should be adopted. In conclusion, to achieve improvement, creating and implementing complex policy initiatives and legislative measures seem unavoidable, both at national and EU levels.

Supplementary Materials: The following are available online at https://www.mdpi.com/article/10.3390/nu13072156/s1. Table S1: Distribution of the study population. Definitions of the variables that were used in the study.

Author Contributions: Conceptualization, N.M. and O.V.; data curation, N.M.; formal analysis, N.M., C.A.S.A., N.K., D.W.N. and O.V.; funding acquisition, O.V.; methodology, N.M., C.A.S.A. and O.V.; supervision, O.V.; writing—original draft, N.M.; writing—review and editing, N.M., C.A.S.A., N.K., D.W.N. and O.V. All authors have read and agreed to the published version of the manuscript.

Funding: This work was supported by the GINOP-2.3.2-15-2016-00005 project, co-financed by the European Union under the European Social Fund and European Regional Development Fund. O.V. receives a fellowship from the Hungarian Academy of Sciences (Premium Postdoctoral Research Program).

Institutional Review Board Statement: The Ethical Committee of the Hungarian Scientific Council on Health has approved the research protocol and methodology (61327-2017/EKU).

Informed Consent Statement: Not applicable.

Data Availability Statement: Microdata of EHIS 2009 and 2014 are available via Eurostat upon request. Microdata of EHIS 2019 is available via Statistical Office of the Slovak Republic upon request.

Acknowledgments: We express our gratitude to Eurostat microdata team and to Statistical Office of the Slovak Republic.

Conflicts of Interest: The authors declare no conflict of interest. The funders had no role in the design of the study; in the collection, analyses, or interpretation of data; in the writing of the manuscript, or in the decision to publish the results.

Disclaimer: The responsibility for all conclusions drawn from the data lies entirely with the authors.

Source of the Data: This study is based on data from Eurostat, European Health Interview survey 2009, 2014, and Statistical Office of the Slovak Republic, European Health Interview Survey, 2019.

References

1. Institute for Health Metrics and Evaluation (IHME). *GBD Compare*; IHME, University of Washington: Seattle, WA, USA, 2019.
2. World Health Organization. Fact Sheets: Diabetes. Available online: https://www.who.int/news-room/fact-sheets/detail/diabetes (accessed on 15 May 2021).
3. International Diabetes Federation (IDF). Type 2 Diabetes. Available online: https://idf.org/aboutdiabetes/type-2-diabetes.html (accessed on 15 May 2021).
4. International Diabetes Federation. *IDF Diabetes Atlas*, 9th ed.; International Diabetes Federation: Brussels, Belgium, 2019.
5. European Parliament. European Parliament News: Parliament Calls for EU Diabetes Strategy. Available online: https://www.europarl.europa.eu/news/en/press-room/20120313IPR40731/parliament-calls-for-eu-diabetes-strategy (accessed on 15 May 2021).
6. World Health Organization. Action Plan for the Prevention and Control of Noncommunicable Diseases in the WHO European Region. In Proceedings of the Regional Committee for Europe 66th Session 2016, Copenhagen, Denmark, 12–15 September 2016.
7. European Commission. Non-Communicable Diseases Overview. Available online: https://ec.europa.eu/health/non_communicable_diseases/overview_en (accessed on 27 March 2021).
8. Eurostat. Glossary: European Health Interview Survey (EHIS). Available online: https://ec.europa.eu/eurostat/statistics-explained/index.php/Glossary:European_health_interview_survey_(EHIS) (accessed on 27 March 2021).
9. World Health Organization. Diabetes Mellitus. Fact Sheet, no. 138. April 2002. Available online: https://www.who.int/health-topics/diabetes (accessed on 13 May 2021).
10. Schwingshackl, L.; Knüppel, S.; Michels, N.; Schwedhelm, C.; Hoffmann, G.; Iqbal, K.; De Henauw, S.; Boeing, H.; Devleesschauwer, B. Intake of 12 food groups and disability-adjusted life years from coronary heart disease, stroke, type 2 diabetes, and colorectal cancer in 16 European countries. *Eur. J. Epidemiol.* **2019**, *34*, 765–775. [CrossRef]
11. Carbone, S.; Del Buono, M.G.; Ozemek, C.; Lavie, C.J. Obesity, risk of diabetes and role of physical activity, exercise training and cardiorespiratory fitness. *Prog. Cardiovasc. Dis.* **2019**, *62*, 327–333. [CrossRef] [PubMed]
12. Seiglie, J.A.; Marcus, M.-E.; Ebert, C.; Prodromidis, N.; Geldsetzer, P.; Theilmann, M.; Agoudavi, K.; Andall-Brereton, G.; Aryal, K.K.; Bicaba, B.W.; et al. Diabetes Prevalence and Its Relationship With Education, Wealth, and BMI in 29 Low- and Middle-Income Countries. *Diabetes Care* **2020**, *43*, 767–775. [CrossRef] [PubMed]
13. Schipf, S.; Ittermann, T.; Tamayo, T.; Holle, R.; Schunk, M.; Maier, W.; Meisinger, C.; Thorand, B.; Kluttig, A.; Greiser, K.H.; et al. Regional differences in the incidence of self-reported type 2 diabetes in Germany: Results from five population-based studies in Germany (DIAB-CORE Consortium). *J. Epidemiol. Commun. Health* **2014**, *68*, 1088–1095. [CrossRef]
14. Du, Y.; Baumert, J.; Paprott, R.; Teti, A.; Heidemann, C.; Scheidt-Nave, C. Factors associated with undiagnosed type 2 diabetes in Germany: Results from German Health Interview and Examination Survey for Adults 2008–2011. *BMJ Open Diabetes Res. Care* **2020**, *8*, e001707. [CrossRef]

15. Rawshani, A.; Svensson, A.-M.; Zethelius, B.; Eliasson, B.; Rosengren, A.; Gudbjörnsdottir, S. Association Between Socioeconomic Status and Mortality, Cardiovascular Disease, and Cancer in Patients with Type 2 Diabetes. *JAMA Intern. Med.* **2016**, *176*, 1146–1154. [CrossRef]

16. Global Burden of Disease Collaborative Network. *Global Burden of Disease Study 2019 (GBD 2019) Results*; Global Burden of Disease Collaborative Network: Seattle, WA, USA, 2020.

17. Braver, N.R.D.; Lakerveld, J.; Rutters, F.; Schoonmade, L.J.; Brug, J.; Beulens, J.W.J. Built environmental characteristics and diabetes: A systematic review and meta-analysis. *BMC Med.* **2018**, *16*, 1–26. [CrossRef]

18. Kumari, M.; Head, J.; Marmot, M. Prospective Study of Social and Other Risk Factors for Incidence of Type 2 Diabetes in the Whitehall II Study. *Arch. Intern. Med.* **2004**, *164*, 1873–1880. [CrossRef]

19. Stringhini, S.; Zaninotto, P.; Kumari, M.; Kivimäki, M.; Batty, G.D. Lifecourse socioeconomic status and type 2 diabetes: The role of chronic inflammation in the English Longitudinal Study of Ageing. *Sci. Rep.* **2016**, *6*, 24780. [CrossRef] [PubMed]

20. Rumball-Smith, J.; Barthold, D.; Nandi, A.; Heymann, J. Diabetes Associated with Early Labor-Force Exit: A Comparison of Sixteen High-Income Countries. *Health Aff.* **2014**, *33*, 110–115. [CrossRef] [PubMed]

21. Magadzire, B.P.; Mathole, T.; Ward, K. Reasons for missed appointments linked to a public-sector intervention targeting patients with stable chronic conditions in South Africa: Results from in-depth interviews and a retrospective review of medical records. *BMC Fam. Pract.* **2017**, *18*, 1–10. [CrossRef] [PubMed]

22. Pazmino, L.; Esparza, W.; Aladro-Gonzalvo, A.; León, E. Impact of Work and Recreational Physical Activity on Prediabetes Condition among U.S. Adults: NHANES 2015–2016. *Int. J. Environ. Res. Public Health* **2021**, *18*, 1378. [CrossRef] [PubMed]

23. Kang, E. Differences in Clinical Indicators of Diabetes, Hypertension, and Dyslipidemia Among Workers Who Worked Long Hours and Shift Work. *Workplace Health Saf.* **2021**, *69*, 268–276. [CrossRef] [PubMed]

24. Vareiro, D.; Bach-Faig, A.; Quintana, B.R.; Bertomeu, I.; Buckland, G.; De Almeida, M.D.V.; Serra-Majem, L. Availability of Mediterranean and non-Mediterranean foods during the last four decades: Comparison of several geographical areas. *Public Health Nutr.* **2009**, *12*, 1667–1675. [CrossRef] [PubMed]

25. Riccardi, G.; Vitale, M.; Vaccaro, O. Are Europeans moving towards dietary habits more suitable for reducing cardiovascular disease risk? *Nutr. Metab. Cardiovasc. Dis.* **2020**, *30*, 1857–1860. [CrossRef] [PubMed]

26. Jiang, Z.; Sun, T.-Y.; He, Y.; Gou, W.; Zuo, L.-S.-Y.; Fu, Y.; Miao, Z.; Shuai, M.; Xu, F.; Xiao, C.; et al. Dietary fruit and vegetable intake, gut microbiota, and type 2 diabetes: Results from two large human cohort studies. *BMC Med.* **2020**, *18*, 1–11. [CrossRef] [PubMed]

27. Cooper, A.; Sharp, S.J.; Lentjes, M.; Luben, R.; Khaw, K.-T.; Wareham, N.J.; Forouhi, N.G. A Prospective Study of the Association Between Quantity and Variety of Fruit and Vegetable Intake and Incident Type 2 Diabetes. *Diabetes Care* **2012**, *35*, 1293–1300. [CrossRef]

28. Neuenschwander, M.; Ballon, A.; Weber, K.S.; Norat, T.; Aune, D.; Schwingshackl, L.; Schlesinger, S. Role of diet in type 2 diabetes incidence: Umbrella review of meta-analyses of prospective observational studies. *BMJ* **2019**, *366*, l2368. [CrossRef]

29. López-Valenciano, A.; Mayo, X.; Liguori, G.; Copeland, R.J.; Lamb, M.; Jimenez, A. Changes in sedentary behaviour in European Union adults between 2002 and 2017. *BMC Public Health* **2020**, *20*, 1–10. [CrossRef]

30. Aune, D.; Norat, T.; Leitzmann, M.; Tonstad, S.; Vatten, L.J. Physical activity and the risk of type 2 diabetes: A systematic review and dose–response meta-analysis. *Eur. J. Epidemiol.* **2015**, *30*, 529–542. [CrossRef]

31. Ross, L.M.; Slentz, C.A.; Zidek, A.M.; Huffman, K.M.; Shalaurova, I.; Otvos, J.D.; Connelly, M.A.; Kraus, V.B.; Bales, C.W.; Houmard, J.A.; et al. Effects of Amount, Intensity, and Mode of Exercise Training on Insulin Resistance and Type 2 Diabetes Risk in the STRRIDE Randomized Trials. *Front. Physiol.* **2021**, *12*. [CrossRef]

32. Bailey, D.P.; Hewson, D.J.; Champion, R.B.; Sayegh, S.M. Sitting Time and Risk of Cardiovascular Disease and Diabetes: A Systematic Review and Meta-Analysis. *Am. J. Prev. Med.* **2019**, *57*, 408–416. [CrossRef] [PubMed]

33. Taheri, S.; Zaghloul, H.; Chagoury, O.; Elhadad, S.; Ahmed, S.H.; El Khatib, N.; Amona, R.A.; El Nahas, K.; Suleiman, N.; Alnaama, A.; et al. Effect of intensive lifestyle intervention on bodyweight and glycaemia in early type 2 diabetes (DIADEM-I): An open-label, parallel-group, randomised controlled trial. *Lancet Diabetes Endocrinol.* **2020**, *8*, 477–489. [CrossRef]

34. Richardson, E.; Zaletel, J.; Ellen, N. *National Diabetes Plans: In Europe What Lessons Are there for the Prevention and Control of Chronic Diseases in Europe?* National Institute of Public Health Ljubljana: Ljubljana, Slovenia, 2016.

35. World Health Organization. Diabetes Country Profiles: Slovakia. Available online: https://www.who.int/diabetes/country-profiles/svk_en.pdf (accessed on 15 May 2021).

36. World Obesity Federation. Global Obesity Observatory. Available online: https://data.worldobesity.org/ (accessed on 15 May 2021).

37. Sisnowski, J.; Street, J.M.; Merlin, T. Improving food environments and tackling obesity: A realist systematic review of the policy success of regulatory interventions targeting population nutrition. *PLoS ONE* **2017**, *12*, e0182581. [CrossRef]

38. Roberts, S.; Pilard, L.; Chen, J.; Hirst, J.; Rutter, H.; Greenhalgh, T. Efficacy of population-wide diabetes and obesity prevention programs: An overview of systematic reviews on proximal, intermediate, and distal outcomes and a meta-analysis of impact on BMI. *Obes. Rev.* **2019**, *20*, 947–963. [CrossRef] [PubMed]

39. Nianogo, R.A.; Arah, O.A. Impact of Public Health Interventions on Obesity and Type 2 Diabetes Prevention: A Simulation Study. *Am. J. Prev. Med.* **2018**, *55*, 795–802. [CrossRef]

nutrients

Article

High Fructose Intake Contributes to Elevated Diastolic Blood Pressure in Adolescent Girls: Results from The HELENA Study

Laurent Béghin [1,*], Inge Huybrechts [2,3], Elodie Drumez [4,5], Mathilde Kersting [6], Ryan W Walker [7], Anthony Kafatos [8], Denes Molnar [9], Yannis Manios [10], Luis A Moreno [11], Stefaan De Henauw [2] and Frédéric Gottrand [1]

[1] Univ. Lille, Inserm, CHU Lille, U1286—INFINITE and CIC-1403, F-59000 Lille, France; Frederic.gottrand@chu-lille.fr

[2] Department of Public Health and Primary Care, Faculty of Medicine and Health Sciences, Ghent University, B-9000 Ghent, Belgium; HuybrechtsI@iarc.fr (I.H.); Stefaan.DeHenauw@ugent.be (S.D.H.)

[3] Dietary Exposure Assessment Group, International Agency for Research on Cancer, F-69000 Lyon, France

[4] Univ. Lille, CHU Lille, ULR 2694—METRICS: Évaluation des Technologies de Santé et des Pratiques Médicales, F-59000 Lille, France; elodie.drumez@chu-lille.fr

[5] CHU Lille, Department of Biostatistics, F-59000 Lille, France

[6] Research Department of Child Nutrition, Pediatric University Clinic, Ruhr-University Bochum, D-44791 Bochum, Germany; mathilde.kersting@ruhr-uni-bochum.de

[7] Department of Environmental Medicine and Public Health, Icahn School of Medicine at Mount Sinai, New York, NY 10029, USA; ryan.walker@mssm.edu

[8] Preventive Medicine and Nutrition Clinic, University of Crete School of Medicine, G-14122 Crete, Greece; kafatos@med.uoc.gr

[9] Department of Pediatrics, University of Pecs, H-7600 Pecs, Hungary; denes.molnar@aok.pte.hu

[10] Department of Nutrition and Dietetics, University of Harakopio, G-10431 Athens, Greece; manios@hua.gr

[11] GENUD (Growth, Exercise, Nutrition and Development) Research Group Escuela Universitaria de Ciencas de la Salud, Universidad de Zaragoza, S-50009 Zaragoza, Spain; lmoreno@unizar.es

* Correspondence: Laurent.beghin@chu-lille.fr

Citation: Béghin, L.; Huybrechts, I.; Drumez, E.; Kersting, M.; Walker, R.W.; Kafatos, A.; Molnar, D.; Manios, Y.; Moreno, L.A.; De Henauw, S.; et al. High Fructose Intake Contributes to Elevated Diastolic Blood Pressure in Adolescent Girls: Results from The HELENA Study. *Nutrients* **2021**, *13*, 3608. https://doi.org/10.3390/nu13103608

Academic Editor: Luisa Cigliano

Received: 12 August 2021
Accepted: 10 October 2021
Published: 15 October 2021

Publisher's Note: MDPI stays neutral with regard to jurisdictional claims in published maps and institutional affiliations.

Abstract: Background: The association between high fructose consumption and elevated blood pressure continues to be controversial, especially in adolescence. The aim of this study was to assess the association between fructose consumption and elevated blood pressure in an European adolescent population. Methods: A total of 1733 adolescents (mean ± SD age: 14.7 ± 1.2; percentage of girls: 52.8%) were analysed from the Healthy Lifestyle in Europe by Nutrition in Adolescence (HELENA) study in eight European countries. Blood pressure was measured using validated devices and methods for measuring systolic blood pressure (SBP) and diastolic blood pressure (DBP). Dietary data were recorded via repeated 24 h recalls (using specifically developed HELENA–DIAT software) and converted into pure fructose (monosaccharide form) and total fructose exposure (pure fructose + fructose from sucrose) intake using a specific fructose composition database. Food categories were separated at posteriori in natural vs. were non-natural foods. Elevated BP was defined according to the 90th percentile cut-off values and was compared according to tertiles of fructose intake using univariable and multivariable mixed logistic regression models taking into account confounding factors: centre, sex, age and z-score–BMI, MVPA (Moderate to Vigorous Physical Activity) duration, tobacco consumption, salt intake and energy intake. Results: Pure fructose from non-natural foods was only associated with elevated DBP (DBP above the 10th percentile in the highest consuming girls (OR = 2.27 (1.17–4.40); $p = 0.015$) after adjustment for cofounding factors. Conclusions: Consuming high quantities of non-natural foods was associated with elevated DBP in adolescent girls, which was in part due to high fructose levels in these foods categories. The consumption of natural foods containing fructose, such as whole fruits, does not impact blood pressure and should continue to remain a healthy dietary habit.

Keywords: pure fructose consumption; adolescent; blood pressure

1. Introduction

The introduction of industrial foods and new manufacturing food process at the beginning of 20th century has modified sources of fructose in humans. For centuries, the main source of fructose in humans was the consumption of natural foods containing fructose such as honey, fruits and vegetables intrinsically rich in the pure monosaccharide form (α-D-glucopyranose) [1]. Since the substitution of sucrose for fructose in many foods and beverages, the consumption of extrinsic fructose has increased in parallel with increasing industrially manufactured or confectionary foods commonly appearing as an added sugar [2,3]. This high use in industrially manufactured or confectionary foods can be attributed to the low cost of fructose compared to sucrose. For nearly two decades, the overconsumption of pure fructose has been reported as a potential unique dietary risk factor because of links to obesity risk, the preeminent epidemic and health concern in the US and worldwide [4,5]. The overconsumption of pure fructose is also associated with risk of chronic metabolic noncommunicable diseases such as hypertension [6], cardiovascular diseases [3,7], type 2 diabetes [8], metabolic syndrome [9], steatohepatitis [10] and inflammatory conditions [11]. In this context, overconsumption of pure fructose became an important public health issue in both the scientific and public domain [12–14].

Arterial hypertension is one of the top five leading worldwide risks of mortality [15]. It is also the most common modifiable risk factor for cardiovascular disease [16], with the tracking of hypertension across adolescence to adulthood being a very important health metric [17]. Previous studies have suggested that high levels of pure fructose intake contribute to elevation in blood pressure [6,18–22]. This observation has been confirmed in animal studies, where consuming a diet comprising 60% of total energy from fructose induces hypertension [23–27]. Few studies have examined the possible deleterious health effects of pure fructose consumption among adolescents [6], despite the observation that adolescents are the highest consumers of fructose [28,29]. To our knowledge, only two studies have assessed the impact of fructose-rich beverages such as caffeinated and/or sugar-sweetened beverages (SSB) on blood pressure in adolescents [18,30], both showing that consumption of these beverages was associated with higher systolic blood pressure. However, results from these studies were highly criticized because they did not take into account of potential confounding factors such as physical activity level, tobacco consumption and salt intake. Moreover, they did not examine the total all food sources containing fructose [31]; they were only focalized on some food categories such as SSB, for example. The impact of fructose consumption on adolescent blood pressure continues to be controversial [32,33], and the relationship between blood pressure and fructose consumption should be viewed as a major public health issue [34].

The purpose of our study was to assess the association of fructose intake from various foods categories on the blood pressure of adolescents. The hypothesis was that high fructose intake from several food sources extrinsically fructose rich could increase blood pressure in adolescents.

2. Materials and Methods

2.1. Sample

Data were derived from the Healthy Lifestyle in Europe by Nutrition in Adolescence (HELENA) study from eight different countries in Northern (Ghent in Belgium, Lille in France, Dortmund in Germany, Stockholm in Sweden), Central (Vienna in Austria) and Southern Europe (Athens in Greece, Roma in Italy, Zaragoza in Spain) between 2006 and 2007, as previously described [35]. Briefly, the HELENA study was a multisite study designed to obtain reliable, comparable data about nutritional habits and patterns, body composition and levels of physical activity and fitness from European adolescents aged 12.5 y to 17.5 y. Details of the sampling procedures, field team preparation, the pilot study and data reliability are presented elsewhere [36]. The study was performed in accordance with the ethical guidelines of the Declaration of Helsinki, good clinical practice and the legislation concerning clinical research in each of the participating countries. The protocol was approved by the appropriate independent ethics committee for each study centre, and written informed consent was obtained from both parents and adolescents [37]. In total, 3865 adolescents were enrolled through their schools, which were randomly selected according to a proportional cluster sampling methodology that took age and socioeconomic status into account [38]. From the initial data set of 3528 analysable adolescents, 1705 were complete for fructose intake and blood pressure analysis.

2.2. Assessment of Daily Fructose Intake

Dietary intake was assessed by two nonconsecutive 24 h recalls performed at any point in the week [39]. The recalls did not necessarily include a weekday and a weekend day for each individual. The 24 h recalls were recorded with a self-administered, computer-based tool: the HELENA Dietary Intake Assessment Tool (HELENA–DIAT) adapted from the YANA-C tool developed with and validated for Flemish adolescents [40]. The HELENA–DIAT tool is based on six meal occasions (breakfast, morning snacks, lunch, afternoon snacks, evening meal and evening snacks) on the previous day. Trained dieticians assisted the adolescents to complete the 24 h recalls when needed. The adolescents selected all foods and beverages consumed at each meal occasion from a standardized food list [41]. Dietary sources of fructose were extracted according to Mesana et al. [42] and Duffey et al. [43] for SSB. Fructose in its pure monosaccharide form was computed using fructose content data from several studies/databases described in Supplementary Table S1 [44–47]. As fructose is also made available through the consumption of sucrose (a disaccharide made of α-D-glucopyranosyl $(1{\rightarrow}2)$-β-D-fructofuranoside), commonly appearing as an added sugar in processed foods [2,3], fructose was also expressed as fructose total exposure. Total fructose exposure was calculated as fructose from monosaccharide form plus fructose from sucrose ($0.5 \times$ sucrose per 100 g), in accordance with Ramne et al. [48]. Pure fructose and total fructose exposure were separately analysed because of their difference in intestinal/blood absorption [9]. Two foods categories were created at posteriori using a classification close to Aeberli et al. [49]: (i) food categories extrinsically fructose rich such as industrial/manufactured/confectionary non-natural food products: sugar-sweetened beverages, nonchocolate confectionary, chocolate, cakes/pies/biscuits, desserts and puddings, breakfast cereals and others sources categories [50], and (ii) intrinsically fructose-rich natural foods: fruit/vegetable juices, honey/jam/syrup and whole fruits.

2.3. Assessment of Anthropometrics

Weight was measured in underwear, with shoes removed, using an electronic scale (SECA 861, SECA, Birmingham, UK) to the nearest 0.1 kg. Height was measured with shoes removed using a telescopic height measuring instrument (SECA 225) to the nearest 0.1 cm. Body Mass Index (BMI) was calculated by dividing body weight (kg) by the square of height (m^2), and BMI z-score was calculated using the lambda, mu and sigma method [51].

2.4. Assessment of Blood Pressure

Blood pressure (BP) measurements were performed following the recommendations for adolescent population [52]. BP was measured in mmHg twice after weight and height measurements were taken. The subjects were seated in a separate, quiet room for 10 min with their backs supported and feet on the ground. Two BP readings were taken with a 10 min interval of quiet rest. The lower of the two measurements was used. Systolic blood pressure (SBP) and diastolic blood pressure (DBP) were measured by the arm blood pressure oscillometric monitor device OMRON HEALTHCARE HEM7001 (OMRON HEALTHCARE, Koyoto, Japan), which has been approved by the British Hypertension Society [53]. Data collection HEM7001 procedures have been described earlier [36]. The use of percentiles is usually used to define reference clinical data in a set of data from a population. Adolescents of this study were separated in 2 groups according to percentiles of blood pressure data: systolic and diastolic, >90th percentile or ≤90th percentile. Elevated BP was defined where systolic or diastolic blood pressure was >90th percentile of the

HELENA population analysed. The cut-off of the 90th percentile to define elevated blood pressure for systolic and diastolic has been described by Flynn, J.T. et al. [54].

2.5. Assessment of Physical Activity

Physical activity was measured using a uniaxial accelerometer (ActiGraph GT1M®, Pensacola, FL, USA) in pure living conditions [55]. The accelerometer recorded activity for 7 consecutive days the same week as daily fructose intake collection. PA collection by the accelerometer was taken off at night. Moderate to vigorous physical activity (MVPA) time spent was computed when PA counts/min were more than 2000 count/min, cut-off defined by Ekelund et al. [56].

2.6. Assessment of Tobacco Consumption

Regular tobacco smoking, defined as the regular consumption of at least one cigarette per day in the past month [57], was assessed using the ad hoc HELENA study questionnaire [36].

2.7. Statistical Analysis

Data are presented as frequency (percentage) for categorical variables and mean ± standard deviation (SD) or median interquartile range (IQR) for continuous variables. Normality of distribution was checked graphically and by using the Shapiro–Wilk test. To assess the potential bias related to missing or incomplete nutrients and BP parameters, the main characteristics of included and excluded adolescents were compared using Student's t-test for continuous variables and Chi-square test for categorical variables. To evaluate the magnitude of differences between analysed and nonanalysed participants, we calculated the absolute standardized differences; a standardized difference >20% denotes a meaningful imbalance. Associations of elevated SBP and DBP values with pure and total fructose exposure (categorized according to tertiles distributions) were investigated with and without adjustment for confounding factors (data of confounding factors are presented in Supplemental Table S2). Mixed logistic regression models were used including elevated blood pressure as dependent variables, fructose exposure and the confounding variables as independent fixed effects and centre as a random effect. To avoid case deletion in multivariate analyses, missing data were imputed by multiple imputations using the regression-switching approach (chained equations, m = 10 imputations) [58]. The imputation procedure was performed under the missing-at-random assumption using all variables, with the predictive mean-matching method for continuous variables and logistic regression (binary, ordinal or multinomial) models for categorical variables. Rubin's rules were used to combine the estimates derived from multiple imputed data sets [58]. Due to established sex differences in prior similar studies of blood pressure in children, analyses were stratified by sex [59,60]. All statistical tests were done at the two-tailed α level of 0.05. Data were analysed with SAS software version 9.4 (SAS Institute Inc., Cary, NC, USA).

3. Results

Main characteristics of the 1705 participants are presented in Supplemental Table S3. Body mass of excluded adolescents from statistical analysis was slightly higher (ASD = 26.6%) than included adolescents. Consequently, a similar difference (ASD = 35.6%) was found for z-score BMI.

Median pure fructose intake was 34.69 g/day in girls (*n* = 901) and 45.29 g/day in boys (*n* = 804) from all food sources and 12.94 g/day in girls and 21.14 g/day in boys from non-natural foods sources only (Table 1). Total fructose exposure was higher and reached 51.24 g/day in girls and 63.62 g/day in boys from all food sources and 24.47 g/day in girls and 35.89 g/day in boys from non-natural foods sources only.

Table 1. Daily fructose intake from various fructose-containing foods in girls and boys.

Girls	Foods	Pure Fructose (g/day)	Total Fructose (g/day)
	Sugar-sweetened beverages	7.98 (2.51 to 17.22)	8.80 (2.76 to 18.98)
	Nonchocolate confectionary	0.12 (0.05 to 0.28)	1.00 (0.40 to 2.41)
	Chocolate	1.53 (0.67 to 2.73)	1.71 (0.75 to 3.05)
	Cakes/pies/biscuits	1.95 (1.07 to 2.90)	8.79 (4.83 to 13.10)
	Desserts and puddings	0.09 (0.06 to 0.19)	0.22 (0.14 to 0.47)
	Breakfast and cereals	0.01 (0.01 to 0.20)	0.13 (0.07 to 1.90)
	Others	0.02 (0.01 to 0.02)	0.02 (0.01 to 0.02)
	Fruit and vegetable juices	5.14 (1.85 to 10.08)	5.86 (2.10 to 11.49)
	Honey/jam/syrup	1.31 (0.37 to 5.28)	1.61 (0.45 to 6.50)
	Fruits	8.60 (4.62 to 13.70)	10.45 (5.61 to 16.64)
	Fructose from all food sources	34.69 ** (25.39 to 46.35)	51.24 ** (39.17 to 65.12)
	Fructose from non-natural foods *	12.94 ** (7.57 to 22.36)	25.47 ** (18.04 to 36.76)
Boys	**Foods**	**Pure Fructose (g/day)**	**Total Fructose (g/day)**
	Sugar-sweetened beverages	15.10 (6.04 to 27.45)	16.64 (6.66 to 30.25)
	Nonchocolate confectionary	0.06 (0.03 to 0.20)	0.55 (0.28 to 1.78)
	Chocolate	1.71 (0.91 to 3.74)	1.91 (1.01 to 4.18)
	Cakes/pies/biscuits	2.12 (1.03 to 3.41)	9.58 (4.64 to 15.40)
	Desserts and puddings	0.05 (0.04 to 0.08)	0.12 (0.09 to 0.19)
	Breakfast and cereals	0.03 (0.02 to 0.28)	0.25 (0.15 to 2.70)
	Others	0.01 (0.01 to 0.02)	0.01 (0.01 to 0.02)
	Fruit and vegetable juices	5.43 (1.94 to 11.19)	6.19 (2.21 to 12.75)
	Honey/jam/syrup	1.40 (0.61 to 5.81)	1.73 (0.76 to 7.15)
	Fruits	8.05 (3.59 to 13.29)	9.78 (4.37 to 16.13)
	Fructose from all food sources	45.29 ** (32.19 to 59.98)	63.62 ** (47.55 to 81.12)
	Fructose from non-natural foods *	21.14 ** (11.69 to 34.60)	35.89 ** (23.61 to 51.55)

Values are medians and interquartile range * Non-natural foods are: sugar-sweetened beverages, nonchocolate confectionary, chocolate, cakes/pies/biscuits, desserts and puddings, breakfast cereals and others sources categories. ** values presented are medians; the sum of medians from each foods categories do not follow an arithmetic computation.

Across the two food classifications "all food sources" or "non-natural", adolescents were classified into low/middle/high tertiles for "pure" and "total fructose" exposure (Table 2).

Table 2. Tertiles of daily pure fructose intake from all food sources and from non-natural foods sources.

Tertiles Levels For Girls	*n* (%)	Ranges for Pure Fructose (g/day)	*n* (%)	Ranges for Total Fructose Exposure (g/day)
All food sources				
Low	284 (31.5%)	7.85 to 27.60	285 (31.6%)	13.78 to 41.88
Middle	311 (34.5%)	27.63 to 41.76	310 (34.4%)	41.92 to 59.38
High	306 (34.0%)	41.90 to 130.86	306 (34.0%)	59.52 to 160.16
Non-natural foods				
Low	296 (32.9%)	1.39 to 9.12	296 (32.8%)	3.80 to 20.18
Middle	305 (33.8%)	9.15 to 18.73	304 (33.7%)	20.22 to 32.01
High	300 (33.3%)	18.76 to 81.50	301 (33.4%)	32.15 to 112.56
Tertiles Levels For Boys	***n* (%)**	**Ranges for Pure Fructose (g/day)**	***n* (%)**	**Ranges for Total Fructose Exposure (g/day)**
All food sources				
Low	258 (32.1%)	9.55 to 35.65	256 (31.8%)	13.17 to 51.76
Middle	270 (33.6%)	35.65 to 53.98	273 (34.0%)	51.77 to 72.70
High	276 (34.3%)	54.00 to 136.75	275 (34.2%)	72.83 to 191.25
Non-natural foods				
Low	267 (33.2%)	2.21 to 14.15	269 (33.5%)	4.56 to 27.55
Middle	272 (33.8%)	14.19 to 28.92	267 (33.2%)	27.61 to 45.01
High	265 (33.0%)	28.9 to 114.79	268 (33.3%)	45.07 to 161.18

Table 3 presents mean ± SD of blood pressure in all adolescents and cut-off values for elevated BP or mildly elevated BP. For girls, cut-offs for 90th and 75th percentile were 125 and 118 mmHg for SBP and 75 and 69 mmHg for DBP, respectively. For boys, cut-offs for 90th and 75th percentile were 137 and 127 mmHg for SBP and 75 and 69 mmHg for DBP respectively.

Table 3. Blood pressure data and threshold for elevated blood pressure.

Blood Pressure for Girls	*n*	**Mean ± SD (mmHg)**
Systolic blood pressure (SBP)	901	111.56 ± 11.21
Elevated SBP above the 90th percentile > 125 (mmHg)	84	132.52 ± 7.27
Mildly elevated above the 75th percentile > 118 (mmHg)	223	125.7 ± 7.1
Mildly elevated above 110 (mmHg)	487	119.6 ± 7.6
Diastolic blood pressure (DBP)	901	64.48 ± 8.50
Elevated DBP above the 90th percentile > 75 (mmHg)	86	80.52 ± 6.10
Mildly elevated above the 75th percentile > 69 (mmHg)	228	75.4 ± 5.7
Mildly elevated above 70 (mmHg)	199	76.1 ± 5.7
Blood Pressure for Boys	*n*	**Mean ± SD (mmHg)**
Systolic blood pressure (SBP)	804	119.61 ± 13.40
Elevated SBP above the 90th percentile > 137 (mmHg)	79	145.32 ± 7.85
Mildly elevated above the 75th percentile > 127 (mmHg)	203	137.1 ± 8.5
Mildly elevated above 110 (mmHg)	600	125.1 ± 10.6
Diastolic blood pressure (DBP)	804	63.99 ± 8.41
Elevated DBP above the 90th percentile > 75 (mmHg)	86	79.21 ± 4.33
Mildly elevated above the 75th percentile > 69 (mmHg)	197	75.0 ± 4.8
Mildly elevated above 70 (mmHg)	170	75.8 ± 4.7

Abbreviations: SBP = systolic blood pressure; DBP = diastolic blood pressure.

Table 4 shows the association between elevated systolic blood pressure and pure and total fructose exposure from various fructose-containing foods for girls and boys. Elevated blood pressure values were not associated with pure and total fructose exposure from fructose-containing foods.

Table 5 shows the association between elevated diastolic blood pressure and pure and total fructose exposure from various fructose-containing foods. Elevated diastolic blood pressure values were not associated with pure and total fructose exposure from various fructose-containing foods.

Table 6 shows the association between elevated systolic blood pressure and pure and total fructose exposure from non-natural foods. Elevated systolic blood pressure values were not associated with pure and total fructose exposure from various fructose-containing foods.

Table 7 shows the association between elevated diastolic blood pressure and pure and total fructose exposure from non-natural foods for girls and boys. Among the highest tertiles of fructose consumption from non-natural foods, we found an association with elevated diastolic blood pressure in girls (OR = 2.27 (1.17–4.40); p = 0.015) that persisted after adjustment for cofounding factors compared to the lowest tertiles of "pure" fructose.

Table 4. Association between elevated systolic blood pressure and pure and total fructose exposure from various fructose-containing foods.

For Girls	Elevated SBP (n = 84)	Model 1 OR (95%CI)	p	Model 2 OR (95%CI)	p
Pure fructose			0.97 *		0.66 *
Low	22/284 (7.7%)	1.00 (ref.)	-	1.00 (ref.)	-
Middle	22/311 (7.1%)	0.72 (0.37 to 1.39)	0.32	0.76 (0.38 to 1.53)	0.45
High	40/306 (13.1%)	0.96 (0.51 to 1.84)	0.91	1.14 (0.55 to 2.39)	0.72
Total fructose exposure			0.48 *		0.86 *
Low	24/285 (8.4%)	1.00 (ref.)	-	1.00 (ref.)	-
Middle	21/310 (6.8%)	0.67 (0.35 to 1.28)	0.23	0.80 (0.40 to 1.60)	0.54
High	39/306 (12.7%)	0.83 (0.44 to 1.57)	0.57	1.06 (0.50 to 2.21)	0.89
For Boys	Elevated SBP (n = 79)	Model 1 OR (95%CI)	p	Model 2 OR (95%CI)	p
Pure fructose			0.12 *		0.035 *
Low	24/258 (9.3%)	1.00 (ref.)	-	1.00 (ref.)	-
Middle	25/270 (9.3%)	0.72 (0.38 to 1.34)	0.30	0.55 (0.27 to 1.09)	0.084
High	30/276 (10.9%)	0.60 (0.31 to 1.14)	0.12	0.44 (0.21 to 0.93)	0.031
Total fructose exposure			0.44 *		0.13 *
Low	21/256 (8.2%)	1.00 (ref.)	-	1.00 (ref.)	-
Middle	28/273 (10.3%)	0.96 (0.51 to 1.80)	0.89	0.64 (0.32 to 1.30)	0.22
High	30/275 (10.9%)	0.78 (0.41 to 1.50)	0.46	0.54 (0.25 to 1.17)	0.12

High SBP is defined as a value greater than the 90th percentile (>125 mmHg for girls and >137 mmHg for boys). Model 1: adjusted for centre. Model 2: adjusted for centre, age, Z-score BMI, MVPA duration, tobacco consumption, salt intake and energy intake and calculated after multiple imputations (m = 10) to handle missing data. * p calculated using fructose levels as ordinal variable. Abbreviations: SBP = systolic blood pressure; OR = odds-ratio; CI = confidence interval.

Table 5. Association between elevated diastolic blood pressure and pure and total fructose exposure from various fructose-containing foods.

For Girls	Elevated DBP (n = 86)	Model 1 OR (95%CI)	p	Model 2 OR (95%CI)	p
Pure fructose			0.12 *		0.038 *
Low	23/284 (8.1%)	1.00 (ref.)	-	1.00 (ref.)	-
Middle	18/311 (5.8%)	0.68 (0.35 to 1.33)	0.26	0.67 (0.34 to 1.34)	0.26
High	45/308 (14.7%)	1.51 (0.81 to 2.70)	0.19	1.93 (0.99 to 3.75)	0.052
Total fructose exposure			0.45 *		0.086 *
Low	23/285 (8.1%)	1.00 (ref.)	-	1.00 (ref.)	-
Middle	23/310 (7.4%)	0.89 (0.47 to 1.66)	0.71	1.10 (0.57 to 2.10)	0.78
High	40/306 (13.1%)	1.24 (0.67 to 2.31)	0.50	1.83 (0.91 to 3.65)	0.089
For Boys	Elevated DBP (n = 86)	Model 1 OR (95%CI)	p	Model 2 OR (95%CI)	p
Pure fructose			0.88 *		0.38 *
Low	25/258 (9.7%)	1.00 (ref.)	-	1.00 (ref.)	-
Middle	29/270 (10.7%)	1.02 (0.57 to 1.83)	0.95	0.87 (0.45 to 1.68)	0.69
High	32/276 (11.6%)	0.96 (0.52 to 1.77)	0.89	0.72 (0.34 to 1.51)	0.39
Total fructose exposure			0.45 *		0.97 *
Low	22/256 (8.6%)	1.00 (ref.)	-	1.00 (ref.)	-
Middle	29/273 (10.6%)	1.11 (0.61 to 2.03)	0.73	0.98 (0.49 to 1.94)	0.95
High	35/275 (12.7%)	1.26 (0.68 to 2.34)	0.46	1.01 (0.47 to 2.16)	0.98

High DBP is defined as a value upper than the 90th percentile (>75 mmHg). Model 1: adjusted for centre. Model 2: adjusted for centre, age, Z-score BMI, MVPA duration, tobacco consumption, salt intake and energy intake and calculated after multiple imputations (m = 10) to handle missing data. * p calculated using fructose levels as ordinal variable. Abbreviations: DBP = diastolic blood pressure; OR = odds-ratio; CI = confidence interval.

Table 6. Association between elevated systolic blood pressure with pure and total fructose exposure from non-natural foods.

For Girls	Elevated SBP (*n* = 84)	Model 1 OR (95%CI)	*p*	Model 2 OR (95%CI)	*p*
Pure fructose			0.79 *		0.70 *
Low	21/296 (7.1%)	1.00 (ref.)	-	1.00 (ref.)	-
Middle	29/305 (9.5%)	1.01 (0.54 to 1.88)	0.98	1.09 (0.56 to 2.10)	0.80
High	34/300 (11.3%)	0.92 (0.49 to 1.76)	0.81	1.15 (0.57 to 2.32)	0.70
Total fructose exposure			0.22 *		0.38 *
Low	26/296 (8.8%)	1.00 (ref.)	-	1.00 (ref.)	-
Middle	26/304 (8.6%)	0.79 (0.43 to 1.44)	0.44	1.03 (0.52 to 2.04)	0.93
High	32/301 (10.6%)	0.67 (0.36 to 1.26)	0.22	1.45 (0.69 to 3.04)	0.33

For Boys	Elevated SBP (*n* = 79)	Model 1 OR (95%CI)	*p*	Model 2 OR (95%CI)	*p*
Pure fructose			0.75 *		0.52 *
Low	19/267 (7.1%)	1.00 (ref.)	-	1.00 (ref.)	-
Middle	30/272 (11.0%)	1.15 (0.61 to 2.18)	0.66	1.04 (0.53 to 2.06)	0.90
High	30/265 (11.3%)	0.93 (0.48 to 1.81)	0.83	0.81 (0.39 to 1.67)	0.56
Total fructose exposure			0.83 *		0.45 *
Low	21/269 (7.8%)	1.00 (ref.)	-	1.00 (ref.)	-
Middle	28/267 (10.5%)	1.02 (0.54 to 1.93)	0.94	0.68 (0.35 to 1.33)	0.26
High	30/268 (11.2%)	0.94 (0.49 to 1.79)	0.85	0.83 (0.39 to 1.76)	0.62

High SBP is defined as a value greater than the 90th percentile (>125 mmHg for girls and >137 mmHg for boys)). Model 1: adjusted for centre. Model 2: adjusted for centre, age, Z-score BMI, MVPA duration, tobacco consumption, salt intake and energy intake and calculated after multiple imputations (m = 10) to handle missing data. * *p* calculated using fructose levels as ordinal variable. Abbreviations: SBP = systolic blood pressure; OR = odds-ratio; CI = confidence interval.

Table 7. Association between elevated diastolic blood pressure with pure and total fructose exposure from non-natural foods.

For Girls	Elevated DBP (*n* = 86)	Model 1 OR (95%CI)	*p*	Model 2 OR (95%CI)	*p*
Pure fructose			0.051 *		0.013 *
Low	19/296 (6.4%)	1.00 (ref.)	-	1.00 (ref.)	-
Middle	26/305 (8.5%)	1.23 (0.65 to 2.33)	0.52	1.36 (0.71 to 2.59)	0.36
High	41/300 (13.7%)	1.82 (0.97 to 3.39)	0.061	2.27 (1.17 to 4.40)	0.015
Total fructose exposure			0.18 *		0.030 *
Low	22/296 (7.4%)	1.00 (ref.)	-	1.00 (ref.)	-
Middle	24/304 (7.9%)	0.97 (0.52 to 1.80)	0.92	2.06 (1.14 to 3.70)	0.016
High	40/301 (13.3%)	1.47 (0.81 to 2.66)	0.21	1.85 (0.91 to 3.79)	0.091

For Boys	Elevated DBP (*n* = 86)	Model 1 OR (95%CI)	*p*	Model 2 OR (95%CI)	*p*
Pure fructose			0.73 *		0.87 *
Low	24/267 (9.0%)	1.00 (ref.)	-	1.00 (ref.)	-
Middle	30/272 (11.0%)	1.06 (0.59 to 1.91)	0.84	1.04 (0.53 to 2.01)	0.91
High	32/265 (12.1%)	1.11 (0.60 to 2.06)	0.73	1.06 (0.52 to 2.16)	0.86
Total fructose exposure			0.71 *		0.86 *
Low	25/269 (9.3%)	1.00 (ref.)	-	1.00 (ref.)	-
Middle	29/267 (10.9%)	1.02 (0.57 to 1.84)	0.94	0.82 (0.42 to 1.59)	0.55
High	32/268 (11.9%)	1.12 (0.61 to 2.04)	0.72	1.17 (0.55 to 2.49)	0.68

High DBP is defined as a value greater than the 90th percentile (>75 mmHg). Model 1: adjusted for centre. Model 2: adjusted for centre, age, Z-score BMI, MVPA duration, tobacco consumption, salt intake and energy intake and calculated after multiple imputations (m = 10) to handle missing data. * *p* calculated using fructose levels as ordinal variable. Abbreviations: DBP = diastolic blood pressure; OR = odds-ratio; CI = confidence interval.

4. Discussion

Fructose intake has quadrupled in the US since the beginning of the 20th century [61,62] and is probably rising across Europe [42], a trend which is predominately driven by the consumption of processed and non-natural food products such as SSB [13]. The median pure and total fructose exposures from our study were similar to the 2007–2010 Dutch food consumption survey (46 g/day) in the same age group [46] and an earlier NHANES study (59 g/day) [63], but greater than a similar study in New Zealander adolescents (21.6 g/day for boys and 18.3 g/day for girls) [46]. In these prior studies, the higher fructose intake observed was explained by SSB consumption [46,63]. Other high-fructose-containing foods contributing to total fructose exposure were honey/jam/syrup, fruits, fruit/vegetable juices and cakes/pies/biscuits, which was also observed in a study of adolescents from Switzerland [49]. In this context, fructose from SSBs could be considered the main source of pure fructose exposure in the typical adolescent diet [62]. Therefore, we consider the data we obtained in the HELENA study to represent habitual pure fructose intake of European adolescent. Most studies showed that medians of fructose intake in subjects more than 15 years old was 50 g/day [9,64,65]. The cut-off of 50 g/day is close to the cut-off of excessive fructose intake used in our study: 41.90 g/day for girls and 54 g/day for boys. The use of tertiles to find the cut-off of excessive fructose (the third tertile) intake in our specific adolescent population permits a well-balanced sample size between groups according to a more powerful statistical analysis.

The effect of fructose intake in the present study was analysed accounting for co-founding factors known to mediate blood pressure [60,66], as demonstrated in similar studies [18,66,67]. Other influencing factors have been analysed such as menarche, presence or absence of menstruation cycle (within one week before BP measurement) and contraceptive use. Data did not show any influence of menarche, presence or absence of menstruation cycle or contraceptive use on girls diastolic BP (using Chi-square tests, data not shown).

In our study, cut-off of 90th percentiles for elevated SBP and DBP were close to the study described by Flynn, J.T. et al. [54]. Adolescence is considered as a vulnerable period of high blood pressure development because the highest peak of blood pressure occurs during puberty [68,69]. As increase of adrenal androgen production occurs earlier in girls than in boys, pure fructose intake levels have more impact on DBP and were significant in girls from this study population [70].

DBP differed among pure fructose intake levels (from non-natural foods) in adolescent girls but not in boys. Indeed, during girls' puberty, there is an increase of adrenal androgen production [71]. Additionally, girls are less physically active than boys at this age [72], which could contribute to elevated DBP. Moreover, presence of polycystic ovary syndrome, which is now frequently associated with an increase of DBP in adolescent girls, could be a possible explanation.

The proposed mechanism by which fructose intake increases blood pressure after an acute load is related to fructose metabolism. After an acute load, maximum blood fructose levels are rapidly achieved at 60 min [65]. Pure fructose is degraded in the liver, resulting in a rapid and transient increase of blood uric acid levels. Importantly, fructose is the only carbohydrate that increases blood uric acid in humans, and uric acid exerts hemodynamic effects (increased oxidative stress, endothelial dysfunction and activation of the renin–angiotensin–aldosterone system) that have been shown to contribute to high blood pressure [6,20,21,23]. Indeed, there are epidemiologic/clinical data and plausible mechanisms that explain increased blood pressure with excessive fructose intake (more than 40 or 50 g/day) [73]. In the high-fructose, non-natural-foods-consumers groups from HELENA study, we observed elevated diastolic blood pressure in adolescent girls. This may be explained by fructose from fruits and vegetables being more slowly absorbed, compared to pure fructose, due to the presence of dietary fibres that slow fructose metabolism [74].

There are several strengths in the present study. Most studies on fructose consumption have focused on SSB intake [18,28,75–77], whereas our study takes into account the contribution of several other food types to fructose exposure. The HELENA study is comprised of a large sample size from eight geographically diverse European cities, and fructose exposure was collected in a "real life" manner from validated dietary instruments. Dietary intake was collected, across all eight countries, using these standardized and validated tools (HELENA–DIAT) and the same food composition database, allowing us to assess both between- and within-individual variability. The large battery of health data, lifestyle measures, anthropometric data and SES variables measured in the HELENA adolescents were obtained through standardized and validated procedures/tools, allowing for analyses to be adjusted for several possible confounding factors. There are some limitations to the present study. Due to the cross-sectional design of the study, causality cannot be determined. Pure fructose consumption was collected from dietary records representing 2 days; thus, it may not completely reflect habitual adolescent consumption patterns and allow total quantification of fructose exposure. Similarly, the fructose content of foods comes from multiple food data composition database or datasets, as a comprehensive dataset on the content of fructose in foods does not exist. Unfortunately, the question about the presence of polycystic ovary syndrome could not been answered. Indeed, confirmation of this diagnostic should include a hormone blood sample analysis and an ovarian ultrasound imaging, which were not performed in the HELENA study. Differences in intake of fructose in the various countries were not assessed in the HELENA study. Some blood sample analyses such as renin or aldosterone were not performed, and only one-third of adolescents' Hb1Ac was analysed [38]. Lastly, our study did not record history of parental hypertension and genetic factors associated with BP; therefore, we could not use these data in our analyses.

5. Conclusions

In conclusion, consuming high quantities of non-natural foods containing extrinsically high fructose is associated with elevated DBP in adolescent girls. The consumption of natural foods containing fructose, such as whole fruits, does not impact blood pressure in our study and should continue to remain a healthy dietary habit [78–80], especially since whole fruit could decrease diastolic blood pressure in girls [81].

Supplementary Materials: The following are available online at https://www.mdpi.com/article/10.3390/nu13103608/s1, Supplemental Table S1. Pure fructose and fructose total exposure intake content in various fructose-containing foods. Supplemental Table S2. Data of confounding factors according to fructose exposure. Supplemental Table S3. Comparison of main characteristics between the included and nonincluded adolescents.

Author Contributions: L.B., I.H. and M.K. designed the data collection instruments, coordinated and supervised data collection, conducted the initial analyses and drafted the initial manuscript. E.D. conducted the initial analyses, statistical analysis and drafted the initial manuscript. D.M., A.K., R.W.W. and Y.M. designed data collection instruments, coordinated and supervised data collection and critically reviewed the manuscript for important intellectual content. S.D.H., L.A.M. and F.G. conceptualized and designed the study, supervised data collection, coordinated the study and critically reviewed the manuscript for important intellectual content. All authors have read and agreed to the published version of the manuscript.

Funding: The HELENA Study took place with the financial support of the European Commission Sixth RTD Framework Programme (Contract FOOD-CT-20056007034). This study was also supported by grants from Spanish Ministry of Science and Innovation (AP 2008-03806).

Institutional Review Board Statement: The study was conducted according to the guidelines of the Declaration of Helsinki and approved by the appropriate independent ethics committee for each study centre.

Informed Consent Statement: Consent was obtained from both parents and adolescents.

Data Availability Statement: Not applicable.

Acknowledgments: The content of this article reflects only the authors' views, and the European Commission is not liable for any use that may be made of the information contained herein. We thank Muriel Beuvry and Anne Gautreau (CIC-1403-CHU-Inserm de Lille, France) for help in typing this manuscript.

Conflicts of Interest: Frédéric Gottrand has received consulting fees from Nestlé. The remaining authors state no conflict of interest. The content of this article reflect only the authors' views, and the European Community is not liable for any use that may be made of the information contained herein.

References

1. Ventura, E.E.; Davis, J.N.; Goran, M.I. Sugar Content of Popular Sweetened Beverages Based on Objective Laboratory Analysis: Focus on Fructose Content. *Obesity* **2011**, *19*, 868–874. [CrossRef] [PubMed]
2. Latasa, P.; Louzada, M.; Steele, E.M.; Monteiro, C.A. Added sugars and ultra-processed foods in Spanish households (1990–2010). *Eur. J. Clin. Nutr.* **2018**, *72*, 1404–1412. [CrossRef] [PubMed]
3. Lelis, D.F.; Andrade, J.M.O.; Almenara, C.C.P.; Broseguini-Filho, G.B.; Mill, J.G.; Baldo, M.P. High fructose intake and the route towards cardiometabolic diseases. *Life Sci.* **2020**, *259*, 118235. [CrossRef] [PubMed]
4. Bray, G.A.; Nielsen, S.J.; Popkin, B.M. Consumption of high-fructose corn syrup in beverages may play a role in the epidemic of obesity. *Am. J. Clin. Nutr.* **2004**, *79*, 537–543. [CrossRef] [PubMed]
5. Sievenpiper, J.L.; de Souza, R.J.; Mirrahimi, A.; Yu, M.E.; Carleton, A.J.; Beyene, J.; Chiavaroli, L.; Di Buono, M.; Jenkins, A.L.; Leiter, L.A.; et al. Effect of Fructose on Body Weight in Controlled Feeding Trials A Systematic Review and Meta-analysis. *Ann. Intern. Med.* **2012**, *156*, 291–304. [CrossRef]
6. Ha, V.; Sievenpiper, J.L.; de Souza, R.J.; Chiavaroli, L.; Wang, D.D.; Cozma, A.I.; Mirrahimi, A.; Yu, M.E.; Carleton, A.J.; Dibuono, M.; et al. Effect of fructose on blood pressure: A systematic review and meta-analysis of controlled feeding trials. *Hypertension* **2012**, *59*, 787–795. [CrossRef]
7. Rippe, J.M.; Angelopoulos, T.J. Fructose-Containing Sugars and Cardiovascular Disease. *Adv. Nutr.* **2015**, *6*, 430–439. [CrossRef]
8. Malik, V.S.; Popkin, B.M.; Bray, G.A.; Despres, J.-P.; Willett, W.C.; Hu, F.B. Sugar-Sweetened Beverages and Risk of Metabolic Syndrome and Type 2 Diabetes. *Diabetes Care* **2010**, *33*, 2477–2483. [CrossRef]
9. Kelishadi, R.; Mansourian, M.; Heidari-Beni, M. Association of fructose consumption and components of metabolic syndrome in human studies: A systematic review and meta-analysis. *Nutrition* **2014**, *30*, 503–510. [CrossRef]
10. Chiu, S.; Sievenpiper, J.L.; de Souza, R.J.; Cozma, A.I.; Mirrahimi, A.; Carleton, A.J.; Ha, V.; Di Buono, M.; Jenkins, A.L.; Leiter, L.A.; et al. Effect of fructose on markers of non-alcoholic fatty liver disease (NAFLD): A systematic review and meta-analysis of controlled feeding trials. *Eur. J. Clin. Nutr.* **2014**, *68*, 416–423. [CrossRef]
11. Wang, D.D.; Sievenpiper, J.L.; de Souza, R.J.; Chiavaroli, L.; Ha, V.; Cozma, A.I.; Mirrahimi, A.; Yu, M.E.; Carleton, A.J.; Di Buono, M.; et al. The Effects of Fructose Intake on Serum Uric Acid Vary among Controlled Dietary Trials. *J. Nutr.* **2012**, *142*, 916–923. [CrossRef] [PubMed]
12. Dornas, W.C.; de Lima, W.G.; Pedrosa, M.L.; Silva, M.E. Health Implications of High-Fructose Intake and Current Research. *Adv. Nutr.* **2015**, *6*, 729–737. [CrossRef] [PubMed]
13. Lustig, R.H. Fructose: Metabolic, Hedonic, and Societal Parallels with Ethanol. *J. Am. Diet. Assoc.* **2010**, *110*, 1307–1321. [CrossRef]
14. Steele, E.M.; Baraldi, L.G.; Louzada, M.L.D.; Moubarac, J.C.; Mozaffarian, D.; Monteiro, C.A. Ultra-processed foods and added sugars in the US diet: Evidence from a nationally representative cross-sectional study. *BMJ Open* **2016**, *6*, e009892. [CrossRef]
15. WHO. Diet, health and physical activity. In *Global Status Report on Non Communicable Diseases*; WHO: Geneva, Switzerland, 2014.
16. Gidding, S.S.; McMahan, C.A.; McGill, H.C.; Colangelo, L.A.; Schreiner, P.J.; Williams, O.D.; Liu, K. Prediction of coronary artery calcium in young adults using the Pathobiological Determinants of Atherosclerosis in Youth (PDAY) risk score: The CARDIA study. *Arch. Intern. Med.* **2006**, *166*, 2341–2347. [CrossRef]
17. Chen, X.; Wang, Y. Tracking of blood pressure from childhood to adulthood—A systematic review and meta-regression analysis. *Circulation* **2008**, *117*, 3171–3180. [CrossRef]
18. Nguyen, S.; Choi, H.K.; Lustig, R.H.; Hsu, C.-Y. Sugar-Sweetened Beverages, Serum Uric Acid, and Blood Pressure in Adolescents. *J. Pediatrics* **2009**, *154*, 807–813. [CrossRef]
19. Feig, D.I.; Soletsky, B.; Johnson, R.J. Effect of allopurinol on blood pressure of adolescents with newly diagnosed essential hypertension—A randomized trial. *JAMA-J. Am. Med Assoc.* **2008**, *300*, 924–932. [CrossRef]
20. Klein, A.V.; Kiat, H. The mechanisms underlying fructose-induced hypertension: A review. *J. Hypertens.* **2015**, *33*, 912–920. [CrossRef] [PubMed]
21. Jayalath, V.H.; Sievenpiper, J.L.; de Souza, R.J.; Ha, V.; Mirrahimi, A.; Santaren, I.D.; Mejia, S.B.; Di Buono, M.; Jenkins, A.L.; Leiter, L.A.; et al. Total Fructose Intake and Risk of Hypertension: A Systematic Review and Meta-Analysis of Prospective Cohorts. *J. Am. Coll. Nutr.* **2014**, *33*, 328–339. [CrossRef]
22. Ahmad, R.; Mok, A.; Rangan, A.M.; Louie, J.C.Y. Association of free sugar intake with blood pressure and obesity measures in Australian adults. *Eur. J. Nutr.* **2020**, *59*, 651–659. [CrossRef]
23. Johnson, R.J.; Perez-Pozo, S.E.; Sautin, Y.Y.; Manitius, J.; Sanchez-Lozada, L.G.; Feig, D.I.; Shafiu, M.; Segal, M.; Glassock, R.J.; Shimada, M.; et al. Hypothesis: Could Excessive Fructose Intake and Uric Acid Cause Type 2 Diabetes? *Endocr. Rev.* **2009**, *30*, 96–116. [CrossRef]

24. Singh, A.K.; Amlal, H.; Haas, P.J.; Dringenberg, U.; Fussell, S.; Barone, S.L.; Engelhardt, R.; Zuo, J.; Seidler, U.; Soleimani, M. Fructose-induced hypertension: Essential role of chloride and fructose absorbing transporters PAT1 and Glut5. *Kidney Int.* **2008**, *74*, 438–447. [CrossRef] [PubMed]

25. Catena, C.; Giacchetti, G.; Novello, M.; Colussi, G.; Cavarape, A.; Sechi, L.A. Cellular mechanisms of insulin resistance in rats with fructose-induced hypertension. *Am. J. Hypertens.* **2003**, *16*, 973–978. [CrossRef]

26. Huang, B.W.; Chiang, M.T.; Yao, H.T.; Chiang, W. The effect of high-fat and high-fructose diets on glucose tolerance and plasma lipid and leptin levels in rats. *Diabetes Obes. Metab.* **2004**, *6*, 120–126. [CrossRef]

27. Chen, H.H.; Chu, C.H.; Wen, S.W.; Lai, C.C.; Cheng, P.W.; Tseng, C.J. Excessive Fructose Intake Impairs Baroreflex Sensitivity and Led to Elevated Blood Pressure in Rats. *Nutrients* **2019**, *11*, 2581. [CrossRef] [PubMed]

28. Vos, M.B.; Kimmons, J.E.; Gillespie, C.; Welsh, J.; Blanck, H.M. Dietary fructose consumption among US children and adults: The Third National Health and Nutrition Examination Survey. *Medscape. J. Med.* **2008**, *10*, 160.

29. Marriott, B.P.; Cole, N.; Lee, E. National Estimates of Dietary Fructose Intake Increased from 1977 to 2004 in the United States. *J. Nutr.* **2009**, *139*, S1228–S1235. [CrossRef] [PubMed]

30. Savoca, M.R.; Evans, C.D.; Wilson, M.E.; Harshfield, G.A.; Ludwig, D.A. The association of caffeinated beverages with blood pressure in adolescents. *Arch. Pediatrics Adolesc. Med.* **2004**, *158*, 473–477. [CrossRef]

31. White, J.S. Sugar-sweetened beverage effect on cardiovascular risk factors lacks significance. *J. Pediatrics* **2010**, *156*, 860–861. [CrossRef]

32. Stanhope, K.L. Sugar consumption, metabolic disease and obesity: The state of the controversy. *Crit. Rev. Clin. Lab. Sci.* **2016**, *53*, 52–67. [CrossRef]

33. Rippe, J.M.; Marcos, A. Controversies about sugars consumption: State of the science. *Eur J. Nutr.* **2016**, *55*, 11–16. [CrossRef]

34. Genovesi, S.; Giussani, M.; Orlando, A.; Orgiu, F.; Parati, G. Salt and Sugar: Two Enemies of Healthy Blood Pressure in Children. *Nutrients* **2021**, *13*, 697. [CrossRef]

35. Moreno, L.A.; De, H.S.; Gonzalez-Gross, M.; Kersting, M.; Molnar, D.; Gottrand, F.; Barrios, L.; Sjostrom, M.; Manios, Y.; Gilbert, C.C.; et al. Design and implementation of the Healthy Lifestyle in Europe by Nutrition in Adolescence Cross-Sectional Study. *Int. J. Obes.* **2008**, *32* (Suppl. 5), S4–S11. [CrossRef]

36. Iliescu, C.; Béghin, L.; Maes, L.; Bourdeaudhuij, I.D.; Libersa, C.; Vereecken, C.; Gonzalez-Gross, M.; Kersting, M.; Molnar, D.; Leclercq, C.; et al. Socioeconomic questionnaire and clinical assessment in the HELENA Cross-Sectional Study: Methodology. *Int. J. Obes.* **2008**, *32* (Suppl. 5), S19–S25. [CrossRef]

37. Béghin, L.; Castera, M.; Manios, Y.; Gilbert, C.C.; Kersting, M.; De, H.S.; Kafatos, A.; Gottrand, F.; Molnar, D.; Sjostrom, M.; et al. Quality assurance of ethical issues and regulatory aspects relating to good clinical practices in the HELENA Cross-Sectional Study. *Int. J. Obes.* **2008**, *32* (Suppl. 5), S12–S18. [CrossRef]

38. Béghin, L.; Huybrechts, I.; Vicente-Rodríguez, G.; De Henauw, S.; Gottrand, F.; Gonzales-Gross, M.; Dallongeville, J.; Sjöström, M.; Leclercq, C.; Dietrich, S.; et al. Main characteristics and participation rate of European adolescents included in the HELENA study. *Arch. Public Health* **2012**, *70*, 1–11. [CrossRef]

39. Biro, G.; Hulshof, K.F.; Ovesen, L.; Amorim Cruz, J.A. Selection of methodology to assess food intake. *Eur. J. Clin. Nutr.* **2002**, *56* (Suppl. 2), S25–S32. [CrossRef]

40. Vereecken, C.A.; Covents, M.; Matthys, C.; Maes, L. Young adolescents' nutrition assessment on computer (YANA-C). *Eur. J. Clin. Nutr.* **2005**, *59*, 658–667. [CrossRef]

41. Vereecken, C.A.; Covents, M.; Sichert-Hellert, W.; Alvira, J.M.; Le Donne, C.; De Henauw, S.; De Vriendt, T.; Phillipp, M.K.; Béghin, L.; Manios, Y.; et al. Development and evaluation of a self-administered computerized 24-h dietary recall method for adolescents in Europe. *Int. J. Obes.* **2008**, *32* (Suppl. 5), S26–S34. [CrossRef] [PubMed]

42. Mesana, M.I.; Hilbig, A.; Androutsos, O.; Cuenca-Garcia, M.; Dallongeville, J.; Huybrechts, I.; De Henauw, S.; Widhalm, K.; Kafatos, A.; Nova, E.; et al. Dietary sources of sugars in adolescents' diet: The HELENA study. *Eur. J. Nutr.* **2018**, *57*, 629–641. [CrossRef]

43. Duffey, K.J.; Huybrechts, I.; Mouratidou, T.; Libuda, L.; Kersting, M.; De, V.T.; Gottrand, F.; Widhalm, K.; Dallongeville, J.; Hallstrom, L.; et al. Beverage consumption among European adolescents in the HELENA study. *Eur. J. Clin. Nutr.* **2012**, *66*, 244–252. [CrossRef]

44. Walker, R.W.; Dumke, K.A.; Goran, M.I. Fructose content in popular beverages made with and without high-fructose corn syrup. *Nutrition* **2014**, *30*, 928–935. [CrossRef]

45. Walker, R.W.; Goran, M.I. Laboratory Determined Sugar Content and Composition of Commercial Infant Formulas, Baby Foods and Common Grocery Items Targeted to Children. *Nutrients* **2015**, *7*, 5850–5867. [CrossRef]

46. Sluik, D.; Engelen, A.I.; Feskens, E.J. Fructose consumption in the Netherlands: The Dutch national food consumption survey 2007-2010. *Eur. J. Clin. Nutr.* **2015**, *69*, 475–481. [CrossRef]

47. Raatz, S.K.; Johnson, L.K.; Picklo, M.J. Consumption of Honey, Sucrose, and High-Fructose Corn Syrup Produces Similar Metabolic Effects in Glucose-Tolerant and -Intolerant Individuals. *J. Nutr.* **2015**, *145*, 2265–2272. [CrossRef]

48. Ramne, S.; Gray, N.; Hellstrand, S.; Brunkwall, L.; Enhorning, S.; Nilsson, P.M.; Engstrom, G.; Orho-Melander, M.; Ericson, U.; Kuhnle, G.G.C.; et al. Comparing Self-Reported Sugar Intake With the Sucrose and Fructose Biomarker From Overnight Urine Samples in Relation to Cardiometabolic Risk Factors. *Front. Nutr.* **2020**, *7*, 62. [CrossRef]

49. Aeberli, I.; Zimmermann, M.B.; Molinari, L.; Lehmann, R.; l'Allemand, D.; Spinas, G.A.; Berneis, K. Fructose intake is a predictor of LDL particle size in overweight schoolchildren. *Am. J. Clin. Nutr.* **2007**, *86*, 1174–1178. [CrossRef]
50. Yeung, C.H.C.; Louie, J.C.Y. Methodology for the assessment of added/free sugar intake in epidemiological studies. *Curr. Opin. Clin. Nutr. Metab. Care* **2019**, *22*, 271–277. [CrossRef]
51. Cole, T.J. The LMS method for constructing normalized growth standards. *Eur. J. Clin. Nutr.* **1990**, *44*, 45–60.
52. The 4th report on the diagnosis, evaluation, and treatment of high blood pressure in children and adolescents. *Pediatrics* **2004**, *114* (Suppl. 2), 555–576. [CrossRef]
53. Topouchian, J.A.; El Assaad, M.A.; Orobinskaia, L.V.; El Feghali, R.N.; Asmar, R.G. Validation of two automatic devices for self-measurement of blood pressure according to the International Protocol of the European Society of Hypertension: The Omron M6 (HEM-7001-E) and the Omron R7 (HEM 637-IT). *Blood Press. Monit.* **2006**, *11*, 165–171. [CrossRef]
54. Flynn, J.T.; Kaelber, D.C.; Baker-Smith, C.M.; Blowey, D.; Carroll, A.E.; Daniels, S.R.; de Ferranti, S.D.; Dionne, J.M.; Falkner, B.; Flinn, S.K.; et al. Clinical Practice Guideline for Screening and Management of High Blood Pressure in Children and Adolescents. *Pediatrics* **2017**, *140*, 1–72. [CrossRef]
55. Buxens, A.; Ruiz, J.R.; Arteta, D.; Artieda, M.; Santiago, C.; Gonzalez-Freire, M.; Martinez, A.; Tejedor, D.; Lao, J.I.; Gomez-Gallego, F.; et al. Can we predict top-level sports performance in power vs endurance events? A genetic approach. *Scand. J. Med. Sci. Sports* **2011**, *21*, 570–579. [CrossRef]
56. Ekelund, U.; Anderssen, S.A.; Froberg, K.; Sardinha, L.B.; Andersen, L.B.; Brage, S. Independent associations of physical activity and cardiorespiratory fitness with metabolic risk factors in children: The European youth heart study. *Diabetologia* **2007**, *50*, 1832–1840. [CrossRef]
57. Malcon, M.C.; Menezes, A.M.B.; Chatkin, M. Prevalence and risk factors for smoking among adolescents. *Rev. Saude Publica* **2003**, *37*, 1–7. [CrossRef]
58. van Buuren, S.; Groothuis-Oudshoorn, K. mice: Multivariate Imputation by Chained Equations in R. *J. Stat. Softw.* **2011**, *45*, 1–67. [CrossRef]
59. Rosner, B.; Cook, N.; Portman, R.; Daniels, S.; Falkner, B. Determination of blood pressure percentiles in normal-weight children: Some methodological issues. *Am. J. Epidemiol.* **2008**, *167*, 653–666. [CrossRef]
60. de Moraes, A.C.; Carvalho, H.B.; Rey-Lopez, J.P.; Gracia-Marco, L.; Béghin, L.; Kafatos, A.; Jimenez-Pavon, D.; Molnar, D.; De, H.S.; Manios, Y.; et al. Independent and combined effects of physical activity and sedentary behavior on blood pressure in adolescents: Gender differences in two cross-sectional studies. *PLoS ONE* **2013**, *8*, e62006. [CrossRef]
61. Popkin, B.M. Patterns of beverage use across the lifecycle. *Physiol. Behav.* **2010**, *100*, 4–9. [CrossRef]
62. Newens, K.J.; Walton, J. A review of sugar consumption from nationally representative dietary surveys across the world. *J. Hum. Nutr. Diet.* **2016**, *29*, 225–240. [CrossRef]
63. Sun, S.Z.; Anderson, G.H.; Flickinger, B.D.; Williamson-Hughes, P.S.; Empie, M.W. Fructose and non-fructose sugar intakes in the US population and their associations with indicators of metabolic syndrome. *Food Chem. Toxicol.* **2011**, *49*, 2875–2882. [CrossRef]
64. Rizkalla, S.W. Health implications of fructose consumption: A review of recent data. *Nutr. Metab.* **2010**, *7*. [CrossRef] [PubMed]
65. Jameel, F.; Phang, M.; Wood, L.G.; Garg, M.L. Acute effects of feeding fructose, glucose and sucrose on blood lipid levels and systemic inflammation. *Lipids Health Dis.* **2014**, *13*, 1–7. [CrossRef] [PubMed]
66. Aleixandre, A.; Miguel, M. Dietary fiber and blood pressure control. *Food Funct.* **2016**, *7*, 1864–1871. [CrossRef] [PubMed]
67. Hu, F.B.; Malik, V.S. Sugar-sweetened beverages and risk of obesity and type 2 diabetes: Epidemiologic evidence. *Physiol. Behav.* **2010**, *100*, 47–54. [CrossRef]
68. Dubach, G.; Loret De Mola, C.; Hardy, R.; Droyfus, J.; Santos, A.C.; Horta, B.L. Early menarche and blood pressure in adulthood: Systematic review and meta-analysis. *J. Public Health* **2018**, *40*, 476–484. [CrossRef]
69. Kwok, M.K.; Leung, G.M.; Schooling, C.M. Pubertal testis volume, age at pubertal onset, and adolescent blood pressure: Evidence from Hong Kong's "Children of 1997" birth cohort. *Am. J. Hum. Biol.* **2017**, *29*, e22993. [CrossRef]
70. Pratt, J.H.; Manatunga, A.K.; Wagner, M.A.; Jones, J.J.; Meaney, F.J. Adrenal androgen excretion during adrenarche. Relation to race and blood pressure. *Hypertension* **1990**, *16*, 462–467. [CrossRef]
71. Szklo, M. Epidemiologic patterns of blood pressure in children. *Epidemiol. Rev.* **1979**, *1*, 143–169. [CrossRef]
72. Béghin, L.; Vanhelst, J.; Drumez, E.; Migueles, J.H.; Androutsos, O.; Widhalm, K.; Julian, C.; Moreno, L.A.; De Henauw, S.; Gottrand, F. Gender influences physical activity changes during adolescence: The HELENA study. *Clin. Nutr.* **2019**, *38*, 2900–2905. [CrossRef]
73. Lowndes, J.; Sinnett, S.; Yu, Z.; Rippe, J. The Effects of Fructose-Containing Sugars on Weight, Body Composition and Cardiometabolic Risk Factors When Consumed at up to the 90th Percentile Population Consumption Level for Fructose. *Nutrients* **2014**, *6*, 3153–3168. [CrossRef] [PubMed]
74. Gaby, A.R. Adverse effects of dietary fructose. *Altern. Med. Rev.* **2005**, *10*, 294–306. [PubMed]
75. Lancaster, K.J. Current Intake and Demographic Disparities in the Association of Fructose-Rich Foods and Metabolic Syndrome. *JAMA Netw. Open* **2020**, *3*, e2010224. [CrossRef] [PubMed]
76. Stanhope, K.L.; Schwarz, J.M.; Keim, N.L.; Griffen, S.C.; Bremer, A.A.; Graham, J.L.; Hatcher, B.; Cox, C.L.; Dyachenko, A.; Zhang, W.; et al. Consuming fructose-sweetened, not glucose-sweetened, beverages increases visceral adiposity and lipids and decreases insulin sensitivity in overweight/obese humans. *J. Clin. Investig.* **2009**, *119*, 1322–1334. [CrossRef] [PubMed]

77. Tappy, L.; Le, K.A. Metabolic effects of fructose and the worldwide increase in obesity. *Physiol. Rev.* **2010**, *90*, 23–46. [CrossRef] [PubMed]
78. Sievenpiper, J.L.; de Souza, R.J.; Jenkins, D.J.A. Sugar: Fruit fructose is still healthy. *Nature* **2012**, *482*, 470. [CrossRef]
79. Angelopoulos, T.J.; Lowndes, J.; Sinnett, S.; Rippe, J.M. Fructose Containing Sugars Do Not Raise Blood Pressure or Uric Acid at Normal Levels of Human Consumption. *J. Clin. Hypertens.* **2015**, *17*, 87–94. [CrossRef]
80. Moore, L.L.; Bradlee, M.L.; Singer, M.R.; Qureshi, M.M.; Buendia, J.R.; Daniels, S.R. Dietary Approaches to Stop Hypertension (DASH) eating pattern and risk of elevated blood pressure in adolescent girls. *Br. J. Nutr.* **2012**, *108*, 1678–1685. [CrossRef]
81. Mansoori, S.; Kushner, N.; Suminski, R.R.; Farquhar, W.B.; Chai, S.C. Added Sugar Intake is Associated with Blood Pressure in Older Females. *Nutrients* **2019**, *11*, 2060. [CrossRef] [PubMed]

Article

The Profiling of Diet and Physical Activity in Reproductive Age Women and Their Association with Body Mass Index

Mamaru Ayenew Awoke [1], Thomas P. Wycherley [2], Arul Earnest [3], Helen Skouteris [4,5] and Lisa J. Moran [1,*]

[1] Monash Centre for Health Research and Implementation, School of Public Health and Preventive Medicine, Monash University, Clayton, VIC 3168, Australia; mamaru.awoke@monash.edu

[2] Alliance for Research in Exercise, Nutrition and Activity (ARENA), Allied Health and Human Performance, University of South Australia, Adelaide, SA 5001, Australia; tom.wycherley@unisa.edu.au

[3] Department of Epidemiology and Preventive Medicine, School of Public Health and Preventive Medicine, Monash University, Melbourne, VIC 3004, Australia; arul.earnest@monash.edu

[4] Health and Social Care Unit, School of Public Health and Preventive Medicine, Monash University, Melbourne, VIC 3004, Australia; helen.skouteris@monash.edu

[5] Warwick Business School, Warwick University, Coventry CV4 7AL, UK

* Correspondence: lisa.moran@monash.edu; Tel.: +61-3-8572-2664

Abstract: Pre-pregnancy, pregnancy and postpartum are critical life stages associated with higher weight gain and obesity risk. Among these women, the sociodemographic groups at highest risk for suboptimal lifestyle behaviours and core lifestyle components associated with excess adiposity are unclear. This study sought to identify subgroups of women meeting diet/physical activity (PA) recommendations in relation to sociodemographics and assess diet/PA components associated with body mass index (BMI) across these life stages. Cross-sectional data (Australian National Nutrition and Physical Activity Survey 2011–2012) were analysed for pre-pregnancy, pregnant and postpartum women. The majority (63–95%) of women did not meet dietary or PA recommendations at all life stages. Core and discretionary food intake differed by sociodemographic factors. In pre-pregnant women, BMI was inversely associated with higher whole grain intake (β = −1.58, 95% CI −2.96, −0.21; p = 0.025) and energy from alcohol (β = −0.08, −0.14, −0.005; p = 0.035). In postpartum women, BMI was inversely associated with increased fibre (β = −0.06, 95% CI −0.11, −0.004; p = 0.034) and PA (β = −0.002, 95% CI −0.004, −0.001; p = 0.013). This highlights the need for targeting whole grains, fibre and PA to prevent obesity across life stages, addressing those most socioeconomically disadvantaged.

Keywords: diet; physical activity; body mass index; dietary guideline; reproductive age women

Citation: Awoke, M.A.; Wycherley, T.P.; Earnest, A.; Skouteris, H.; Moran, L.J. The Profiling of Diet and Physical Activity in Reproductive Age Women and Their Association with Body Mass Index. *Nutrients* **2022**, *14*, 2607. https://doi.org/10.3390/nu14132607

Academic Editor: Panagiota Mitrou

Received: 26 April 2022
Accepted: 21 June 2022
Published: 23 June 2022

Publisher's Note: MDPI stays neutral with regard to jurisdictional claims in published maps and institutional affiliations.

1. Introduction

Reproductive age women are at higher risk of longitudinal weight gain and developing obesity [1]. Data from longitudinal studies reports that women gain on average up to 0.7 kg per year, and there are greater rates of weight gain in women aged 18–50 years compared to women aged 50 and over [1]. Reproductive life stages, including preconception, pregnancy and postpartum, are critical windows that drive weight gain and maternal adiposity [2]. Nearly 50% of women enter pregnancy with overweight or obesity [3] or gain weight above the Institute of Medicine guidelines' recommendation during pregnancy [4], and postpartum women retain an extra 0.5–3 kg on average during each pregnancy [5]. Overweight and obesity in preconception and during pregnancy increase the risk of maternal complications and adverse birth outcomes [4,6]. Maternal obesity at conception increases the time to conceive, reduces fertility and increases the risk of future comorbidities, such as type-2 diabetes and cardiovascular diseases, including hypertension [6]. Furthermore, higher pre-pregnancy body mass index (BMI) is a strong predictor of excessive gestational and pregnancy complications [7]. Excessive gestational weight gain additionally drives post-partum weight retention (PPWR), which further increases risks for subsequent pregnancies and exacerbates maternal obesity [8].

Diet and physical activity (PA) are key modifiable risk factors in weight gain and obesity, and optimal diet and regular PA are inversely associated with weight gain and obesity [9,10]. Optimal diet and a higher level of PA can therefore prevent weight gain and obesity [11]. Suboptimal diet and PA have been reported in adults at the population level [12,13]. As a specific high-risk population for weight gain and future obesity, women in pre-pregnancy, during pregnancy and postpartum also have unhealthy dietary patterns and poor diet quality [14]. For example, only 7–10% of pregnant and postpartum women meet population-level recommended intakes of healthy core foods [15,16], and 80% of pregnant women are insufficiently active, which persists into postpartum [16–18]. This may be related to barriers such as fatigue or a lack of motivation, and confidence and time. Women may also prioritise family commitments (e.g., parenting or household responsibilities) over their personal lifestyles [19,20]. All these barriers to a healthy lifestyle and sociodemographic factors are potentially associated with increased adiposity in pre-pregnancy, excessive gestational weight gain during pregnancy and PPWR. However, there is limited and conflicting research on sociodemographic factors associated with meeting population-level diet and PA recommendations in women across the reproductive life stages [21].

National guidelines broadly recommend targeting unhealthy diet and sedentary behaviour for management of overweight and obesity in the general population [22]. Women at key reproductive life stages may also benefit from targeting specific diet and PA components to prevent excess adiposity. Identifying both specific diet and PA components and specific groups of reproductive age women could contribute to future interventions for preventing weight again and obesity. This would also contribute to the evidence base for tailoring intervention strategies to improve healthy eating and increase PA in specific high-risk groups of women. We hypothesise that women across reproductive life stages have inadequate diets and PA levels, which may be disparately linked with sociodemographic characteristics, including age, ethnicity, geographic location, marital status, employment and educational and socioeconomic disadvantages. We also hypothesise that higher intakes of specific core foods and increased PA will be associated with lower BMI, but increased total energy intake and higher energy from total discretionary foods will be associated with higher BMI.

The aims of this study were: (i) to identify women who meet and do not meet diet and PA population-level recommendations based on sociodemographic factors and (ii) to assess the key diet and PA components associated with BMI in women across the reproductive life stages.

2. Materials and Methods

2.1. Data Source and Study Participants

We used data from the National Nutrition and PA Survey (NNPAS) component of the 2011–2012 Australian Health Survey (AHS) conducted by the Australian Bureau of Statistics (ABS) between May 2011 and June 2012. This national survey was designed to provide detailed information on the health and wellbeing of the Australian population. A stratified multistage sampling of urban and rural private dwellings was obtained to ensure a representative sample of Australians (N = 12,153). Detailed information on participant recruitment, the survey design, data collection and response rates have been previously reported in the Australian Health Survey User Guide [23].

Data were collected using a face-to-face interview from randomly selected people in each selected household (one adult ≥ 18 years and one child aged 2–17 years where applicable). Ethical approval was not required because this study was based on secondary data using Confidentialised Unit Record Files (CURF).

This sub-study was a cross-sectional analysis limited to reproductive age women in various key reproductive life stages (pre-pregnancy, pregnancy and postpartum) (Figure 1)

(N = 2492). Key reproductive life stages were identified based on proxy questions on 'female life stages' and 'number of children' in the household (household type, 'HHTYP' variable). The question 'female life stages' has responses: 1: Have never menstruated. 2: Currently pregnant. 3: Currently breastfeeding. 4: Currently experiencing menopause. 5: Post menopause. 6: None of these apply. 9: Not applicable. To identify pre-pregnancy and postpartum, responses 4 and 5 were excluded. Then, female life stage responses 1 or 6 or 9 (have never menstruated OR none of these apply OR not applicable) AND household type responses: 1 (person living alone), 2 (couple only), 5 (unrelated persons aged 15+ only), 6 (all other households) AND age 18–48 (to exclude lower limit of peri-menopause [24]) were classified as pre-pregnancy. Female life stage responses: 3 or 6 or 9 (current breastfeeding OR none of these apply OR not applicable) AND household type responses: 3 (couple family with children) or 4 (one parent family with children) AND age 18–48 were classified as postpartum women; and those who responded currently pregnant taken as pregnant women (Table S1). We note that by this definition, pre-pregnant women were all reproductive age women who were not pregnant or postpartum at the time of the survey.

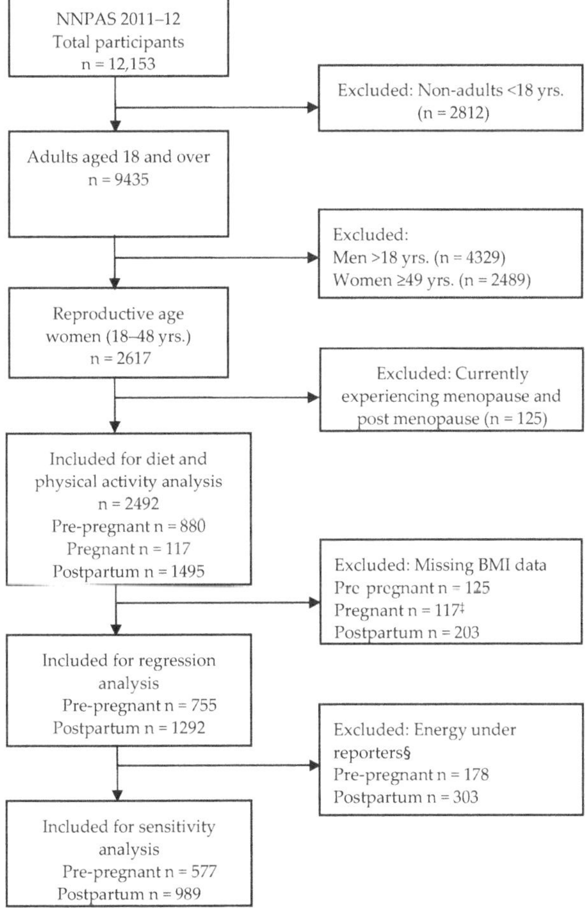

Figure 1. Flow diagram of study participant inclusion for analysis. NNPA, National Nutrition and Physical activity Survey. ‡ Weight/BMI measurement from pregnant women was not taken. § Energy under reporters based on Goldberg cut-off (EI:BMR < 0.9).

2.2. Variables and Measures

2.2.1. Dietary Assessment

Dietary information was collected face-to-face using 24 h dietary recall administered by trained interviewers. ABS used an Automated Multiple-Pass Method (AMPM) [25] developed by the Agricultural Research Service of the United States Department of Agriculture to capture all foods, beverages and dietary supplements. The nutrients and energy (kJ) intake were calculated from each food and beverage consumed using the 2011–2013 Australian Food and Nutrient (AUSNUT) food composition database developed by Food Standards Australia New Zealand (FSANZ). Individual foods were each given an eight-digit food code and classified into food classification groups using the AUSNUT 2011–2013 database. Two-day and 24 h dietary recall were collected; the second day was collected via a telephone interview conducted 8 days or more after the first interview. The first day's dietary recall response rate was 98% (n = 12,153), and the second day's recall response rate was 64% (n = 7735). The first day of dietary recall was used for all analyses to retain a larger sample size and ensure national representativeness consistent with previous studies [13,26,27].

Daily serves of the five core food groups and total daily energy from discretionary foods/beverages were calculated. Details of the five core food group serving size definitions and daily recommended intake are presented in Table S2. The usual daily intakes of fruit and vegetables (serves per day), grain/cereal foods (serves/day), whole grains (serves/day, g/day, as half proportion of grains), dairy products (serves per day), lean meats and alternatives (serves per day), total energy intake (kJ/d), energy from macronutrients (carbohydrate (%E), protein (%E), total fat (%E), saturated fat (%E), polyunsaturated fat (%E), monounsaturated fat (%E)) and fibre (g/day) were included in the analyses. Discretionary foods and beverages, including percentages of energy from total discretionary foods/beverages, sugar sweetened beverages (SSBs), saturated fats, alcohol intake and added sugar, were included in the analyses. The ABS classified discretionary foods and beverages in the NNPAS using the discretionary flag list, which was based on food grouping level (five-digit code, e.g., 11,501 soft drinks non-cola, 11,503 soft drinks cola) or individual food level (eight-digit codes where the flag is assigned to individual food codes within the five-digit subgroup). A list of discretionary choices and respective food codes with examples are presented in Table S3.

2.2.2. PA

Self-reported PA levels were assessed using the Active Australia Survey, which has been validated against accelerometers in middle-aged women [28]. Respondents reported the estimated time spent in walking, moderate-intensity activity (e.g., gentle swimming, social tennis doubles, golf) and vigorous PA (e.g., jogging, fast cycling, circuit training, competitive tennis) in the past week. The reported durations (excluding the number of sessions) of these activities were summed (sum of minutes) to estimate the total time spent in PA. We used only the duration of PA reported during the previous week to ensure comparability with international guidelines. Total minutes of PA was dichotomised as meeting ($\geq$150 min/week) or not meeting the guidelines (<150 min/week) according to the 2014 Australia's PA and Sedentary Behaviour Guidelines for Adults [29]. Furthermore, the reported durations for moderate and vigorous activity (multiplied by two) were summed to estimate the total time spent in moderate–vigorous PA (MVPA), which was used both as a continuous variable and dichotomised as $\geq$150 MVPA minutes/week or $\leq$ 150 MVPA minutes/week. In multivariable analysis, we included total PA, as it includes all types of activity, such as walking and moderate to vigorous activities, which can be common across life stages.

2.2.3. Covariates

Covariates in the analyses included age (in years), marital status (married vs. not married), country of birth (Australian born, mainly English-speaking country born, other countries), educational level (bachelor/graduate diploma, certificates/advanced diploma or other no non-school qualifications), socio-economic index for areas (SEIFA) or index of relative socio-economic disadvantage (IRSD) (in quintiles: quintile one corresponds to the lowest scores for the most disadvantaged areas and quintile five represents the highest scores for the most advantaged areas), remoteness (inner regional Australia, major cities and other (outer regional/remote)), household income (in quintiles considered as continuous in regression analyses) and health behaviours, such as smoking status (current smoker, ex-smoker or never smoked) and self-assessed health (excellent/very good, good, fair and poor). Participants were also asked whether they were currently on a diet: responses included currently on a diet to lose weight, currently on a diet for health reasons, currently on a diet to lose weight and for health reasons, not currently on a diet or not applicable. Responses were dichotomized to currently on a diet for any reason and not currently on diet.

2.2.4. Dependent Variable

Anthropometric measures (weight and height) of the respondents were taken during the interview using a digital scale (maximum 150 kg and recoded to the nearest 0.1 kg) and a stadiometer (maximum 210 cm and recorded to the nearest 0.1 cm) respectively. Participants were encouraged to remove their shoes and heavy clothing before measurements were taken. Height measurements were repeated on a random 10% sample of respondents to validate the measurement, and if the second measurement of height or waist varied by more than one centimetre, then a third reading was taken. Body mass index (BMI, kg/m^2) was calculated from measured weight and height as weight in kilograms divided by the square of height in metres. Anthropometric data of women who were pregnant at the time of the survey were not obtained. According to WHO categories, BMI was defined as underweight ($<18.5 \ kg/m^2$), normal weight ($18.5–24.9 \ kg/m^2$), overweight ($25.0–29.9 \ kg/m^2$) and obese ($\geq30 \ kg/m^2$).

2.3. Statistical Analysis

Descriptive statistics were used to estimate the proportion and mean consumption of dietary intake and PA across reproductive life stages (pre-pregnancy, pregnancy and postpartum). Pearson Chi-square tests were used to determine differences between categorical variables and student's t-test for continuous variables.

Univariable and multivariable linear regression analyses were performed to investigate diet, PA and sociodemographic factors associated with BMI in pre-pregnancy and postpartum. For multivariable regression, residuals were checked and met the normality assumption. All estimates (proportion, means, standard error, beta-coefficients and 95% CI) were population weighted to take into account sampling weights and sampling design of the survey by applying replicate weights. Jack knife replicate weights were used to obtain unbiased standard errors and coefficient estimates. The analysis was based on complete case data, and codes followed recommendations [30] to account for the complex survey design.

The backward stepwise regression technique was used to select most appropriate variables, removing the least significant variables one by one (variable with the highest *p*-value in the model) and continued until a parsimonious model was reached ($p < 0.05$). The variables were assessed for multicollinearity through the variance inflation factor (VIF) and tolerance statistics (VIF > 10) to exclude the redundant explanatory variables. Collinear variables were excluded, as they showed linear relationship with the other independent variables. All statistical analyses were performed using STATA SE version 16.1 (StataCorp LLC, College Station, TX, USA). Statistical significance was considered at *p*-value ≤ 0.05.

2.4. Sensitivity Analysis

Under-reporting is common in nutrition surveys, as people tend to underestimate their food intakes [31], which would affect the overall results. The most utilised method to identify under-reporters is to compare each person's basal metabolic rate (BMR) with their reported energy intake (EI) and apply Goldberg cut-off values to examine whether the EI reported is plausible. We employed this approach to identify under-reporters, and a sensitivity analysis was performed in women only with plausible energy intakes (excluding under-reporters) consistent with previous studies [32,33]. Briefly, BMR is the amount of energy needed for an individual's minimum set of body functions required for life over a 24 h period. This was calculated in kilojoules per 24 h based on individual's age, sex and weight without activity level adjustment. The ratio of energy intake (EI) to BMR (EI:BMR) was used to identify under-reporters (implausibly low energy intakes) using the Goldberg cut-off limit of 0.9 for EI:BMR. This is for data below the 95% confidence limit for an individual, allowing for daily variation in energy intakes and errors in EI:BMR computation. After excluding 481 women (pre-pregnancy n = 178 and postpartum n = 303) with implausibly low energy intake, n = 577 pre-pregnant and n = 989 postpartum women were included for sensitivity analysis in the multivariable model (Figure 1). Approximately a quarter of the total analytical sample, with similar proportions of pre-pregnant (23.6%) and postpartum women (23.5%), were excluded in the sensitivity analysis.

3. Results

3.1. Sociodemographic Characteristics

Socio-demographic characteristics across reproductive life stages are presented in Table 1. The mean ages of study participants were 31.2 ± 8.4, 29.3 ± 5.3 and 33.6 ± 8.7 years for pre-pregnant, pregnant and postpartum women, respectively. One in five (20.7%) and 19.3% of pre-pregnant women were overweight and obese, respectively, whereas 25.4% and 22.3% of postpartum women were overweight and obese, respectively. The majority of the participants (64.8% pre-pregnant, 66.7% pregnant and 74.9% postpartum) were Australian born. Next, 42.0% of pre-pregnant, 33.8% of pregnant and 29.4% of postpartum women had high education levels (bachelor's degree or graduate diploma); 38.6% of pre-pregnant, 69.6% of pregnant and 52.2% of postpartum women were married; 58.6% of prepregnant, 55.6% of pregnant and 60.3% of postpartum women reported never smoking; and 90.0% of pre-pregnant, 96.0% of pregnant and 90.6% of postpartum women reported 'excellent/very good/good' self-rated health.

Table 1. Sociodemographic characteristics across reproductive life stages (n = 2492).

	Pre-Pregnancy (n = 880)	Pregnancy (n = 117)	Postpartum (n = 1495)
	n (%) [a]	n (%) [a]	n (%) [a]
Age, mean (SD)	31.2 ± 8.4	29.3 ± 5.3	33.6 ± 8.7
BMI, kg/m^2	25.4 ± 6.21	NA	26.2 ± 5.61
BMI (WHO categories)			
Under/normal weight (<25 kg/m^2)	412 (60.3)	NA	639 (52.3)
Overweight (25–<30 kg/m^2)	180 (20.7)	NA	336 (25.4)
Obese (≥ 30 kg/m^2)	163 (19.3	NA	317 (22.3)
Dieting			
Not currently on diet	132 (13.6)	1 (0.18)	242 (16.6)
Currently on diet	748 (86.4)	116 (99.8)	1253 (85.1)

Table 1. *Cont.*

	Pre-Pregnancy (n = 880)	Pregnancy (n = 117)	Postpartum (n = 1495)
	n (%) [a]	n (%) [a]	n (%) [a]
Country of birth			
Australia	645 (64.8)	87 (66.7)	1113 (74.9)
English speaking countries	80 (10.7)	7 (9.7)	145 (9.8)
Others	155 (24.5)	23 (15.3)	237 (15.3)
Remoteness area			
Major cities	596 (79.4)	67 (61.7)	972 (72.6)
Inner regional	150 (14.1)	26 (22.4)	286 (18.2)
Other	134 (6.50)	24 (15.9)	237 (9.16)
Marital status			
Married	313 (38.6)	83 (69.6)	880 (52.2)
Not married	567 (61.4)	34 (30.4)	615 (47.8)
Non-school educational level			
Bachelor/Graduate diploma	362 (42.0)	41 (33.8)	447 (29.4)
Certificates/Advanced diploma	276 (33.3)	37 (31.9)	523 (35.7)
No non-school qualification	232 (24.7)	39 (34.3)	514 (35.0)
Household income [b]			
Q1 (lowest)	72 (8.21)	17 (17.4)	260 (14.0)
Q2	97 (15.2)	15 (13.0)	265 (17.2)
Q3	127 (17.0)	25 (25.4)	297 (21.9)
Q4	242 (28.3)	27 (23.3)	301 (24.0)
Q5 (highest)	261 (31.4)	28 (20.9)	216 (22.8)
Occupation			
Professional	320 (34.2)	37 (29.6)	369 (26.3)
Assoc. Professional	296 (33.0)	30 (24.0)	474 (33.8)
Clerical trade	135 (16.2)	7 (7.94)	177 (13.6)
Other	129 (16.5)	40 (38.5)	475 (26.3)
SEIFA quintile			
Q1 (lowest)	151 (15.0)	18 (18.6)	277 (18.6)
Q2	162 (18.4)	36 (25.2)	274 (17.6)
Q3	176 (23.7)	20 (14.7)	316 (21.6)
Q4	158 (20.5)	21 (26.0)	260 (16.9)
Q5 (highest)	233 (22.3)	22 (15.4)	368 (25.4)
Smoking status			
Current smoker	201 (23.3)	12 (9.5)	295 (16.3)
Ex-smoker	169 (18.2)	42 (35.0)	380 (23.8)
Never smoked	510 (58.6)	68 (55.6)	820 (60.3)
Self-assessed health			
Excellent/very good/good	789 (90.0)	111 (96.0)	1341 (90.6)
Fair/poor	91 (10.0)	6 (4.0)	154 (9.4)

[a] Weighted percentage (replicate weight accounted); BMI, body mass index; NA, not applicable as BMI data were not collected from pregnant women. SEIFA, socio-economic index of disadvantage; quintile one represents the most disadvantaged areas, and quintile five represents the least disadvantaged areas (higher quintiles correspond to areas with lower levels of disadvantage areas where fewer individuals have low incomes, low educational attainment or work in unskilled occupations); [b] Equivalised household income (weekly, AUD)—an indicator of the economic resources available to each member of a household to indicate the situation of individuals and households.

3.2. Proportion of Reproductive Age Women Meeting Recommended Intakes of Core Foods, Discretionary Choices and PA

The mean intakes of core food groups, discretionary foods, energy from macronutrients and PA across reproductive life stages are shown in Tables 2 and 3; and the proportions of women who met and did not meet population-level dietary recommendations are shown in Figure 2. Similar mean proportions of total daily energy were from discretionary foods, beverages and SSBs in pre-pregnant (33.4% and 3.77% respectively), pregnant (29.1% and 4.94% respectively) and postpartum (31.5% and 3.51% respectively) women (Table 2).

Table 2. Core food groups and energy from discretionary foods in women across reproductive life stages.

	Pre-Pregnancy (n = 880)	Pregnancy (n = 117)	Postpartum (n = 1495)	Population Level Recommendations
	Mean ± SD	Mean ± SD	Mean ± SD	
Vegetables, legumes/beans (serve/day)	2.90 ± 2.67	2.95 ± 2.97	2.95 ± 2.97	≥5 serves/day
Fruit (serves/day)	1.32 ± 1.58	1.77 ± 1.88	1.31 ± 1.51	≥2 serves/day
Grain/cereals foods (serve/day)	3.97 ± 2.75	5.21 ± 3.34	4.01 ± 2.67	≥6 serves/day
Milk, yoghurt, cheese and alternatives (serve/day)	1.32 ± 1.11	1.97 ± 1.80	1.39 ± 1.13	≥2.5 serves/day
Meats and alternatives (serve/day)	1.56 ± 1.45	1.24 ± 1.12	1.60 ± 1.46	≥2.5 serves/day
Whole grains (serves/day)	1.26 ± 1.57	1.61 ± 1.80	1.19 ± 1.55	≥3 serves/day
Fibre (g/day)	20 ± 11.20	23.3 ± 12.4	20.5 ± 10.9	≥25 g/day
DF (%E)	33.4 ± 22.4	29.1 ± 18.9	31.5 ± 19.4	<2.5 serves/day
SSBs (%E)	3.77 ± 7.21	4.94 ± 9.70	3.51 ± 6.81	<2.5 serves/day

Proportion is weighted (population weight and survey design accounted); DF, discretionary foods; SSBs, sugar sweetened beverages; %E, percentage energy.

Table 3. Energy and macronutrient intake and PA in women across reproductive life stages.

	Pre-Pregnancy (n = 880)	Pregnancy (n = 117)	Postpartum (n = 1495)	Population Level Recommendations
	Mean ± SD	Mean ± SD	Mean ± SD	
Total energy (kJ)	7781.4 ± 3118.6	8683.3 ± 4037.6	7637.2 ± 2880.8	8700 kJ
Protein (%E)	17.8 ± 6.71	17.0 ± 5.48	18.3 ± 5.76	15–25%
CHO (%E)	44.4 ± 11.2	49.4 ± 10.8	44.4 ± 10.7	45–65%
Total fat intake (%E)	31.0 ± 9.14	30.6 ± 8.32	31.8 ± 8.79	20–35%
Saturated and trans-fat (%E)	12.1 ± 4.88	12.3 ± 4.62	12.5 ± 4.74	<10%
Trans-fat intake (%E)	0.56 ± 0.36	0.59 ± 0.34	0.57 ± 0.36	-
Monosaturated fat intake (%E)	11.8 ± 4.20	11.3 ± 3.48	12.2 ± 4.14	-
Added sugar (%E)	10.3 ± 8.92	11.8 ± 11.44	9.44 ± 7.79	<10%
Sodium (mg/day)	2249 ± 1334.0	2295.5 ± 1190.8	2222 ± 1153.2	2000 mg/day [a]
Alcohol (%E)	3.92 ± 9.61	0.07 ± 0.83	2.49 ± 6.46	<10 standard drink/week [b]
Total PA (min/week) [1]	255.4 ± 253.1	149.2 ± 204.6	201.8 ± 224.8	≥150 min/day
MVPA (min/week) [2]	143.9 ± 248.0	30.6 ± 80.1	115.2 ± 210.7	≥150 min/day

Proportion is weighted (population weight and survey design accounted); CHO, carbohydrates; MVPA, moderate–vigorous physical activity; PA, physical activity; %E, percentage energy. [1] Total minutes of physical activity undertaken in last week (includes walking for transport + walking for fitness + moderate + vigorous time but does not include sessions). [2] Moderate–vigorous physical activity derived from time spent in moderate and vigorous intensity activities (moderate time + 2 times vigorous time). [a] Reference for sodium guideline [34]. [b] Reference for alcohol guideline [35].

Approximately one in ten women across all life stages obtained much of their daily energy from added sugars, and this portion was slightly higher in pregnant women (11.8%).

Despite reproductive age women not meeting the recommended serves of core foods, energy intake from macronutrients was generally in the optimal range across reproductive life stages (Table 3) (acceptable macronutrient distribution range: 15–25% of energy from protein, 20–35% of energy from fat and 45–65% of energy from carbohydrates) [34]. Similar proportions of women met the recommended daily intakes of both fruit and vegetables (5.12% of pre-pregnant, 2.14% of pregnant and 4.35% of postpartum women), vegetables (15.9% for pre-pregnant and 15.0% for pregnant and postpartum women) and meat and alternatives (17.9% of pregnant and 22.0% of pre-pregnant and postpartum women) (Figure 2). The proportions of women who met recommended intakes of fruit, grains/cereals and dairy or alternatives were higher for pregnant women (37.1%, 33.3% and 27.7%, respectively) than pre-pregnant (26.0%, 19.6% and 13.4%, respectively) and postpartum (24.9%, 18.5% and 14.9%, respectively) women (Figure 2). The proportion of women meeting PA guidelines (total activity in minutes) was lower for pregnant women (31.0%) than for pre-pregnant (59.4%) and postpartum (48.8%) women. Similarly, a low proportion of pregnant women (6.9%) had ≥150 min MVPA/week compared to pre-pregnant (30.6%) and postpartum women (25.4%) (Figure 3).

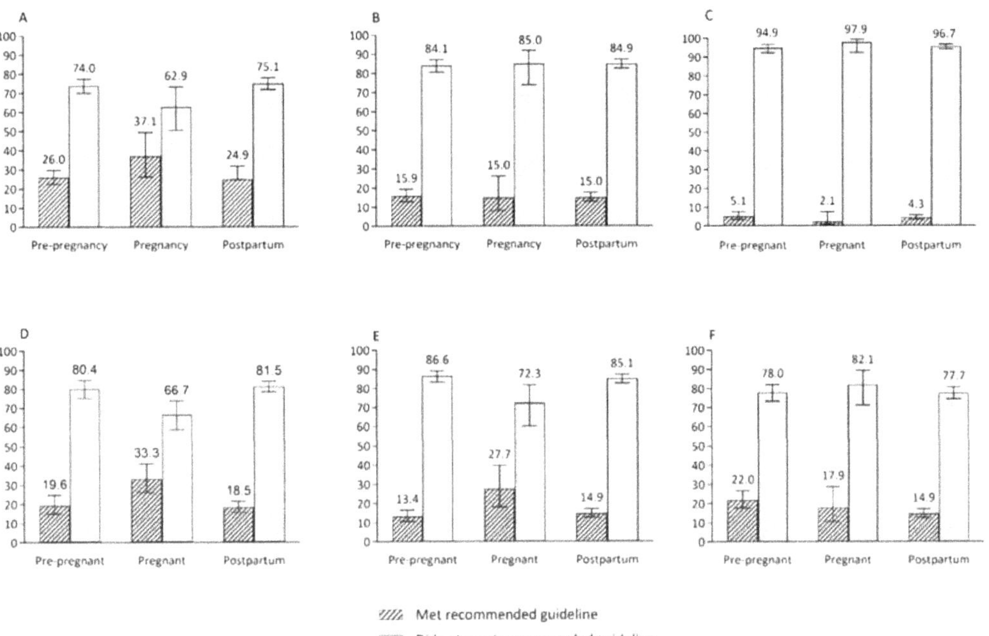

Figure 2. The proportions of women who met and do not meet the population-level recommended intakes of core food groups across reproductive life stages. (**A**) Fruit, (**B**) vegetables and legumes, (**C**) fruit and vegetable combined, (**D**) grain (cereal) foods, (**E**) milk and alternatives, (**F**) meat and alternatives.

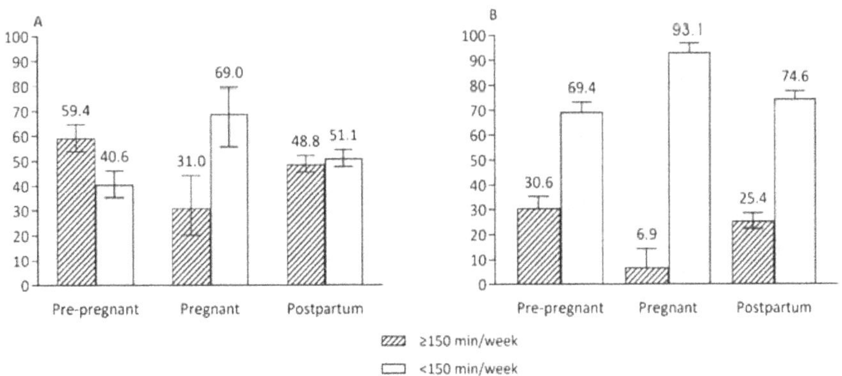

Figure 3. The proportions of women who spent time performing physical activity at least 150 min per week or more across reproductive life stages. (**A**) Total minutes of physical activity per week, (**B**) moderate–vigorous intensity activity per week. Error bars represent 95% CI of proportions.

Differences in sociodemographic characteristics for women meeting or not meeting population recommendations for diet and PA pre-pregnancy and postpartum are reported in Tables S4a–g. These data are not presented for pregnant women due to the small sample size. For pre-pregnant women, the recommended intakes of vegetables; fruit; dairy or alternatives; and meat or alternatives, did not differ by sociodemographic factors. For post-partum women, the recommended intakes of dairy or alternatives and meat or alternatives did not differ by sociodemographic factors. Pre-pregnant women born in Australia were less likely to meet the recommended intake of grains/cereals and more likely to have intake of discretionary foods above the recommended level (>2.5 serves/day). Those with a higher education and SEIFA were more likely to meet the PA recommendations. Postpartum women with a higher education were more likely to meet the recommended intakes of vegetables and fruit; those born in Australia were less likely to meet the recommended intakes of fruit and grains/cereals and discretionary foods; those with professional jobs were more likely to meet the recommended intake of fruit, and those with a higher education and SEIFA were more likely to meet the PA recommendations.

3.3. Diet and PA Variables Associated with BMI

In multivariable analysis among pre-pregnant women, BMI was inversely associated with higher intake of whole grains ($\beta = -1.58$, 95% CI -2.96, -0.21; $p = 0.025$) and with increased energy from alcohol ($\beta = -0.08$, 95% CI -0.14, -0.005; $p = 0.035$) (Table 4). However, no associations were found among core foods (fruit, vegetable, grain/cereal foods, dairy, meat and/or alternatives), total energy intake, energy from discretionary foods/beverages, SSBs and PA. With regard to sociodemographic factors, in pre-pregnant women, age ($\beta = 0.22$, 95% CI 0.15, 0.29; $p < 0.001$), being born in other county ($\beta = -3.20$, 95% CI -4.52, -1.88; $p < 0.001$), being a current smoker ($\beta = -1.40$, 95% CI -2.76, -0.04; $p = 0.044$), excellent/very good/good health ($\beta = -2.89$, 95% CI -5.51, -0.28; $p = 0.030$) and currently on a diet ($\beta = 2.25$, 95% CI 0.25, 4.24; $p = 0.028$) were independently associated with BMI (Table 4).

Table 4. Associations between diet and physical activity and BMI in pre-pregnant women (N = 755).

	Unadjusted Model		Adjusted Model	
	β (95% CI)	*p* Value	β (95% CI)	*p* Value
Age, year	0.20 (0.14, 0.25)	<0.001	0.22 (0.15, 0.29)	**<0.001**
Country of birth				
Australia (ref.)				
Other English-speaking country	−0.20 (−2.05, 1.66)	0.832	0.34 (−1.10, 1.79)	0.639
Others	−3.04 (−4.11, −1.96)	<0.001	−3.20 (−4.52, −1.88)	**<0.001**
Remoteness area				
Major cities (ref.)				
Inner regional	1.77 (0.22, 3.32)	0.026	0.36 (−1.04, 1.76)	0.613
Other	3.01 (0.83, 5.20)	0.008	1.73 (−0.37, 3.83)	0.104
Marital status				
Not married (ref.)				
Married	0.30 (−0.95, 1.55)	0.633	−0.51 (−1.79, 0.77)	0.432
Education				
Bachelor/Graduate diploma (ref.)				
Certificates/Advanced diploma	0.99 (−0.42, 2.41)	0.164	−0.97 (−2.62, 0.67)	0.242
No non-school qualification	2.09 (0.34, 3.85)	0.020	−0.56 (−2.18, 1.06)	0.491
Household income (cont. decile)	−0.08 (−0.33, 0.18)	0.553	..	..
Occupation				
Clerical trade (ref.)				
Professional	−0.82 (−2.54, 0.91)	0.347	−0.77 (−2.33, 0.79)	0.327
Assoc. Professional	−0.90 (−2.50, 0.70)	0.265	−1.08 (−2.55, 0.40)	0.150
Other	0.05 (−2.18, 2.27)	0.967	−0.54 (−2.50, 1.42)	0.584

Table 4. *Cont.*

	Unadjusted Model		Adjusted Model	
	β (95% CI)	*p* Value	β (95% CI)	*p* Value
SEIFA				
1st (highest disadvantage) (ref.)				
2nd quintile	−0.52 (−2.36,1.32)	0.575	−0.65 (−2.65, 1.36)	0.521
3rd quintile	−2.10 (−3.98, −0.22)	0.029	−1.16 (−2.90, 0.57)	0.184
4th quintile	−1.55 (−3.79,0.69)	0.171	−1.48 (−3.53, 0.57)	0.15
5th quintile (least disadvantage)	−2.18 (−3.98, −0.38)	0.018	−1.85 (−3.71, 0.01)	0.051
Smoking status				
Never smoked (ref.)				
Current smoker	0.55 (−1.09, 2.18)	0.505	−1.40 (−2.76, −0.04)	**0.044**
Ex-smoker	1.73 (0.05, 3.41)	0.043	0.53 (−1.08, 2.15)	0.512
Self-assessed health				
Fair/poor (ref.)				
Excellent/very good/good	−2.77 (−5.57, 0.03)	0.052	−2.89 (−5.51, −0.28)	**0.030**
Dieting				
Not currently on diet (ref)				
Currently on diet	2.41 (0.21, 4.61)	0.032	2.25 (0.25, 4.24)	**0.028**
Vegetables, legumes/beans (categorical)				
<1 serves/day (ref.)				
≥1 to <3 serves/day	−0.43 (−1.82, 0.96)	0.537	..	..
≥3 to <5 serves/day	0.49 (−1.24, 2.21)	0.574	..	..
≥5 serves/day	−1.56 (−3.32, 0.21)	0.083	..	..
Vegetables, legumes/beans (binary)				
<5 serves/day (ref.)				
≥5 serves/day	−1.48 (−2.93, −0.03)	0.046	−1.06 (−2.54, 0.42)	0.156
Fruit (categorical)				
<1 serves/day (ref.)				
≥1 to <2 serves/day	−1.55 (−3.86, 0.72)	0.007	..	..
≥2 to <3 serves/day	−0.79 (−2.50, 0.93)	0.363	..	..
≥3 serves/day	−1.13 (−3.19, 0.94)	0.279	..	..
Fruit (binary)				
<2 serves/day (ref.)				
≥2 serves/day	−0.47 (−1.84, 0.90)	0.494	0.39 (−1.04, 1.81)	0.588
Grain (cereal) foods (categorical)				
Zero or none (ref.)				
>0 to <2 serves/day	−1.62 (−4.71, 1.47)	0.299	..	..
≥2 to <4 serves/day	−2.48 (−5.67, 0.71)	0.125	..	..
≥4 to <6 serves/day	−2.42 (−5.73, 0.89)	0.149	..	..
≥6 serves/day	−2.78 (−6.09, 0.53)	0.098	..	..
Grain (cereal) foods (binary)				
<6 serves/day (ref.)				
≥6 serves/day	−0.64 (−2.07, 0.79)	0.373	0.74 (−0.92, 2.39)	0.376
Whole grain (categorical)				
<1 serves/day (ref.)				
≥1 to <2 serves/day	−0.64 (−2.34, 1.07)	0.459	−0.12 (−1.69, 1.44)	0.874
≥2 to <3 serves/day	−1.88 (3.43, −0.33)	0.019	−1.63 (−3.44, 0.17)	0.075
≥3 serves/day	−2.10 (−3.78, −0.41)	0.016	−1.58 (−2.96, −0.21)	**0.025**
Whole grains (binary)				
<48 g/day (ref.)				
≥48 g/day	−1.65 (−2.96, −0.34)	0.014	..	..

Table 4. *Cont.*

	Unadjusted Model		Adjusted Model	
	β (95% CI)	*p* Value	β (95% CI)	*p* Value
Whole grain (half of total grain intake)				
<50% (ref.)				
≥50%	−1.84 (−3.12, −0.56)	0.006	..	..
Milk and/or alternatives (categorical)				
<0.5 serves/day (ref.)				
≥0.5 to <1.5 serves/day	−1.23 (−3.19, 0.73)	0.214	..	..
≥1.5 to <2.5 serves/day	−1.08 (−3.05, 0.89)	0.280	..	..
≥2.5 serves/day	−1.66 (−4.22, 0.89)	0.199	..	..
Milk and/or alternatives (binary)				
<2.5 serves/day (ref.)				
≥2.5 serves/day	−0.86 (−2.89, 1.17)	0.401	..	..
Meats and/or alternatives (categorical)				
<0.5 serves/day (ref.)				
≥0.5 to <1.5 serves/day	−1.86 (−3.54, −0.17)	0.032	..	..
≥1.5 to <2.5 serves/day	−0.38 (−1.93, 1.17)	0.624	..	..
≥2.5 serves/day	−0.58 (−2.32. 1.17)	0.510	..	..
Meats and/or alternatives (binary)				
<2.5 serves/day (ref.)				
≥2.5 serves/day	0.22 (−1.12, 1.55)	0.748	0.87 (−0.34, 2.08)	0.157
Fibre (g/day)	−0.07 (−0.11, −0.03)	0.001	..	..
CHO (%E)	0.01 (−0.04, 0.07)	0.639	..	..
Protein (%E)	0.11 (0.03, 0.20)	0.010	..	..
Total fat (%E)	−0.03 (−0.07, 0.02)	0.197	−0.06 (−0.14, 0.03)	0.179
Trans-fat (%E)	0.84 (−0.66, 2.34)	0.268	..	..
Saturated and trans-fat (%E)	−0.05 (−0.17, 0.07)	0.414	..	..
Monosaturated fat (%E)	−0.04 (−0.18, 0.10)	0.545	..	..
Polyunsaturated fat (%E)	−0.14 (−0.47, 0.18)	0.381	..	..
Alcohol (%E)	−0.03 (0.09, 0.03)	0.352	−0.08 (−0.14, −0.005)	**0.035**
DF (%E)	0.02 (−0.01, 0.05)	0.221	0.01 (−0.03, 0.05)	0.475
SSBs (%E)	0.09 (−0.007, 0.18)	0.069	0.04 (−0.06, 0.15)	0.378
Added sugar intake (%E)	0.05 (−0.03, 0.13)	0.211	..	..
Total energy (MJ)	−0.24 (−0.42, −0.05)	0.013	−0.16 (−0.37, 0.05)	0.135
Total PA (minutes/wk) cont [1]	−0.003 (−0.004, −0.001)	0.005	−0.002 (−0.004, 0.0002)	0.080
Total PA (minutes/wk) [1]				
Did not meet recommended guideline (ref.)				
Met recommended guidelines	−1.38 (−2.86, 0.09)	0.066	..	..

Data were analysed using linear regression with data presented from univariable and multivariable linear regression analyses. Collinear diet variables (VIF > 10) were excluded (protein correlated with meat and alternatives, carbohydrate with alcohol, monosaturated fat, polyunsaturated fat, trans-fat saturated fat + trans-fat correlated with total fat, added sugar correlated with SSBs). β, beta-coefficient; CHO, carbohydrate; DF, discretionary foods; MJ, megajoules; MVPA, moderate–vigorous physical activity; PA, physical activity; SEIFA, socio-economic index of disadvantage for areas; SSBs, sugar sweetened beverage, %E, percentage energy. [1] Total minutes of physical activity undertaken in last week (includes walking for transport + walking for fitness + Moderate + Vigorous time but does not include sessions). Used as continuous and dichotomized as whether physical activity last week met 150 min recommended guidelines.

In postpartum women, BMI was inversely associated with increased fibre intake (β = −0.06, 95% CI −0.11, −0.004; *p* = 0.034) and each minute increase in PA per week (β = −0.002, 95% CI −0.004, −0.001; *p* = 0.013) (Table 5). There were no significant associations between fruit, vegetable, whole grain, dairy, meat and/or alternatives, total energy intake and energy from discretionary foods and BMI in postpartum. Higher socioeconomic disadvantage (β = −1.73, 95% CI −3.12, −0.05; *p* = 0.017; Q5 vs. Q1) and excellent/very good/good health condition (β = −2.30, 95% CI −4.06, −0.54; *p* = 0.011) were inversely associated with BMI, whereas currently on diet (β = 2.82, 95% CI 1.76, 3.89; *p* < 0.001) and increased age (β = 0.13, 95% CI 0.08, 0.19; $p < 0.001$) were positively associated with BMI (Table 5).

Table 5. Associations between diet and physical activity and BMI in postpartum women (N = 1292).

	Unadjusted Model		Adjusted Model	
	β (95% CI)	p Value	β (95% CI)	p Value
Age, year	0.12 (0.07, 0.16)	<0.001	0.13(0.08, 0.19)	**<0.001**
Country of birth				
Australia (ref.)				
Other English-speaking country	0.18 (−1.28, 1.65)	0.802	−0.01(−1.59, 1.58)	0.993
Others	−0.45 (−1.67, 0.77)	0.461	−0.98 (−2.31, 0.36)	0.148
Remoteness				
Major cities (ref.)				
Inner regional	1.24 (0.07, 2.40)	0.038	0.40 (−0.77, 1.58)	0.496
Other	1.87 (0.45, 3.28)	0.011	1.16 (−37, 3.83)	0.129
Marital status				
Not married (ref.)				
Married	0.77 (−0.15, 1.70)	0.101	..	..
Education				
Bachelor/Graduate diploma (ref.)	1.36 (0.18, 2.55)	0.025	−0.40 (−1.58, 0.78)	0.500
Certificates/Advanced diploma	0.50 (−0.61, 1.61)	0.373	0.83 (−0.20, 1.87)	0.491
No non-school qualification				
Household income (cont. decile)	−0.03 (−0.20, 0.14)	0.703	..	..
Occupation				
Clerical trade (ref.)				
Professional	−0.59 (−2.25, 1.07)	0.482	−0.58 (−2.34, 1.18)	0.511
Assoc. Professional	−0.51 (−2.23, 1.21)	0.559	−0.92 (−2.40, 0.55)	0.216
Other	−0.23 (−2.08, 1.63)	0.807	−0.19 (−1.89, 1.51)	0.828
SEIFA				
1st (highest disadvantage) (ref.)				
2nd quintile	−0.76 (−2.49,0.97)	0.383	−0.37 (−1.99, 1.24)	0.644
3rd quintile	−1.36 (−3.46, 0.74)	0.199	−1.07 (−3.06, 0.92)	0.286
4th quintile	−1.94 (−3.67, −0.22)	0.028	−1.62 (−3.28, 0.05)	0.058
5th quintile (least disadvantage)	2.30 (−3.75, −0.86)	0.002	−1.73 (−3.12, −0.32)	**0.017**
Smoking status				
Never smoked (ref.)				
Current smoker	1.22 (0.20, 2.25)	0.020	0.15 (−0.84, 1.15)	0.761
Ex-smoker	1.49 (0.55, 2.43)	0.002	0.85 (−0.14, 1.84)	0.092
Self-assessed health				
Fair/poor (ref.)				
Excellent/very good/good	−3.16 (−4.92, −1.41)	0.001	−2.30 (−4.06, −0.54)	**0.011**
Dieting				
Not currently on diet (ref)				
Currently on diet	2.81 (1.75, 3.88)	<0.001	2.82 (1.76, 3.89)	**<0.001**
Vegetables, legumes (categorical)				
<1 serves/day (ref.)				
≥1 to <3 serves/day	−0.59 (−1.76, 0.57)	0.311	..	..
≥3 to <5 serves/day	−0.90 (−2.33, 0.53)	0.211	..	..
≥5 serves/day	−0.85 (−2.31, 0.60)	0.246	..	..
Vegetables, legumes/beans (binary)				
<5 serves/day (ref.)				
≥5 serves/day	−0.36 (−1.68, 0.96)	0.587	1.04 (−0.53, 2.62)	0.189
Fruit (categorical)				
<1 serves/day(ref.)				
≥1 to <2 serves/day	0.73 (−0.40, 1.86)	0.201	..	..
≥2 to <3 serves/day	−0.24 (−1.65, 1.16)	0.731	..	..
≥3 serves/day	−1.49 (−2.94, −0.05)	0.043	..	..

Table 5. *Cont.*

	Unadjusted Model		Adjusted Model	
	β (95% CI)	*p* Value	β (95% CI)	*p* Value
Fruit (binary)				
<2 serves/day (ref.)				
≥2 serves/day	−1.09 (−2.07, 0.11)	0.030	−0.29 (−1.44, 0.86)	0.617
Grain (cereal) foods (categorical)				
Zero or none (ref.)				
>0 to <2 serves/day	−0.95 (−3.40, 1.51)	0.442	..	..
≥2 to <4 serves/day	−1.51 (−3.561, 0.49)	0.137	..	..
≥4 to <6 serves/day	−1.36 (−3.53, 0.80)	0.212	..	..
≥6 serves/day	−1.93 (−4.14, 0.28)	0.086		
Grain (cereal) foods (binary)				
<6 serves (ref.)				
≥6 serves	−0.66 (−1.72, 0.41)	0.222	..	..
Whole grain (categorical)				
<1 serves/day (ref.)				
≥1 to <2 serves/day	−1.34 (−2.35, −0.31)	0.459	−0.85 (−1.86, 0.16)	0.097
≥2 to <3 serves/day	−0.52 (−1.91, 0.87)	0.460	−0.40 (−1.80, 1.00)	0.569
≥3 serves/day	−1.90 (−3.11, −0.69)	0.003	−1.08 (−2.54, 0.38)	0.144
Whole grain (binary)				
<48 g/day (ref.)				
≥48 g/day	−0.81 (−1.69, 0.73)	0.071	..	..
Whole grain (half of total grains intake)				
<50% (ref.)				
≥50%	−0.95 (−1.99, 0.09)	0.072	..	..
Milk and/or alternatives (categorical)				
<0.5 serves/day (ref.)				
≥0.5 to <1.5 serves/day	0.64 (−0.40, 1.68)	0.221	..	..
≥1.5 to <2.5 serves/day	−0.16 (−1.46, 1.14)	0.808	..	..
≥2.5 serves	0.04 (−1.24, 1.32)	0.953	..	..
Milk and/or alternatives (binary)				
<2.5 serves/day (ref.)				
≥2.5 serves/day	−0.20 (−1.28, 0.88)	0.715	−0.48 (−1.53, 0.57)	0.366
Meats and/or alternatives (categorical)				
<0.5 serves/day (ref.)				
≥0.5 to <1.5 serves/day	−0.96 (−2.01, 0.09)	0.071	..	..
≥1.5 to <2.5 serves/day	−0.69 (−1.82, 0.44)	0.228	..	..
≥2.5 serves/day	−0.62 (−1.85, 0.62)	0.322	..	..
Meats and alternatives (binary)				
<2.5 serves/day (ref.)				
≥2.5 serves/day	−0.08 (−1.12, 0.97)	0.880	−0.29 (−1.48, 0.90)	0.630
Fibre (g/day)	−0.06 (−0.09, −0.03)	0.001	−0.06 (−0.11, −0.004)	**0.034**
CHO (%E)	0.01 (−0.03, 0.05)	0.687	..	..
Protein (%E	0.05 (−0.02, 0.132)	0.154	..	..
Total fat (%E)	−0.03 (−0.07, 0.02)	0.197	−0.04 (−0.09, 0.008)	0.097
Trans-fat (%E)	0.40 (−0.76, 1.56)	0.492	..	..
Saturated and trans-fat (%E)	0.02 (−0.10, 0.06)	0.628	..	..
Monosaturated fat (%E)	−0.04 (−0.18, 0.10)	0.545	..	..
Polyunsaturated fat (%E)	−0.14 (−0.31, 0.27)	0.100	..	..
Alcohol (%E)	0.01 (−0.06, 0.08)	0.820	−0.03 (−0.09, 0.04)	0.460

Table 5. *Cont.*

	Unadjusted Model		Adjusted Model	
	β (95% CI)	*p* Value	β (95% CI)	*p* Value
DF (%E)	0.007 (−0.01, 0.03)	0.437	−0.006 (−0.03,0.02)	0.673
SSBs (%E)	0.04 (−0.05, 0.13)	0.340	..	..
Added sugar intake (%E)	0.06 (−0.01, 0.12)	0.075	..	..
Total energy (MJ)	−0.11 (−0.28, 0.04)	0.144	0.15 (0.09, 0.39)	0.209
Total PA (minutes/wk) cont [1]	−0.003 (−0.004, −0.001)	0.005	−0.002 (−0.004, −0.001)	0.013
Total PA (minutes/wk) [1]				
Did not meet recommended guideline (ref.)				
Met recommended guidelines	−0.96 (−1.95, 0.03)	0.057	..	..

Data were analysed using linear regression with data presented from univariable and multivariable linear regression analyses. Collinear diet variables were excluded (protein correlate with meat and alternatives, carbohydrate with alcohol, monosaturated fat, polyunsaturated fat, trans-fat saturated fat + trans-fat correlated with total fat, added sugar correlated with SSBs). β, beta-coefficient; CHO, carbohydrate; DF, discretionary foods; MJ, megajoules; MVPA, moderate-vigorous physical activity; PA, physical activity; SEIFA, socio-economic index of disadvantage for areas; SSBs, sugar sweetening beverages, %E, percentage energy. [1] Total minutes of physical activity undertaken in last week (includes walking for transport + walking for fitness + Moderate + Vigorous time but do not include sessions). Used as continuous and dichotomized as whether physical activity last week met 150 min recommended guidelines. The difference in the number of variables included in the multivariable analysis for pre-pregnant and postpartum women is due to model selection strategies using backward elimination techniques.

3.4. Results from Sensitivity Analysis

Multivariable linear regression analysis results of diet and PA and BMI after excluding implausible energy reporters in pre-pregnant and postpartum women are presented in Table 6. In pre-pregnant women, the associations between whole grains and energy from alcohol and BMI were not maintained in the sensitivity analysis. Other diet variables showed similar associations in terms of directionality without substantial differences in the magnitudes of estimates (<20% relative change) from the main analysis. In postpartum women, the association between fibre and BMI was no longer statistically significant, but the total energy intake's association with BMI became statistically significant. The inverse association between increased PA and BMI was maintained after excluding implausible energy reporters.

Table 6. Sensitivity analysis of the associations between diet and physical activity and BMI after excluding implausible energy reporters among pre-pregnant and postpartum women.

	Pre-Pregnancy (n = 577) [§]		Postpartum (n = 989) [§]	
	Adjusted β (95% CI)	*p* Value	Adjusted β (95% CI)	*p* Value
Age, year	0.20 (0.13, 0.27)	**<0.001**	0.14 (0.08., 0.19)	**<0.001**
Country of birth				
Australia (ref.)				
Other English-speaking country	0.39 (−0.96, 1.74)	0.567	0.18 (−1.61, 1.97)	0.841
Others	−1.66 (−2.70, −0.62)	**0.002**	−0.68 (−2.11, 0.75)	0.154
Remoteness				
Major cities (ref.)				
Inner regional	0.80 (−90, 2.51)	0.350	0.03 (−1.11, 1.17)	0.952
Other	1.70 (−0.67, 4.07)	0.157	1.18 (−0.46, 2.82)	0.154
Marital status				
Not married (ref.)				
Married	−0.35 (−1.44, 0.74)	0.517	...	..
Education				
Bachelor/Graduate diploma (ref.)				
Certificates/Advanced diploma	−0.75 (−2.54, 1.05)	0.408	−0.68 (−1.86, 0.50)	0.254
No non-school qualification	−0.28 (−1.93, 1.36)	0.731	1.08 (−0.07, 2.22)	0.064

Table 6. *Cont.*

	Pre-Pregnancy (n = 577) [§]		Postpartum (n = 989) [§]	
	Adjusted β (95% CI)	*p* Value	Adjusted β (95% CI)	*p* Value
Occupation				
Clerical trade (ref.)				
Professional	−0.31 (−1.99, 1.36)	0.710	−0.83 (−2.74, 1.09)	0.392
Assoc. Professional	−0.35 (−1.92, 1.22)	0.659	−1.46 (−3.20, 0.29)	0.100
Other	−1.51 (−2.96, −0.06)	**0.041**	−0.69 (−2.71, 1.33)	0.498
SEIFA				
1st (highest disadvantage) (ref.)				
2nd quintile	0.18 (−2.15, 2.50)	0.880	−0.48 (−2.27, 1.30)	0.591
3rd quintile	−0.42 (−2.49, 1.64)	0.684	−1.13 (−3.04, 0.78)	0.241
4th quintile	−0.62 (−2.75, 1.51)	0.564	−1.77 (−3.66, 0.12)	0.066
5th quintile (least disadvantage)	−0.49 (−2.52, 1.54)	0.631	−1.49 (−3.10, 0.12)	0.68
Smoking status				
Never smoked (ref.)				
Current smoker	−1.55 (−2.95, −0.16)	0.030	0.33 (−0.96, 1.62)	0.612
Ex-smoker	0.47 (−1.12, 2.06)	0.558	0.74 (−0.56, 2.04)	0.262
Self-assessed health				
Fair/poor (ref.)				
Excellent/very good/good	−2.22 (−5.09, 0.64)	0.125	−1.04 (−2.66, 0.58	0.203
Dieting				
Not currently on diet (ref)				
Currently on diet	2.28 (0.39, 4.17)	**0.019**	3.24 (1.79, 4.71)	**<0.001**
Vegetables, legumes/beans (binary)				
<5 serves/day (ref.)				
≥5 serves/day	−1.23 (−2.46, 0.003)	0.051	0.63 (−0.72, 1.99)	0.354
Fruit (binary)				
<2 serves/day (ref.)				
≥2 serves	0.86 (−0.31, 2.04)	0.146	−0.74 (−2.07, 0.59)	0.270
Grain (cereal) foods (binary)				
<6 serves (ref.)				
≥6 serves/day	1.44 (−0.21,3.09)	0.086	..	...
Whole grain (serve, categorical)				
<1 serves/day (ref.)				
≥1 to <2 serves/day	0.48 (−1.07, 2.03)	0.537	−1.10 (−2.17, −0.03)	0.044
≥2 to <3 serves/day	−0.58 (−2.44, 1.28)	0.537	−0.22 (−1.59, 1.15)	0.748
≥3 serves/day	−1.11 (−2.42, 0.20)	0.096	−0.78 (−2.35, 0.79)	0.324
Milk and/or alternatives (binary)				
<2.5 serves/day (ref.)				
≥2.5 serves/day	..	..	−0.26 (−1.38, 0.85)	0.641
Meats and alternatives (binary)				
<2.5 serves/day (ref.)				
≥2.5 serves/day	1.09 (−0.08, 2.25)	0.067	−0.66 (−1.88, 0.57)	0.287
Fibre (g/day)	..	...	−0.04 (−0.09, 0.01)	0.144
Total fat (%E)	−0.01 (−0.09, 0.07)	0.724	−0.04 (−0.09, 0.02)	0.211
Alcohol (%E)	−0.04 (−0.11, 0.03)	0.231	−0.01 (−0.08, 0.06)	0.737
DF (%E)	0.03 (−0.01, 0.07)	0.153	−0.0003 (−0.03, 0.03)	0.865
SSBs (%E)	0.06 (−0.06, 0.17)	0.303	..	..
Total energy (MJ)	−0.01 (−0.23, 0.21)	0.927	0.33 (0.09, 0.58)	**0.007**
Total PA (minutes/wk) cont [1].	−0.001(−0.003, 0.001)	0.208	−0.002 (−0.004, −0.0001)	**0.041**

Data were analysed using linear regression. Only data from multivariable linear regression analyses are presented. β, beta-coefficient; DF, discretionary foods; MJ, megajoules; MVPA, moderate-vigorous physical activity; PA, physical activity; SEIFA, socio-economic index of disadvantage for areas; SSBs, sugar sweetened beverages, %E, percentage energy. [§] Variables included here are similar to variables that were included in the main analyses and were selected based on backward stepwise elimination processes. 1 Total minutes of physical activity undertaken in last week (includes walking for transport + walking for fitness + Moderate + Vigorous time but do not include sessions).

4. Discussion

4.1. Main Findings

We report here for the first time on diet and PA and their associations with BMI in a nationally representative sample of Australian women across key reproductive life stages. We confirm women across life stages failed, on average, to meet population-level recommended intakes of key core foods. A higher proportion of daily energy from discretionary foods persisted in pre-pregnant, pregnant and postpartum women. Sociodemographic factors, including country of birth, education, occupation and socioeconomic disadvantage areas, were associated with core and discretionary food intake and PA in pre-pregnancy and postpartum women. An inverse association was observed for both higher whole grain intake and higher energy from alcohol and BMI in pre-pregnant women, whereas increased fibre intake and PA were inversely associated with BMI in postpartum women.

4.2. Meeting Recommended Intakes of Core Foods, Discretionary Choices and PA

Our findings of failure to meet population-level recommendations for core foods, discretionary foods and PA are consistent with previous studies in pre-pregnant, pregnant [21,36] and postpartum women [37,38] and in the general population [13]. Failure to meet dietary or PA recommendations may be explained by factors, including lack of awareness, limited resources for accessing healthy foods for low-socioeconomic-status women, [36], cultural influences on food choice, lack of social support, exposure to fast food outlets [39] and other psychosocial barriers [40]. Low proportions of pregnant and postpartum women met the PA guidelines, which is consistent with previous studies [41–43]. Several barriers may prevent pregnant women from engaging in PA, including perceived mother–baby safety concerns, fatigue, lack of motivation and lack of social support [44,45]; and barriers such as time limitations, lack of childcare, lack of partner support and family responsibilities may prevent PA by postpartum women [20].

4.3. Socioeconomic Factors

Consistent with prior research in the general population [46–48], a range of sociodemographic characteristics were associated with meeting the recommended intakes of PA guidelines in pre-pregnant and postpartum women. For women pre-pregnancy, those born in Australia were less likely to have an optimal intake of grains/cereals, and those with higher education and the least socioeconomically disadvantaged were more likely to meet PA guidelines. Conversely, the recommended intakes of vegetables, fruit, dairy or alternatives and meat or alternatives did not differ by sociodemographic factors in pre-pregnant women, as previously reported [21]. Postpartum women with higher education and living in socioeconomically advantaged areas were more likely to have the recommended intakes of vegetables and fruits and the recommended level of PA. While research is limited in postpartum women, this finding is consistent with previous studies reporting socioeconomic disadvantage as being a strong determinant of fruit and vegetable intake [46] and PA [48] in the general population. A disparity in overall food and nutrient intake has been previously reported between Australian-born and overseas-born women, with overseas born women having higher intakes of cereals/beans but less vegetable/legume, dairy and meat intakes than Australian-born women [49]. This suggests that future interventions could potentially target grains/cereals and PA for pregnant women; and vegetables, fruit, discretionary foods and PA for postpartum women, specifically those from different ethnic and socioeconomic backgrounds.

4.4. Dietary Components and BMI

A higher whole grain intake ($\geq$3 servings/day) in pre-pregnant women and increased fibre intake in postpartum were associated with decreased BMI (kg/m^2). The association between wholegrains and BMI in pre-pregnancy is consistent with previous meta-analyses in the general population reporting $\geq$3 servings/day whole grains was associated with a lower BMI and less central adiposity [50]. In postpartum women, a 1 g/d increase in fibre was associated with a 0.06 kg/m^2 lower BMI and 0.15 kg lower postpartum weight gain [51]; and fibre intake below the recommendation (<29 g/day) increased the risk of PPWR by 24% [52]. These findings may be related to the effects of whole grains [53] and fibre [54] on satiety and fullness and the subsequent inhibitory effect on energy intake. Given the mean wholegrains and fibre intakes were ~1.2 serves/day and ~20 g/day, respectively (compared to broad guidelines of ≥3 serves/day [55] and population recommendations of 25–30 g/day [34], respectively), it is imperative to target both fibre and wholegrains for optimising weight management in women at key reproductive life-stages.

We report an inverse association between BMI and energy from alcohol in pre-pregnant women. While there is a lack of research currently on the association between alcohol and obesity in pre-pregnant and postpartum women, there are inconsistent findings in the general population [56–59]. This may be partly attributed to variations in frequency, amount or types of alcohol, and variations in lifestyle and dietary habits or energy intake for drinkers and non-drinkers [60]. Our analysis was adjusted for factors such as dieting and total energy intake, and indicates an independent relationship between alcohol and BMI in pre-pregnant women. The link between alcohol intake and BMI is likely complex and modifiable across life stages due to physiological variations. Furthermore, given that a large proportion of pregnancies are unplanned [61] and population recommendations are to stop alcohol intake when trying to conceive [35], the contribution of alcohol to both BMI and adverse pregnancy outcomes must be considered.

We observed no significant association between fruit or vegetable intake and BMI in pre-pregnant and postpartum women. This is in contrast to prior studies reporting inverse associations between the 'vegetables and meat' pattern and BMI in preconception [14] and the fruit and vegetable index and weight gain in young women [62], and systematic reviews reporting an inverse association between fruit and vegetable intake and BMI in the general population [63]. This discrepancy could be due to the lack of consistent adjustment for important confounders, including total energy intake and the low intake of fruit and vegetables. In contrast to reports in the general population [64,65], we also report no association between energy from discretionary foods or SSBs and BMI in pre-pregnant and postpartum women. These disparate results are unclear, but may be due to the use of different analysis approaches. We used energy from discretionary foods, unlike prior studies that assessed individual discretionary foods [65]. Total energy intake was not also associated with BMI in pre-pregnant and postpartum women, in contrast to prior studies of postpartum women [66]. This may be partly explained by energy misreporting, particularly in women with higher BMIs [67], or other factors, such as the relatively high rate of dieting. In sensitivity analysis, however, increased energy intake was associated with BMI in postpartum women, even after exclusion of energy misreporters. Here, for our main analysis, we reported results without exclusion of energy misreporters, as these are consistent with prior reports from this large national survey [68].

4.5. PA and BMI

We report here a modest but significant inverse association between PA and BMI in postpartum women, which is consistent with some [69,70] but not all [71] observational studies. Conversely, we found no significant association between PA and BMI in pre-pregnant women, which is in contrast to longitudinal studies in reproductive-aged women that reported an inverse association between a higher level of MVPA and weight gain [72] or overweight and obesity [73]. However, these studies did not consistently adjust for important confounders, such as energy intake and other dietary factors. Differences in study design (cross-sectional vs. longitudinal) may partly explain the inconsistent reports. It is difficult to explain the finding here that PA is more closely associated with BMI only in postpartum women, but this may be related to the smaller sample size for the pre-pregnant population. Given over half of women currently undertake suboptimal amounts of PA [16] and the benefits of PA for psychological and physical wellbeing [74] and weight management [69], there is a need for further research to elucidate the mixed findings of PA and BMI in free-living pre-pregnant and postpartum women.

4.6. Strength and Limitations

This study has several strengths. Given the use of a subsample from a nationally representative survey, the results can be generalisable to the Australian population of reproductive age women. Height and weight were collected based on measured data, which give accurate BMI status and more reliable estimates than self-reported data. Furthermore, our analysis followed rigorous methods by accounting for sampling weight and survey design, resulting in unbiased estimates. The analysis also adjusted for several important confounders, such as dieting. However, the limitations of this study should be acknowledged. First, self-reported dietary data based on 24 h recall may be subject to recall bias or misreporting due to social desirability bias [75], which affects the results towards the null. However, the use of AMPM aids to minimize recall bias by maximising the recall of foods and accounting for intrapersonal variability [25]. Second, although 24 h recall gives a good estimation of dietary intake at the population level, dietary data based on one day recall may not reflect the usual intake of foods and nutrients. Third, we used a proxy method to identify reproductive life stages using prespecified terms 'female life stages' and 'number of children' in the household that may not definitely separate pre-pregnancy and postpartum women. Despite the lack of certainty on the definitions of the specific reproductive life stages, this dataset allowed the use of robust methods of dietary assessment by 24 h recall. Fourth, the sample size for pre-pregnant and pregnant women was relatively small, which may have reduced the power in the multivariable analysis for pre-pregnant women. Finally, the cross-sectional study design precluded the assessment of any causal relationships and may also explain inconsistencies with previous longitudinal studies. We also note that some food group recommendations are different for lactating women. We did not differentiate postpartum women based on breastfeeding status, as the subgroup sample sizes, particularly for sociodemographic analysis, would have been too small.

5. Conclusions

Our findings showed that women across reproductive life stages, in particular, those from lower socioeconomic groups and those born in Australia, failed to meet population-level diet and PA recommendations. In pre-pregnant women, whole grains and energy from alcohol were inversely associated with BMI, and fibre and PA were associated with BMI in postpartum women. This study suggests that the lifestyle components of whole grains, fibre, alcohol and PA; and sociodemographic groups of country of birth, education and socioeconomic disadvantage, should be targeted in future interventions to prevent weight gain or obesity in women across reproductive life stages. The findings, however, should be interpreted with caution due to the indirect definitions of reproductive life stages.

Supplementary Materials: The following supporting information can be downloaded at: https://www.mdpi.com/article/10.3390/nu14132607/s1. Table S1: Definitions of reproductive life stages. Table S2: Core food components, servings and daily recommended intakes according to the Australian Dietary Guideline (ADG 2013). Table S3: Discretionary food groups (food flag) based on the food code assigned in the food classifications system based on Australian Bureau of Statistics. Table S4 a–g: Dietary intake and physical activity by sociodemographic characteristics in pre-pregnant and postpartum women.

Author Contributions: Conceptualization, L.J.M., M.A.A. and T.P.W.; statistical analyses and initial draft of the manuscript, M.A.A.; manuscript review, interpretation of results, edits and supervision, H.S., A.E., T.P.W. and L.J.M. All authors have read and agreed to the published version of the manuscript.

Funding: No funding support was obtained for this study. M.A.A was funded by the Monash International Tuition Scholarship and Monash Graduate Scholarship; L.J.M. is funded by the National Heart Foundation Future Leader Fellowship.

Informed Consent Statement: Informed consent was obtained from all subjects involved in the study.

Data Availability Statement: Publicly available datasets were analysed in this study and can be requested from the Australian Bureau of Statistics: the National Nutrition and Physical Activity Survey 2011–2012 in de-identified format.

Conflicts of Interest: The authors declare no conflict of interest.

References

1. Lucke, J.; Waters, B.; Hockey, R.; Spallek, M.; Gibson, R.; Byles, J.; Dobson, A. Trends in women's risk factors and chronic conditions: Findings from the Australian Longitudinal Study on Women's Health. *Women's Health* **2007**, *3*, 423–432. [CrossRef] [PubMed]
2. Ziauddeen, N.; Roderick, P.J.; Macklon, N.S.; Alwan, N.A. The duration of the interpregnancy interval in multiparous women and maternal weight gain between pregnancies: Findings from a UK population-based cohort. *Sci. Rep.* **2019**, *9*, 9175. [CrossRef] [PubMed]
3. National Research Council; Institute of Medicine. *Weight Gain during Pregnancy: Reexamining the Guidelines*; National Academies Press: Washington, DC, USA, 2010.
4. Goldstein, R.F.; Abell, S.K.; Ranasinha, S.; Misso, M.L.; Boyle, J.A.; Harrison, C.L.; Black, M.H.; Li, N.; Hu, G.; Corrado, F. Gestational weight gain across continents and ethnicity: Systematic review and meta-analysis of maternal and infant outcomes in more than one million women. *BMC Med.* **2018**, *16*, 153. [CrossRef] [PubMed]
5. Gore, S.A.; Brown, D.M.; West, D.S. The role of postpartum weight retention in obesity among women: A review of the evidence. *Ann. Behav. Med.* **2003**, *26*, 149–159. [CrossRef]
6. Poston, L.; Caleyachetty, R.; Cnattingius, S.; Corvalán, C.; Uauy, R.; Herring, S.; Gillman, M.W. Preconceptional and maternal obesity: Epidemiology and health consequences. *Lancet Diabetes Endocrinol.* **2016**, *4*, 1025–1036. [CrossRef]
7. Marchi, J.; Berg, M.; Dencker, A.; Olander, E.; Begley, C. Risks associated with obesity in pregnancy, for the mother and baby: A systematic review of reviews. *Obes. Rev.* **2015**, *16*, 621–638. [CrossRef]
8. Nehring, I.; Schmoll, S.; Beyerlein, A.; Hauner, H.; von Kries, R. Gestational weight gain and long-term postpartum weight retention: A meta-analysis. *Am. J. Clin. Nutr.* **2011**, *94*, 1225–1231. [CrossRef]
9. Seifu, C.N.; Fahey, P.P.; Hailemariam, T.G.; Frost, S.A.; Atlantis, E. Dietary patterns associated with obesity outcomes in adults: An umbrella review of systematic reviews. *Public Health Nutr.* **2021**, *24*, 6390–6414. [CrossRef]
10. World Health Organization. Overweight and Obesity. Available online: https://www.who.int/news-room/fact-sheets/detail/obesity-and-overweight (accessed on 2 November 2021).
11. Farpour-Lambert, N.J.; Ells, L.J.; Martinez de Tejada, B.; Scott, C. Obesity and weight gain in pregnancy and postpartum: An evidence review of lifestyle interventions to inform maternal and child health policies. *Front. Endocrinol.* **2018**, *9*, 546. [CrossRef]
12. Bennie, J.A.; Pedisic, Z.; van Uffelen, J.G.; Gale, J.; Banting, L.K.; Vergeer, I.; Stamatakis, E.; Bauman, A.E.; Biddle, S.J. The descriptive epidemiology of total physical activity, muscle-strengthening exercises and sedentary behaviour among Australian adults–results from the National Nutrition and Physical Activity Survey. *BMC Public Health* **2015**, *16*, 73. [CrossRef]
13. Fayet-Moore, F.; McConnell, A.; Cassettari, T.; Tuck, K.; Petocz, P.; Kim, J. Discretionary intake among Australian adults: Prevalence of intake, top food groups, time of consumption and its association with sociodemographic, lifestyle and adiposity measures. *Public Health Nutr.* **2019**, *22*, 1576–1589. [CrossRef] [PubMed]
14. Cuco, G.; Fernandez-Ballart, J.; Sala, J.; Viladrich, C.; Iranzo, R.; Vila, J.; Arija, V. Dietary patterns and associated lifestyles in preconception, pregnancy and postpartum. *Eur. J. Clin. Nutr.* **2006**, *60*, 364–371. [CrossRef] [PubMed]
15. Wen, L.M.; Flood, V.M.; Simpson, J.M.; Rissel, C.; Baur, L.A. Dietary behaviours during pregnancy: Findings from first-time mothers in southwest Sydney, Australia. *Int. J. Behav. Nutr. Phys. Act.* **2010**, *7*, 13. [CrossRef] [PubMed]
16. van der Pligt, P.; Olander, E.K.; Ball, K.; Crawford, D.; Hesketh, K.D.; Teychenne, M.; Campbell, K. Maternal dietary intake and physical activity habits during the postpartum period: Associations with clinician advice in a sample of Australian first time mothers. *BMC Pregnancy Childbirth* **2016**, *16*, 27. [CrossRef]
17. Carmen, A.-P.; Pablo, L.-C.; Rocío, O.-R.; Juan, M.-M.; Aurora, B.-C.; José, J.J.-M. Compliance with leisure-time physical activity recommendations in pregnant women. *Acta Obstet. Gynecol. Scand.* **2011**, *90*, 245–252.
18. Pereira, M.A.; Rifas-Shiman, S.L.; Kleinman, K.P.; Rich-Edwards, J.W.; Peterson, K.E.; Gillman, M.W. Predictors of change in physical activity during and after pregnancy: Project Viva. *Am. J. Prev. Med.* **2007**, *32*, 312–319. [CrossRef]
19. Carter-Edwards, L.; Østbye, T.; Bastian, L.A.; Yarnall, K.S.; Krause, K.M.; Simmons, T.-J. Barriers to adopting a healthy lifestyle: Insight from postpartum women. *BMC Res. Notes* **2009**, *2*, 161. [CrossRef]
20. Makama, M.; Awoke, M.A.; Skouteris, H.; Moran, L.J.; Lim, S. Barriers and facilitators to a healthy lifestyle in postpartum women: A systematic review of qualitative and quantitative studies in postpartum women and healthcare providers. *Obes. Rev.* **2021**, *22*, e13167. [CrossRef]

21. Olmedo-Requena, R.; Gomez-Fernandez, J.; Mozas-Moreno, J.; Lewis-Mikhael, A.-M.; Bueno-Cavanillas, A.; Jimenez-Moleon, J.-J. Factors associated with adherence to nutritional recommendations before and during pregnancy. *Women Health* **2018**, *58*, 1094–1111. [CrossRef]

22. NHMRC. Australian Dietary Guidelines, Canberra: National Health and Medical Research Council. Available online: https://www.nhmrc.gov.au/adg (accessed on 16 July 2021).

23. Australian Bureau of Statistic. Australian Health Survey: Nutrition First Results-Foods and Nutrients. Users' Guide, 2011–2013 (cat. no. 4363.0.55.001). Canberra, ACT (2013). Available online: https://www.abs.gov.au/statistics/health/health-conditions-and-risks/australian-health-survey-nutrition-first-results-foods-and-nutrients/latest-release (accessed on 16 July 2021).

24. Johnston, J.M.; Colvin, A.; Johnson, B.D.; Santoro, N.; Harlow, S.D.; Merz, C.N.B.; Sutton-Tyrrell, K. Comparison of SWAN and WISE menopausal status classification algorithms. *J. Women's Health* **2006**, *15*, 1184–1194. [CrossRef]

25. Bliss, R.M. Researchers produce innovation in dietary recall. *Agric. Res.* **2004**, *52*, 10–13.

26. Blumfield, M.; McConnell, A.; Cassettari, T.; Petocz, P.; Warner, M.; Campos, V.; Lê, K.-A.; Minehira, K.; Marshall, S.; Fayet-Moore, F. Balanced carbohydrate ratios are associated with improved diet quality in Australia: A nationally representative cross-sectional study. *PLoS ONE* **2021**, *16*, e0253582. [CrossRef] [PubMed]

27. Seifu, C.N.; Fahey, P.P.; Atlantis, E. Unhealthy Diet Pattern Mediates the Disproportionate Prevalence of Obesity among Adults with Socio-Economic Disadvantage: An Australian Representative Cross-Sectional Study. *Nutrients* **2021**, *13*, 1363. [CrossRef] [PubMed]

28. Brown, W.J.; Burton, N.W.; Marshall, A.L.; Miller, Y.D. Reliability and validity of a modified self-administered version of the Active Australia physical activity survey in a sample of mid-age women. *Aust. N. Z. J. Public Health* **2008**, *32*, 535–541. [CrossRef]

29. Australian Government Department of Health. Australia's Physical Activity and Sedentary Behaviour Guidelines. Available online: http://www.health.gov.au/internet/main/publishing.nsf/Content/health-pubhlth-strateg-phys-act-guidelines (accessed on 23 July 2021).

30. Birrell, C.L.; Steel, D.G.; Batterham, M.J.; Arya, A. How to use replicate weights in health survey analysis using the National Nutrition and Physical Activity Survey as an example. *Public Health Nutr.* **2019**, *22*, 3315–3326. [CrossRef]

31. Livingstone, M.B.E.; Black, A.E. Markers of the validity of reported energy intake. *J. Nutr.* **2003**, *133*, 895S–920S. [CrossRef]

32. Johnson, B.J.; Bell, L.K.; Zarnowiecki, D.; Rangan, A.M.; Golley, R.K. Contribution of discretionary foods and drinks to Australian children's intake of energy, saturated fat, added sugars and salt. *Children* **2017**, *4*, 104. [CrossRef]

33. Dutch, D.C.; Golley, R.K.; Johnson, B.J. Diet Quality of Australian Children and Adolescents on Weekdays versus Weekend Days: A Secondary Analysis of the National Nutrition and Physical Activity Survey 2011–2012. *Nutrients* **2021**, *13*, 4128. [CrossRef]

34. Australian Government and New Zealand Ministry of Health. *Nutrient Reference Values for Australia and New Zealand*; Commonwealth of Australia: Canberra, Australia, 2005. Available online: https://www.nrv.gov.au/nutrients (accessed on 11 October 2021).

35. National Health Medical Research Council. *Australian Alcohol Guidelines: Health Risks and Benefits*; National Health and Medical Research Council: Canberra, Australia, 2001.

36. Caut, C.; Leach, M.; Steel, A. Dietary guideline adherence during preconception and pregnancy: A systematic review. *Matern. Child Nutr.* **2020**, *16*, e12916. [CrossRef]

37. George, G.C.; Milani, T.J.; Hanss-Nuss, H.; Freeland-Graves, J.H. Compliance with dietary guidelines and relationship to psychosocial factors in low-income women in late postpartum. *J. Am. Diet. Assoc.* **2005**, *105*, 916–926. [CrossRef]

38. Kay, M.C.; Wasser, H.; Adair, L.S.; Thompson, A.L.; Siega-Riz, A.M.; Suchindran, C.M.; Bentley, M.E. Consumption of key food groups during the postpartum period in low income, non-Hispanic black mothers. *Appetite* **2017**, *117*, 161–167. [CrossRef] [PubMed]

39. Williams, L.K.; Thornton, L.; Crawford, D.; Ball, K. Perceived quality and availability of fruit and vegetables are associated with perceptions of fruit and vegetable affordability among socio-economically disadvantaged women. *Public Health Nutr.* **2012**, *15*, 1262–1267. [CrossRef] [PubMed]

40. Malek, L.; Umberger, W.; Zhou, S.J.; Makrides, M. Understanding drivers of dietary behavior before and during pregnancy in industrialized countries. In *Health and Nutrition in Adolescents and Young Women: Preparing for the Next Generation*; Karger: Basel, Switzerland, 2015; Volume 80, pp. 117–140.

41. Santos, P.C.; Abreu, S.; Moreira, C.; Lopes, D.; Santos, R.; Alves, O.; Silva, P.; Montenegro, N.; Mota, J. Impact of compliance with different guidelines on physical activity during pregnancy and perceived barriers to leisure physical activity. *J. Sports Sci.* **2014**, *32*, 1398–1408. [CrossRef]

42. Borodulin, K.; Evenson, K.R.; Herring, A.H. Physical activity patterns during pregnancy through postpartum. *BMC Women's Health* **2009**, *9*, 32. [CrossRef] [PubMed]

43. Durham, H.A.; Morey, M.C.; Lovelady, C.A.; Brouwer, R.J.N.; Krause, K.M.; Østbye, T. Postpartum physical activity in overweight and obese women. *J. Phys. Act. Health* **2011**, *8*, 988–993. [CrossRef] [PubMed]

44. Coll, C.V.; Domingues, M.R.; Gonçalves, H.; Bertoldi, A.D. Perceived barriers to leisure-time physical activity during pregnancy: A literature review of quantitative and qualitative evidence. *J. Sci. Med. Sport* **2017**, *20*, 17–25. [CrossRef] [PubMed]

45. Harrison, A.L.; Taylor, N.F.; Shields, N.; Frawley, H.C. Attitudes, barriers and enablers to physical activity in pregnant women: A systematic review. *J. Physiother.* **2018**, *64*, 24–32. [CrossRef]

46. Ball, K.; Lamb, K.E.; Costa, C.; Cutumisu, N.; Ellaway, A.; Kamphuis, C.B.; Mentz, G.; Pearce, J.; Santana, P.; Santos, R. Neighbourhood socioeconomic disadvantage and fruit and vegetable consumption: A seven countries comparison. *Int. J. Behav. Nutr. Phys. Act.* **2015**, *12*, 68. [CrossRef]

47. Giskes, K.; Avendaňo, M.; Brug, J.; Kunst, A. A systematic review of studies on socioeconomic inequalities in dietary intakes associated with weight gain and overweight/obesity conducted among European adults. *Obes. Rev.* **2010**, *11*, 413–429. [CrossRef]

48. O'Donoghue, G.; Kennedy, A.; Puggina, A.; Aleksovska, K.; Buck, C.; Burns, C.; Cardon, G.; Carlin, A.; Ciarapica, D.; Colotto, M. Socio-economic determinants of physical activity across the life course: A "Determinants of DIet and Physical ACtivity" (DEDIPAC) umbrella literature review. *PLoS ONE* **2018**, *13*, e0190737. [CrossRef]

49. Liu, H.; Hall, J.J.; Xu, X.; Mishra, G.D.; Byles, J.E. Differences in food and nutrient intakes between Australian-and Asian-born women living in Australia: Results from the Australian Longitudinal Study on Women's Health. *Nutr. Diet.* **2018**, *75*, 142–150. [CrossRef] [PubMed]

50. Harland, J.I.; Garton, L.E. Whole-grain intake as a marker of healthy body weight and adiposity. *Public Health Nutr.* **2008**, *11*, 554–563. [CrossRef] [PubMed]

51. Alderete, T.L.; Wild, L.E.; Mierau, S.M.; Bailey, M.J.; Patterson, W.B.; Berger, P.K.; Jones, R.B.; Plows, J.F.; Goran, M.I. Added sugar and sugar-sweetened beverages are associated with increased postpartum weight gain and soluble fiber intake is associated with postpartum weight loss in Hispanic women from Southern California. *Am. J. Clin. Nutr.* **2020**, *112*, 519–526. [CrossRef] [PubMed]

52. Drehmer, M.; Camey, S.A.; Nunes, M.A.; Duncan, B.B.; Lacerda, M.; Pinheiro, A.P.; Schmidt, M.I. Fibre intake and evolution of BMI: From pre-pregnancy to postpartum. *Public Health Nutr.* **2013**, *16*, 1403–1413. [CrossRef]

53. Sanders, L.M.; Zhu, Y.; Wilcox, M.L.; Koecher, K.; Maki, K.C. Effects of Whole Grain Intake, Compared with Refined Grain, on Appetite and Energy Intake: A Systematic Review and Meta-Analysis. *Adv. Nutr.* **2021**, *12*, 1177–1195. [CrossRef]

54. Howarth, N.C.; Saltzman, E.; Roberts, S.B. Dietary fiber and weight regulation. *Nutr. Rev.* **2001**, *59*, 129–139. [CrossRef]

55. U.S. Department of Health and Human Services; U.S. Department of Agriculture. *2015–2020 Dietary Guidelines for Americans*, 8th ed.; USDA; HHS: Washington, DC, USA, 2015.

56. Wang, L.; Lee, I.-M.; Manson, J.E.; Buring, J.E.; Sesso, H.D. Alcohol consumption, weight gain, and risk of becoming overweight in middle-aged and older women. *Arch. Intern. Med.* **2010**, *170*, 453–461. [CrossRef]

57. Wannamethee, S.G.; Field, A.E.; Colditz, G.A.; Rimm, E.B. Alcohol intake and 8-year weight gain in women: A prospective study. *Obes. Res.* **2004**, *12*, 1386–1396. [CrossRef]

58. O'Donovan, G.; Stamatakis, E.; Hamer, M. Associations between alcohol and obesity in more than 100,000 adults in England and Scotland. *Br. J. Nutr.* **2018**, *119*, 222–227. [CrossRef]

59. Poudel, P.; Ismailova, K.; Andersen, L.B.; Larsen, S.C.; Heitmann, B.L. Adolescent wine consumption is inversely associated with long-term weight gain: Results from follow-up of 20 or 22 years. *Nutr. J.* **2019**, *18*, 56. [CrossRef]

60. Ruf, T.; Nagel, G.; Altenburg, H.-P.; Miller, A.; Thorand, B. Food and nutrient intake, anthropometric measurements and smoking according to alcohol consumption in the EPIC Heidelberg study. *Ann. Nutr. Metab.* **2005**, *49*, 16–25. [CrossRef] [PubMed]

61. Bearak, J.; Popinchalk, A.; Alkema, L.; Sedgh, G. Global, regional, and subregional trends in unintended pregnancy and its outcomes from 1990 to 2014: Estimates from a Bayesian hierarchical model. *Lancet Glob. Health* **2018**, *6*, e380–e389. [CrossRef]

62. Aljadani, H.M.; Patterson, A.; Sibbritt, D.; Taylor, R.M.; Collins, C.E. Frequency and variety of usual intakes of healthy foods, fruit, and vegetables predicts lower 6-year weight gain in young women. *Eur. J. Clin. Nutr.* **2020**, *74*, 945–952. [CrossRef] [PubMed]

63. Mytton, O.T.; Nnoaham, K.; Eyles, H.; Scarborough, P.; Mhurchu, C.N. Systematic review and meta-analysis of the effect of increased vegetable and fruit consumption on body weight and energy intake. *BMC Public Health* **2014**, *14*, 886. [CrossRef]

64. Malik, V.S.; Pan, A.; Willett, W.C.; Hu, F.B. Sugar-sweetened beverages and weight gain in children and adults: A systematic review and meta-analysis. *Am. J. Clin. Nutr.* **2013**, *98*, 1084–1102. [CrossRef] [PubMed]

65. Sui, Z.; Wong, W.K.; Louie, J.C.Y.; Rangan, A. Discretionary food and beverage consumption and its association with demographic characteristics, weight status, and fruit and vegetable intakes in Australian adults. *Public Health Nutr.* **2017**, *20*, 274–281. [CrossRef] [PubMed]

66. Boghossian, N.S.; Yeung, E.H.; Lipsky, L.M.; Poon, A.K.; Albert, P.S. Dietary patterns in association with postpartum weight retention. *Am. J. Clin. Nutr.* **2013**, *97*, 1338–1345. [CrossRef]

67. Tam, K.W.; Veerman, J.L. Prevalence and characteristics of energy intake under-reporting among Australian adults in 1995 and 2011 to 2012. *Nutr. Diet.* **2019**, *76*, 546–559. [CrossRef]

68. Grech, A.; Rangan, A.; Allman-Farinelli, M. Macronutrient composition of the Australian Population's diet; trends from three National Nutrition Surveys 1983, 1995 and 2012. *Nutrients* **2018**, *10*, 1045. [CrossRef]

69. Ha, A.V.V.; Zhao, Y.; Binns, C.W.; Pham, N.M.; Nguyen, P.T.H.; Nguyen, C.L.; Chu, T.K.; Lee, A.H. Postpartum physical activity and weight retention within one year: A prospective cohort study in Vietnam. *Int. J. Environ. Res. Public Health* **2020**, *17*, 1105. [CrossRef]

70. Fadzil, F.; Shamsuddin, K.; Puteh, S.E.W.; Tamil, A.M.; Ahmad, S.; Hayi, N.S.A.; Samad, A.A.; Ismail, R.; Shauki, N.I.A. Predictors of postpartum weight retention among urban Malaysian mothers: A prospective cohort study. *Obes. Res. Clin. Pract.* **2018**, *12*, 493–499. [CrossRef]

71. Oken, E.; Taveras, E.M.; Popoola, F.A.; Rich-Edwards, J.W.; Gillman, M.W. Television, walking, and diet: Associations with postpartum weight retention. *Am. J. Prev. Med.* **2007**, *32*, 305–311. [CrossRef]

72. Byambasukh, O.; Vinke, P.; Kromhout, D.; Navis, G.; Corpeleijn, E. Physical activity and 4-year changes in body weight in 52,498 non-obese people: The Lifelines cohort. *Int. J. Behav. Nutr. Phys. Act.* **2021**, *18*, 75. [CrossRef]
73. Pavey, T.G.; Peeters, G.G.; Gomersall, S.R.; Brown, W.J. Long-term effects of physical activity level on changes in healthy body mass index over 12 years in young adult women. *Mayo Clin. Proc.* **2016**, *91*, 735–744. [CrossRef]
74. Larson-Meyer, D.E. Effect of postpartum exercise on mothers and their offspring: A review of the literature. *Obes. Res.* **2002**, *10*, 841–853. [CrossRef]
75. Lafay, L.; Mennen, L.; Basdevant, A.; Charles, M.; Borys, J.; Eschwege, E.; Romon, M. Does energy intake underreporting involve all kinds of food or only specific food items? Results from the Fleurbaix Laventie Ville Sante (FLVS) study. *Int. J. Obes.* **2000**, *24*, 1500–1506. [CrossRef]

 *nutrients*

Article

Impact of Dietary Trajectories on Obesity and Dental Caries in Preschool Children: Findings from the Healthy Smiles Healthy Kids Study

Narendar Manohar [1,2,3], Andrew Hayen [4], Jane A. Scott [5], Loc G. Do [6,7], Sameer Bhole [8,9] and Amit Arora [1,2,8,10,11,*]

[1] School of Health Sciences, Western Sydney University, Penrith, NSW 2751, Australia; drnarendar@gmail.com
[2] Health Equity Laboratory, Campbelltown, NSW 2560, Australia
[3] Australian College of Physical Education, Sydney Olympic Park, NSW 2127, Australia
[4] School of Public Health, Faculty of Health, University of Technology Sydney, Ultimo, NSW 2007, Australia; andrew.hayen@uts.edu.au
[5] Discipline of Nutrition and Dietetics, School of Population Health, Curtin University, Perth, WA 6102, Australia; jane.scott@curtin.edu.au
[6] School of Dentistry, Faculty of Health and Behavioural Sciences, University of Queensland, Brisbane, QLD 4072, Australia; l.do@uq.edu.au
[7] Australian Research Centre for Population Oral Health, University of Adelaide, Adelaide, SA 5005, Australia
[8] Oral Health Services, Sydney Local Health District and Sydney Dental Hospital, Surry Hills, NSW 2010, Australia; sameer.bhole@health.nsw.gov.au
[9] Sydney Dental School, Faculty of Medicine and Health, The University of Sydney, Surry Hills, NSW 2010, Australia
[10] Translational Health Research Institute, Western Sydney University, Campbelltown, NSW 2560, Australia
[11] Discipline of Child and Adolescent Health, Faculty of Medicine and Health, Sydney Medical School, The University of Sydney, Westmead, NSW 2145, Australia
* Correspondence: a.arora@westernsydney.edu.au

Citation: Manohar, N.; Hayen, A.; Scott, J.A.; Do, L.G.; Bhole, S.; Arora, A. Impact of Dietary Trajectories on Obesity and Dental Caries in Preschool Children: Findings from the Healthy Smiles Healthy Kids Study. *Nutrients* **2021**, *13*, 2240. https://doi.org/10.3390/nu13072240

Academic Editors: Panagiota Mitrou and Gloria Lena Vega

Received: 14 May 2021
Accepted: 25 June 2021
Published: 29 June 2021

Publisher's Note: MDPI stays neutral with regard to jurisdictional claims in published maps and institutional affiliations.

Abstract: This study examines the impact of longitudinal dietary trajectories on obesity and early childhood caries (ECC) in preschool children in Australia. Mother–infant dyads from the Healthy Smiles Healthy Kids study were interviewed at 4 and 8 months, and 1, 2, and 3 years of age. Children underwent anthropometric and oral health assessments between 3 and 4 years of age. Multivariable logistic regression and negative binomial regression analysis were performed for the prevalence of overweight and obesity, and the number of tooth surfaces with dental caries, respectively. The intake of core, discretionary, and sugary foods showed distinct quadratic ($n = 3$) trajectories with age. The prevalence of overweight or obesity was 10% ($n = 72$) and that of early childhood caries (ECC) was 33% (mean decayed, missing, and filled tooth surfaces (dmfs) score: 1.96). Children with the highest trajectories of discretionary foods intake were more likely to be overweight or obese (adjusted OR: 2.51, 95% CI: 1.16–5.42). Continued breastfeeding beyond 12 months was associated with higher dmfs scores (adjusted IRR: 2.17, 95% CI: 1.27–3.73). Highest socioeconomic disadvantage was the most significant determinant for overweight or obesity (adjusted OR: 2.86, 95% CI: 1.11–7.34) and ECC (adjusted IRR: 2.71, 95% CI: 1.48–4.97). Targeted health promotion interventions should be designed to prevent the incidence of two highly prevalent conditions in preschool children.

Keywords: diet; dietary trajectories; dietary patterns; overweight; obesity; early childhood caries; ECC; dental caries; health risk; preschool children

1. Introduction

Obesity and early childhood caries (ECC) are the two common and important health problems affecting Australian children [1,2]. In Australia, 20% of children aged 2–4 years are either overweight or obese [3] and 34% have ECC by the age of 5-years [4]. Both these conditions can have detrimental long-term health consequences [5,6] and also have a tendency of progression into adulthood [1,5].

Childhood obesity and ECC are multi-factorial in origin with a diverse range of risk factors [7]. Recent evidence shows that these two conditions share common risk factors including low socioeconomic status (SES), poor diet, and other social–environmental factors [8,9]. Hence, considering the evident association between these two conditions, the Common Risk Factor Approach (CRFA) seems to be the most suitable interventional strategy as it seeks to target risk factors that are common to both the conditions [10].

Poor diet, comprising of energy-dense, low-nutritious foods, and an earlier introduction of such foods can contribute to the rising prevalence of childhood obesity and ECC [11,12]. For example, earlier research from the Healthy Smiles Healthy Kids (HSHK) study [13–15] identified that 95% of infants were introduced to discretionary foods, and more specifically, almost 43% of infants were introduced to sugar beverages (SSBs) before the recommended age of 52 weeks [16]. The research team further noted that over 13% and over 76% of Sydney infants were introduced to solid foods before 17- and 26-weeks post-partum, respectively [15]. Another Australian cohort study found that a high consumption of sugary drinks led to obesity and ECC in young children [9]. In addition to sugars, poor diet comprises of other constituents that are considered to be obesogenic (e.g., ultra-processed foods high in saturated and trans fats and/or salts) and cariogenic (e.g., acidic foods or beverages) [17,18]. Contrariwise, an individual's diet also consists of foods that can have an obesity-protective effect (e.g., fruits and vegetables) [19]. Therefore, in order to understand the disease epidemiology and prevention of obesity and dental caries in early life, it is important to examine the impact of whole diet, rather than single nutrients or foods. Measuring multiple dietary elements together and their patterns assists in identifying the synergistic and correlational nature of individual foods and nutrients. Moreover, they may be suggestive of how individuals eat and in what frequency [20]. Furthermore, measuring diet patterns longitudinally (from early infancy to preschool age) will help with defining which foods and/or nutrients are amenable to change and at what stage of life [21], particularly if those foods or nutrients are specifically related to disease outcomes.

Dietary patterns in children have been examined using statistical approaches such as factor analysis, principal component analysis (PCA), and cluster analysis [22–24]. In recent years, the Group-Based Trajectory Modelling (GBTM) approach has emerged for examining longitudinal dietary patterns [25]. The GBTM identifies clusters of individuals who follow similar trajectories over time [26]. This study is innovative and novel, since it examines the dietary intakes evolving over the first three years of life using advanced statistical approaches such as GBTM and their impact on two subsequent health outcomes. This may provide a greater understanding of the diet–disease relationship, which in the case of the present study is obesity and ECC.

Dietary patterns and their association with later health outcomes in children have been investigated previously [27,28]. However, studies exploring the association between dietary patterns and obesity amongst children [28,29] as well as the relationship between dietary patterns and ECC [28,30] have shown inconsistent findings. A recent Australian cohort study did not find any meaningful association between two dietary patterns (i.e., healthy and unhealthy), body mass index (BMI) scores, and ECC in toddlers [28]. Whilst a Singapore-based study [31] involving a multi-ethnic Asian children population, using multi-level mixed modelling to examine the dietary trajectories between 6 and 12 months of age and ECC at a later age (i.e., age 2 and 3 years), demonstrated an inverse relationship between healthy diet patterns and ECC [31]. It is suggested that inconsistencies in the limited existing evidence may be due to discrepancies in the methodological approaches used to examine dietary patterns [28]. Moreover, to the best of our knowledge, no study has used GBTM to examine the longitudinal dietary patterns of core, discretionary, and added-sugar foods in infancy and early childhood and their association with later health outcomes. Therefore, the aim of this study was to examine the association between healthy and unhealthy dietary trajectories and obesity and ECC in preschool children. Moreover, the use of GBTM to characterise dietary patterns in early life would assist in providing a more detailed picture of diet and its evolution with age. Hence, the objectives of this study were:

1. To investigate the impact of longitudinal dietary trajectories on obesity and ECC in Australian preschool children.
2. To ascertain the impact of sociodemographic, socioeconomic, behavioural, and biological factors on obesity and ECC in Australian preschool children.

2. Methods

2.1. Data Source

This study is a secondary analysis of data collected from an ongoing birth cohort study called Healthy Smiles Healthy Kids (HSHK) [32], which has followed socioeconomically diverse families based in South West Sydney (SWS) since 2009, as described in detail in earlier studies [15,32,33]. In terms of recruitment, Child and Family Health Nurses (CFHNs) recruited mother–infant dyads (n = 1035) between October 2009 and February 2010 from public hospitals located within the Sydney and South Western Sydney Local Health Districts (LHDs) (formerly known as Sydney South West Area Health Service). The mothers were provided information on the study at the first post-natal visit, and written consent was obtained. Interpreter services and written material in the native language of non-English speaking participants (i.e., Arabic, Assyrian, Hindi, Cambodian, Cantonese, Mandarin, Vietnamese, and Samoan) were also arranged to facilitate participation.

2.2. Data Collection

2.2.1. Dietary Data

Children's dietary data were periodically collected via telephone interviews at five age points i.e., 4 months, 8 months, 1 year, 2 years, and 3 years, respectively. The dietary questionnaire was adapted from the Iowa Fluoride study [34], the NSW Child Health Questionnaire [35], the National Child Oral Health Survey [36], the Perth Infant Feeding Studies (PIFS I and II) [37,38], and the HSHK pilot study [39]. Children's dietary habits, in terms of consumption of 32 individual food and drink items in the preceding seven days, were recorded using a short food frequency questionnaire (FFQ) (Table S1). At every interview, mothers were asked an open-ended question "In the past 7 days, how often was your baby/child fed each of the following foods and/or drinks?". A numerical response was recorded to represent the number of times the specified food and/or drink was consumed in a week.

For dietary trajectory analyses, the 32 listed food and drink items were broadly categorised into 'core' and 'discretionary' foods groups based on the 2013 Australian Dietary Guidelines [16,40]. The same categorisation method was used in previously published research [14]. The core foods group (n = 12 items) comprised of dairy, grains, fruits, vegetables, and meat and its alternatives; whilst the discretionary foods group (n = 20 items) comprised of foods with added fats and/or salt, and foods with added sugars. Additionally, the discretionary foods group was further categorised into sugary foods group (n = 18 items). The frequency (continuous data) of each item in the five individual core foods subgroups were summed to give the 'total of the core food group intake', and the frequency of each item in the two discretionary food groups were summed to give the 'total of the discretionary food group intake'. This same method was used for sugar-containing items to give the 'total of the sugary food group intake'. The focus of this study was on dietary trajectories of core, discretionary, and sugary foods, respectively. Children were included in the study if they had diet data available for at least three interview age points along with the clinical outcome data.

2.2.2. Other Predictors

Information on sociodemographic characteristics (including maternal age, marital status, country of birth, education level, employment status at 12 months postpartum, parity, and child gender), area-level socioeconomic status (SES), behavioural factors including maternal smoking practices during pregnancy, and biological factors including infant birth weight were collected via telephone interview at baseline (i.e., 8 weeks postpartum) using the adapted study questionnaire. Data on child age, breastfeeding duration, and age at which complementary (solid) foods were first introduced were periodically recorded via telephone interviews at 4 months, 8 months, 1 year, 2 years, and 3 years age points. Furthermore, children's physical activity (time spent playing outdoors) was recorded at the 3-year interview [41].

2.2.3. Anthropometric Data

Once the children reached the age of 3 years, they were invited (accompanied by their mothers) for a dental assessment. At the time of the dental visit, children's weight and height were measured by trained and experienced health professionals using standardised methodology and equipment [42]. For weight measurement, a calibrated digital scale placed on a flat, hard surface was used, with children wearing light clothing and no shoes. The measured weight was recorded to the nearest 100 g. For the height measurement, a portable SECA stadiometer was used with a vertical backboard and movable headboard. The child's head, back, buttocks, and heels were in contact with the vertical backboard. The measured height was recorded to the nearest 1 mm. For weight and height assessment, two measurements were taken by the examiner, and if the two measures for weight differed by more than 50 g, and/or if the two measures for height differed by more than 5 mm, then a third measurement was taken. For anthropometric outcome, age and gender-specific BMI z-score was calculated using the World Health Organization's (WHO) (Geneva, Switzerland) AnthroPlus software program version 2.0 [43]. Children were categorised (based on the WHO age and gender specific cut-offs) as healthy (BMI z-score ≥ -2 and $\leq +2$ standard deviations (SD)), overweight (BMI z-score $> +2$ and $\leq +3$ SD) or obese (BMI z-score $> +3$ SD) [44]. For analytical purposes, overweight and obese categories were combined into a single category 'overweight or obese' (BMI z-score $> +2$ SD) due to the small number of obese children.

2.2.4. Dental Data

Children's dental examinations were conducted between the ages of 3 and 4 years. The examinations were conducted by trained and experienced dental therapists in clinical settings using standardised protocols [45,46]. A standard dental index—decayed, missing, and filled tooth surfaces (dmfs)—was used to record the ECC prevalence [47]. For this study, ECC was characterised as the 'presence of one or more decayed (non-cavitated or cavitated lesions), missing (due to decay), or filled tooth surfaces in any primary (baby) teeth in children less than 6 years of age' [48].

2.3. Statistical Analyses

Statistical analyses were performed using Stata Statistical Software version 15.0 (Stata-Corp, College Station, TX, USA). Continuous data were presented as mean and SD and categorical data were presented as frequency and percentages.

2.4. Dietary Pattern Analyses

2.4.1. Group-Based Trajectory Modelling

Dietary patterns were examined using the Group-Based Trajectory Modelling (GBTM) analysis. A plug-in (PROC TRAJ) in Stata was used to construct the dietary trajectories. The GBTM uses finite mixture modelling and creates meaningful subgroups comprising of individuals who follow statistically similar trajectories. It statistically identifies (rather than assuming a priori) groups of distinctive trajectories that are summarised by a finite set of different polynomial functions of age or time, as determined by maximum likelihood estimation. The maximisation uses a general quasi-Newtown method. GBTM allows the trajectories to emerge from the data itself rather than establishing trajectories on the basis of an individual trait or traits. In terms of trajectory groups, this method determines the form and numbers that best fit the data. GBTM predicts the trajectory of each group, the form of each trajectory, estimates the probability for each individual for group membership, and assigns them to the group for which they have the highest probability.

For GBTM, Bayesian information criteria (BIC) are often used for selecting the model (number of trajectory groups) that best represents the heterogeneity in the trajectories of the study sample. However, the BIC does not always clearly identify the ideal number of groups. Therefore, the objective of model selection is centred around summarising the data features in as parsimonious manner as possible. In this study analysis, a Poisson-based model was used due to the continuous distribution (count data) of the food frequency data at each age point. GBTM analysis comprises of a two-step process: (1) select the number of groups; and (2) determine the order of the polynomial defining each group's trajectory (i.e., zero-order, linear, cubic, and quadratic). A series of 2- to 6-group models were fitted, starting with zero-order specifications for the trajectory shapes and moving to linear, cubic, and quadratic specifications until the best-fitting model (which was parsimonious and analytically tractable) was established.

Prior to dietary analyses, the 32 food and drink items were categorised into meaningful groups (described earlier in the methods section) that had clinical relevance to the outcomes of obesity and ECC, respectively. Three food groups, namely core, discretionary, and sugary foods were used as input variables for GBTM. Trajectories of core foods were constructed due to their protective effect against obesity [19]. Trajectories of discretionary foods were constructed in relation to the obesity outcome, since foods high in saturated fats, salts, and sugars are known to be obesogenic [17]. Meanwhile, trajectories of sugary foods were generated in relation to the ECC outcome, since sugars are known to be one of the principal determinants of ECC [49]. Dietary data available for at least three interview periods were included in the analyses. In summary, dietary trajectories for 'core' and 'discretionary' foods were constructed for a total of 738 children, since they had the anthropometric data, whilst dietary trajectories for 'sugary' foods were constructed for a total of 718 participants, since they had ECC outcome data.

2.4.2. Predictors of Anthropometric Measures and ECC

Multi-level multivariable regression modelling was used for primary outcomes of overweight/obesity and ECC, respectively. Regression modelling for both outcomes was based on conceptual models; it was guided by prior evidence that certain dietary, biological, sociodemographic, socioeconomic, and behavioural factors were the likely candidate predictors [17,19,49,50]. Binary logistic regression was used to investigate the associations of dietary patterns (specifically 'core' and 'discretionary' foods) and other predictors with child weight status. Additionally, to investigate the associations of sugary foods dietary patterns and other predictors with the presence of dental caries; the countfit-function was used [51] to compare negative binomial and zero-inflated negative binomial regression. The negative binomial regression was the best fit, and hence, it was used for this analysis.

For each primary outcome, a series of models was generated. The first model was generated with only the diet trajectory groups. Then, demographic factors were sequentially entered into the model to check for effect size and random variations. Furthermore, socioeconomic factors at the individual and area level were sequentially entered into the models followed by behavioural and biological factors, respectively. For this study, each research question is answered by presenting the full model with factors at all the levels. The fixed effects were presented as odds ratios (OR) and 95% confidence intervals (95% CI) for child weight status, and incidence rate ratios (IRR) and 95% CI for the prevalence of ECC. A significance level of 5% was used for all analyses.

2.5. Ethics Approval and Participant Consent

Ethics approval to conduct this study was given by the former Sydney South West Area Health Service—RPAH Zone (ID number X08-0115), Liverpool Hospital, University of Sydney and Western Sydney University. All participants signed a written consent form prior to study commencement.

3. Results

Fifteen hundred mothers were invited to participate in the HSHK study, of whom 1035 formally agreed to participate (69% response rate). Participating ($n = 1035$) and non-participating mothers ($n = 465$) were compared on certain sociodemographic characteristics and chosen method of infant feeding. Both cohorts had no significant differences in terms of maternal age (Chi-square (X^2) = 4.75, $p = 0.153$), educational level ($X^2 = 6.65$, $p = 0.328$), and method of infant feeding ($X^2 = 2.46$, $p = 0.813$). Before the baseline interview, a further 101 mothers either opted out or were non-contactable. Hence, in total, 934 mothers completed the interviews (62.2% response rate), and of these, 738 had the anthropometric outcome data (21% attrition rate), whilst 718 participants had the ECC outcome data (23% attrition rate). No differences in the age, education level, and method of infant feeding were evident between mothers who withdrew from the study and those who completed all the interviews, including the clinical assessments (data not reported).

'Core' and 'discretionary' foods intake trajectories were constructed for 738 children having anthropometric data (52% males and 48% females), while 'sugary' foods intake trajectories were constructed for 718 children having ECC data (52% males and 48% females). The mean ($\pm$SD) age of children was 3.57 ($\pm$0.25) years at the time of clinical assessment. Most children were of healthy weight (90%), with 7% being overweight, 3% having obesity, and 0.27% being underweight. Eleven percent ($n = 38$) of females and 9% ($n = 34$) of males were overweight or obese in the study sample. In relation to the prevalence of dental caries, 33% ($n = 239$) of children had ECC with a mean dmfs score of 1.96. There was no difference in the caries experience of male ($n = 124$) and female ($n = 115$) children. Furthermore, the majority of mothers were married and/or partnered, university educated, and non-smokers during pregnancy. The characteristics of participants in relation to the anthropometric and ECC outcomes are shown in Tables 1 and 2, respectively.

Table 1. Participant characteristics—diet trajectories, demographic, socioeconomic, behavioural, and biological characteristics based on anthropometric groups ($n = 738$).

Characteristics	Healthy [a] ($n = 666$)	Overweight and Obese [b] ($n = 72$)
Diet trajectories		
Core foods		
Lowest (Gradual increase with late decrease)	149 (89.76%)	17 (10.24%)
Medium (Rapid increase with late decrease)	288 (88.62%)	37 (11.38%)
Highest (Rapid increase with early decrease)	229 (92.71%)	18 (7.29%)
Discretionary foods		
Lowest (Low and gradual rising)	282 (93.69%)	19 (6.31%)
Medium (Moderate and stable)	297 (89.46%)	35 (10.54%)
Highest (High and late declining)	87 (82.86%)	18 (17.14%)
Demographic factors		
Child age (in years) Mean $\pm$SD	3.57 ± 0.25	3.58 ± 0.27
Child gender		
Male	350 (91.15%)	34 (8.85%)
Female	316 (89.27%)	38 (10.73%)
Maternal age (in years) Mean $\pm$ SD	31.54 ± 5.02	30.64 ± 6.58
Maternal marital status		
Married	616 (90.32%)	66 (9.68%)
Single	50 (89.89%)	6 (10.71%)

Table 1. *Cont.*

Characteristics	Healthy [a] (*n* = 666)	Overweight and Obese [b] (*n* = 72)
Maternal country of birth		
Australia-born	309 (9.09%)	34 (9.91%)
English speaking country	36 (90%)	4 (10%)
Non-English-speaking country	321 (90.42%)	34 (9.58%)
Number of children in household		
1	334 (90.27%)	36 (9.73%)
2	210 (90.52%)	22 (9.48%)
≥3	122 (89.71%)	14 (10.29%)
Individual-level socioeconomic status		
Maternal education		
University	314 (92.90%)	24 (7.10%)
College/TAFE	119 (89.47%)	14 (10.53%)
Completed 12	132 (88.59%)	17 (11.41%)
Left school < 12	101 (85.59%)	17 (14.41%)
Maternal work status		
Not working	339 (88.74%)	43 (11.26%)
Working	298 (91.98%)	26 (8.02%)
Area-level socioeconomic status		
Index of relative socioeconomic advantage and disadvantage		
Deciles 9–10	162 (95.29%)	8 (4.71%)
Deciles 7–8	123 (92.48%)	10 (7.52%)
Deciles 5–6	18 (90%)	2 (10%)
Deciles 3–4	157 (88.70%)	20 (11.30%)
Deciles 1–2	206 (86.55%)	32 (13.45%)
Behavioural factors		
Breastfeeding duration		
<17 weeks	236 (87.41%)	34 (12.59%)
17–25 weeks	71 (89.87%)	8 (10.13%)
26–51 weeks	165 (90.16%)	18 (9.84%)
≥52 weeks	193 (94.15%)	12 (5.85%)
Age of introduction of solid foods		
<17 weeks	63 (85.14%)	11 (14.86%)
17–25 weeks	362 (90.05%)	40 (9.95%)
≥26 weeks	232 (91.70%)	21 (8.30%)
Infant feeding at 4-weeks age		
Only BF	436 (92.77%)	34 (7.23%)
Only FF	87 (83.65%)	17 (16.35%)
Both BF and FF	143 (87.20%)	21 (12.80%)
Outdoor physical activity		
≥180 min	414 (89.61%)	48 (10.39%)
<180 min	239 (91.22%)	23 (8.78%)
Maternal smoking during pregnancy		
No	636 (90.60%)	66 (9.40%)
Yes	30 (83.33%)	6 (16.67%)
Biological factors		
Infant birthweight		
Normal/High	634 (90.44%)	67 (9.56%)
Low	32 (86.49%)	5 (13.51%)

[a,b] The total of the categories might not always add up to 738 due to missing or incomplete data for some items. Index of relative socioeconomic advantage and disadvantage: deciles 9–10 = least disadvantaged; deciles 7–8 = low disadvantaged; deciles 5–6 = moderately disadvantaged; deciles 3–4 = highly disadvantaged; deciles 1–2 = most disadvantaged. SD: standard deviation. *n*: sample size. BF: breastfeeding. FF: formula feeding.

Table 2. Participant characteristics—diet trajectories, demographic, socioeconomic, and behavioural characteristics based on early childhood caries (ECC) groups (*n* = 718).

Characteristics	ECC − ve [a] (*n* = 479)	ECC + ve [b] (*n* = 239)
Diet trajectories		
Sugary foods		
Lowest (Low and gradual rising)	219 (67.80%)	104 (32.20%)
Medium (Moderate and stable)	194 (65.54%)	102 (34.46%)
Highest (High and late declining)	66 (66.67%)	33 (33.33%)
Demographic factors		
Child age (in years) Mean ± SD	3.56 ± 0.25	3.59 ± 0.25
Child gender		
Male	248 (66.67%)	124 (33.33%)
Female	231 (66.76%)	115 (33.24%)
Maternal age (in years) Mean ±SD	31.54 ± 5.09	31 ± 5.49
Maternal marital status		
Married	442 (66.77%)	220 (33.23%)
Single	37 (66.07%)	19 (33.93%)
Number of children in household		
1	257 (71.59%)	102 (28.41%)
2	146 (64.60%)	80 (35.40%)
≥3	76 (57.14%)	57 (42.86%)
Individual-level socioeconomic status		
Maternal education		
University	223 (69.25%)	99 (30.57%)
College/TAFE	97 (75.78%)	31 (24.22%)
Completed 12	94 (62.67%)	56 (37.33%)
Left school < 12	65 (55.08%)	53 (44.92%)
Maternal work status		
Not working	233 (62.47%)	140 (37.53%)
Working	226 (72.20%)	87 (27.80%)
Area-level socioeconomic status		
Index of relative socioeconomic advantage and disadvantage		
Deciles 9–10	122 (78.21%)	34 (21.79%)
Deciles 7–8	88 (69.84%)	38 (30.16%)
Deciles 5–6	16 (76.19%)	5 (23.81%)
Deciles 3–4	116 (65.91%)	60 (34.09%)
Deciles 1–2	137 (57.32%)	102 (42.68%)
Behavioural factors		
Breastfeeding duration		
<17 weeks	176 (66.17%)	90 (33.83%)
17–25 weeks	54 (71.05%)	22 (28.95%)
26–51 weeks	129 (72.88%)	48 (27.12%)
≥52 weeks	119 (60.10%)	79 (39.90%)
Maternal smoking during pregnancy		
No	453 (66.42%)	229 (33.58%)
Yes	26 (72.22%)	10 (27.78%)

[a,b] The total of the categories might not always add up to 718 due to missing or incomplete data for some items. Index of relative socioeconomic advantage and disadvantage: deciles 9–10 = least disadvantaged; deciles 7–8 = low disadvantaged; deciles 5–6 = moderately disadvantaged; deciles 3–4 = highly disadvantaged; deciles 1–2 = most disadvantaged. SD: standard deviation. *n*: sample size. ECC: early childhood caries.

3.1. Dietary Pattern Trajectories

3.1.1. Core Foods

Three distinct core foods trajectories were identified (Figure 1): trajectory 1 (Lowest consumers—gradual increase with late decrease) comprising 22.9% of the sample; trajectory 2 (Medium consumers—rapid increase with late decrease) comprising 43.6%; and trajectory 3 (Highest consumers—rapid increase with early decrease) comprising 33.5%. The resulting patterns indicate that children's core foods intake increased between 4 months and 2 years of age, with frequency for all patterns decreasing between 2 and 3 years of age. From the age of 1 to 2 years, children with the highest consumption began to decrease their intake of core foods, while children in the lower consumption trajectories continued to increase their consumption until 2 to 3 years, after which a downward decline in core foods consumption was observed. Overall, the medium and highest trajectories seemed to converge with advancing age, whilst the lowest trajectory remained distinct at the 3-year age point (Figure 1).

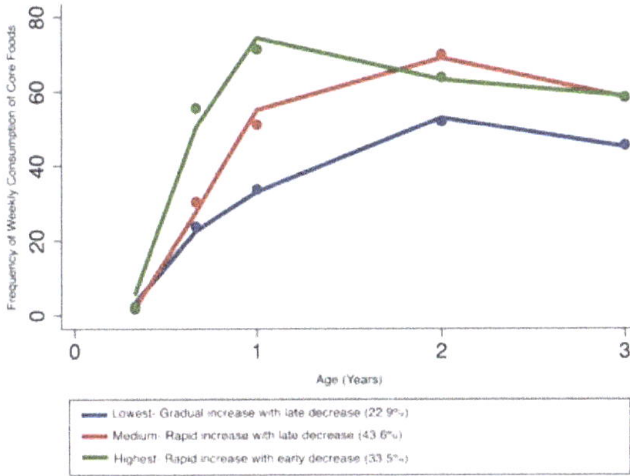

Figure 1. Trajectories of core foods consumption in infancy and early childhood.

3.1.2. Discretionary Foods

Three distinct discretionary foods trajectories were identified (Figure 2): trajectory 1—'Lowest consumers—Low and gradual rising' comprising 40.8% of the sample; trajectory 2—'Medium consumers—Moderate and gradual rising' comprising 44.8%; and trajectory 3—'Highest consumers—High and late declining' comprising 14.4%. The resulting patterns indicate that children's discretionary foods intake steadily increased between 4 months and 3 years of age. Between 2 and 3 years of age, children who had the lowest and medium trajectories continued to have slightly higher or stable intakes respectively, whilst children who had the highest trajectories began a downward trend in discretionary foods consumption. Overall, all three trajectories remained distinctive with advancing age (Figure 2).

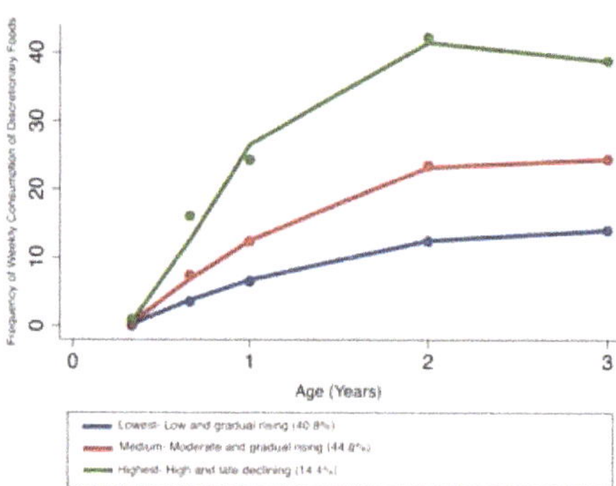

Figure 2. Trajectories of discretionary foods consumption in infancy and early childhood.

3.1.3. Sugary Foods

Three distinct sugary foods trajectories were identified (Figure 3): trajectory 1—'Lowest consumers—Low and gradual rising' comprising 41.3% of the sample; trajectory 2—'Medium consumers—Moderate and stable' comprising 45.3%; and trajectory 3—'Highest consumers—High and late declining' comprising 13.4% of the total sample. The resulting patterns indicate that children's sugary foods intake steadily increased between 4 months and 3 years of age. Between the ages of 2 and 3 years, children who had the lowest and medium trajectories continued to have slightly higher or stable intakes, respectively, whilst children who had the highest trajectories tended to begin a downward trend in sugar foods consumption. Overall, all the three trajectories remained distinctive with advancing age (Figure 3).

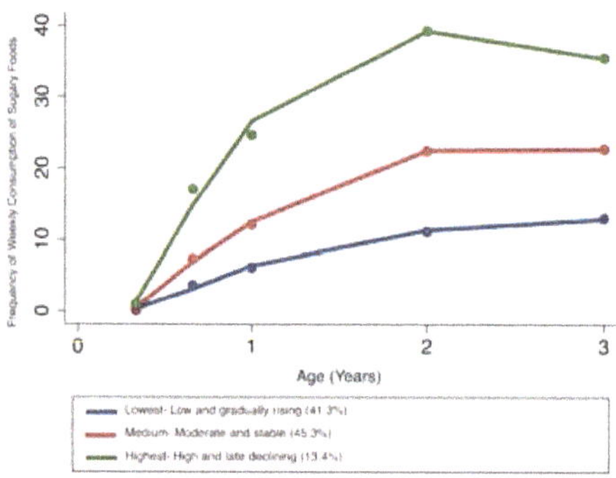

Figure 3. Trajectories of sugary foods consumption in infancy and early childhood.

3.2. Impact of Dietary Pattern Trajectories and Other Predictors on Overweight/Obesity

After adjustment of covariates, the highest trajectory of discretionary foods intake compared with the lowest trajectory was independently associated with overweight or obesity (OR = 2.51, 95% CI: 1.16–5.42; p = 0.019). Additionally, low area-level SES (deciles 1–2: most disadvantaged), compared with highest area-level SES (deciles 9–10: least disadvantaged) was associated with overweight or obesity (OR = 2.86, 95% CI: 1.11–7.34; p = 0.029). There was no independent association between core food intake trajectories and overweight or obesity (Table 3).

Table 3. Final model—impact of diet trajectories, demographic, socioeconomic, behavioural, and biological factors on overweight or obesity in early childhood.

	Adjusted OR	95% CI		*p*-Value	Overall *p*-Value
Diet trajectories					
Core foods					
Lowest (Gradual increase with late decrease)	ref				
Medium (Rapid increase with late decrease)	1.26	0.64	2.47	0.505	0.559
Highest (Rapid increase with early decrease)	0.82	0.39	1.72	0.603	
Discretionary foods					
Lowest (Low and gradual rising)	ref				
Medium (Moderate and stable)	1.55	0.83	2.89	0.165	0.022
Highest (High and late declining)	2.51	1.16	5.42	0.019	
Demographic factors					
Child age (in years)	1.03	0.37	2.86	0.958	0.973
Child gender					
Male	ref				0.496
Female	1.21	0.72	2.06	0.468	
Maternal age (in years)	1.00	0.95	1.06	0.856	0.803
Maternal marital status					
Married	ref				0.521
Single	0.68	0.24	1.89	0.457	
Maternal country of birth					
Australia-born	ref				
English speaking country	1.23	0.38	3.99	0.734	0.252
Non-English-speaking country	0.71	0.38	1.31	0.275	
Number of children in household					
1	ref				
2	0.80	0.43	1.48	0.481	0.346
≥3	0.67	0.31	1.45	0.314	
Individual-level socioeconomic status					
Maternal education					
University	ref				
College/TAFE	1.11	0.52	2.39	0.777	
Completed 12	1.06	0.51	2.24	0.861	0.707
Left school < 12	1.19	0.51	2.80	0.678	
Maternal work status					
Not working	ref				0.381
Working	0.77	0.43	1.37	0.375	
Area-level socioeconomic status					
Index of relative socioeconomic advantage and disadvantage					
Deciles 9–10	ref				
Deciles 7–8	1.65	0.60	4.49	0.328	
Deciles 5–6	1.94	0.35	10.83	0.449	0.030
Deciles 3–4	2.25	0.87	5.81	0.092	
Deciles 1–2	2.86	1.11	7.34	0.029	
Behavioural factors					
Breastfeeding duration					
26–51 weeks	ref				
<17 weeks	0.63	0.29	1.38	0.257	
17–25 weeks	0.65	0.23	1.89	0.433	0.121
≥52 weeks	0.55	0.25	1.22	0.143	

Table 3. *Cont.*

	Adjusted OR	95% CI		*p*-Value	Overall *p*-Value
Age of introduction of solid foods					
<17 weeks	ref				
17–25 weeks	0.97	0.41	2.28	0.948	0.492
≥26 weeks	0.88	0.34	2.25	0.786	
Infant feeding at 4-weeks age					
Only BF	ref				
Only FF	1.96	0.81	4.75	0.137	0.072
Both BF and FF	1.85	0.96	3.56	0.067	
Outdoor physical activity					
≥180 mins	ref				
<180 min	0.89	0.51	1.58	0.714	0.730
Maternal smoking during pregnancy					
No	ref				
Yes	1.07	0.31	3.69	0.908	0.822
Biological factors					
Infant birthweight					
Normal/High	ref				
Low	1.31	0.45	3.78	0.621	0.730

Index of relative socioeconomic advantage and disadvantage: deciles 9–10 = least disadvantaged; deciles 7–8 = low disadvantaged; deciles 5–6 = moderately disadvantaged; deciles 3–4 = highly disadvantaged; deciles 1–2 = most disadvantaged. AOR: adjusted odds ratio. 95% CI: 95% confidence interval. ref: reference category. BF: breastfeeding. FF: formula feeding.

3.3. Impact of Dietary Pattern Trajectories and Other Predictors on ECC

No statistically significant or clinically meaningful association was found between trajectories of sugary foods intake and ECC after adjusting for covariates (Table 4). In regard to other predictors, low area-level SES (deciles 3–4: highly disadvantaged—IRR = 2.02, 95% CI: 1.08–3.77; p = 0.027 and deciles 1–2: most disadvantaged—IRR = 2.71, 95% CI: 1.48–4.97; p = 0.001), compared with highest area-level SES (deciles 9–10: least disadvantaged) was associated with ECC. Furthermore, a longer duration of breastfeeding (≥52 weeks) was associated with ECC (IRR = 2.17, 95% CI: 1.27–3.73; p = 0.005) compared with breastfeeding for 26 to 51 weeks (Table 4).

Table 4. Final model—impact of diet trajectories, demographic, socioeconomic, and behavioural factors on early child-hood caries.

	Adjusted IRR	95% CI		*p*-Value	Overall *p*-Value
Diet trajectories					
Sugary foods					
Lowest (Low and gradual rising)	ref				
Medium (Moderate and stable)	1.30	0.85	2.00	0.228	0.737
Highest (High and late declining)	0.90	0.47	1.70	0.747	
Demographic factors					
Child age (in years)	0.87	0.39	1.97	0.747	0.941
Child gender					
Male	ref				
Female	0.86	0.57	1.29	0.473	0.246
Maternal age (in years)	0.98	0.94	1.03	0.464	0.755
Maternal marital status					
Married	ref				
Single	1.34	0.59	3.00	0.482	0.352

Table 4. *Cont.*

	Adjusted IRR	95% CI		*p*-Value	Overall *p*-Value
Number of children in household					
1	ref				
2	1.11	0.69	1.78	0.654	0.563
≥3	1.53	0.84	2.79	0.164	
Individual-level socioeconomic status					
Maternal education					
University	ref				
College/TAFE	0.75	0.43	1.33	0.329	
Completed 12	1.07	0.60	1.89	0.817	0.195
Left school < 12	1.75	0.91	3.37	0.092	
Maternal work status					
Not working	ref				
Working	0.78	0.52	1.18	0.241	0.126
Area-level socioeconomic status					
Index of relative socioeconomic advantage and disadvantage					
Deciles 9–10	ref				
Deciles 7–8	1.64	0.86	3.13	0.132	
Deciles 5–6	0.67	0.19	2.28	0.521	
Deciles 3–4	2.02	1.08	3.77	0.027	0.005
Deciles 1–2	2.71	1.48	4.97	0.001	
Behavioural factors					
Breastfeeding duration					
26–51 weeks	ref				
<17 weeks	1.23	0.72	2.08	0.448	
17–25 weeks	0.99	0.47	2.06	0.976	0.008
≥52 weeks	2.17	1.27	3.73	0.005	
Maternal smoking during pregnancy					
No	ref				
Yes	0.51	0.17	1.56	0.242	0.223

Index of relative socioeconomic advantage and disadvantage: deciles 9–10 = least disadvantaged; deciles 7–8 = low disadvantaged; deciles 5–6 = moderately disadvantaged; deciles 3–4 = highly disadvantaged; deciles 1–2 = most disadvantaged. Adjusted IRR: adjusted incidence rate ratio. 95% CI: 95% confidence interval. ref: reference category.

4. Discussion

This study augments the existing literature on childhood nutrition by exploring the longitudinal trajectories of dietary intake in Australian children as they transition from infancy to early childhood using an innovative statistical approach such as GBTM and their association with two highly prevalent childhood health issues.

In the present study, the prevalence of overweight and obesity was 10% while the prevalence of ECC was 33%. A similar birth cohort study based in South Australia reported 8.2% prevalence of overweight and obesity and 8.8% prevalence of ECC in children aged 24–36 months [28]. In an Australia-wide context, the prevalence of overweight and obesity in the present study is significantly lower (10% vs. 20%) [3], while the prevalence of ECC is identical (33% vs. 34%) [4].

This study investigated the association of longitudinal dietary trajectories with overweight or obesity and ECC at 3–4 years of age. A positive association between high frequency of discretionary foods intake (poor quality diet) and overweight or obesity was observed. However, no association was found between high frequency of sugary foods intake and ECC. Furthermore, no evidence of an inverse association between trajectories of core foods intake (good quality diet) and overweight or obesity was found. Nevertheless, the risk of childhood overweight or obesity and ECC was predicted by area-level SES, and the risk of ECC was predicted by child breastfeeding practices in expected directions.

The frequent consumption of energy-dense, nutrient-poor discretionary foods was prospectively associated with overweight and obesity. This finding is consistent with previous literature [52,53]. For instance, a previous systematic review identified a positive relationship between energy-dense, high-fat, and low-fibre dietary patterns, and later overweight and obesity risk based on high-quality prospective studies [54]. Another recent review including longitudinal cohort studies suggested that energy-dense diets increase the risk of obesity in childhood [55]. However, inconsistent findings have been reported, and a recent Australian cohort study did not find any association between unhealthy diet patterns and obesity in children aged 24–36 months [28]. Frequent consumption of discretionary foods can displace the intake of healthier core foods, provide excessive energy leading to weight gain, and ultimately contribute to incidence of chronic health conditions [56]. Hence, understanding the contribution of discretionary nutrients (such as saturated fat, salt, and added sugars) to total energy intake will assist in the identification of targets to focus our efforts on preschool children to improve their nutrient intakes, diet quality, and subsequent health.

Previous literature related to associations between unhealthy dietary patterns and anthropometric outcomes mostly includes older children and/or adolescents and/or is predominantly based on cross-sectional analysis using explanatory dietary pattern methods (e.g., factor analysis, PCA, and cluster analysis) [22–24]. To the best of our knowledge, this is the first Australian cohort study that uses the newly emerging GBTM approach to examine children's diet patterns (or trajectories) in the first three years of life and investigates their impact on overweight/obesity between 3 and 4 years of age.

For ECC, the present study did not find an association between sugary foods trajectories and ECC. This outcome is not surprising considering the inconsistent findings about the association between sugary diet patterns and dental caries measures reported in earlier studies [28,31]. A probable explanation for the lack of association in the present study might be that the dietary items, particularly sugary items, within the FFQ were limited, and the questionnaire assessed the dietary intake on a weekly basis. Moreover, the dietary assessment was self-reported. Nonetheless, a recent Australian cohort study also did not find any association between sugar-containing dietary patterns in the first 12 months and ECC at 24–36 months of age [28]. It was suggested that the lack of association was because 24–36 months of age might be too early to detect the impact of poor diet [28]. Likewise, another recent cohort study did not find any evidence for sugary diet patterns subsequently causing dental caries in Asian toddlers [31].

In relation to the methods of dietary assessment, FFQ is one of the most widely used instruments in epidemiological studies because of its relative simplicity, time efficiency, and cost-effectiveness [57]. However, a major disadvantage of FFQ is the reliance on self-reporting, which is influenced by participants' honesty, education level, memory, and cognitive capability in terms of dietary intake. Such factors are suggested to significantly underestimate the true energy intake [58], thereby possibly affecting the true diet–disease relationship. On the contrary, objective data, for example SES (including education level, income, and employment status) is a composite indicator of health and is suggested to influence the dietary intake and diet quality, as well as the health of individuals. Thus, using such data helps to better explain the social inequalities in nutrition and health [59].

In the present study, beside dietary patterns, the two health outcomes were associated with certain socioeconomic and behavioural factors. Familial socioeconomic disadvantage was found to be a common risk factor for both overweight or obesity and ECC in the study sample. A recent systematic review identified familial wealth as one of the most important predictors for both obesity and ECC risk in preschool children [8]. The present study findings are consistent with previous literature that also found an inverse association between parental SES and children's weight status and dental caries experience [60,61].

Poor parental income influences children's dietary choices, since low-calorie, nutrient-rich foods (namely fruits, vegetables, and whole-grain cereals) are likely to be more expensive, therefore possibly leading to higher consumption of an energy-dense, nutrient-poor diet, which is relatively inexpensive [62]. Furthermore, a low family income may lead to limited access to a low-sugar diet, fluoridated toothpaste, and professional preventative measures [61], and subsequent poor dietary and lifestyle choices [62].

Breastfeeding is known to have considerable health benefits for both child and the mother [63]. However, this study found that a longer duration of breastfeeding (i.e., beyond 12 months of age) increases ECC risk. This finding adds to the current body of evidence about the causal role of breastfeeding on ECC. Previous systematic reviews [64,65] reported that the risk of ECC increases with breastfeeding beyond 12 months of age. Although a recent Australian cohort study [66] did not find an association between breastfeeding beyond 12 months of age and ECC, the authors concluded that the size and direction of the relationship was suggestive of a higher risk. Meanwhile, a second national Australian study found that this effect was modified by fluoride, and an association between ECC and breastfeeding beyond 24 months was only evident in children without access to a fluoridated water supply [67]. Breastfeeding is known to be protective against caries in the first 6 to 12 months of life compared to no breastfeeding, but ECC prevalence tends to increase if breastfeeding continues beyond the first year of life [64,65]. The exact age at which breastfeeding begins to have a cariogenic effect is not known, which is possibly because studies have used different cut-offs, e.g., 12 months [68], 18 months [69], and 24 months and beyond [70]. Human breast milk is suggested to be more cariogenic than bovine milk but less than infant formula [71]. In the first 12 months, infants are usually fed either breastmilk or formula, both of which have almost the same carbohydrate content [65]. After 12 months, infants are usually weaned onto cow's milk, which has a significantly lower carbohydrate content than both formula and human milk. Hence, a prolonged contact of teeth with human milk combined with intake of sugars in the diet leads to an acidogenic oral environment and subsequent demineralisation of tooth (or teeth) due to the bacterial fermentation of sugars [72]. Furthermore, these elements can be modified by various risk factors such as SES, maternal education, maternal smoking, parity, sugar intake, fluoride exposure, and oral hygiene practices [73]. Some studies have also shown that children who are breastfed for longer durations tend to consume cariogenic foods more frequently [68,74].

Strengths and Limitations

This study used the longitudinal data from the HSHK birth cohort study, thereby overcoming the possibility of reverse directionality between dietary patterns and overweight/obesity and ECC outcomes [28]. Furthermore, oral health related birth-cohort studies such as HSHK are rare, and this study provides an opportunity to explore the common risk factors of two highly prevalent health issues affecting children at present. This study examines the frequency of intake of core foods, discretionary, and added-sugar foods over a longitudinal period in early years of life and its impact on obesity and ECC. Earlier studies have primarily focussed on assessing dietary patterns in older children or adults [24,25]. The dietary intake was able to be estimated at multiple and regularly spaced time points over the first three years of life, thus allowing for the influence of dietary patterns to possibly manifest into examinable changes in weight status and teeth at 3 to 4 years of age. The dietary patterns were generated using the newly emerging Group-Based Trajectory Modelling (GBTM) in longitudinal and clinical research, which provides a comprehensive picture of the evolution of children's diet over time. Additionally, the use of multiple meaningful dietary patterns (i.e., core, discretionary, and sugar-based) to assess relationships with obesity and ECC outcomes is unique and a major strength of the present study. The attrition rate of the study sample by the time of clinical assessment (i.e., at 3–4 years of age) was considerably lower than those of other birth cohort studies [28,75]. Various sociodemographic, socioeconomic, behavioural, and biological predictors were also assessed, which would assist in identifying potential target groups for preventing two health outcomes using the common risk factor approach in health promotion.

This study also has some limitations that need to be acknowledged. First, the dietary intake was based on parent reports through interviews and food frequency questionnaire (FFQ; therefore, there is a possibility of underreporting, inaccurate dietary recall, and/or social desirability bias. However, FFQ are commonly used in longitudinal studies because they are cost-effective and relatively quick to complete [57]. This assisted in maintaining good retention in the present study. In addition, since the data were longitudinal, the chances of heaping of data and recall bias are minimised. Second, a FFQ adapted from well-established literature was used in the study; however, it was not able to capture the whole diet. Moreover, the short FFQ consisted of essential core and discretionary foods listed in the Australian dietary guidelines; however, the items' list is limited, particularly in relation to foods with free sugars (a strong determinant of ECC) [66], which might have produced different dietary trajectories. Similarly, in regard to core foods, refined cereals could not be distinguished from unrefined cereals; therefore, popular cereals that are high in sugar were categorised as core foods rather than discretionary foods [76]. Additionally, only the frequency of intake of foods was recorded rather than their frequency, amount, and relative percent of total calories, so we could not capture actual dietary intake, since the amount might vary within and between individuals, and total calories might vary between different foods. Another potential limitation is that the influence of parental BMI on children's weight status could not be evaluated, considering that this is one of the most important determinants of childhood overweight and obesity. Furthermore, in relation to the predictors of ECC, the frequency and timing of breastfeeding (day- and night-time), and role of oral hygiene practices such as supervised tooth brushing, frequency of toothbrushing, and use of fluoridated toothpaste could not be evaluated, which might have produced different results.

5. Conclusions

In summary, this is one of the first studies to describe the early life dietary patterns using Group-Based Trajectory Modelling (GBTM) and examine their impact on two highly prevalent chronic diseases in childhood—obesity and early childhood caries (ECC). An independent association was found between the highest trajectory of discretionary foods intake and being overweight or obese. However, an association between trajectories of sugary foods intake and ECC could not be established. Further research to investigate the impact of longitudinal trajectories of free-sugars intake on ECC is warranted. Familial socioeconomic disadvantage was identified to be a common risk factor for both health conditions, thus justifying the concept of common risk factor approach (CRFA) in disease epidemiology and prevention. Additionally, breastfeeding beyond 12 months was found to be a significant predictor of ECC in this sample; however, other factors related to breastfeeding (frequency and timing) and oral hygiene practices were unadjusted. In conclusion, these study findings have identified target groups to implement preventative and interventional strategies against the rising obesity and dental caries burden in early childhood.

Supplementary Materials: The following are available online at https://www.mdpi.com/article/10.3390/nu13072240/s1, Table S1: List of dietary items (*n* = 32) recorded in the short food frequency questionnaire.

Author Contributions: N.M., A.A., L.G.D., and J.A.S. conceived the study. A.A., L.G.D., and J.A.S. developed the study questionnaire. A.A. and S.B. were involved in data collection. N.M. performed the analysis and interpreted the results with assistance from A.H., J.A.S., L.G.D., and A.A.; N.M. prepared the first draft, and all authors critically revised the manuscript. All authors have read and agreed to the published version of the manuscript.

Funding: This project is supported by NHMRC Grants (1069861, 1033213, 1134075), NSW Health, Australian Dental Research Foundation, Western Sydney University, and Oral Health Foundation.

Institutional Review Board Statement: The study was conducted according to the guidelines of the Declaration of Helsinki. Ethics approval to conduct this study was given by the former Sydney South West Area Health Service—RPAH Zone (ID number X08-0115, 4 September 2008), Liverpool Hospital, University of Sydney, and Western Sydney University.

Informed Consent Statement: All participants signed a written consent form prior to study commencement. All research participants consented to use their de-identified data for publishing in scientific publications.

Data Availability Statement: The data of this study cannot be shared publicly due to the presence of sensitive (confidential) participants' information.

Acknowledgments: We would like to thank the staff from Sydney and South Western Sydney Local Health Districts for their continued support with the project. The authors are grateful to the Child and Family Health Nurses for recruiting families to this birth cohort study. We would like to thank the families who are a part of the Healthy Smiles Healthy Kids study for their continued commitment to this ongoing cohort study. We would also like to acknowledge the Australian College of Physical Education (ACPE) for its support. Lastly, we would like to acknowledge Jonathan Broadbent and Bobby Jones for their expert statistical assistance.

Conflicts of Interest: The authors declare no conflict of interest.

References

1. Ho, N.; Olds, T.; Schranz, N.; Maher, C. Secular trends in the prevalence of childhood overweight and obesity across Australian states: A meta-analysis. *J. Sci. Med. Sport* **2017**, *20*, 480–488. [CrossRef] [PubMed]
2. Kassebaum, N.J.; Bernabé, E.; Dahiya, M.; Bhandari, B.; Murray, C.; Marcenes, W. Global burden of untreated caries: A systematic review and metaregression. *J. Dent. Res.* **2015**, *94*, 650–658. [CrossRef]
3. Australian Institute of Health and Welfare. *A Picture of Overweight and Obesity in Australia 2017. Cat. No. PHE 216*; AIHW: Canberra, Australia, 2017.
4. Ha, D.H.; Roberts-Thomson, K.F.; Arrow, P.; Peres, K.G.; Do, L.G. Children's oral health status in Australia, 2012–2014. In *Oral Health of Australian Children: The National Child Oral Health Study 2012–14*; Do, L.G., Spencer, A.J., Eds.; Adelaide University Press: Adelaide, Australia, 2016; pp. 86–152.
5. Sanders, R.H.; Han, A.; Baker, J.S.; Cobley, S. Childhood obesity and its physical and psychological co-morbidities: A systematic review of Australian children and adolescents. *Eur. J. Pediatr.* **2015**, *174*, 715–746. [CrossRef]
6. Martins-Júnior, P.A.; Vieira-Andrade, R.G.; Corrêa-Faria, P.; Oliveira-Ferreira, F.; Marques, L.S.; Ramos-Jorge, M.L. Impact of Early Childhood Caries on the Oral Health-Related Quality of Life of Preschool Children and Their Parents. *Caries Res.* **2013**, *47*, 211–218. [CrossRef]
7. Hooley, M.; Skouteris, H.; Boganin, C.; Satur, J.; Kilpatrick, N. Body mass index and dental caries in children and adolescents: A systematic review of literature published 2004 to 2011. *Syst. Rev.* **2012**, *1*, 57. [CrossRef] [PubMed]
8. Manohar, N.H.A.; Fahey, P.; Arora, A. Obesity and Dental Caries in Early Childhood: A Systematic Review and Meta-Analyses. *Obes. Rev.* **2020**, *21*, e12960. [CrossRef]
9. Hooley, M.; Skouteris, H.; Millar, L. The relationship between childhood weight, dental caries and eating practices in children aged 4–8 years in Australia, 2004–2008. *Pediatr. Obes.* **2012**, *7*, 461–470. [CrossRef] [PubMed]
10. Sheiham, A.; Watt, R.G. The common risk factor approach: A rational basis for promoting oral health. *Community Dent. Oral Epidemiol.* **2000**, *28*, 399–406. [CrossRef]
11. Hayden, C.; Bowler, J.O.; Chambers, S.; Freeman, R.; Humphris, G.; Richards, D.; Cecil, J.E. Obesity and dental caries in children: A systematic review and meta-analysis. *Community Dent. Oral Epidemiol.* **2013**, *41*, 289–308. [CrossRef] [PubMed]
12. Koh, G.A.; Scott, J.A.; Oddy, W.H.; Graham, K.I.; Binns, C.W. Exposure to non-core foods and beverages in the first year of life: Results from a cohort study. *Nutr. Diet.* **2010**, *67*, 137–142. [CrossRef]
13. Irvine, V.; John, J.R.; Scott, J.A.; Hayen, A.; Do, L.G.; Bhole, S.; Ha, D.; Kolt, G.S.; Arora, A. Factors Influencing the Early Introduction of Sugar Sweetened Beverages among Infants: Findings from the HSHK Birth Cohort Study. *Nutrients* **2020**, *12*, 3343. [CrossRef] [PubMed]
14. Manohar, N.; Hayen, A.; Bhole, S.; Arora, A. Predictors of Early Introduction of Core and Discretionary Foods in Australian Infants—Results from HSHK Birth Cohort Study. *Nutrients* **2020**, *12*, 258. [CrossRef]
15. Arora, A.; Manohar, N.; Hector, D.; Bhole, S.; Hayen, A.; Eastwood, J.; Scott, J.A. Determinants for early introduction of complementary foods in Australian infants: Findings from the HSHK birth cohort study. *Nutr. J.* **2020**, *19*, 1–10. [CrossRef] [PubMed]
16. National Health and Medical Research Council. *Eat. for Health—Australian Dietary Guidelines*; National Health and Medical Research Council: Canberra, Australia, 2013.
17. Poti, J.M.; Braga, B.; Qin, B. Ultra-processed food intake and obesity: What really matters for health—processing or nutrient content? *Curr. Obes. Rep.* **2017**, *6*, 420–431. [CrossRef]

18. Moynihan, P.; Petersen, P.E. Diet, nutrition and the prevention of dental diseases. *Public Health Nutr.* **2004**, *7*, 201–226. [CrossRef]

19. Dhandevi, P.; Jeewon, R. Fruit and vegetable intake: Benefits and progress of nutrition education interventions-narrative review article. *Iran. J. Public Health* **2015**, *44*, 1309.

20. Moeller, S.M.; Reedy, J.; Millen, A.E.; Dixon, L.B.; Newby, P.; Tucker, K.L.; Krebs-Smith, S.M.; Guenther, P.M. Dietary patterns: Challenges and opportunities in dietary patterns research: An Experimental Biology workshop, 1 April 2006. *J. Am. Diet. Assoc.* **2007**, *107*, 1233–1239. [CrossRef]

21. Ambrosini, G.L.; Emmett, P.M.; Northstone, K.; Jebb, S.A. Tracking a dietary pattern associated with increased adiposity in childhood and adolescence. *Obesity* **2014**, *22*, 458–465. [CrossRef]

22. Bell, L.; Golley, R.; Daniels, L.; Magarey, A. Dietary patterns of Australian children aged 14 and 24 months, and associations with socio-demographic factors and adiposity. *Eur. J. Clin. Nutr.* **2013**, *67*, 638–645. [CrossRef]

23. Lioret, S.; Betoko, A.; Forhan, A.; Charles, M.-A.; Heude, B.; de Lauzon-Guillain, B. Dietary Patterns Track from Infancy to Preschool Age: Cross-Sectional and Longitudinal Perspectives. *J. Nutr.* **2015**, *145*, 775–782. [CrossRef] [PubMed]

24. Kerr, J.A.; Gillespie, A.N.; Gasser, C.E.; Mensah, F.K.; Burgner, D.; Wake, M. Childhood dietary trajectories and adolescent cardiovascular phenotypes: Australian community-based longitudinal study. *Public Health Nutr.* **2018**, *21*, 2642–2653. [CrossRef] [PubMed]

25. Gasser, C.E.; Kerr, J.A.; Mensah, F.K.; Wake, M. Stability and change in dietary scores and patterns across six waves of the Longitudinal Study of Australian Children. *Br. J. Nutr.* **2017**, *117*, 1137–1150. [CrossRef] [PubMed]

26. Nagin, D.S.; Odgers, C.L. Group-based trajectory modeling in clinical research. *Annu. Rev. Clin. Psychol.* **2010**, *6*, 109–138. [CrossRef] [PubMed]

27. Smithers, L.G.; Golley, R.K.; Brazionis, L.; Lynch, J.W. Characterizing whole diets of young children from developed countries and the association between diet and health: A systematic review. *Nutr. Rev.* **2011**, *69*, 449–467. [CrossRef]

28. Bell, L.K.; Schammer, C.; Devenish, G.; Ha, D.; Thomson, M.W.; Spencer, J.A.; Do, L.G.; Scott, J.A.; Golley, R.K. Dietary patterns and risk of obesity and early childhood caries in Australian toddlers: Findings from an australian cohort study. *Nutrients* **2019**, *11*, 2828. [CrossRef]

29. Marshall, S.; Burrows, T.; Collins, C.E. Systematic review of diet quality indices and their associations with health-related outcomes in children and adolescents. *J. Hum. Nutr. Diet.* **2014**, *27*, 577–598. [CrossRef]

30. Chaffee, B.W.; Feldens, C.A.; Rodrigues, P.H.; Vítolo, M.R. Feeding practices in infancy associated with caries incidence in early childhood. *Community Dent. Oral Epidemiol.* **2015**, *43*, 338–348. [CrossRef]

31. Hu, S.; Sim, Y.F.; Toh, J.Y.; Saw, S.M.; Godfrey, K.M.; Chong, Y.S.; Yap, F.; Lee, Y.S.; Shek, L.P.; Tan, K.H.; et al. Infant dietary patterns and early childhood caries in a multi-ethnic Asian cohort. *Sci. Rep.* **2019**, *9*, 852. [CrossRef]

32. Arora, A.; Scott, J.; Bhole, S.; Do, L.; Schwarz, E.; Blinkhorn, A. Early childhood feeding practices and dental caries in preschool children: A multi-centre birth cohort study. *BMC Public Health* **2011**, *11*, 28. [CrossRef]

33. Arora, A.; Manohar, N.; Hayen, A.; Bhole, S.; Eastwood, J.; Levy, S.; Scott, J.A. Determinants of breastfeeding initiation among mothers in Sydney, Australia: Findings from a birth cohort study. *Int. Breastfeed. J.* **2017**, *12*, 39. [CrossRef]

34. Rankin, S.J.; Levy, S.M.; Warren, J.J.; Gilmore, J.E.; Broffitt, B. Relative validity of an FFQ for assessing dietary fluoride intakes of infants and young children living in Iowa. *Public Health Nutr.* **2011**, *14*, 1229–1236. [CrossRef] [PubMed]

35. Centre for Epidemiology and Evidence. *2009–2010 Summary Report from the New South Wales Child. Health Survey*; NSW Ministry of Health: Sydney, Australia, 2012.

36. Ha, D.H.; Amarasena, N.; Crocombe, L.A. *The Dental Health of Australia's Children by Remoteness: Child. Dental Health Survey Australia 2009*; Australian Institute of Health and Welfare: Canberra, Australia, 2013.

37. Scott, J.A.; Binns, C.W.; Graham, K.I.; Oddy, W.H. Temporal changes in the determinants of breastfeeding initiation. *Birth* **2006**, *33*, 37–45. [CrossRef] [PubMed]

38. Scott, J.A.; Binns, C.W.; Aroni, R.A. Breast-feeding in Perth: Recent trends. *Aust. N. Z. J. Public Health* **1996**, *20*, 210–211. [CrossRef]

39. Arora, A.; Gay, M.; Thirukumar, D. Parental choice of infant feeding behaviours in South West Sydney: A preliminary investigation. *Health Educ. J.* **2012**, *71*, 461–473. [CrossRef]

40. National Health and Medical Research Council. *Eat for Health Educator Guide*; National Health and Medical Research Council: Canberra, Australia, 2013.

41. Australian Department of Health. *Australia's Physical Activity and Sedentary Behaviour Guidelines and the Australian 24-Hour Movement Guidelines*; Australian Department of Health: Canberra, Australia, 2019.

42. World Health Organization. *The Use and Interpretation of Anthropometry: Report of a World Health Organization (WHO) Expert Committee*; World Health Organization: Geneva, Switzerland, 1995.

43. World Health Organization. *Application Tools: WHO AnthroPlus Software*; World Health Organization: Geneva, Switzerland, 2014.

44. De Onis, M.; Lobstein, T. Defining obesity risk status in the general childhood population: Which cut-offs should we use? *Int. J. Paediatr. Obes.* **2010**, *5*, 458–460. [CrossRef] [PubMed]

45. World Health Organization. *Oral Health Survey: Basic Methods*, 5th ed.; World Health Organization: Geneva, Switzerland, 2013.

46. National Institute of Dental and Craniofacial Research (NIDCR). *National Health and Nutrition Examination Survey Dental Examiners Procedures Manual*; National Center for Health Statistics: Hyattsville, MD, USA, 2002.

47. Palmer, J.; Anderson, R.; Downer, M. Guidelines for prevalence studies of dental caries. *Community Dent. Health* **1984**, *1*, 55–66.

48. American Academy of Pediatric Dentistry. Policy on Early Childhood Caries (ECC): Classifications, Consequences, and Preventive Strategies. *Pediatr. Dent.* **2016**, *38*, 52–54.
49. Moynihan, P. Sugars and dental caries: Evidence for setting a recommended threshold for intake. *Adv. Nutr.* **2016**, *7*, 149–156. [CrossRef]
50. Gonzalez-Casanova, I. Individual, family, and community predictors of overweight and obesity among colombian children and adolescents. *Prev. Chronic. Dis* **2014**, *11*, E134. [CrossRef]
51. Long, J.S.; Freese, J. *Regression Models for Categorical Outcomes Using Stata*; Stata Press: College Station, TX, USA, 2005.
52. Chung, A.; Peeters, A.; Gearon, E.; Backholer, K. Contribution of discretionary food and drink consumption to socio-economic inequalities in children's weight: Prospective study of Australian children. *Int. J. Epidemiol.* **2018**, *47*, 820–828. [CrossRef]
53. Zhang, J.; Wang, H.; Wang, Y.; Xue, H.; Wang, Z.; Du, W.; Su, C.; Zhang, J.; Jiang, H.; Zhai, F. Dietary patterns and their associations with childhood obesity in China. *Br. J. Nutr.* **2015**, *113*, 1978–1984. [CrossRef]
54. Ambrosini, G.L. Childhood dietary patterns and later obesity: A review of the evidence. *Proc. Nutr. Soc.* **2014**, *73*, 137–146. [CrossRef]
55. Pérez-Escamilla, R.; Obbagy, J.E.; Altman, J.M.; Essery, E.V.; McGrane, M.M.; Wong, Y.P.; Spahn, J.M.; Williams, C.L. Dietary energy density and body weight in adults and children: A systematic review. *J. Acad. Nutr. Diet.* **2012**, *112*, 671–684. [CrossRef] [PubMed]
56. Johnson, B.J.; Bell, L.K.; Zarnowiecki, D.; Rangan, A.M.; Golley, R.K. Contribution of discretionary foods and drinks to Australian children's intake of energy, saturated fat, added sugars and salt. *Children* **2017**, *4*, 104. [CrossRef] [PubMed]
57. Shim, J.-S.; Oh, K.; Kim, H.C. Dietary assessment methods in epidemiologic studies. *Epidemiol. Health* **2014**, *36*, e2014009. [CrossRef]
58. Archer, E.; Lavie, C.J.; Hill, J.O. The failure to measure dietary intake engendered a fictional discourse on diet-disease relations. *Front. Nutr.* **2018**, *5*, 105. [CrossRef]
59. Alkerwi, A.A.; Vernier, C.; Sauvageot, N.; Crichton, G.E.; Elias, M.F. Demographic and socioeconomic disparity in nutrition: Application of a novel Correlated Component Regression approach. *BMJ Open* **2015**, *5*, e006814. [CrossRef]
60. Dinsa, G.D.; Goryakin, Y.; Fumagalli, E.; Suhrcke, M. Obesity and socioeconomic status in developing countries: A systematic review. *Obes. Rev.* **2012**, *13*, 1067–1079. [CrossRef]
61. Schwendicke, F.; Dörfer, C.; Schlattmann, P.; Page, L.F.; Thomson, W.; Paris, S. Socioeconomic inequality and caries: A systematic review and meta-analysis. *J. Dent. Res.* **2015**, *94*, 10–18. [CrossRef]
62. Kant, A.K.; Graubard, B.I. Family income and education were related with 30-year time trends in dietary and meal behaviors of American children and adolescents. *J. Nutr.* **2013**, *143*, 690–700. [CrossRef]
63. Victora, C.G.; Bahl, R.; Barros, A.J.; França, G.V.; Horton, S.; Krasevec, J.; Murch, S.; Sankar, M.J.; Walker, N.; Rollins, N.C. Breastfeeding in the 21st century: Epidemiology, mechanisms, and lifelong effect. *Lancet* **2016**, *387*, 475–490. [CrossRef]
64. Cui, L.; Li, X.; Tian, Y.; Bao, J.; Wang, L.; Xu, D.; Zhao, B.; Li, W. Breastfeeding and early childhood caries: A meta-analysis of observational studies. *Asia. Pac. J. Clin. Nutr.* **2017**, *26*, 867.
65. Tham, R.; Bowatte, G.; Dharmage, S.C.; Tan, D.J.; Lau, M.X.; Dai, X.; Allen, K.J.; Lodge, C.J. Breastfeeding and the risk of dental caries: A systematic review and meta-analysis. *Acta Paediatr.* **2015**, *104*, 62–84. [CrossRef] [PubMed]
66. Devenish, G.; Mukhtar, A.; Begley, A.; Spencer, A.J.; Thomson, W.M.; Ha, D.; Do, L.; Scott, J.A. Early childhood feeding practices and dental caries among Australian preschoolers. *Am. J. Clin. Nutr.* **2020**, *111*, 821–828. [CrossRef] [PubMed]
67. Ha, D.H.; Spencer, A.J.; Peres, K.G.; Rugg-Gunn, A.J.; Scott, J.A.; Do, L.G. Fluoridated Water Modifies the Effect of Breastfeeding on Dental Caries. *J. Dent. Res.* **2019**, *98*, 755–762. [CrossRef]
68. Feldens, C.A.; Giugliani, E.R.J.; Vigo, Á.; Vítolo, M.R. Early Feeding Practices and Severe Early Childhood Caries in Four-Year-Old Children from Southern Brazil: A Birth Cohort Study. *Caries Res.* **2010**, *44*, 445–452. [CrossRef] [PubMed]
69. Tanaka, K.; Miyake, Y.; Sasaki, S.; Hirota, Y. Infant feeding practices and risk of dental caries in Japan: The Osaka maternal and child health study. *Pediatr. Dent.* **2013**, *35*, 267–271.
70. Peres, K.G.; Nascimento, G.G.; Peres, M.A.; Mittinty, M.N.; Demarco, F.F.; Santos, I.S.; Matijasevich, A.; Barros, A.J. Impact of prolonged breastfeeding on dental caries: A population-based birth cohort study. *Pediatrics* **2017**, *140*, e20162943. [CrossRef]
71. Peres, R.C.R.; Coppi, L.C.; Volpato, M.C.; Groppo, F.C.; Cury, J.A.; Rosalen, P.L. Cariogenic potential of cows', human and infant formula milks and effect of fluoride supplementation. *Br. J. Nutr.* **2008**, *101*, 376–382. [CrossRef]
72. Erickson, P.R.; Mazhari, E. Investigation of the role of human breast milk in caries development. *Pediatr. Dent.* **1999**, *21*, 86–90.
73. Selwitz, R.H.; Ismail, A.I.; Pitts, N.B. Dental caries. *Lancet* **2007**, *369*, 51–59. [CrossRef]
74. Hallonsten, A.L.; Mejàre, I.; Birkhed, D.; Håkansson, C.; Lindvall, A.M.; Edwardsson, S.; Koch, G. Dental caries and prolonged breast-feeding in 18-month-old Swedish children. *Int. J. Paediatr. Dent.* **1995**, *5*, 149–155. [CrossRef]
75. Johnson, S.; Carpenter, L.; Amezdroz, E.; Dashper, S.; Gussy, M.; Calache, H.; de Silva, A.M.; Waters, E. Cohort Profile: The VicGeneration (VicGen) study: An Australian oral health birth cohort. *Int. J. Epidemiol.* **2017**, *46*, 29–30. [CrossRef] [PubMed]
76. Dunford, E.K.; Louie, J.C.Y.; Walker, K.Z.; Gill, T.P. Nutritional quality of Australian breakfast cereals-are they improving? *Appetite* **2012**, *59*, 464–470. [CrossRef]

Review

Dietary Interventions to Prevent Childhood Obesity: A Literature Review

Ana Rita Pereira [1] and Andreia Oliveira [2,3,4,*]

[1] Faculty of Health Sciences (Nutrition Sciences), University Fernando Pessoa, Rua Carlos da Maia 296, 4200-150 Porto, Portugal; 35460@ufp.edu.pt
[2] EPIUnit—Instituto de Saúde Pública da Universidade do Porto (Institute of Public Health of the University of Porto), Rua das Taipas 135, 4050-600 Porto, Portugal
[3] Laboratory for Integrative and Translational Research in Population Health (ITR), Rua das Taipas 135, 4050-600 Porto, Portugal
[4] Department of Public Health and Forensic Sciences, and Medical Education, Faculty of Medicine, University of Porto, Alameda Professor Hernâni Monteiro, 4200-319 Porto, Portugal
* Correspondence: acmatos@ispup.up.pt; Tel.: +351-222-061-820

Abstract: Several dietary interventions have been conducted to prevent/reduce childhood obesity, but most of them are known to have failed in tackling the obesity epidemic. This study aimed to review the existing literature on dietary interventions for the prevention of childhood obesity and their effectiveness. A literature search was conducted using PubMed Central®. Only articles published between 2009 and 2021, written in English, conducted in humans, and including children and/or adolescents (<18 years old) were considered. The majority of studies were school-based interventions, with some addressing the whole community, and including some interventions in the food sector (e.g., taxation of high fat/sugar foods, front-of-pack labelling) and through mass media (e.g., restrictions on food advertising for children) that directly or indirectly could help to manage childhood obesity. Most of the programs/interventions conducted focus mainly on person-based educational approaches, such as nutrition/diet education sessions, allied to the promotion of physical activity and lifestyles to students, parents, and school staff, and less on environmental changes to offer healthier food choices. Only a few trials have focused on capacity building and macro-policy changes, such as the adaptation of the built environment of the school, serving smaller portion sizes, and increasing the availability and accessibility of healthy foods and water in schools, and restricting the access to vending machines, for example. Overall, most of the intervention studies showed no consistent effects on changing the body mass index of children; they have only reported small weight reductions, clinically irrelevant, or no effects at all. Little is known about the sustainability of interventions over time.

Keywords: pediatric obesity; children; dietary interventions; diet; prevention

Citation: Pereira, A.R.; Oliveira, A. Dietary Interventions to Prevent Childhood Obesity: A Literature Review. *Nutrients* **2021**, *13*, 3447. https://doi.org/10.3390/nu13103447

Academic Editor: Panagiota Mitrou

Received: 22 July 2021
Accepted: 24 September 2021
Published: 28 September 2021

1. Introduction

In the recent past, there was a shift from prevailing infectious diseases to a high prevalence of chronic and degenerative diseases associated with lifestyle choices [1]. Obesity is one of the conditions that has dramatically increased all over the world, and children, in particular, are a cause of public health concern [2].

The prevalence of overweight and obesity has increased substantially over the past four decades, and an epidemiological transition from underweight to overweight and obesity has been described throughout the world [3]. This alarming rise has been observed in all regions, including developing countries, with an increase of overweight and obesity prevalence from 1980 to 2013 of 8.1% to 12.9% (in boys) and 8.4% to 13.4% (in girls) [2]. These increases have been also reported in developed countries, among children and adolescents, with 23.8% of boys and 22.6% of girls having either overweight or obesity in

2013 [2]. Although the prevalence is clearly higher in developed countries at all ages, the differences between sexes are small. Nonetheless, the prevalence of childhood obesity in the United States and some European countries has apparently reached a plateau [4], but it continues at high rates.

Obesity is a complex, multifactorial disease. Although genetics may be an important etiological factor for obesity development, genes do not fully explain the huge and fast increase of obesity at the population level [4,5]. It is believed that this obesity epidemic may be due to gene–environment interactions [6], enhanced by an increasingly permissive obesogenic environment, with different levels of determinants [1,7]. There are micro-environmental settings, such as schools, workplaces, homes, and neighborhoods, and these are influenced by macro-environmental sectors, such as the health system and the food industry that may be key settings to tackle the obesity epidemic [7]. Now, more than ever, individuals are embedded in a more permissive environment with concern to eating habits and are more likely to adopt sedentary behaviors. It has been recognized many years ago in the Ottawa Charter that it is very important to promote supportive environments [8]. In the case of children, the family and school are included in a wider proximal context [7].

It is well known that diet and other habits are shaped at the earlier stages of life and maintained through adulthood [9]. With the current increasing rates of childhood obesity, there has been a growing amount of research focusing on the determinants of obesity in children and their families, and several studies have described possible dietary/nutritional interventions to prevent childhood obesity. It is known that interventions that are mostly based on educational, behavioral, or pharmacological measures are not very effective in preventing and treating obesity [10,11].

This study aims to review the existing literature on dietary interventions for the prevention of childhood obesity and to assess their effectiveness.

2. Materials and Methods

A literature search was conducted using the PubMed Central® search engine, the most comprehensive dataset for biomedical literature. The search expression used for this search included the mesh terms "(pediatric obesity) OR (childhood obesity) AND (primary prevention) AND "diet". Due to the extensive amount of published data, we limited the timeline to have only articles from 2009 up to 2019. An update search was then performed to include studies published between 2019 and 2021. Only articles written in English, conducted in humans, and including children and/or adolescents (<18 years old) were included. This search yielded 538 articles, of which we excluded 513, including 25 in this study. The literature search had three stages, the search for the titles, then abstracts, and finally the full-text papers were searched and retrieved (when deemed of interest). Some articles were discarded because they did not report measures to prevent obesity in children (n = 192) or because these measures were implemented only in adults (n = 321).

Additional papers (n = 27) were included in this review from a snowball process or searched to put into context the dietary interventions for the prevention of childhood obesity, totaling 52 references. Figure 1 presents the flowchart of the studies' selection.

In this literature review, dietary interventions to prevent childhood obesity were grouped and described into four levels: school-based interventions, community-based interventions, interventions through mass media, and food sector interventions.

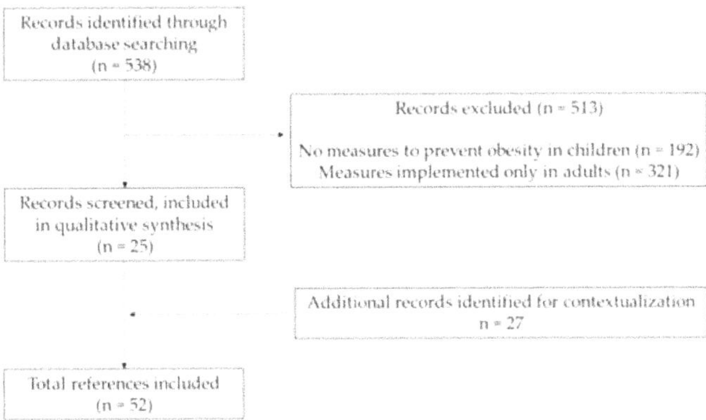

Figure 1. Flowchart of studies' selection.

3. Results

To prevent obesity in children there is a need to take multidimensional actions at different levels, including the individual, familial, institutional, and environmental levels. At the moment, these types of multilevel interventions seem to be the most promising ones to actually prevent/manage obesity. In particular, children are very influenced by social and environmental conditions, so at these ages, community-based interventions, changing the supportive environment, seem to play an especially important role [12].

Table 1 provides a descriptive summary of the dietary interventions to tackle childhood obesity, described in detail below.

3.1. School-Based Interventions

The Ballabeina study is a cluster-randomized controlled single-blinded trial that took place in some preschools in Switzerland, designed to study the effect of a multidimensional lifestyle intervention on aerobic fitness and adiposity, mainly in migrant preschoolers with the duration of over one school year [13]. This study included 652 preschool children with a mean age of 5.1 years. The interventions comprised a physical activity program, lessons on nutrition, media use, and sleep, and adaptation of the built environment of the preschool. The dietary intervention included weekly nutrition lessons given by a dietician; the students could learn about balanced nutrition and healthy nutritional behaviors in a didactic way. These lessons were centered on five messages: "drink water", "eat fruit and vegetables", "eat regularly", "make clever choices", and "turn your screen off when you eat", which were developed in collaboration with the Swiss Society for Nutrition. These messages were also described on funny cards that children could get with a task to implement the message at home. After 4 months of intervention, the results showed no differences between the groups in the children's body mass index (BMI). However, an increase in aerobic fitness by the end of the intervention was reported, and children in the intervention group also showed beneficial effects in the percentage of body fat (-1.1%), and their motor agility, when compared with the children in the control group. It was also possible to observe benefits in reported physical activity, media use (less screen time in boys), and eating habits, such as an increase in fruit and vegetable consumption in the intervention group [13].

In the Netherlands, a school-based trial was implemented including students from the ages of 12–14 years old (n = 1108), within a multidimensional health promotion intervention [14,15]. There were 10 intervention and eight control secondary schools included. The intervention included an educational component, with classes in biology and physical education, and a computer-based information program; and an environmental component, with propositions such as serving smaller portion sizes in the canteen and healthier food options, or restricting the access to vending machines. There were also posters affixed to create more awareness about which foods were healthier and which were not. With a twenty-month follow-up, it was observed in the intervention group a reduction in body composition measures, such as skinfold thickness, lower consumption of sugar-containing beverages at 12 months, and less screen time (but only in boys) [14,15].

A school-based obesity-prevention trial in Chile evaluated the effect of weekly physical activity classes and classes on healthy nutrition for parents and students from 1st to 8th grade; 2141 schools were in the intervention group and 945 in the control group [16]. Some environmental changes were also made, including instructing school kiosks to offer healthier options to students and still remain lucrative. The results showed a reduction in BMI z-scores in boys after 6 months of intervention and better physical fitness in both genders. On the other hand, the modifications in the kiosk's food availability did not seem to change the students' food choices [16].

The school-based Healthier Options for Public Schoolchildren (HOPS) is a randomized trial implemented over two school years (2004–2005 and 2005–2006) that included six elementary schools (4588 children aged 6 to 13 years; 48% Hispanic) in Osceola, Florida. Interventions implemented included modifications in the school menu, school gardens, and physical activity [17]. Complementarily, there were healthy nutrition and physical activity lessons for the students and parents through monthly newsletters. After 2 years, it was possible to observe a higher percentage of students who maintained a normal weight (under the 85th percentile of BMI-for-age) in the intervention group (52.1%) than in the control group (40.7%). Students in the intervention group had also improved academic performance compared to the control group [17].

The "Shape up Somerville" (SUS) is a non-randomized controlled trial conducted over two school years (September 2003–June 2005) in 1178 children in grades 1–3 (average of 8 years old) attending public school in three different communities from Somerville, Massachusetts, United States [18]. This intervention included more physical activity opportunities around the school, such as information on safe routes to school and walking to the school bus; modifications inside the school space, such as new equipment for physical activities; and a dietary intervention. This included taste tests of fruit and vegetables during lunchtime, where children could vote on whether they would like to see those fruits or vegetables on the monthly school menu; new vegetarian recipes and fresh fruit were made available every day for breakfast and lunch; colorful educational posters with nutrition and health information were displayed in the school cafeterias, and food service staff was trained. Additionally, there was an approval of restaurants according to SUS guidelines which offer low-fat dairy products, some dishes in smaller portion sizes, fruits and vegetables as side dishes, and have visible signs highlighting healthier options. After 1 year, results showed that the BMI z-scores were 0.06 lower in the intervention group than in the control group [18]. There was a decrease in overweight and obesity and an increase in remission in both sexes in the intervention group, but the comparison groups were not randomly assigned.

A randomized cluster controlled trial was performed in Mexico on 532 school-aged children from the 2nd and 3rd grades, with an average age of 8.5 ± 0.73 years at baseline (280 children in the intervention and 252 in the control group; each arm with one public and one private school, totaling four) [19]. It aimed to make these children and their parents reduce their sedentary behaviors, consumption of soft drinks and high-fat and salt-containing snack foods and increase their consumption of fruits and vegetables. The intervention consisted of sessions for discussing healthy lifestyles dedicated to the school board and teachers, conducted by nutritionists and physical activity professionals. There were also interactive lessons provided by nutrition graduated students for the children with the intent of increasing their fruit and vegetable intake, physical activity practice, and reducing their intake of soda and high-fat and salt-containing snacks, while simultaneously lowering their TV watching time. There were also nutrition sessions for parents run by nutrition professionals, with the intent of educating them about healthy eating. The results showed that by the sixth month of the intervention, there was a greater decrease in BMI in the intervention group than in the control group (difference of $-0.82\ \text{kg/m}^2$ in children BMI), although this was not sustained in the long-term, after 18 and 24 months [19].

A Multicomponent School Nutrition Policy Initiative on the prevention of overweight and obesity among children was conducted in 1349 students in grades four through six from 10 schools in a US city [20]. This initiative included the following interventions: school self-assessment, in which the schools suggested strategies such as limiting the use of food as reward/punishment, promoting active recess, and serving breakfasts in the classrooms to ensure the students eat a healthy meal; training in nutrition education for the school staff; nutrition education classes for the children; nutrition policies in the intervention schools, such as changing the foods that were sold and served according to the Dietary Guidelines for Americans to meet the nutritional standards; social marketing, such as giving raffle tickets to students who purchased or brought from home healthy snacks and beverages; and parent outreach through nutrition educators in home and school association meetings, report card nights, parent education meetings, and weekly nutrition workshops [20]. The results of this intervention were a 50% reduction in the incidence of overweight. There were significantly fewer children in the intervention schools (7.5%) than in the control schools (14.9%) who had become overweight after 2 years. However, there were no differences in the incidence or prevalence of obesity, nor in the remission of being overweight or obesity after 2 years of follow-up [20].

Donnelly et al. [21] conducted a 2-year trial in students from grades three to five in two school districts in rural Nebraska aiming to reduce obesity and improve physical and metabolic fitness. The intervention consisted of nutrition education, modified school lunches, and increased physical activity. The meals were planned with the kitchen staff according to the Lunchpower! Program. This program consists of energy, fat, and sodium reduced lunches, in agreement with the Healthy People 2000 objectives [22]. According to this, the fat content is restricted to 30% of the total energy intake, the sodium is limited to 1000 mg, the cholesterol to 100 mg, and the dietary fiber is increased to 8 to 10 g per day. There were also nutrition classes given by the teacher, after being trained. These classes included basic nutrition, nutrition for proper growth and development, the relationship between diet and health, healthy food choices, how to reduce fat in the diet, snack alternatives, and food safety. After two years of intervention, the control school showed significantly higher total energy (9%) and total fat (25%). The control school also showed considerably greater values for sodium and smaller for fiber. After the first year of intervention, there were no significant differences between the control and intervention schools in nutrition knowledge. However, after two years of intervention, the intervention school reduced by 45% their wrong answers about nutrition knowledge. Concerning physical activity, the control school practiced significantly more sports outside school compared to the intervention school. After 2 years of the intervention, neither the control nor intervention schools showed significant increases in aerobic capacity. Both schools showed no significant changes in the percentage of body fat, but a significant increase in BMI [21].

The DECIDE-Children Study [23] is a cluster-randomized controlled trial conducted in 1200 Chinese students from four primary schools (8–10 years old). The intervention consisted of health education activities for the parents; supervision and encouragement of the children as a way of increasing their physical activity practice outside of school; school policies to prevent obesity and health education activities for the children. There was also the development of an app called 'Eat Wisely, Move Happily' that aids in diffusing information, monitoring the children's behavior, managing their weight, and giving feedback for the teachers and parents. Since this study is ongoing, the results of this intervention are not yet available [23].

In 2020, a multicenter randomized controlled trial [24] was conducted in 4846 Chinese school children aged 7 to 13 years, in which the intervention consisted of the development of a nutrition handbook that was given to all students; nutrition and health courses to the students, parents, teachers, and health workers about the proportion of the meals, how to choose healthy foods, and how to reduce eating out, unhealthy fast food, sugar-sweetened beverages, and snacks; and displaying informative posters around the school. Courses on physical activity for the parents and physical activity classes for the students were also given. There were no significant improvements in the overall diversity of the food consumption in the intervention group; however, there were some improvements in the diversity of the foods consumed at breakfast and a decrease in the consumption of some unhealthy foods [24]. No effects on children's BMI were studied.

The Abriendo Caminos Program [25] was implemented in several schools in Illinois, California, Iowa, Texas, and Puerto Rico targeting families of parents and one child aged 6–18 years old (n = 500). This randomized control trial consisted of workshops, presentations and activities on nutrition education, family wellness, and physical activity. There are still no known results from this study.

Another randomized control trial called Healthy Start [26] was conducted in Denmark and targeted school children aged 2 to 6 years (n = 3722) and consisted of guiding families on how to improve their children's diet and physical activity practices, reduce stress, and improve sleep quantity and quality. Activities included cooking classes, games focused on exercise and motor skills development, and access to a website that provided recipe inspiration and ideas. The clinical effects of this intervention on children's growth and body composition measures were small [26].

The FIVALIN Project [27] is a quasi-experimental study conducted in 810 children aged 8–12 years and 600 parents in Barcelona. This study consisted of workshops on health education and sports educational sessions. Educational materials, mobile messages to remind parents to attend the workshops, with the date and hour, and videos were sent to families to reinforce the health behaviors encouraged during the workshops and sports educational sessions. This study is ongoing; therefore, there are still no known results.

The CHIRPY DRAGON Intervention [28] was a cluster-randomized controlled trial led in Chinese school children with a mean age of 6.15 years (n = 1641). This school- and family-based obesity prevention program consisted of workshops and family activities to promote physical activity and healthy eating behaviors, and school support to improve physical activity and healthy food provision. After 12 months of intervention, the BMI z- scores of children in the intervention group decreased, along with an increase in the consumption of fruit and vegetables, and a decrease in the consumption of sugar-sweetened beverages and unhealthy snacks. Screen time also decreased and physical activity increased in this group [28].

The Kids in Action [29] was a controlled trial conducted with children aged 9–12 years from four primary schools in Amsterdam. The study consisted of meetings with the children to develop interventions that targeted their physical activity and healthy eating habits. This intervention consists of environmental changes, organizational changes, or educational approaches, and depending on the type of intervention, the executors could be dieticians, sports coaches, or supermarkets in the community. There are no results from this study yet.

In 2018, an education-based intervention study called The ABC of Healthy Eating Project (including 464 students) was conducted in Poland [30]. This study included students aged 11–13 years. The intervention group received a diet and lifestyle-related educational program and both the intervention and the control group partook in school activities with the theme of nutrition and healthy lifestyles. There are still no known results.

3.2. Community-Based Interventions

The MOVE/me *Muevo* was a randomized community trial implemented in 30 recreation centers in San Diego County in a total of 541 families with children between the ages of 5 and 8 years to prevent and control childhood obesity [31]. This program consisted of activities at the recreation centers and participants' homes, as well as phone calls from health coaches and emailing tip sheets. The intervention families had "Family Health Coaches" who addressed the following nutrition behaviors: increase the consumption of fruit and vegetables through modifications in meal and snack purchases and preparation; decrease the consumption of sugar-sweetened beverages through changes in food purchases and setting limits; increase healthy food portions by modifying the food consumption behaviors; reduce eating out and when eating out, choosing healthier options; increase the availability and accessibility of healthy foods and beverages at home; reduce screen time and avoid eating in front of the television, and increase the number of meals eaten as a family. After 2 years, there were no significant differences between the control and intervention groups concerning BMI or waist circumference [31]. Some changes were observed in the dietary domain, namely a reduction in fat and sugary beverages, which means that it was easier for the participants to adopt healthier behaviors in this field, compared to the more complex and multidimensional attitudes of physical activity [31].

The "Romp & Chomp" is a community-based trial carried out in Australia in children aged 1–5 years old (n = 12,000) and their families [32]. There were changes regarding the provision of water in childcare centers, childcare policies regarding healthy eating and physical activity, and skills in physical activity and nutrition were taught to the childcare professionals. Amongst the nutrition interventions, there were the following: a collaboration with Dental Health Services Victoria, which provided some resources (lunch boxes and drink bottles, and some marketing material for the kindergarten children); training of the staff as a way to support nutrition messages and healthy eating choices for children aged 5 years; support from dental health professionals to the kindergartens, as a way to engage with parents on the topic of healthy eating and with the intent of providing support for the staff to implement health and nutrition policies; access to a dietitian and other allied health professionals through e-mails, phone calls, and site visits; production and distribution of promotional materials (balloons, stickers, posters, postcards). After 3 years of intervention, the 3.5 years old subsample showed considerably lower mean weight, BMI, and z-score BMI, and the 2 and 3.5 years old children showed a considerably lower prevalence of overweight and obesity when compared with baseline values. The intervention group also showed a considerably lower intake of packaged snacks and fruit juice [32].

The Aventuras Para Niños Study is a community-based intervention to promote healthy eating and physical activity and prevent excess weight gain in Latino children [33]. It was performed in thirteen elementary schools, with randomization to assign them to either a family-only intervention, a community-only, or a family+community intervention. In the family-only intervention, professionals would either call the families or make home visits as a way of discussing ways to pass through the difficulties of maintaining a healthy diet and being physically active, by showing them how to prepare healthy meals at home, as well as presenting them with the benefits of encouraging their children to eat healthily and practice physical activity. The community-only intervention included improving the schools' playgrounds, implementing salad bars, as well as community parks, and displaying water bottles in classrooms for the students. It also included the implementation of better physical education equipment and healthy menus for the children, all of this combined with spreading media messages through posters, news, and point-of-choice messages in grocery stores with healthy messages. The family+community intervention included all of the interventions above. The results showed no noteworthy main effects for the family or community interventions. Therefore, it is possible that not any real effects for the family or community interventions were observed in the BMI z-scores of the children compared with either of those circumstances alone. Despite the lack of significant effects on children's BMI z-scores, there were several obesity-related behaviors in these children that were changed by the family intervention, such as the increased consumption of fruits and vegetables [33].

The EPODE (Ensemble Prevenons l'Obesité Des Enfants/Together Let's Prevent Childhood Obesity) aims to reduce childhood obesity through a societal process that consists of childhood settings, local environments, and family norms becoming more supportive and making it easier for children to adopt healthy lifestyles by enjoying healthy eating, active play, and recreation [34]. This program was launched in 2004 in 10 French pilot communities, and targeted children aged 1–12 years, their families, and various local stakeholders who have the power to initiate micro-changes in these children and their families through local initiatives focusing on better and balanced eating habits and the regular practice of physical activity. Recently, there have been some other programs, inspired by the EPODE methodology, such as the Healthy Weight Communities in Scotland or the JOGG program in the Netherlands.

The Pacific Obesity Prevention in Communities (OPIC) Project was carried out in four countries, Australia, Fiji, New Zealand, and Tonga, over 30 months, between 2004 and 2009 [35]. This was a complex community-based intervention that included 18,000 secondary-school children (aged 12–18 years) from eight ethnic and cultural groups, 60 multi-professional research staff, 300 stakeholders and partner organizations, and 27 higher degree research students. The interventions varied across sites, but all sites included targeting reductions in the consumption of high-sugar content drinks and energy-dense snacks and increasing physical activity. The authors state that the project may have positive effects on diet and physical activity, but the effects on childhood obesity are not clearly described [36].

3.3. Interventions through Mass Media

Some interventions to tackle childhood obesity through mass media have been based on restrictions on food advertising to children. It has been shown that restricting the number of hours spent watching television (TV) can be an effective approach to reduce the prevalence of childhood obesity, and reducing the meals in front of a TV has been shown to be as important as increasing physical activity [37]. Energy-dense foods and drinks and fast-food companies often target children in their advertisements, since they are very easily influenced at young ages, namely through TV commercials. Thus, reducing the time spent in front of the TV might be a useful strategy to try to reduce the childhood obesity prevalence. Sweden has banned TV commercials/advertisements to children under 12 and TV advertising to children. Norway, Denmark, Austria, Ireland, Australia, and Greece have also imposed some restrictions on advertising to children [38], as well as Portugal [39].

3.4. Food Sector Interventions

Food taxation is a primordial prevention measure taken that is currently being applied in several countries, such as some parts of the USA and Canada [40], to reduce the intake of unhealthy foods and, in the long term, their health effects such as obesity. Some examples are high-volume foods with low nutritional value, such as soft drinks, confectionery, and snack foods. Portugal has also adopted the taxation of sugar-sweetened beverages as an intervention to reduce its high consumption in the country [41]. There was a decrease of 6.58 million liters of sugar-sweetened beverages sold per year, which translates into a decrease in consumption of 21% compared to the baseline consumption data from the National Dietary Survey [41]. The number of cases of obesity prevented by taxing sugar-sweetened beverages was studied, concluding that there was a higher impact on adolescents (0.012%), preventing 0.76 cases of obesity yearly [41].

According to Teng et al. studies suggest that the implementation of sugar-sweetened beverages taxes worldwide has proven effective in reducing sugar-sweetened beverages purchases and intake [42]. Evidence also shows that the taxation of sugar-sweetened beverages might be an effective tool to reduce the consumption of sugar-sweetened beverages and an important component to prevent obesity [42]. Roberts et al. suggest that a fiscal strategy could very likely reduce the purchase of high-sugar content products, even if in the short term [43].

Another measure currently being taken is the addition of logos or some type of labeling to alert the consumers to the healthier products, making it easier for them to choose healthy foods. Although it is not directly focused on childhood obesity, it may have indirect effects. Anastasiou et al. reported that food labeling may affect the consumer's dietary intake;

however, results are inconclusive [44]. It is uncertain if using health-related claims is beneficial or damaging. Nonetheless, other than health-related claims, negative effects derived from food labeling seem highly unlikely according to the evidence. Therefore, food labeling should continue to be promoted in policies and education programs [44].

An example of this intervention is the "Pick the Trick" Program, conducted in Australia and New Zealand, providing foods with symbols for the consumers that make it easier to identify the healthier choices [45]. In Europe, the WHO European Food and Nutrition Action Plan 2015–2020 identifies the introduction of interpretative, consumer-friendly labeling on the front of packages as a priority policy issue [46]. Although the majority of countries in the region (n = 15) have some form of front-of-pack labeling, fewer countries have interpretive systems which provide judgments about the relative healthfulness of foods. Among other future policies, there is the intention of the application of a single front-of-pack labeling system in all countries. A WHO report summarizes the existing evidence on the development processes and effectiveness of front-of-pack food labeling policies in the WHO European region [47].

The portion sizes have also been getting increasingly larger over the past four decades in most high-income countries [48,49]. Despite this increase in portion sizes, few countries report measures to reduce them. Most measures are focused on information to consumers rather than changes in the food and drink environment [50].

Table 1. Summary of dietary interventions on childhood obesity and their main characteristics and results.

Author (Study Title) (Reference)	Country, Year	Type of Intervention	Intervention Description	Target Audience	Results
School-Based Interventions					
Niederer I, et al. (Ballabeina study) [13]	Switzerland, 2009	Cluster randomized controlled single-blinded trial	Lessons on nutrition (balanced nutrition and healthy nutritional behaviors in a didactic way), physical activity program, media use, and sleep, and adaptation of the built environment of the preschool.	Preschool children (mean 5.1 years) (n = 652), the parents, and the teachers	No differences in children's BMI were found between groups. However, the intervention group had a reduction in body fat percentage, better motor agility, as well as benefits in reported physical activity, media use, and eating habits.
Singh AS, et al. [14,15]	Netherlands, 2006	School-based trial	Educational component (classes in biology and physical education, and a computer-based information program); and an environmental component (e.g., serving smaller portion sizes in the canteen and healthier options, restricting access to vending machines, and food awareness by posters).	Students from the ages of 12–14 years (n = 1108)	With a twelve-month follow-up, a reduction in the skinfold thickness of the intervention groups was found, as well as lower consumption of sugar-containing beverages, and less screen time (but only in boys).
Kain J, et al. [16]	Chile, 2004	School-based obesity-prevention trial	Weekly classes on physical activity and healthy nutrition for parents and students. Some environmental changes were also made (e.g., school kiosks were instructed to offer healthier choices and at the same time remain lucrative).	Parents and students from 1st to 8th grade; 2141 schools in the intervention group and 945 in the control group.	After 6 months, there was a reduction of body mass index (BMI) z-scores in boys and better physical fitness in both genders. On the other hand, the modifications in the kiosk's food availability did not seem to change the students' food choices.

Table 1. *Cont.*

Author (Study Title) (Reference)	Country, Year	Type of Intervention	Intervention Description	Target Audience	Results
Hollar D, et al. (Healthier Options for Public Schoolchildren (HOPS)) [17]	Florida, US, 2004–2006	Randomized trial	Modifications in the school menu, school gardens, and physical activity; monthly newsletters with healthy nutrition and physical activity lessons for the students and parents.	6 elementary schools (4588 children aged 6 to 13 years; 48% Hispanic)	After 2 years, a higher percentage of students who maintained a normal weight (<85th percentile of BMI-for-age) was found in the intervention group (52.1%) when comparing with the control group (40.7%). Students in the intervention group had improved academic performance.
Economos CD, et al. (Shape up Somerville) [18]	United States, (September 2003–June 2005)	Non-randomized controlled trial	Dietary intervention (e.g., promotion of fresh fruit and vegetables and taste tests, posters with nutritional and health information, training of food staff, modification of food offers in restaurants according to the study guidelines); increase of physical activity opportunities around the school (e.g., information on safe routes); modifications inside the school space (e.g., new equipment).	1178 children (average 7.92 years) attending public school in three different communities from Somerville, Massachusetts	After 1 year, the BMI z-scores were 0.06 lower in the intervention group than in the control group. There was a decrease in overweight and obesity and an increase in remission in both sexes in the intervention group. The study design did not include randomization of the intervention.
Bacardí-Gascon M, et al. [19]	Mexico, 2012	Randomized cluster controlled trial	Sessions discussing healthy lifestyles to the school board and the teachers; interactive lessons to the children to increase fruit and vegetables intake and physical activity practice, and reduce soda and high fat and salt-containing snacks intake, while simultaneously decreasing TV watching time; healthy eating sessions to parents.	532 school-aged children from 2nd and 3rd grade	By the sixth month, there was a greater decrease in BMI in the intervention group than in the control group (difference of -0.82 kg/m^2 in children BMI), although it was not sustained after 18 and 24 months of intervention.
Foster GD, et al. [20]	USA, 2008	Multicomponent School Nutrition Policy Initiative	School self-assessment (e.g., strategies like limiting the use of food as reward/punishment, promoting active recess, and serving breakfasts in classrooms); training of school staff and children in nutrition education; nutrition policies (e.g., changing sold foods); social marketing; school association meetings/workshops.	1349 students in grades 4 through 6 from 10 schools	There were significantly fewer children in the intervention schools (7.5%) than in the control schools (14.9%) who became overweight after 2 years, but no differences after 2 years of follow-up.
Donnelly JE, et al. [21]	Nebraska, USA, 1996	2-year trial	Nutrition education (basic nutrition, diet, and general health, nutrition for growth and development, healthy food choices, snack alternatives, food safety), modified school lunches (meals planned according to the Lunchpower! Program aiming to reduce energy, fat, and sodium lunches), and increased physical activity.	Students from grades 3 to 5 in two school districts in rural Nebraska (n = 338)	After 2 years of the intervention, both schools showed no significant changes in the body fat percentage, but a significant increase in the BMI. The control school showed significantly higher total energy, total fat and sodium intake, and lower fiber intake.

Table 1. *Cont.*

Author (Study Title) (Reference)	Country, Year	Type of Intervention	Intervention Description	Target Audience	Results
Liu Z, et al. (The DECIDE-Children study) [23]	China, 2019	Cluster-randomized controlled trial	Health education activities for parents and children; supervision and encouragement of children's physical activity practice outside of school; school policies to prevent obesity. Development of an app called 'Eat Wisely, Move Happily' that aids in diffusing information, monitoring the children's behavior, and managing their weight.	4-grade primary schools (8–10 years old) (n = 1200)	No known results.
Xu H, et al. [24]	China, 2020	Multicenter randomized controlled trial	Development of a nutrition handbook that was given to all students; nutrition and health courses to students, parents, teachers, and health workers (e.g., meals proportion, how to choose healthy foods, reduce eating out and unhealthy foods); informative posters around the school; course on physical activity for parents, and physical activity classes for students.	4846 school children aged 7–13 years	The effects on children's BMI were studied. There were some improvements in the diversity of the foods consumed at breakfast and a decrease in the consumption of some unhealthy foods.
Hannon BA, et al. (*Abriendo Caminos* Program) [25]	Illinois, California, Iowa, Texas, and Puerto Rico, 2019	Randomized control trial	Workshop presentations and activities on nutrition education, family wellness, and physical activity.	Families of parents and 1 child aged 6–18 years (n = 500)	No known results.
Olsen NJ, et al. (Healthy Start) [26]	Denmark, 2020	Randomized controlled trial	Guidance on how to improve the child's diet and physical activity, quantity and quality of sleep, and reduce their stress. Cooking classes, games focused on exercise and motor skills development, access to a website with recipes.	Children aged 2 to 6 years (n = 3722)	The clinical effects of this intervention in the children's growth and body composition were small.
Homs C, et al. (FIVALIN project) [27]	Barcelona, 2021	Quasi-experimental design	Workshops on health education and sports educational sessions.	810 children aged 8–12 years and 600 parents	No known results.
Li B, et al. (The CHIRPY DRAGON intervention) [28]	China, 2019	Cluster-randomized controlled trial	Workshops and family activities to promote physical activity and healthy eating behaviors; school support to improve physical activity and healthy food provision.	School children with a mean age of 6.15 years (n = 1641)	There was a decrease in the BMI z-scores of the children in the intervention group, along with an increase in the consumption of fruit and vegetables, and a decrease in the consumption of sugar-sweetened beverages and unhealthy snacks. There was also a decrease in screen time and an increase in physical activity in this group.
Anselma M, et al. (Kids in Action) [29]	Amsterdam, 2019	Controlled trial	Meetings with children to develop interventions that targeted their physical activity and healthy eating habits. These interventions consist of environmental changes, organizational changes, or educational approaches.	Children aged 9–12 years from four primary schools	No known results.

Table 1. *Cont.*

Author (Study Title) (Reference)	Country, Year	Type of Intervention	Intervention Description	Target Audience	Results
Hamulka J, et al. (The ABC of Healthy Eating Project) [30]	Poland, 2018	Education-based intervention study	Diet and lifestyle-related programs for the intervention group and school activities with the theme of nutrition and healthy lifestyles for both the intervention and the control group.	School children aged 11–13 years. (464 students)	No known results.
Community-based interventions					
Elder JP, et al. (MOVE/me Muevo) [31]	San Diego County, USA, 2014	Randomized community trial	Activities and phone calls from health coaches on how to increase the consumption of fruit and vegetables; decrease the consumption of sugar-sweetened beverages; increase healthy food portions; reduce eating out and do healthier options when eating out; increase the availability and accessibility of healthy foods and beverages at home; reduce the screen time and avoid eating in front of the TV, and increase the number of family meals.	541 families with children between the ages of 5 and 8 years old	After 2 years, there were no significant differences between the control and the intervention group concerning BMI or waist circumference. Some changes were observed in dietary intake, namely a reduction in fat and sugary beverages in the intervention group.
De Silva-Sanigorski A, et al. (Romp & Chomp) [32]	Australia, 2020	Community-based trial	Changes in the provision of water in childcare centers, childcare policies regarding healthy eating and physical activity; teaching of skills in physical activity and nutrition to the childcare professionals; production and distribution of promotional materials (balloons, stickers, posters, postcards).	Children aged 1–5 y (n = 12,000) and their families	After 3 years of intervention, the 3.5 years old subsample showed considerably lower mean weight, BMI, and z-score BMI, and the 2 and 3.5 years old children showed a considerably lower prevalence of overweight and obesity when compared with the baseline values. The intervention group also showed a considerably lower intake of packaged snacks and fruit juice.
Crespo NC, et al. (The Aventuras Para Niños Study) [33]	Southern California, 2003	Randomized Community-based trial	Three arms: family-only, community-only, or family+community intervention. In the family-only intervention, professionals call/make home visits to discuss how to maintain a healthy diet, prepare meals, and be physically active. The community-only intervention included improving the school's playgrounds, implementing salad bars, as well as community parks, displaying water bottles in the classrooms for the students, better physical education equipment and healthy menus for the children, all of this combined with spreading media messages through posters, news and point-of-choice messages in grocery stores, with health messages. The family+community included all described.	811 predominantly Mexican immigrant/Mexican-American mothers with children in kindergarten through second grade	No noteworthy main effects nor interactions for the family or community interventions were found, including on BMI z-scores. Despite the lack of significant effects on the children's BMI z-score, there were multiple obesity-related behaviors in these children that were changed by the family intervention, like increased consumption of fruit and vegetables.

Table 1. *Cont.*

Author (Study Title) (Reference)	Country, Year	Type of Intervention	Intervention Description	Target Audience	Results
Borys JM, et al. (EPODE (Ensemble Prevenons l'Obesité Des Enfants/Together Let's Prevent Childhood Obesity) [34]	France, 2004	Community-based intervention	Changes in local environments, childhood settings, and family norms to make them more supportive and aid the adoption of healthy lifestyles in children.	Children aged 1–12 years, and their families, as well as a wide variety of local stakeholders in 10 French pilot communities	No known results.
Swinburn BA, et al. and Schultz JT, et al. (Pacific Obesity Prevention in Communities (OPIC) project) [35,36]	Australia, Fiji, New Zealand, and Tonga, over 30 months, between 2004 and 2009	Community-based intervention	Interventions aiming to reduce the consumption of high sugar content drinks and energy-dense snacks and increase physical activity.	18 000 children 12–18 years, 300 stakeholders, 60 multi-professional research staff, 27 research students.	The authors state that the project can produce positive effects in diet and physical activity, but effects on childhood obesity are not clearly described.
Interventions through mass media					
World Health Organization and Assembly of the Republic (TV ban/restrictions of food commercials to kids in several countries [38] and Portugal) [39]	Sweden, Norway, Denmark, Austria, Ireland, Australia, Greece, and Portugal, 2019	Mass-media based-intervention	Sweden has banned TV food commercials for children under the age of 12 and TV food advertising for children. Norway, Denmark, Austria, Ireland, Australia, and Greece have also made some restrictions on commercials for children. Portugal approved a law to restrict advertising to children for foodstuffs and beverages of high energy value, salt, sugar, and saturated fatty acids content.	Children	No efficacy results. However, energy-dense foods and drinks and fast-food companies often target children in their advertisements, since they are very easily influenced at this age, namely through TV commercials.
Interventions through the Food Sector					
Goiana-da-Silva F, et al. (Taxation of sugar-sweetened beverages) [11]	Portugal, 2017	Food sector intervention	Taxation of sugar-sweetened beverages as an intervention to reduce its high consumption in the country.	Community	Decrease of 6.58 million liters per year, which translates into a decrease in consumption of 21% compared to the baseline consumption data of IAN-AF 2015–2016. The number of cases of obesity prevented had a higher impact in adolescents (0.012%), preventing 0.76 cases of obesity yearly, followed by an impact of 0.062% in adults aged 18 to <65 years, and the children showed an impact of 0.049%. These data show that Portugal achieved its goal, decreasing sales of sugar-sweetened beverages.
Young L, et al. ("Pick the Trick" program) [45]	Australia and New Zealand	Food sector intervention	Providing foods with symbols for the consumers making it easier to identify the healthier choices.	Community	No known results
Kelly B, et al. and Nielsen S, et al. (WHO front-of-pack labeling system) [47]	WHO-E Food and Nutrition Action Plan 2015–2020	Food sector intervention	Among other future policies, there is the intention of application of a single front-of-pack labeling system in all countries.	Community	No known results

4. Discussion

This study aimed to review the most recent literature on dietary interventions for the prevention of childhood obesity. It describes different levels of interventions: the school level, the community level, the mass media, and the food sector level, and provides an overview of their effectiveness (the ability to show consistent results overtime on decreasing children's BMI), which stand out from previous reviews.

Given the complexity and multifactorial nature of obesity, it is consensual that there is a need to take actions at multidimensional levels, including individual, familial, institutional, and environmental. The majority of the studies included in this review aiming to reduce/manage childhood obesity were school-based interventions, with some addressing the whole community, and some including distal interventions through the food sector and mass media, which may have an indirect effect on childhood obesity by changing food behaviors.

Children are highly influenced by social and environmental conditions, so at these ages, the modification of the environment is expected to play an important role. However, most of the programs/interventions conducted focus mainly on person-based educational approaches, such as nutrition/diet education sessions combined with the promotion of physical activity and lifestyles to students, parents, and school staff, and less on environmental changes that facilitate healthier behavioral choices. Only some trials [13,14,16–18,20,21,30,31] have focused on capacity building and macro-policy changes, such as the adaptation of the built environment of the school, serving smaller portion sizes and increasing the availability and accessibility of healthy foods and water in schools, and restricting access to vending machines, for example.

Multidimensional intervention studies are usually difficult to evaluate and highly depend on the complexity of evaluation designs (e.g., only outcome evaluation vs. complex evaluation including process, impact, and outcome). Moreover, especially in the multidimensional community-based programs, it is hard to distinguish which part of the intervention was the most effective.

Overall, most of the intervention studies did not show consistent effects on changing children's BMI. A large number of studies, mainly based on school interventions, did not show very effective results, which may be a reflection of the difficulties experienced trying to obtain significant results when relying only on school-based interventions. In fact, the small weight reductions described in most studies could be clinically irrelevant. It is difficult to figure why the interventions taken until now to prevent/reduce childhood obesity have failed to provide substantial results in terms of effectiveness. The ineffectiveness of some interventions may be due to insubstantial evaluation, or because studies were too short to detect appropriate outcomes, or, simply, because they do not work [51]. Another possible explanation is the lack of interventions at multiple levels of determinants, especially environmental changes (distal level). Importantly, little is known about the sustainability of interventions over time [51]. However, other positive results, such as the change of dietary behaviors or physical activity performance have been described and should not be discarded.

Actions to prevent childhood obesity need to be taken in multiple settings and incorporate a variety of approaches and involve a wide range of stakeholders [51]. Complex interventions focused on environmental changes and the strengthening of individuals and communities as well as macro-policy changes seem to be promising strategies to reduce childhood obesity without increasing socioeconomic inequalities [52]. The best approach should include the family context and contemplate early life determinants. An approach that could be much more effective to prevent obesity is a combination of interventions that promote healthier diets and increase physical activities through society, rather than an approach focused solely on school environments [52]. Focusing on mass media campaigns and political actions to prevent obesity by influencing people's eating choices and the increase of physical activities might be an effective approach to this problem [52].

Overall, sustained interventions are likely to be required at several levels, at an individual level in schools and community settings to effect behavioral change, and in sector changes involving different stakeholders [51].

5. Conclusions

Most dietary interventions to tackle childhood obesity focus mainly on person-based educational approaches and less on environmental changes to offer healthier behavioral choices. Most of them failed to reduce childhood obesity.

The creation of environments supportive of healthier behaviors seems to be the best approach to mitigate the challenge that is childhood obesity. Complex and multilevel interventions focused on environmental changes and the strengthening of individuals and communities, including family, as well as macro-policy changes will have the potential to tackle childhood obesity without increasing socioeconomic inequalities.

Author Contributions: A.R.P. drafted the manuscript, played an important role in interpreting the result, and approved its final version. A.O. conceived the work that led to the submission, played an important role in interpreting the results, revised the manuscript, and approved its final version. All authors have read and agreed to the published version of the manuscript.

Funding: The authors acknowledge FEDER from the Operational Programme Factors of Competitiveness—COMPETE and national funding from the Foundation for Science and Technology—FCT (Portuguese Ministry of Education and Science) under the projects "Appetite regulation and obesity in childhood: a comprehensive approach towards understanding genetic and behavioral influences" (POCI-01-0145-FEDER-030334; PTDC/SAUEPI/30334/2017) and "Appetite and adiposity—evidence for gene-environment interplay in children" (IF/01350/2015), and the Investigator Contract (IF/01350/2015—Andreia Oliveira).

Institutional Review Board Statement: Not applicable.

Informed Consent Statement: Not applicable.

Data Availability Statement: Not applicable.

Conflicts of Interest: The authors declare no conflict of interest.

References

1. Popkin, B.M. Impact on Body Composition. *Proc. Nutr. Soc.* **2011**, *70*, 82–91. [CrossRef]
2. Ng, M.; Fleming, T.; Robinson, M.; Thomson, B.; Graetz, N.; Margono, C.; Mullany, E.C.; Biryukov, S.; Abbafati, C.; Abera, S.F.; et al. Global, regional, and national prevalence of overweight and obesity in children and adults during 1980–2013: A systematic analysis for the Global Burden of Disease Study 2013. *Lancet* **2014**, *384*, 766–781. [CrossRef]
3. Null, N. Trends in adult body-mass index in 200 countries from 1975 to 2014. A pooled analysis of 1698 population-based measurement studies with 19·2 million participants. *Lancet* **2016**, *387*, 1377–1396.
4. Wabitsch, M.; Moss, A.; Kromeyer-Hauschild, K. Unexpected plateauing of childhood obesity rates in developed countries. *BMC Med.* **2014**, *12*, 1–5. [CrossRef] [PubMed]
5. Wardle, J.; Carnell, S.; Haworth, C.M.A.; Plomin, R. Evidence for a strong genetic influence on childhood adiposity despite the force of the obesogenic environment. *Am. J. Clin. Nutr.* **2008**, *87*, 398–404. [CrossRef] [PubMed]
6. Llewellyn, C.; Wardle, J. Behavioral susceptibility to obesity: Gene-environment interplay in the development of weight. *Physiol. Behav.* **2015**, *152*, 494–501. [CrossRef]
7. Davison, K.K.; Birch, L.L. Childhood overweight: A contextual model and recommendations for future research. *Obes. Rev.* **2001**, *2*, 159–171. [CrossRef]
8. Thompson, S.R.; Watson, M.C.; Tilford, S. The Ottawa Charter 30 years on: Still an important standard for health promotion. *Int. J. Health Promot. Educ.* **2018**, *56*, 73–84. [CrossRef]
9. Issanchou, S. Determining Factors and Critical Periods in the Formation of Eating Habits: Results from the Habeat Project. *Ann. Nutr. Metab.* **2017**, *70*, 251–256. [CrossRef]
10. Plachta-Danielzik, S.; Kehden, B.; Landsberg, B.; Rosario, A.S.; Kurth, B.-M.; Arnold, C.; Graf, C.; Hense, S.; Ahrens, W.; Müller, M.J. Attributable Risks for Childhood Overweight: Evidence for Limited Effectiveness of Prevention. *Pediatrics* **2012**, *130*, e865–e871. [CrossRef]
11. Kopelman, P.; Jebb, S.A.; Butland, B. Executive summary: Foresight 'tackling obesities: Future choices' project. *Obes. Rev.* **2007**, *8*. [CrossRef]

12. Bleich, S.N.; Segal, J.; Wu, Y.; Wilson, R.; Wang, Y. Systematic review of community-based childhood obesity prevention studies. *Pediatrics* **2013**, *132*. [CrossRef] [PubMed]

13. Niederer, I.; Kriemler, S.; Zahner, L.; Bürgi, F.; Ebenegger, V.; Hartmann, T.; Meyer, U.; Schindler, C.; Nydegger, A.; Marques-Vidal, P.; et al. Influence of a lifestyle intervention in preschool children on physiological and psychological parameters (Ballabeina): Study design of a cluster randomized controlled trial. *BMC Public Health* **2009**, *9*, 94. [CrossRef] [PubMed]

14. Singh, A.S.; Chin, A.; Paw, M.J.M.; Kremers, S.P.J.; Visscher, T.L.S.; Brug, J.; Van Mechelen, W. Design of the Dutch Obesity Intervention in Teenagers (NRG-DOiT): Systematic development, implementation and evaluation of a school-based intervention aimed at the prevention of excessive weight gain in adolescents. *BMC Public Health* **2006**, *6*, 1–15. [CrossRef] [PubMed]

15. Singh, A.S.; Chin, A.; Paw, M.J.M.; Brug, J.; van Mechelen, W. Dutch obesity intervention in teenagers: Effectiveness of a school-based program on body composition and behavior. *Arch. Pediatr. Adolesc. Med.* **2009**, *163*, 309. [CrossRef] [PubMed]

16. Kain, J.; Uauy, R.; Vio, A.F.; Cerda, R.; Leyton, B. School-based obesity prevention in Chilean primary school children: Methodology and evaluation of a controlled study. *Int. J. Obes.* **2004**, *28*, 483–493. [CrossRef]

17. Hollar, D.; Messiah, S.E.; Lopez-Mitnik, G.; Hollar, T.L.; Almon, M.; Agatston, A.S. Effect of a two-year obesity prevention intervention on percentile changes in body mass index and academic performance in low-income elementary school children. *Am. J. Public Health* **2010**, *100*, 646–653. [CrossRef] [PubMed]

18. Economos, C.D.; Hyatt, R.R.; Must, A.; Goldberg, J.P.; Kuder, J.; Naumova, E.; Collins, J.J.; Nelson, M.E. Shape Up Somerville two-year results: A community-based environmental change intervention sustains weight reduction in children. *Prev. Med.* **2013**, *57*, 322–327. [CrossRef]

19. Bacardí-Gascon, M.; Pérez-Morales, M.E.; Jiménez-Cruz, A. A six month randomized school intervention and an 18-month follow-up intervention to prevent childhood obesity in Mexican elementary schools. *Nutr. Hosp.* **2012**, *27*, 755–762. [PubMed]

20. Foster, G.D.; Sherman, S.; Borradaile, K.E.; Grundy, K.M.; Vander Veur, S.S.; Nachmani, J.; Karpyn, A.; Kumanyika, S.; Shults, J.; Healy, C.M.; et al. A Policy-Based School Intervention to Prevent Overweight and Obesity. *Pediatrics* **2008**, *121*, e794–e802. [CrossRef]

21. Donnelly, J.E.; Jacobsen, D.J.; Whatley, J.E.; Hill, J.O.; Swift, L.L.; Cherrington, A.; Polk, B.; Tran, Z.V.; Reed, G. Nutrition and Physical Activity Program to Attenuate Obesity and Promote Physical and Metabolic Fitness in Elementary School Children. *Obes. Res.* **1996**, *4*, 229–243. [CrossRef]

22. U.S. Department of Health and Human Services. *Healthy People 2000: National Health Promotion and Disease Prevention Objectives*; DHHS Publication No. (PHS) 91-50212; U.S. Government Printing Office, Public Health Service: Washington, DC, USA, 1990; pp. 93–134.

23. Liu, Z.; Wu, Y.; Niu, W.-Y.; Feng, X.; Lin, Y.; Gao, A.; Zhang, F.; Fang, H.; Gao, P.; Li, H.-J.; et al. A school-based, multi-faceted health promotion programme to prevent obesity among children: Protocol of a cluster-randomised controlled trial (the DECIDE-Children study). *BMJ Open* **2019**, *9*, e027902. [CrossRef]

24. Xu, H.; Ecker, O.; Zhang, Q.; Du, S.; Liu, A.; Li, Y.; Hu, X.; Li, T.; Guo, H.; Li, Y.; et al. The effect of comprehensive intervention for childhood obesity on dietary diversity among younger children: Evidence from a school-based randomized controlled trial in China. *PLoS ONE* **2020**, *15*, e0235951. [CrossRef] [PubMed]

25. Hannon, B.; Teran-Garcia, M.; Nickols-Richardson, S.M.; Musaad, S.M.; Villegas, E.M.; Hammons, A.; Wiley, A.; Fiese, B.H. Implementation and Evaluation of the Abriendo Caminos Program: A Randomized Control Trial Intervention for Hispanic Children and Families. *J. Nutr. Educ. Behav.* **2019**, *51*, 1211–1219. [CrossRef]

26. Olsen, N.J.; Ängquist, L.; Frederiksen, P.; Lykke Mortensen, E.; Lilienthal Heitmann, B. Primary prevention of fat and weight gain among obesity susceptible healthy weight preschool children. Main results from the "Healthy Start" randomized controlled intervention. *Pediatr. Obes.* **2021**, *16*, 1–15. [CrossRef] [PubMed]

27. Homs, C.; Berruezo, P.; Según, G.; Estrada, L.; de Bont, J.; Riera-Romaní, J.; Carrillo-Álvarez, E.; Schröder, H.; Milà, R.; Gómez, S.F. Family-based intervention to prevent childhood obesity among school-age children of low socioeconomic status: Study protocol of the FIVALIN project. *BMC Pediatr.* **2021**, *21*, 1–14. [CrossRef] [PubMed]

28. Li, B.; Pallan, M.; Liu, W.J.; Hemming, K.; Frew, E.; Lin, R.; Martin, J.; Zanganeh, M.; Hurley, K.; Cheng, K.K.; et al. The CHIRPY DRAGON intervention in preventing obesity in Chinese primary-school–aged children: A cluster-randomised controlled trial. *PLoS Med.* **2019**, *16*, e1002971. [CrossRef]

29. Anselma, M.; Altenburg, T.; Chinapaw, M. Kids in Action: The protocol of a Youth Participatory Action Research project to promote physical activity and dietary behaviour. *BMJ Open* **2019**, *9*, 1–10. [CrossRef]

30. Hamulka, J.; Wadolowska, L.; Hoffmann, M.; Kowalkowska, J.; Gutkowska, K. Effect of an education program on nutrition knowledge, attitudes toward nutrition, diet quality, lifestyle, and body composition in polish teenagers. The ABC of healthy eating project: Design, protocol, and methodology. *Nutrients* **2018**, *10*, 1439. [CrossRef]

31. Elder, J.P.; Crespo, N.C.; Corder, K.; Ayala, G.X.; Slymen, D.J.; Lopez, N.V.; Moody, J.S.; McKenzie, T.L. Childhood obesity prevention and control in city recreation centres and family homes: The MOVE/me Muevo Project. *Pediatr. Obes.* **2014**, *9*, 218–231. [CrossRef]

32. De Silva-Sanigorski, A.; Elea, D.; Bell, C.; Kremer, P.; Carpenter, L.; Nichols, M.; Smith, M.; Sharp, S.; Boak, R.; Swinburn, B. Obesity prevention in the family day care setting: Impact of the Romp & Chomp intervention on opportunities for children's physical activity and healthy eating. *Child Care Health Dev.* **2011**, *37*, 385–393.

33. Crespo, N.C.; Elder, J.P.; Ayala, G.X.; Slymen, D.J.; Campbell, N.R.; Sallis, J.F.; McKenzie, T.L.; Baquero, B.; Arredondo, E.M. Results of a Multi-level Intervention to Prevent and Control Childhood Obesity among Latino Children: The Aventuras Para Niños Study. *Ann. Behav. Med.* **2012**, *43*, 84–100. [CrossRef] [PubMed]
34. Borys, J.-M.; Le Bodo, Y.; Jebb, S.A.; Seidell, J.; Summerbell, C.; Richard, D.; De Henauw, S.; Moreno, L.A.; Romon, M.; Visscher, T.L.S.; et al. EPODE approach for childhood obesity prevention: Methods, progress and international development. *Obes. Rev.* **2011**, *13*, 299–315. [CrossRef]
35. Swinburn, B.A.; Millar, L.; Utter, J.; Kremer, P.; Moodie, M.; Mavoa, H.; Snowdon, W.; McCabe, M.; Malakellis, M.; De Courten, M.; et al. The Pacific Obesity Prevention in Communities project: Project overview and methods. *Obes. Rev.* **2011**, *12* (Suppl. 2), 3–11. [CrossRef] [PubMed]
36. Schultz, J.T.; Moodie, M.; Mavoa, H.; Utter, J.; Snowdon, W.; McCabe, M.P.; Millar, L.; Kremer, P.; Swinburn, B.A. Experiences and challenges in implementing complex community-based research project: The Pacific Obesity Prevention in Communities project. *Obes. Rev.* **2011**, *12* (Suppl. 2), 12–19. [CrossRef]
37. Robinson, T.N. Reducing children's television viewing to prevent obesity: A randomized controlled trial. *JAMA* **1999**, *282*, 1561–1567. [CrossRef]
38. World Health Organization. Marketing of Foods High in Fat, Salt and Sugar to Children: Update 2012–2013. 2013. Available online: https://www.euro.who.int/__data/assets/pdf_file/0019/191125/e96859.pdf (accessed on 2 September 2021).
39. Assembley of the Republic. Law no 30/2019. Diário da República n.o 79/2019, Série I de 2019-04-23. Available online: https://dre.pt/application/conteudo/122151046 (accessed on 2 September 2021).
40. Jacobson, M.F.; Brownell, K.D. Small Taxes on Soft Drinks and Snack Foods to Promote Health | Center for Science in the Public Interest. *Am. J. Public Health* **2000**, *90*, 854–857.
41. Goiana-Da-Silva, F.; Severo, M.; Silva, D.C.E.; Gregório, M.J.; Allen, L.N.; Muc, M.; Nunes, A.M.; Torres, D.; Miraldo, M.; Ashrafian, H.; et al. Projected impact of the Portuguese sugar-sweetened beverage tax on obesity incidence across different age groups: A modelling study. *PLoS Med.* **2020**, *17*, e1003036. [CrossRef]
42. Teng, A.M.; Jones, A.C.; Mizdrak, A.; Signal, L.; Genç, M.; Wilson, N. Impact of sugar-sweetened beverage taxes on purchases and dietary intake: Systematic review and meta-analysis. *Obes. Rev.* **2019**, *20*, 1187–1204. [CrossRef]
43. Roberts, K.E.; Ells, L.J.; McGowan, V.J.; Machaira, T.; Targett, V.C.; Allen, R.E.; Tedstone, A.E. A rapid review examining purchasing changes resulting from fiscal measures targeted at high sugar foods and sugar-sweetened drinks. *Nutr. Diabetes* **2017**, *7*, 302. [CrossRef] [PubMed]
44. Anastasiou, K.; Miller, M.; Dickinson, K. The relationship between food label use and dietary intake in adults: A systematic review. *Appetite* **2019**, *138*, 280–291. [CrossRef] [PubMed]
45. Young, L.; Swinburn, B. Impact of the Pick the Tick food information programme on the salt content of food in New Zealand. *Health Promot. Int.* **2002**, *17*, 13–19. [CrossRef] [PubMed]
46. WHO. European Food and Nutrition Action Plan 2015–2020 (2014). Denmark: WHO Regional Office for Europe. 2014. Available online: https://www.euro.who.int/__data/assets/pdf_file/0003/294474/European-Food-Nutrition-Action-Plan-20152020-en.pdf (accessed on 2 September 2021).
47. Kelly, B.; Jewell, J. What Is the Evidence on the Policy Specifications, Development Processes and Effectiveness of Existing Front-of-Pack Food Labelling Policies in the WHO European Region. 2018. Available online: https://www.euro.who.int/__data/assets/pdf_file/0007/384460/Web-WHO-HEN-Report-61-on-FOPL.pdf (accessed on 2 September 2021).
48. Nielsen, S.; Popkin, B. Patterns and trends in food portion sizes, 1977–1998. *JAMA* **2003**, *289*, 450–453. [CrossRef] [PubMed]
49. Young, L.R.; Nestle, M. Reducing portion sizes to prevent obesity. *Am. J. Prev. Med.* **2012**, *43*, 565–568. [CrossRef]
50. World Health Organisation. Global Nutrition Policy Review 2016–2017: Country Progress in Creating Enabling Policy Environments for Promoting Healthy Diets and Nutrition. 2018. Available online: https://www.who.int/publications/i/item/9789241514873 (accessed on 2 September 2021).
51. World Health Organization. Population-Based Approaches to Childhood Obesity Prevention. 2012. Available online: https://apps.who.int/iris/bitstream/handle/10665/80149/9789241504782_eng.pdf;jsessionid=C20C82A9D40706BE6D43BDAB600599FF?sequence=1 (accessed on 2 September 2021).
52. Oliveira, A.; Durão, C.; Lopes, C. Social and health behaviour determinants of obesity. In *Recent Advances in Obesity: Understanding Obesity—From Its Origins to Impact on Life*; Monteiro, R., Martins, M., Eds.; Bentham Science Publishers: Sharjah, United Arab Emirates, 2020.

nutrients

Article

Promoting Health and Food Literacy through Nutrition Education at Schools: The Italian Experience with MaestraNatura Program

Beatrice Scazzocchio [1,*,†], Rosaria Varì [1,†], Antonio d'Amore [1], Flavia Chiarotti [2], Sara Del Papa [1], Annalisa Silenzi [1], Annamaria Gimigliano [3], Claudio Giovannini [1] and Roberta Masella [1,*]

[1] Center for Gender-Specific Medicine, Istituto Superiore di Sanità, Viale Regina Elena 299, 00161 Rome, Italy; rosaria.vari@iss.it (R.V.); antonio.damore@iss.it (A.d.); sara.delpapa8@gmail.com (S.D.P.); annalisa.silenzi@iss.it (A.S.); claudio.giovannini@iss.it (C.G.)

[2] Center for Behavioral Sciences and Mental Health, Istituto Superiore di Sanità, Viale Regina Elena 299, 00161 Rome, Italy; flavia.chiarotti@iss.it

[3] AMG Edutainment, Viale Egeo 14, 00144 Rome, Italy; annamaria.gimigliano@gmail.com

* Correspondence: beatrice.scazzocchio@iss.it (B.S.); roberta.masella@iss.it (R.M.)

† These authors contributed equally to this work.

Abstract: MaestraNatura is an innovative nutrition education program aimed at both enhancing awareness about the importance of a healthy food–lifestyle relationship and the ability to transfer the theoretical principles of nutrition guidelines to everyday life. The educational contents of the program resulted from the analysis of the answers to a questionnaire submitted to students aged 6–13 in order to assess their degree of knowledge about nutritional facts. Educational paths were specifically designed and implemented to address the main knowledge gaps identified through the analysis of the answers and were then tested for teachers' satisfaction in a sample of 56 schools in the north, centre, and south of Italy, involving 790 classes, 600 teachers, and 15,800 students. The results showed an approval rating from teachers from 90% to 94%. Said paths were designed for primary (6–10 years old) and first-level secondary (11–13 years old) school students. In addition, in a pilot study carried out in nine Educational Institutes located in an area close to Rome (Lazio region), a specific path was tested for effectiveness in increasing students' knowledge about fruit and vegetables by conducting questionnaires before (T0) and after (T1) the didactic activities. Results showed a significant increase in right answers at T1 with respect to T0 ($z = 2.142$, $p = 0.032$). Fisher's exact probability test showed an answer variability depending on the issue considered. In conclusion, this work could be considered as a first necessary step toward the definition of new educational program, aimed at increasing food literacy and favouring a healthier relationship with food, applicable in a widespread and effective manner, also outside of Italy.

Keywords: health literacy; education; food; students; healthy lifestyle

Citation: Scazzocchio, B.; Varì, R.; d'Amore, A.; Chiarotti, F.; Del Papa, S.; Silenzi, A.; Gimigliano, A.; Giovannini, C.; Masella, R. Promoting Health and Food Literacy through Nutrition Education at Schools: The Italian Experience with MaestraNatura Program. *Nutrients* **2021**, *13*, 1547. https://doi.org/10.3390/nu13051547

Academic Editor: Panagiota Mitrou

Received: 30 March 2021
Accepted: 2 May 2021
Published: 4 May 2021

Publisher's Note: MDPI stays neutral with regard to jurisdictional claims in published maps and institutional affiliations.

1. Introduction

Chronic non-communicable diseases (NCDs) are the leading cause of death worldwide [1,2]. The main NCDs risk factors are unhealthy lifestyles, mostly bad eating habits and physical inactivity [3,4]. For this reason, the increasing incidence of NCDs is a huge challenge for a sustainable health system that intends to make universal prevention its main tool for protecting people's health [5]. The need for preventive interventions and policies in the nutritional field concerns all age groups, because the negative effects associated with inadequate lifestyles affect all age groups worldwide. In Italy, although overweight and obesity in children have been reported to follow a decreasing trend from 2008 to 2016 [6], they still affect 30–40% of children under 18 years of age [7]. As any other behaviour, the development of the eating behaviour starts very early in life in response to a range of personal, social, economic, and environmental factors. Once acquired, eating behaviours are very hard to change [8–10].

Health literacy (HL) is defined as "the degree to which individuals can obtain, process, and understand the basic health information and services they need to make appropriate health decisions" [11]. HL is a determinant of health as it favours the adoption of correct lifestyles, the adherence to therapies, and the appropriate access to health services [12]. Furthermore, there is a growing interest about food literacy (FL) defined as a set of skills and knowledge related to food, which enables people to make informed choices about food and nutrition for improving their own health [13]. The big challenge, therefore, is to start very early with nutrition education programs to encourage the adoption of adequate lifestyles. School appears to be the most eligible setting to implement strategies aimed at improving students' diets and food choices that play a pivotal role in promoting health [14,15].

Studies carried out in the last years suggest that the most successful school-based nutrition education interventions must be intensive, long lasting and comprehensive, and take account of environmental changes at school as well as family involvement and support. Research also suggests such interventions should be theory-based and incorporated into regular school curricula and activities [16,17].

Despite the profusion and widespread dissemination of guidelines for a healthy nutrition both in the US and Europe, the prevalence rates of overweight and obesity have been increasing [18,19], which is evidence of the scarce influence that mere information can exert on the modification of behaviour patterns. In 2008, the European Parliament resolution on the "White Paper on nutrition, overweight and obesity related health issues" indicates a multilevel and comprehensive approach to be the best way to fight obesity among the EU population. It has been pointed out the need of European programs on research, health, education, and lifelong learning, as an important step in an overall strategy to address diet-related chronic diseases in Europe. The resolution also highlighted the importance of actions aimed at improving the HL of citizens and the need of a broader educational strategy, for example by means of lessons on diet and health in primary schools [20]. In line with this vision, since 2013, the EU has promoted a school program to favour the consumption of fruit in children by distributing fresh fruit to the primary schools accordingly with the EU 'School fruit and vegetable scheme'. The scheme also suggests the need of educational measures, including lessons, farm visits, school gardens, tasting and cooking [21]. However, much still needs to be done to reach the full potential of food and nutrition education [22]. In particular, a new paradigm is needed that goes beyond school class-based transmission of basic, generic information about food and nutrition to promote an active, hands-on learning and skill development to deal with food and nutrition in real life settings. To the best of our knowledge, this is the first study showing a nutrition education program that can be spread and easily adapted everywhere, and that allows to standardize the intervention going beyond the traditional frontal lesson which belongs to the category of passive learning together with reading, listening, and watching movies, characterized by low percentage of knowledge retention [23]. The purpose of this study was to design an effective, innovative nutrition education program, namely MaestraNatura (MN), aimed at increasing HL, and FL in particular, among primary and first-level secondary school students (6–13 years old). Final objective having students develop a balanced relationship with food together with the ability to transfer the theoretical principles contained in dietary guidelines to the actual context of a daily diet.

2. Methods

2.1. Ethical Aspects

The data were collected according to the parental written informed consent obtained for the participation of their children, in agreement with both ethical and legal (personal data protection) requirements of the Italian law. The study was explained to the participants before the start, by meeting with the teachers and providing leaflets to the parents through the schools.

The data collected were pseudonymised soon after the data quality control assigning a univocal numerical code to each subject in order to allow the connection of data collected on the same subject before and after the educational plan. The study was not a clinical trial nor did it gather any genetic data or biological samples. For these reasons, it was not compulsory to submit the study to the IRB review. The study was part of a larger educational initiative (innovative protocols for food education in primary and lower secondary schools) approved by the Ministry of Health, General Direction Food Safety and Nutrition, which supported the program and subsequently the study.

2.2. Study Design

The methodological approach was built upon the active participation of students in experimental activities at school, and the involvement of parents—or rather the family as a whole—in practical applications (cooking). The project activities were organized in three phases. The first phase of the study was devoted to exposing major misconceptions about food by administering a simple multiple-choice questionnaire to the students aged 8–13 years (Table 1, Panel A and B). To children aged 6–7 years, only Panel B of the questionnaire was administered and the questions were asked orally by the teachers. We also interviewed teachers and parents for criticism and expectations about a nutrition education program and used the collected information to set up the didactic plans, which were then submitted to the teachers for assessment. These steps were repeated a few times for content quality improvement by modifying inadequate parts and strengthening others. This phase was implemented in the Lazio region, and involved 25 schools, 200 classes, 230 teachers, and about 4000 students (aged 6–13). During the second phase, we tested the new education program for acceptance among teachers, and extended the activities to another five regions, including schools in northern (Piemonte, Veneto), central (Lazio) and southern (Campania, Basilicata) Italy. Seven towns of varying size (Torino, Verona, Padova, Roma, Benevento, Potenza, Avigliano) were involved, for a total of 56 schools, 790 classes, 600 teachers, and 15,800 students (aged 6–13) (Figure 1). The adherence and the level of satisfaction to the activities proposed by the program, were evaluated by interviewing the teachers. To evaluate the overall judgement on the project, a 0–5 scale was used; the question score > 2 was considered as positive. The developed contents of the didactic paths, as well as additional information, experiments, and recipes, were provided through a web platform (www.maestranatura.org, accessed on 3 May 2021) that teachers, students and parents could easily access. The web platform was structured to become an actual learning management system. The choice of using an information technology (IT) tool to disseminate contents allowed for cost reduction and real-time updating, not to mention the sharing of experience.

Finally, in order to test the effectiveness of the implemented didactic paths in increasing food knowledge, a pilot study was carried out with 1235 students (aged 8–10) attending 61 primary classes (3rd, 4th, and 5th classes) at 9 Educational Institutes (EI) located in a small town close to Rome, testing the educational path "It Is Easy To Say Vegetables". We chose these classes because the contents of the path were in line with their curricular programs. The path included two didactic power point presentations on "Seeds & Fruits" and "Food Chains", and four experimental and practical activities aimed at increasing knowledge and familiarity with vegetables (Seed's germination and dissemination; Draw a vegetable identity card; Create a vegetable garden from kitchen waste; Extract chlorophyll from leaves). In addition, some recipes on seasonable vegetables were provided in order to favour interactions between children and parents, and to induce the students to experiment new tastes. All the planned activities were carried out throughout the school year. To evaluate the effectiveness of the path in increasing the students' knowledge about vegetables and fruit, they were asked to fill out the panel B of the multiple-choice questionnaire (Table 1, Panel B) at the beginning and at the end of the school year.

Table 1. Questionnaire for the students.

Panel A: What do you know about food?
Question

1. What is the function of fats in our body?
2. What is the function of sugar in our body?
3. What is the function of water in our body?
4. What is the function of proteins in our body?
5. What is the function of vitamins in our body?
6. What is in milk?
7. What is in a steak?
8. Which of the following foods provide more energy?
9. Which of the following foods is a fruit?
10. Which of the following foods is a vegetable?
11. Which is the energy source for plants?

Panel B: What do you know about vegetables?
Question

Q01	Tomatoes are . . .
Q02	Fennel is . . .
Q03	Onions are . . .
Q04	Courgettes are . . .
Q05	Aubergines are . . .
Q06	Carrots are . . .
Q07	Stems develop into . . .
Q08	Roots allow the plant to . . .
Q09	Flowers are useful to . . .
Q10	What do seeds need to develop?

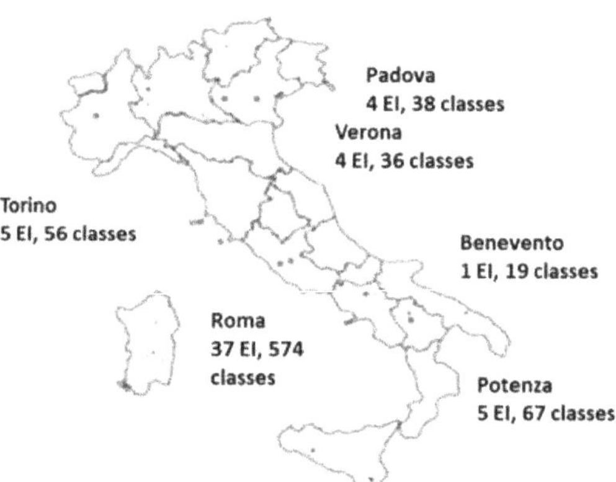

Figure 1. Geographic localization of the Educational Institutes (EI) and classes participating in the implementation of MaestraNatura program.

2.3. Statistical Analysis

Categorical data are presented as absolute and percent frequencies for any question of the questionnaire, both at the beginning (time 0, T0) and at the end (time 1, T1) of the study. Percentages of correct response across the questions of the questionnaire are summarized by median, minimum and maximum, separately at T0 and T1. For any child and for any question, the responses given at T1 have also been classified as changed (from wrong, W0 to correct C1 or, conversely, from correct C0 to wrong W1) or unchanged (correct or wrong at both T0 and T1). For any question and at any time, differences between sexes or classes in the percentage of correct response were assessed by the Fisher exact probability test, because the low expected frequencies in some of the contingency tables make the chi-square test not always applicable. The same test was used for each question to assess if the percentage of "improving" children (W0 to C1) was different from the percentage of "worsening" children" (C0 to W1).

Finally, the percentages of correct answer at T0 were compared to the percentages of correct answer at T1 across the questions of the questionnaire using the Wilcoxon matched-pairs signed ranks test.

For all tests, $p < 0.05$ was considered statistically significant. The significance levels were reported both with and without Bonferroni's correction, which was applied to take into account the increase in Type I error probability due to multiple tests, where appropriate. STATA 16.0 was used for the analyses.

3. Results

3.1. Preliminary Assessment of the Main Knowledge Gaps in Nutritional Facts

The results of the questionnaire (Table 1) highlighted relevant knowledge gaps on the origin and function of foods, metabolism, the role of water and macro/micronutrients, and the energy issues. For instance, 37% of the students (11–13 years old) correctly identified tomatoes as fruit, while only 24% and 16% of them identified pumpkin and courgette, respectively, as the fruit of the plant. Among the students of primary school classes (8–10 years old), 29% correctly identified courgettes but defined radish a fruit instead of a root. The analysis of the answers showed that, in general, the students seemed to have relied more on intuition than knowledge, giving their responses on the basis of colour and shape. For instance, the courgette, green and oblong, was very frequently classified as a stem. The situation was a little better with potatoes and carrots: 25% of the students aged 6–9 years and 40–50% of those aged 10–13 years correctly identified potatoes as tubers and carrots as roots. Regarding the origin of food, about 49% of students of the second and third classes (7–8 years old) and 40% of the students of the fourth and fifth classes (9–10 years old) believed that yoghurt does not derive from milk but rather from vegetables. More than half (60%) of the primary school students (8–10 years old) thought that milk does not contain water, protein being its main component over water and sugar. Finally, only 5% of the students (11–13 years old) correctly answered the questions on the energy, whereas most of them affirmed that steaks and vitamins (in pills) are foods that provide energy faster.

3.2. Definition of the Nutrition Education Contents

These findings helped us define what topics had to be discussed in detail and what basic concepts were worthy of consideration. In conclusion, at the end of the first two years of activity, we implemented a very effective, innovative nutrition education program aimed at increasing HL, and FL in particular, in primary and first level secondary school students (6–13 years old). The didactic paths specifically designed for each class of primary (6–10 years old) and first-level secondary schools (11–13 years old) are shown in Table 2. The nutrition education contents were defined according to WHO [24] and national guidelines [25] while those about sustainable diet were from FAO and United Nations [26,27]. In primary school classes, from the first to the fourth (6–9 years old), the following topics are addressed: the handling and processing of food, the discovery of water as an essential element for life, the identification of the different parts of the plant and their functions together with the knowledge about variety and seasonality of vegetables. In the fifth class, the differences between food and nutrients, and food groups are introduced in order to start learning how to combine food in a balanced daily menu. In the first class of first-level secondary school (11–12 years old), food waste, environmental footprint and sustainable diet issues are discussed in depth. Finally, in the second class of first-level secondary school (12–13 years old), the digestive process and the different organs of the human body that take part in it are discussed along with a further study of nutrients and their function in human metabolism.

Table 2. Didactic contents, experimental and practical activities, and objectives of the MN educational paths.

School Class Educational Path (hours)	Didactic Contents	Experimental Activities	Activities at Home	Objectives
I primary "The miracle of life" (8 h)	"The miracle of life"	Look at the bean germination Combine the card showing flower, fruit, leaves, seed, with the correct food Recognize the plant by touching or smelling it	Let's prepare: Carrot oil, Vanilla extract, Cinnamon and apple cake, Sweetness with rose petals, Mint syrup	Encouraging the manipulation and transformation of food; discovering new flavours and food
II primary "Microorganisms, friends or enemies?" (8 h)	"Organic or Inorganic?" "Food storage methods"	Let's breed yeast Let's turn must into wine Let's make vinegar and yogurt Let's observe milk curdling	Let's prepare: Bread with brewer's yeast Home-made sourdough Bread with sourdough	Encouraging the handling and transformation of food; refining the sensitivity towards "genuine flavours" in children and families; introducing elements of food hygiene.
III primary "Water's superpowers" (8 h)	"Water's superpowers" "Leftovers and food waste"	Which substances dissolve in water? Experiments on solubility Experiments on surface tension and capillarity How do clouds form? Experiments on the water cycle	How does water freeze? Let's prepare icicles Anti-waste recipes: reuse of leftovers Homemade cosmetics and detergents	Introducing children to the knowledge of water, an essential element for life; sensitizing children and families on food waste issues
IV primary "It's too easy to say vegetables" (10 h)	"Seeds and fruits" "Food chains"	Seeds germination and dissemination Draw a poster classifying vegetables based on: (a) edible parts; (b) seasonality; (c) family Vegetable slicing and creation of a vegetable garden from kitchen waste Extract chlorophyll from leaves	Compile the identity card of seeds, underground drums, roots, leaves, fruit from the garden Let's cook seasonal vegetables: educating taste	Recognizing the different parts of the plant and their function; learning which part of the most common vegetables we eat; discovering the variety of seasonal vegetables; developing awareness of the importance of consuming a varied diet, rich in fruit and vegetables to preserve health; reducing children's unwillingness to eat vegetables.

Table 2. *Cont.*

School Class Educational Path (hours)	Didactic Contents	Experimental Activities	Activities at Home	Objectives
V primary "Why do we have to eat?" (10 h)	"Why do we have to eat?" "Discover the egg" "Discover milk"	What's in the egg? Is this egg fresh? What's in milk? What food group does it belong to? Plan a daily menu	Let's go cooking: with and without eggs with and without milk	Understanding the importance of food in maintaining human well-being; Understanding the origin of foods and how technology affects their availability. Distinguishing between food and nutrients; learning how to classify foods in food groups. Learning how to combine foods to plan a balanced daily menu
I secondary "Mindful eating: don't de-vour the planet" (12 h)	"Food waste" "Food sustainability" "Food storage meth-ods" "Milk: from the sta-ble to the table"	Preparing milk products: Let's make yoghurt Let's make curd Let's make butter Contrasting food waste: how does mould form? Read food labels properly How should we store food? Put the food in the fridge	Let's go cooking: Brioche, sponge cake, breakfast cake, anti-waste recipes	Understanding what food waste is and what we can do to reduce it. Learning the meaning of environmental footprint and the importance of having a sustainable diet.
II secondary "We are what we eat" (14 h)	"The digestive process" "There is no perfect food" "Recognizing nutrients"	Simulating the digestion process Discover macronutrients providing energy What's inside? How many times a day, how many times a week? Plan a weekly menu	Let's go cooking: cream, crème caramel, meringues	Identifying human organs and their functions in digestive and metabolic processes; understanding the importance of healthy behaviours; identifying human organs and explaining their functions by means of models; drawing up a scientific report describing all the phases of an experiment; using tools and measurement units with confidence. Recognizing nutrients in food Learning the importance of a healthy, varied diet; learning how to combine foods in meals throughout the week to maintain the right variety of foods and the proper daily and weekly frequencies of consumption

On the whole, MN sets the student as main target, the school as the ideal gateway for the educational program, the teacher as a major player to promote and support the didactic activities, and parents as indispensable actors for transferring the basic concepts of healthy diet guidelines from theory to practice.

The MN program features didactic contents that taken together represent a complete teaching plan strictly connected to the different science programs specifically designed for each class. The contents are meant to be distributed gradually along the entire scholastic path, to allow for a progressive development of the scientific issues by adapting them to the age of children.

Each educational path has the following contents:

(a) Units for teachers that comprise texts and power point presentations on the themes of the course.

(b) Practical activities and experiments to be carried out in the classroom with (i) illustrations and explanations facilitating the achievement of the expected results; (ii) precautions to be taken; (iii) the list of the materials and the time required to carry out the experiment.

(c) Practical activities as homework with the involvement of household adults.

(d) Concept maps that sum up the basic concepts covered by the didactic paths in a simplified schematic view to provide the full picture of a complex process in a simple way.

(e) Questionnaires with a variable number of multiple-choice answers to evaluate the acquired knowledge.

After defining all the educational plans, we evaluated the completing rate for the proposed activities and the approval rating among the teachers involved in the study. The percentage of teachers' positive judgements on the proposed educational contents and practical activities was significantly high (over 90%) (Table 3). The same table shows the percent of fully carried out activities. As regard the overall judgement of the projects, the 100% of teachers expressed a positive judgement. The program also received some criticism mostly regarding the management of printed contents, the composition of some of the kits to carry out the experiments, and the interaction with parents. This survey allowed us to improve the program and reorganize the distribution of contents and questionnaires, which were then transferred to the e-learning platform.

Table 3. Percentage of completing activities and teachers' satisfaction for the proposed activities.

Activity	% of Completing Activities (Mean ± SD)	% of Teachers' Positive Judgement (Mean ± SD)
Experimental laboratory	71 ± 20	90 ± 4
Activities at home	86 ± 5	94 ± 4
Training	96 ± 3	92 ± 3
Overall judgement on the project		100

3.3. Pilot Study to Test One of the Educational Paths

3.3.1. Assessment of Students' Basal Knowledge on Fruit and Vegetables

A pilot study was carried out in a small town close to Rome to assess the increase in knowledge on fruit and vegetables after attending the educational path 'It's easy to say vegetables'. Students were given a multiple-choice questionnaire (Table 1, Panel B) at the beginning and at the end of the didactic activities. Out of the 1235 students enrolled in the study, 826 (67%) filled in the questionnaire at T0. Upon analysing the students' answers to the questions, it became evident that the level of basal knowledge varied considerably depending on the subject of the question. The percentage of correct answers varied from a minimum of 22% (WHAT DO SEEDS NEED TO DEVELOP?) to a maximum of 81% (ROOTS

ALLOW THE PLANT TO ...). There were no gender differences regarding the level of basal knowledge, other than for few topics, such as tomato and fennel, for which the percentage of boys providing the correct answers was significantly higher than for girls (data not shown). Finally, there were not differences in the level of knowledge with students' age, unless for some topics (fennel, flower, seeds) for which the percentage of correct answers significantly increased with age ($p < 0.001$, $p = 0,007$, $p = 0.055$, respectively) (Table 4).

Table 4. Assessment of students' basal knowledge on fruit and vegetables.

Question On	Total ($n = 826$)		3rd Class ($n = 314$)		4th Class ($n = 276$)		5th Class ($n = 236$)		p
	n	%	n	%	n	%	n	%	
tomato	391	47.3	161	51.3	122	44.2	108	45.8	0.199
fennel	214	25.9	52	16.6	88	31.9	74	31.4	<0.001 *
courgette	260	31.5	105	33.4	82	29.7	73	30.9	0.606
onion	513	62.1	198	63.1	160	58.0	155	65.7	0.183
aubergine	271	32.8	109	34.7	86	31.2	76	32.2	0.636
carrot	579	70.1	210	66.9	195	70.7	174	73.7	0.214
stem	407	49.3	163	51.9	151	54.7	93	39.4	0.001 *
root	673	81.5	263	83.8	221	81.1	189	80.1	0.414
flower	457	55.3	153	48.7	158	57.3	146	61.9	0.007
seed	180	21.8	54	17.2	61	22.1	65	27.5	0.055

Percentages of correct answers to panel B questionnaire collected at the beginning of the educational activities (T0). Data are shown in total and by school class. Significance level p refers to Fisher's exact probability test; * $p < 0.05$ when applying Bonferroni's correction.

3.3.2. Evaluation of the Knowledge Increase at the End of Didactical Activities

Out of the 826 filling in the questionnaire at T0, 246 students (30%) answered the questionnaire at T1. At the end of the activities the total knowledge increased, with a median increase of about 12% (range from -13.8% to 32.9% for flower and tomato, respectively). The Wilcoxon test applied to the answers to all the questions showed a significant overall improvement of right answers at T1 with respect to T0 ($z = 2.142$, $p = 0.032$). Specifically, for seven questions out of 10 there was an increase in the right answer ranging from $+7.7$ to $+32.9\%$, for two questions the percentage was quite similar at T0 and T1 (change equal to $+2.4$ and -2.4%), and for one out of 10 questions there was a decrease of -13.8% (Table 5).

Table 5. Evaluation of the knowledge increase at the end of didactical activities.

Question on	T0 ($n = 246$)		T1 ($n = 246$)		T1-T0	Wilcoxon p
	n	%	n	%	%	
tomato	116	47.2	197	80.1	32.9	
fennel	61	24.8	96	39.0	14.2	
courgette	75	30.5	118	48.0	17.5	
onion	142	57.7	136	55.3	−2.4	
aubergine	70	28.5	123	50.0	21.5	
carrot	159	64.6	204	82.9	18.3	
stem	133	54.1	159	64.6	10.6	
root	202	82.1	208	84.6	2.4	
flower	135	54.9	101	41.1	−13.8	
seed	51	20.7	70	28.5	7.7	
						0.032

Percentages of correct answers collected at the beginning (T0) and at the end (T1) of the educational activities. T1-T0 represents the percentage of knowledge increase for each question. Significance level p refers to Wilcoxon matched-pairs signed ranks test $p < 0.05$.

With respect to the assessment of the efficacy in increasing knowledge in the 246 children, Fisher's exact probability test showed that the percentage of "improving" children ((W0 to C1)/W0) was significantly higher than the percentage of "worsening" children ((C0 to W1)/C0) for questions 1, 6, 7, and 8 ($p < 0.001$); the two percentages did not differ for questions 3, 4, and 5, while the percentage of "improving" children was significantly lower than that of "worsening" children for questions 2 ($p < 0.020$), 9 ($p < 0.047$), and 10 ($p < 0.001$) (Figure 2).

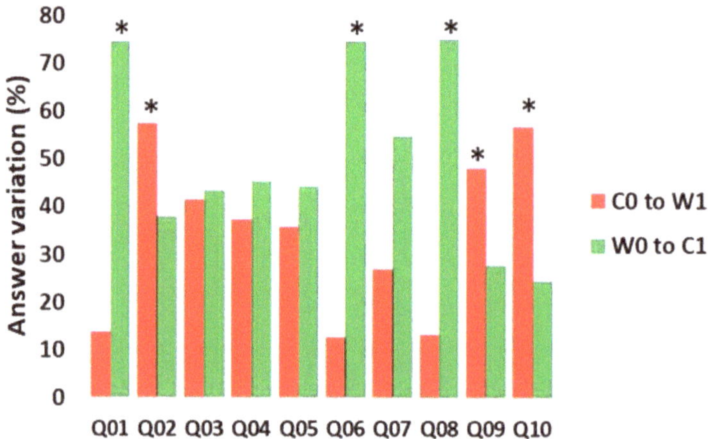

Figure 2. Answer variation to the questionnaires filled in at the end of the didactic path. Data represent the answer variation (%) of children that move from wrong to correct answer (W0 to C1) computed as [(W0 to C1)/W0] and from correct to wrong answer (C0 toW1) computed as [(C0 to W1)/C0]. C0, C1: correct answer at T0 or T1, respectively; W0, W1: wrong answer at T0 or T1, respectively. Q01–Q10: questions showed in Table 1, Panel B. Significance level p refers to Fisher's exact probability test; * $p < 0.05$ when applying Bonferroni's correction.

4. Discussion

MN stands as an innovative nutrition education program that uses food as a didactic tool to stimulate scientific thought and the students' awareness on the importance of healthy dietary habits. MN tries to connect different fields of knowledge, such as biology, physics, chemistry, history, ecology, environmental sciences, anthropology, and taste education, to eventually achieve adequate levels of HL. Actually, food can be considered from many different points of view: as a tool to observe natural and chemical–physical phenomena, as a main determinant of the planet survival because of its impact on the environment, as a product and expression of human culture, and as a vehicle for social relationships starting from very early life when it represents the strongest bond between a child and their parents.

MN program is a model of mixed-mode learning that appears to be especially suitable these days. In view of the information spread through old and new media and technologies, in fact, we need to balance information redundancy and the increasing speed of learning processes with the growth in learning difficulties and functional illiteracy that in Italy involves about 20% of the young population (16–24 years old) [28]. A large part of the adult population is not able to understand the complexity of reality, and similar difficulties affect children that might react with diverse cognitive disorders. The survey conducted among Italian students aged 6–13 years to define their knowledge about food and nutrition showed serious shortcomings. A certain degree of heterogeneity in the students' responses was found depending on the issue considered and the age of participant; taken as a whole, the results of the preliminary survey were rather discouraging. Nevertheless, we started from those results to build up a comprehensive, multifaceted educational program to provide an answer to those needs by using a completely different teaching approach. Firstly, MN proposes a systemic-constructivist approach that aims to facilitate the comprehension of complexity. To face a problem from a systemic point of view is to seek connections with similar problems in different areas and, only later, to explore the distinctive characteristics of the starting problem [29]. This approach simplifies the knowledge process without trivializing it and seeks to be effective in leading to a progressive, self-generating learning typical of the constructivist approach [30]. It also favours cooperative learning, a successful teaching strategy in which small groups of students with different levels of ability engage in a variety of learning activities to improve their understanding of a subject [31]. Finally, the MN method takes into consideration Dale's cone of learning that indicates the active learning techniques as the best base to support the acquiring of knowledge [23]. Considering purely the health aspect of nutrition education, the traditional approach embraced by most interventions is that it is enough to feed scientific information on food and nutrition to the population in order to induce a behavioural change and the adoption of a healthy lifestyle in any person. However, this approach has been questioned because of its poor effectiveness. For this reason, the American Dietetic Association stated that new nutrition education interventions must adopt methodologies able to produce effective changes in dietary habits and not to disseminate solely nutrition information [32]. Furthermore, an interesting paper reviewing the intervention programs carried out in the last years to counteract childhood obesity reports that the most effective ones were those conducted at school in the age range 6–12 and focused to modify one behaviour at a time. Didactic contents promoting healthy nutrition, physical activity, food processing, and safety, among others, appeared to be more effective when originally included in the scholastic plan and delivered in multiple sessions during the school year [33]. Finally, to create an environment facilitating behavioural changes (e.g., offering healthy food like fruit and nuts to the students) and to support parents in improving their relationship with their children also by sharing time together, e.g., cooking, are highlighted as relevant points to get good results [34].

The MN program takes into account these suggestions and introduces some novelties in the traditional educational approach. First of all, the educational path direction: it does not start from the food pyramid to guide behaviour towards healthy eating. On the contrary, the understanding of the food pyramid is the endpoint to be reached after gaining knowledge of nutritional facts in order to understand the guidelines for healthy eating. Focusing on health in general, MN avoids emphasis on concepts like "healthy body weight" or "good/bad" foods and promotes the whole person without neglecting children's psychological and emotional aspects [35], which should be reconciled in integrated intervention programs aimed at the prevention of obesity and eating disorders [36]. Another distinctive aspect is that knowledge is acquired through experience in the eight-year period of primary and secondary school (6–13 years old). Finally, an ambitious objective we set is to define the real effectiveness of educational contents in increasing FL. In this regard, the pilot study carried out to evaluate the efficacy of the path 'It Is Easy To Say Vegetables!' in improving the knowledge on fruit and vegetables, provided interesting and encouraging results. The significant increase in right answers obtained after the path execution with respect to those obtained before, indeed, support the educational effectiveness of the path tested. However, by analysing the answer variability, the students failed more the answers to the questions about the plant functionality than those about plant recognizing. The variability found in some answers, evidenced by Fisher's test, led us to rethink the wording of the questions and to improve the discussion of specific issues. Furthermore, we found a drop out of 70% in completing the questionnaire at T1. This could have been due to several factors. However, the most relevant aspect, in our opinion, was the delay in presenting the project to the teachers with respect to the school times. This determined either the withdrawal of a number of teachers that felt unable to complete the program, or the difficulty in administering the questionnaire at T1 because the end of the school year was approaching. The work herein presented represents the first step of an on-going activity to increase Food

Literacy, and lay the foundation for a healthier relationship with food and the daily diet. In particular, the progressive journey towards a better knowledge of the vegetable world might reduce children's reluctance to eat vegetables and favour awareness regarding the importance of consuming a balanced diet rich in fruits and vegetables. Such knowledge can provide children with a scientific basis to understand how to combine food in daily and weekly menus to maintain the right balance among nutrients and reinforce the concept that consuming a healthy and varied diet is the main tool to preserve human health. In conclusion, our study, although developed for Italian students, can be suitable for application in other countries, as it deals with public health issues and nutrition principles and recommendations widely accepted and shared. Moreover, the program is not limited to specific traditional or geographical contexts and can, thus, be implemented as an effective preventive action on public health.

Author Contributions: Conceptualization, B.S., R.V., and R.M.; data curation, A.d., F.C., S.D.P., A.S., A.G., and C.G.; formal analysis, F.C.; investigation, B.S., R.V., and R.M.; supervision, B.S., R.V., and R.M.; writing—original draft, B.S., R.V., and R.M.; writing—review & editing, A.d. and C.G. All authors have read and agreed to the published version of the manuscript.

Funding: This study was supported by the Italian Ministry of Health, General Direction Food Safety and Nutrition.

Institutional Review Board Statement: Not applicable.

Informed Consent Statement: Informed consent was obtained from all subjects involved in the study.

Data Availability Statement: The data presented in this study are available on request from the corresponding author.

Acknowledgments: The authors would like to thank Patrizia De Sanctis for her valuable help with data collection, Monica Brocco for English language editing and all the participating teachers and the Educational Institutes for their contribution.

Conflicts of Interest: The authors declare to have no conflict of interest.

References

1. Noncommunicable Diseases Country Profiles 2018. Available online: https://www.who.int/nmh/publications/ncd-profiles-2018/en/ (accessed on 3 May 2021).
2. WHO-Noncommunicable Diseases Progress Monitor 2020. Available online: https://www.who.int/publications/i/item/ncd-progress-monitor-2020) (accessed on 3 May 2021).
3. Vos, T.; Lim, S.S.; Abbafati, C.; Abbas, K.M.; Abbasi, M.; Abbasifard, M.; Abbasi-Kangevari, M.; Abbastabar, H.; Abd-Allah, F.; Abdelalim, A.; et al. Global burden of 369 diseases and injuries in 204 countries and territories, 1990–2019: A systematic analysis for the Global Burden of Disease Study 2019. *Lancet* **2020**, *396*, 1204–1222. [CrossRef]
4. Murray, C.J.L.; Abbafati, C.; Abbas, K.M.; Abbasi, M.; Abbasi-Kangevari, M.; Abd-Allah, F.; Abdollahi, M.; Abedi, P.; Abedi, A.; Abolhassani, H.; et al. Five insights from the Global Burden of Disease Study 2019. *Lancet* **2020**, *396*, 1135–1159. [CrossRef]
5. *WHO-Tackling NCDs: 'Best Buys' and Other Recommended Interventions for the Prevention and Control of Non-Communicable Diseases*; WHO: Geneva, Switzerland, 2017.
6. Lauria, L.; Spinelli, A.; Buoncristiano, M.; Nardone, P. Decline of childhood overweight and obesity in Italy from 2008 to 2016: Results from 5 rounds of the population-based surveillance system. *BMC Public Health* **2019**, *19*, 618. [CrossRef] [PubMed]
7. Obesity and the Economics of Prevention: Fit Not Fat—Italy Key Facts. Available online: https://www.oecd.org/els/health-systems/obesityandtheeconomicsofpreventionfitnotfat-italykeyfacts.html (accessed on 3 May 2021).
8. Siega-Riz, A.M.; Deming, D.M.; Reidy, K.C.; Fox, M.K.; Condon, E.; Briefel, R.R. Food Consumption Patterns of Infants and Toddlers: Where Are We Now? *J. Am. Diet. Assoc.* **2010**, *110*, S38–S51. [CrossRef] [PubMed]
9. Mennella, J.A.; Nicklaus, S.; Jagolino, A.L.; Yourshaw, L.M. Variety is the spice of life: Strategies for promoting fruit and vegetable acceptance during infancy. *Physiol. Behav.* **2008**, *94*, 29–38. [CrossRef] [PubMed]
10. Birch, L.L.; Doub, A.E. Learning to eat: Birth to age 2 y. *Am. J. Clin. Nutr.* **2014**, *99*, 723S–728S. [CrossRef] [PubMed]
11. Broucke, S.V.D. Health literacy: A critical concept for public health. *Arch. Public Health* **2014**, *72*, 1–10. [CrossRef]
12. Rowlands, G.; Trezona, A.; Russell, S.; Lopatina, M.; Pelikan, J.; Paasche-Orlow, M.; Drapkina, O.; Kontsevaya, A.; Sørensen, K. *What Is the Evidence on the Methods, Frameworks and Indicators Used to Evaluate Health Literacy Policies, Programmes and Interventions at the Regional, National and Organizational Levels?* WHO: Geneva, Switzerland, 2019.
13. Truman, E.; Lane, D.; Elliott, C. Defining food literacy: A scoping review. *Appetite* **2017**, *116*, 365–371. [CrossRef] [PubMed]

14. WHO-Life Skills Education School Handbook—Noncommunicable Diseases: Introduction. 2020. Available online: https://www.who.int/publications/i/item/97-8924-000484-9 (accessed on 3 May 2021).

15. WHO-Life Skills Education School Handbook—Noncommunicable Diseases: Approaches for School. 2020. Available online: https://www.who.int/publications/i/item/9789240005020 (accessed on 3 May 2021).

16. Waters, E.; de Silva-Sanigorski, A.; Hall, B.J.; Brown, T.; Campbell, K.J.; Gao, Y.; Armstrong, R.; Prosser, L.; Summerbell, C.D. Interventions for preventing obesity in children. *Cochrane Database Syst. Rev.* **2011**. [CrossRef] [PubMed]

17. Liu, Z.; Li, Q.; Maddison, R.; Ni Mhurchu, C.; Jiang, Y.; Wei, D.-M.; Cheng, L.; Cheng, Y.; Wang, D.; Wang, H.-J. A School-Based Comprehensive Intervention for Childhood Obesity in China: A Cluster Randomized Controlled Trial. *Child. Obes.* **2019**, *15*, 105–115. [CrossRef] [PubMed]

18. Hales, C.M.; Carroll, M.D.; Fryar, C.D.; Ogden, C.L. *Prevalence of Obesity among Adults and Youth: United States, 2015–2016*; NCHS Data Brief; DHHS Publication: Hyattsville, MD, USA, 2017; pp. 1–8.

19. Marques, A.; Peralta, M.; Naia, A.; Loureiro, N.; De Matos, M.G. Prevalence of adult overweight and obesity in 20 European countries, 2014. *Eur. J. Public Health* **2018**, *28*, 295–300. [CrossRef] [PubMed]

20. White Paper on Nutrition, Overweight and Obesity-Related Health Issues. Available online: https://eur-lex.europa.eu/LexUriServ/LexUriServ.do?uri=OJ:C:2010:008E:0097:0105:EN:PDF (accessed on 3 May 2021).

21. School Fruit, Vegetables and Milk Scheme. Available online: https://ec.europa.eu/info/food-farming-fisheries/key-policies/common-agricultural-policy/market-measures/school-fruit-vegetables-and-milk-scheme/school-scheme-explained_it (accessed on 3 May 2021).

22. Wolfenden, L.; Nathan, N.K.; Sutherland, R.; Yoong, S.L.; Hodder, R.K.; Wyse, R.J.; Delaney, T.; Grady, A.; Fielding, A.; Tzelepis, F.; et al. Strategies for enhancing the implementation of school-based policies or practices targeting risk factors for chronic disease. *Cochrane Database Syst. Rev.* **2017**. [CrossRef] [PubMed]

23. Dale, E. *Audiovisual Methods in Teaching*; Dryden Press: New York, NY, USA, 1969.

24. Food Based Dietary Guidelines in the WHO European Region. Available online: https://www.euro.who.int/__data/assets/pdf_file/0017/150083/E79832.pdf (accessed on 3 May 2021).

25. Linee Guida per una Sana Alimentazione. Available online: https://www.crea.gov.it/documents/59764/0/Dossier+LG+2017_CAP10.pdf/627ccb4d-4f80-cc82-bd3a-7156c27ddd4a?t=1575530729812) (accessed on 3 May 2021).

26. Sustainable Healthy Diets Guiding Principles. Available online: http://www.fao.org/3/ca6640en/ca6640en.pdf (accessed on 3 May 2021).

27. Sustainable Diets for Healthy People and a Healthy Planet. Available online: https://www.unscn.org/uploads/web/news/document/Climate-Nutrition-Paper-EN-WEB.pdf (accessed on 3 May 2021).

28. Di Francesco, G.; Amendola, M.; Mineo, S. I Low Skilled in Italia-Evidenze dall'indagine PIAAC Sulle Competenze Degli Adulti. Osservatorio Isfol n. 1-2/2016. Available online: http://isfoloa.isfol.it/bitstream/handle/123456789/1262/Oss_12_2016_DiFrancesco_Amendola_Mineo.pdf?sequence=1 (accessed on 3 May 2021).

29. Von Bertalanffy, L. *General System Theory: Foundations, Development, Applications*, Revised edition; George Braziller Inc.: New York, NY, USA, 1976.

30. Reusser, K. Co-constructivism in Educational Theory and Practice. In *Book: International Encyclopedia of the Social & Behavioral Sciences*; Elsevier B.V.: Amsterdam, The Netherlands, 2001; pp. 2058–2062. [CrossRef]

31. Le, H.; Janssen, J.; Wubbels, T. Collaborative learning practices: Teacher and student perceived obstacles to effective student collaboration. *Camb. J. Educ.* **2016**, *48*, 103–122. [CrossRef]

32. Schwartz, N.E. Communicating Nutrition and Dietetics Issues: Balancing diverse perspectives. *J. Am. Diet. Assoc.* **1996**, *96*, 1137–1139. [CrossRef]

33. Manios, Y.; Kafatos, A. Health and Nutrition Education in Elementary Schools: Changes in health knowledge, nutrient intakes and physical activity over a six year period. *Public Health Nutr.* **1999**, *2*, 445–448. [CrossRef]

34. Briggs, M.; Mueller, C.G.; Fleischhacker, S.; American Dietetic, A.; School Nutrition, A.; Society for Nutrition, E. Position of the American Dietetic Association, School Nutrition Association, and Society for Nutrition Education: Comprehensive School Nutrition Services. *J. Am. Diet. Assoc.* **2010**, *110*, 1738–1749. [CrossRef] [PubMed]

35. Spear, B.A. Does dieting increase the risk for obesity and eating disorders? *J. Am. Diet. Assoc.* **2006**, *106*, 523–525. [CrossRef] [PubMed]

36. O'Dea, J.A. Prevention of child obesity: 'First, do no harm'. *Health Educ. Res.* **2004**, *20*, 259–265. [CrossRef] [PubMed]

Article

Unfavorable Dietary Quality Contributes to Elevated Risk of Ischemic Stroke among Residents in Southwest China: Based on the Chinese Diet Balance Index 2016 (DBI-16)

Yingying Wang [1,†], Xu Su [2,†], Yun Chen [1], Yiying Wang [2], Jie Zhou [2], Tao Liu [2,*], Na Wang [1,*] and Chaowei Fu [1]

[1] Key Laboratory of Public Health Safety, NHC Key Laboratory of Health Technology Assessment, School of Public Health, Fudan University, Shanghai 200032, China; 17211020095@fudan.edu.cn (Y.W.); 18211020001@fudan.edu.cn (Y.C.); fcw@fudan.edu.cn (C.F.)

[2] Guizhou Province Center for Disease Prevention and Control, Guiyang 550004, China; susuxuxu@163.com (X.S.); wyy123789123789@163.com (Y.W.); zhoujie19872014@163.com (J.Z.)

* Correspondence: liutaombs@163.com (T.L.); na.wang@fudan.edu.cn (N.W.)

† These authors contributed equally to this work.

Abstract: Background: Little is known about the effects of dietary quality on the risk of ischemic stroke among Southwest Chinese, and evidence from prospective studies is needed. We aimed to evaluate the associations of ischemic stroke with dietary quality assessed by the Chinese Diet Balance Index 2016 (DBI-2016). Methods: The Guizhou Population Health Cohort Study (GPHCS) recruited 9280 residents aged 18 to 95 years from 12 areas in Guizhou Province, Southwest China. Baseline investigations, including information collections of diet and demographic characteristics, and anthropometric measurements were performed from 2010 to 2012. Dietary quality was assessed by using DBI-2016. The primary outcome was incident ischemic stroke diagnosed according to the International Classification of Diseases 10th revision (ICD-10) until December 2020. Data analyzed in the current study was from 7841 participants with complete information of diet assessments and ischemic stroke certification. Cox proportional hazards models were used to estimate the risk of ischemic stroke associated with dietary quality. Results: During a median follow-up of 6.63 years (range 1.11 to 9.53 years), 142 participants were diagnosed with ischemic stroke. Participants with ischemic stroke had a more excessive intake of cooking oils, alcoholic beverages, and salt, and had more inadequacy in meats than those without ischemic stroke. ($p < 0.05$). Compared with participants in the lowest quartile (Q1), those in the highest quartile (Q4) of the higher bound score (HBS) and of the dietary quality distance (DQD) had an elevated risk for ischemic stroke, with the corresponding hazard ratios (HRs) of 3.31 (95%CI: 1.57–6.97) and 2.26 (95%CI: 1.28–4.00), respectively, after adjustment for age, ethnic group, education level, marriage status, smoking and waist circumference, and the medical history of diabetes and hypertension at baseline. In addition, excessive intake levels (score 1–6) of cooking oils, excessive intake levels (score 1–6) of salt, and inadequate intake levels (score −12 to −7) of dietary variety were positively associated with an increased risk for ischemic stroke, with the multiple HRs of 3.00 (95%CI: 1.77–5.07), 2.03 (95%CI: 1.33–3.10) and 5.40 (95%CI: 1.70–17.20), respectively. Conclusions: Our results suggest that unfavorable dietary quality, including overall excessive consumption, excessive intake of cooking oils and salt, or under adequate dietary diversity, may increase the risk for ischemic stroke.

Keywords: dietary quality; ischemic stroke; Chinese Diet Balance Index 2016; DBI-16; Southeast China

Citation: Wang, Y.; Su, X.; Chen, Y.; Wang, Y.; Zhou, J.; Liu, T.; Wang, N.; Fu, C. Unfavorable Dietary Quality Contributes to Elevated Risk of Ischemic Stroke among Residents in Southwest China: Based on the Chinese Diet Balance Index 2016 (DBI-16). *Nutrients* **2022**, *14*, 694. https://doi.org/10.3390/nu14030694

Academic Editor: Panagiota Mitrou

Received: 31 December 2021
Accepted: 3 February 2022
Published: 7 February 2022

Publisher's Note: MDPI stays neutral with regard to jurisdictional claims in published maps and institutional affiliations.

1. Introduction

Stroke is the second leading cause of death and disability worldwide, and has been the first leading cause of death and the leading cause of all-age disability-adjusted life years (DALYs) in China [1,2]. Notably, nationwide studies and periodic governmental reports revealed a great burden of stroke in China, with an increasing prevalence and incidence in the past decade [3,4]. According to the latest annual report in 2019, China currently has 21 million patients with stroke [5]. Ischemic stroke, defined as the permanent infarction of cerebral tissues due to abrupt decreases in cerebral blood flow, accounts for more than 70% of prevalences among all sub-types of stroke. A large prospective cohort study based on half a million Chinese adults showed that the incidence of ischemic stroke during 7.2-year follow-up was 5.86 per 1000 person-years [6].

Dietary intake is a modifiable lifestyle behavior closely associated with most non-communicable diseases (NCDs), including cardiovascular diseases (CVDs) [7–9]. There has been an increasing interest in using specific indexes to evaluate dietary quality and their effects on NCDs, especially in some developed countries, such as the Healthy Eating Index (HEI) and the Diet Quality Index (DQI) developed for Americans [10,11], and the Mediterranean Diet Score (MDS) used for the residents in Northern Europe [12]. With reference to the methods of HEI and DQI, the Chinese Dietary Balance Index (DBI) has been designed to assess the overall diet quality in the Chinese population [13], according to the most recent Dietary Guidelines for Chinese residents [14]. Although associations of DBI with diabetes, hypertension, and cardiometabolic risk factors have been reported in previous cross-sectional studies among subgroups in China [15–17], data on the relationship between DBI and ischemic stroke is inconclusive, and evidence from prospective studies are warranted.

In this current study, we aimed to explore the associations between diet quality assessed by DBI and risk of incident ischemic stroke based on a prospective cohort in Guizhou Province in Southwest China, a region where economy and culture are relatively underdeveloped and with a great disease burden of ischemic stroke [18], in order to provide some evidence on further dietary intervention to manage and prevent ischemic stroke.

2. Methods

2.1. Study Design and Participants

The Guizhou Population Health Cohort Study (GPHCS) is one of few large population-based prospective cohort studies in Southwest China, which was established from 12 areas (five urban districts and seven rural counties) in Guizhou Province between 20 November 2010 and 19 December 2012. A multistage proportional stratified cluster sampling method was used to obtain a representative sample of the general population in Guizhou Province. The inclusive criteria were as follows: (1) age of 18 years or above; (2) living in the study regions for more than six months and having no plan to move out; (3) completing survey questionnaire and blood sampling; (4) signing the written informed consent. A total of 9280 local residents were enrolled in the cohort. All participants were followed up for major chronic diseases and vital status through a repeated investigation conducted between 2016 and 2020. All deaths were confirmed by the record from Death Registration Information System and Basic Public Health Service System. Ethics approval was obtained from the ethics review board of Guizhou Province (No.S2017-02). All participants provided written informed consent at enrollment.

In this study, we excluded participants with a history of ischemic stroke, haemorrhagic stroke, myocardial infarction or other cardiovascular diseases, missing data of diet consumption or cardiovascular diseases at baseline, loss to follow-up, and death, leaving 7841 participants for the analyses (Figure 1).

2.2. Outcome Definition

The primary outcome was the first onset of ischemic stroke (I63) diagnosed according to the International Classification of Diseases 10th revision (ICD-10). All reported events were reviewed and integrated centrally by trained clinical staff. Each participant was followed up until the first occurrence of the corresponding outcome, death, or loss to follow-up, which occurred first before 31 December 2020. The incidence rate was calculated as the number of incident cases divided by follow-up person-years.

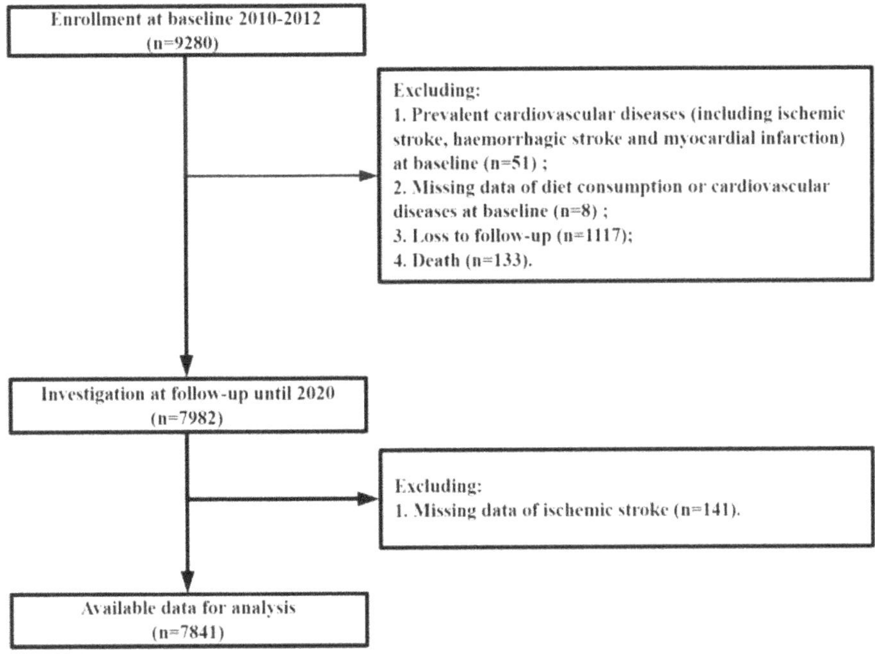

Figure 1. Flow chart of the study.

2.3. Dietary Data Collection

The information of dietary intake for each participant was assessed using a semi-quantitative food frequency questionnaire (FFQ), both at the individual level and at the household level. The individual FFQ covered 23 items of foods and beverages which were commonly consumed, including cereals, tubers, livestock meats, poultry meats, aquatic products, vegetables, fruits, eggs, dairy products, soybean products, etc. A commonly used unit or portion size was specified for each food item, participants were required to answer their usual consuming frequency (daily, weekly, monthly, yearly, or never) of each specific food or beverages over the past one year and the amount of consumption at each time. The daily intake on average for each food item was then calculated according to the product of the intake frequency and the amount consumed at each time (in gram per day, g/day). House condiment consumption, such as cooking oils, salt, sugar, sauces, etc., was determined by evaluating all condiments consumed by all household members for one month. The total amount of condiments consumed in the household divided by the number of members usually eating at home was used to assess individual consumption of condiments.

2.4. Dietary Intake Assessment

The dietary quality among the participants at baseline was assessed by the Chinese Diet Balance Index 2016 (DBI-16) [18], a revised version from the Chinese Diet Balance Index 2007 (DBI-07) [19]. DBI-2016 comprises 14 subgroups of 8 components from the Dietary Guidelines for Chinese residents [14], including: (1) cereal; (2) vegetable and fruit; (3) dairy and soybean; (4) animal food (red meats/products/poultry/game, fish/shrimp, and egg); (5) empty energy foods (cooking oils, and alcoholic beverage); (6) condiments (addible sugar, and salt); (7) diet variety; and (8) drinking water. A score of 0 for each DBI-16 component means that the individual has reached the recommended intake amounts of the corresponding food group. Positive scores (ranging 1 to 12) indicate the excessive intake level of cereals, red meat/products/poultry/game, eggs, cooking oils, alcoholic beverage, addible sugar, salt, while negative scores (ranging -12 to -1) indicate the inadequate intake level of cereals, vegetables, fruits, dairy, soybeans, red meat/products/poultry/game, fish/shrimps, eggs, diet variety, and drinking water. Considering the difference of nutrient requirements in energy consumption, the scoring of these 14 food subgroups was based on 11 levels of energy intake. Scoring details of DBI-16 are shown in Table S1.

Based on the scores for each DBI-16 component, three indicators of diet quality were calculated: (1) the lower bound score (LBS), an indicator for inadequate food intake, was computed by adding all the negative scores; (2) the higher bound score (HBS), an indicator for excessive food intake, was calculated by adding all the positive scores; (3) the diet quality distance, an indicator of unbalanced food intake, is calculated by adding the absolute values of both positive and negative scores [13]. The ranges of LBS, HBS, and DQD were 0 to 60, 0 to 40, and 0 to 84, respectively. For simplicity, each indicator was further divided into five levels to reflect the diet quality: (1) no problem, a score of 0; (2) almost no problem, less than 20% of the total score; (3) low-level problem, between 20% and 40% of the total score; (4) moderate level problem, between 40% and 60% of the total score; and (5) high-level problem, greater than 60% of the total score.

2.5. Other Variables Collection

A standardized questionnaire was used to collect the information of demographic characteristics, lifestyles, and medical history, including age, sex, area, ethnic group, education level, family income, marriage status, occupation status, physical activity, smoking or not, alcohol drinking or not, medical history of diabetes, hypertension and dyslipidemia, use of medications and nutraceuticals. Smoking was defined as smoking at least one cigarette a day for 12 months or more. Alcohol drinking was defined as drinking at least three times a week for 12 months or more. Medication use was defined as taking medications for diabetes, hypertension, dyslipidemia, or obesity regularly. Nutraceutical intake was defined as intaking some common nutraceuticals (such as vitamins or minerals), or foods with health-care functions (such as wine, tea) at least one time a week for 12 months or more. The physical activity level was calculated as the product of the duration and frequency of each activity, weighted by an estimate of the metabolic equivalent (MET) of that activity and summed for all activities performed, with the result expressed as the average MET hours per day.

Height, weight, and waist circumference were determined by trained technicians, using calibrated instruments with standard protocols and recorded to the nearest 0.1 cm or 0.1 kg. Waist-to-height ratio (WHtR) was calculated as waist circumference in centimeters divided by height in meters. Body mass index (BMI) was calculated as weight in kilograms divided by the square of height in meters. Systolic blood pressure (SBP) and diastolic blood pressure (DBP) were measured from the left arm after the participant rested in a seated position. All participants provided a 10-mL blood sample after an overnight fast of at least 10 h, they were also required to undergo an oral glucose tolerance test (OGTT), and the plasma was obtained at 2 h during the test. Fasting plasma glucose (FPG), 2-h postload glucose (2h-PG) and Hemoglobin A1c (HbA1c) were determined by the glucose oxidase methods (Roche Diagnostics, Mannheim, Germany). Serum triglycerides (TG), total cholesterol (CHOL), low-density lipoprotein cholesterol (LDL-C), and high-density lipoprotein cholesterol (HDL-C) were measured by enzymatic methods (Roche Diagnostics, Mannheim, Germany).

Diabetes was defined as those above the threshold of glycemia (FPG $\geq$ 6.1 mmol/L or 2h-PG $\geq$ 7.8 mmol/L), having a reported diabetes history, or experiencing anti-diabetes medications [20]. Hypertension was defined as an abnormal level of current blood pressure (SBP > 140 mmHg or DBP > 90 mmHg), having a reported hypertension history, or experiencing anti-hypertension medications [20]. Dyslipidemia was defined as an abnormal level of current blood lipids (TG $\geq$ 1.7 mmol/L, CHOL $\geq$ 5.2 mmol/L, LDL $\geq$ 3.4 mmol/L, HDL < 1.0 mmol/L), having a reported dyslipidemia history, or experiencing anti-dyslipidemia medications [20]. General overweight or obesity was defined as BMI $\geq$ 24 kg/m^2, central obesity was defined as WC $\geq$ 85 cm for females or $\geq$ 90 cm for males, and obesity status was defined as having either of these two types of obesity [21].

2.6. Statistical Analysis

Continuous variables were expressed as means and standard deviations (mean $\pm$ SD) and compared by using the Student's *t*-test. Categorical variables were presented as frequencies and percentages (*n*, %) and compared by using the Chi-square test. Considering that the proportional hazards assumption showed no strong evidence of departure, cox proportional hazards models were used to estimate hazard ratios (HRs) and 95% confidence intervals (95%CIs) for ischemic stroke by the components of DBI-16 and the indicators of diet quality. The level of statistical significance was defined as α = 0.05 of two-side probability. All analyses were performed using the R program (version 4.0.4, R Foundation for Statistical Computing, Vienna, Austria), and all figures were performed by using GraphPad Prism software (version 9, GraphPad Prism, San Diego, CA, USA).

3. Results

3.1. Descriptions of Study Population

The baseline characteristics of all 7841 participants in this study are shown in Table 1. The mean age was 44.18 $\pm$ 14.97 years at enrollment, and 47.4% (*n* = 3719) were male. Of these, 67.1% (*n* = 5258) of participants were rural residents, and more than half were Han Chinese (58.5%, *n* = 4589) and farmers (57.3%, *n* = 4490). The majority had an education level below junior middle school or no formal educated (86.7%, *n* = 6799).

Table 1. Baseline Characteristics of participants according to ischemic stroke status.

Characteristics	All (*n* = 7841)	Non-Ischemic Stroke (*n* = 7699)	Ischemic Stroke (*n* = 142)	*p* Value
Age (*n*, %)				<0.001
18–40 years	3252 (41.5)	3231 (42.0)	21 (14.8)	
41–60 years	3322 (42.4)	3256 (42.3)	66 (46.5)	
$\geq$60 years	1267 (16.1)	1212 (15.7)	55 (38.7)	
Sex (*n*, %)				0.754
Male	3719 (47.4)	3654 (47.5)	65 (45.8)	
Female	4122 (52.6)	4045 (52.5)	77 (54.2)	
Area (*n*,%)				0.136
Urban	2583 (32.9)	2545 (33.1)	38 (26.8)	
Rural	5258 (67.1)	5154 (66.9)	104 (73.2)	
Ethnic group (*n*,%)				0.003
Ethnic Han	4589 (58.5)	4488 (58.3)	101 (71.1)	
Minority	3252 (41.5)	3211 (41.7)	41 (28.9)	
Education (*n*,%)				0.038
No formal education	1606 (20.5)	1565 (20.3)	41 (28.9)	
Junior middle school and below	5193 (66.2)	5111 (66.4)	82 (57.7)	
Senior high school and above	1042 (13.3)	1023 (13.3)	19 (13.4)	
Family income (*n*, %)				0.010
<3000 RMB/person	1664 (32.6)	1625 (32.6)	39 (33.9)	
3000–10,000 RMB/person	2129 (41.8)	2080 (41.7)	49 (42.6)	
$\geq$10,000 RMB/person	1306 (25.6)	1279 (25.7)	27 (23.5)	
Marriage (*n*,%)				0.001
Married/Cohabit	6340 (80.9)	6226 (80.9)	114 (80.3)	
Unmarried/Single	744 (9.5)	740 (9.6)	4 (2.8)	
Divorced/Widowed/Separated	757 (9.7)	733 (9.5)	24 (16.9)	

Table 1. *Cont.*

Characteristics	All (*n* = 7841)	Non-Ischemic Stroke (*n* = 7699)	Ischemic Stroke (*n* = 142)	*p* Value
Occupation (*n*, %)				0.284
Farmers	4490 (57.3)	4407 (57.2)	83 (58.5)	
Others	2092 (26.7)	2061 (26.8)	31 (21.8)	
Unemployed or retired	1259 (16.1)	1231 (16.0)	28 (19.7)	
Smoking (*n*, %)				0.375
No	5856 (74.7)	5755 (74.7)	101 (71.1)	
Yes	1985 (25.3)	1944 (25.3)	41 (28.9)	
Alcohol drinking (*n*, %)				0.403
No	6038 (77.0)	5924 (76.9)	114 (80.3)	
Yes	1803 (23.0)	1775 (23.1)	28 (19.7)	
Diabetes (*n*, %)				0.015
No	7162 (91.7)	7042 (91.8)	120 (85.7)	
Yes	648 (8.3)	628 (8.2)	20 (14.3)	
Hypertension (*n*, %)				<0.001
No	5835 (74.4)	5756 (74.8)	79 (55.6)	
Yes	2006 (25.6)	1943 (25.2)	63 (44.4)	
Dyslipidemia (*n*, %)				0.581
No	3353 (42.8)	3296 (42.8)	57 (40.1)	
Yes	4488 (57.2)	4403 (57.2)	85 (59.9)	
Obesity (*n*, %)				0.376
No	4794 (64.1)	4716 (64.1)	78 (60.0)	
Yes	2688 (35.9)	2636 (35.9)	52 (40.0)	
Medication use (*n*, %)				<0.001
No	6919 (88.2)	6812 (88.5)	107 (75.4)	
Yes	922 (11.8)	887 (11.5)	35 (24.6)	
Nutraceutical intake (*n*, %)				0.044
No	6944 (88.7)	6810 (88.6)	134 (94.4)	
Yes	883 (11.3)	875 (11.4)	8 (5.6)	
MET (per day, mean ± SD)	109.82 ± 122.62	109.87 ± 122.68	107.17 ± 120.00	0.795
WC (cm, mean ± SD)	7661 ± 9.46	76.57 ± 9.46	78.67 ± 9.63	0.013
WHtR	5.52 ± 10.15	5.50 ± 10.15	6.83 ± 9.76	0.182
BMI (kg/m^2, mean ± SD)	22.90 ± 3.36	22.89 ± 3.36	23.26 ± 3.25	0.203
FPG (mmol/L, mean ± SD)	5.25 ± 1.26	5.25 ± 1.25	5.40 ± 1.51	0.158
2h-PG (mmol/L, mean ± SD)	5.79 ± 2.25	5.79 ± 2.25	6.12 ± 2.52	0.088
SBP (mmHg, mean ± SD)	125.09 ± 20.87	124.90 ± 20.72	135.33 ± 25.97	<0.001
DBP (mmHg, mean ± SD)	78.24 ± 11.90	78.16 ± 11.85	82.56 ± 13.88	<0.001
TG (mmol/L, mean ± SD)	1.76 ± 1.57	1.75 ± 1.56	1.89 ± 1.92	0.324
CHOL (mmol/L, mean ± SD)	4.79 ± 1.32	4.79 ± 1.31	4.85 ± 1.55	0.64
HDL-C (mmol/L, mean ± SD)	1.45 ± 0.56	1.45 ± 0.56	1.41 ± 0.63	0.405
LDL-C (mmol/L, mean ± SD)	2.66 ± 1.18	2.66 ± 1.18	2.54 ± 1.30	0.239

Abbreviation: SD, standard deviation; MET, metabolic equivalent of task; WC, waist circumference; WHtR, waist-to-height ratio; BMI, body mass index; FPG, fasting plasma glucose; 2h-PG, 2-h postload glucose; SBP, systolic blood pressure; DBP, diastolic blood pressure, TG, triglyceride; CHOL, total cholesterol; HDL-C, high-density lipoprotein cholesterol; LDL-C, low-density lipoprotein cholesterol.

During a median follow-up of 6.63 years (range 1.11 to 9.53 years), 142 participants were diagnosed with ischemic stroke. Compare with participants without ischemic stroke, those with incident ischemic stroke were more likely to be older, Han Chinese, with lower economic level, and less likely to be formally educated, and married (divorced/widowed/separated). Those with ischemic stroke also tended to have prevalent diabetes and hypertension (*p* < 0.001).

3.2. Assessments of Dietary Quality

The distributions of scores for the DBI-16 components are presented in Table 2. Overall, 0.3% to 98.9% of participants have reached the recommended dietary intakes (score = 0) of the DBI-16 components, and the majority (over 90%) consumed appropriate amounts of addible sugar and alcoholic beverages. Inadequate intakes (score < 0) were most commonly observed in dairy, fish, fruits, eggs, vegetables and soybeans, with the corresponding proportions among all participants of 99.7%, 97.5%, 95.5%, 83.5%, 62.0%, and 54.4%, respectively. Over 84.8 % of individuals had a dietary variety below the recommended level. By contrast, excessive intakes (score > 0) in cereals, cooking oils, salt and meats were also observed among 72.5%, 64.1%, 60.8%, and 46.6% of participants, respectively. Participants with ischemic stroke had more excessive intake in cooking oils, alcoholic beverages, and salt, and were more likely to have inadequate intake in meats than those without ischemic stroke. ($p < 0.05$).

Table 2. Distributions of scores for the DBI-16 components and the percentages of participants with each score.

Components	Score Range [a]	Group	(−12)–(−11)	(−10)–(−9)	(−8)–(−7)	(−6)–(−5)	(−4)–(−3)	(−2)–(−1)	0	(1)–(2)	(3)–(4)	(5)–(6)	(7)–(8)	(9)–(10)	(11)–(12)	p Value [b]
Cereals	(−12)–(12)	Non-Ischemic stroke	0.6	0.8	1.4	2.2	4.8	1.7	16.0	1.1	18.2	8.8	5.6	2.9	36.0	0.296
		Ischemic stroke	0	1.4	0.7	0.7	5.6	1.4	16.2	2.8	13.4	5.6	7.7	2.1	42.3	
Vegetables	(−6)–(0)	Non-Ischemic stroke				4.2	26.3	31.5	38.0							0.249
		Ischemic stroke				1.4	23.2	35.9	39.4							
Fruits	(−6)–(0)	Non-Ischemic stroke				49.0	39.4	7.1	4.5							0.325
		Ischemic stroke				46.5	45.8	4.9	2.8							
Dairy	(−6)–(0)	Non-Ischemic stroke				91.3	6.2	2.3	0.3							0.167
		Ischemic stroke				88.7	5.6	4.9	0.7							
Soybeans	(−6)–(0)	Non-Ischemic stroke				31.0	12.8	10.6	45.6							0.065
		Ischemic stroke				24.6	14.1	16.9	44.4							
Red meats/products, Poultry/game	(−4)–(4)	Non-Ischemic stroke					5.2	28.3	19.8	15.6	31.1					0.004
		Ischemic stroke					12.0	32.4	16.2	14.1	25.4					
Fish/shrimps	(−4)–(0)	Non-Ischemic stroke					85.8	11.7	2.5							0.546
		Ischemic stroke					88.7	9.9	1.4							
Eggs	(−4)–(4)	Non-Ischemic stroke					49.4	34.1	13.3	1.4	1.8					0.896
		Ischemic stroke					49.3	36.6	12.0	0.7	1.4					
Cooking oils	(0)–(6)	Non-Ischemic stroke							36.2	22.4	12.4	29.0				<0.001
		Ischemic stroke							19.7	19.0	16.2	45.1				
Alcoholic beverages	(0)–(6)	Non-Ischemic stroke							93.8	3.2	1.3	1.7				0.009
		Ischemic stroke							91.5	1.4	4.2	2.8				
Addible sugar	(0)–(6)	Non-Ischemic stroke							98.9	0.7	0.1	0.3				0.915
		Ischemic stroke							99.3	0.7	0	0				
Salt	(0)–(6)	Non-Ischemic stroke							39.5	41.0	5.1	14.4				0.004
		Ischemic stroke							27.5	55.6	3.5	13.4				
Diet variety	(−12)–(0)	Non-Ischemic stroke	0	0.1	4.2	9.3	21.1	50.0	15.3							0.644
		Ischemic stroke	0	0	4.9	6.3	21.8	54.9	12.0							

[a] Score range of total score is −60 to 44; [b] p value for chi-square test for the proportions of the scores for each food group.

The DBI-16 also revealed that 57.3%, 35.1%, and 2.3% of participants had a low, moderate, and high level of under intake (indicated by LBS), respectively; 43.8%, 16.8%, and 0.9% of them had a low to high level of over intake (indicated by HBS), respectively; 50.6%, 44.1%, and 3.9% of them had a low to high-level problem of overall unbalance (indicated by DQD), respectively (Table 3). The ischemic stroke patients had a higher median HBS, higher prevalence of moderate level of over intake (HBS, 25.4%) and higher overall unbalance (DQD, 48.6%) as compared with those without ischemic stroke. Moreover, the distributions of dietary quality across two groups divided by area and ethnicity are shown in Figure 2.

Table 3. Distribution of dietary quality and the percentages of participants with each category.

Diet Quality	Indicator	Score Range	Group	Mean ± SD	Distribution of Dietary Quality (%) [a]				
					No Problem	Almost No Problem	Low Level Problem	Moderate Level Problem	High Level Problem
Under intake	LBS	0–60	Non-Ischemic stroke	22.68 ± 6.82	0	5.3	57.2	35.2	2.3
			Ischemic stroke	22.42 ± 7.09	0	5.6	62.0	29.6	2.8
Over intake	HBS	0–40	Non-Ischemic stroke	11.84 ± 6.68	2.7	36.0	43.8	16.6	0.9
			Ischemic stroke	13.30 ± 6.43	2.1	28.9	42.9	25.4	0.7
Overall unbalance	DQD	0–84	Non-Ischemic stroke	34.51 ± 8.35	0	1.4	50.7	44.0	3.9
			Ischemic stroke	35.72 ± 7.81	0	0	47.2	48.6	4.2

[a] Distribution of the lower bound score (LBS): No problem: 0; Almost no problem: 1–12; Low level: 13–24; Moderate level: 25–36; High level: 37–60. Distribution of the higher bound score (HBS): No problem: 0; Almost no problem: 1–9; Low level: 10–18; Moderate level: 19–27; High level: 28–44. Distribution of the diet quality distance (DQD): No problem: 0; Almost no problem: 1–17; Low level: 18–34; Moderate level: 35–50; High level: 51–84.

3.3. Association Analyses of Ischemic Stroke with Dietary Quality Indicators and DBI-16 Components

The results of Cox regression analyses were shown in Table 4. The hazard ratios (HRs) for ischemic stroke were progressively elevated with increasing HBS. Compared with participants in the lowest quartile (Q1) of HBS, those in the highest quartile (Q4) had a 3.15-fold (95%CI: 1.50–6.63) increased risk for ischemic stroke, after adjustment for age, sex, area, ethnic group, education level, marriage status, smoking, diabetes, hypertension, dyslipidemia, and obesity status; (Model 2), and the association slightly increased after additional adjustment for medication use and nutraceutical intake at baseline (HR: 3.31, 95%CI: 1.57–6.97, Model 3). A similar result was observed for the highest DQD quartile (Q4), with the corresponding multiple-adjusted HR of 2.19 (95%CI: 1.24–3.86) based on Model 2 and 2.26 (95%CI: 1.28–4.00) based on Model 3.

Table 4. Hazard ratios (HRs) and 95% confidence intervals (95%CIs) for ischemic stroke by diet quality indicators and DBI-16 components according to Cox regression models.

Indicators	No (*n*)	Cases (*n*)	Incident Density (Cases per 1000 PYs)	HR (95%CI) [a]		
				Model 1	Model 2	Model 2
LBS [b]						
Quartile 1 (Q1)	1761	34	2.78	1.00	1.00	1.00
Quartile 2 (Q2)	1907	41	3.06	1.04 (0.66–1.64)	1.13 (0.68–1.89)	1.13 (0.68–1.89)
Quartile 3 (Q3)	2008	29	2.07	0.78 (0.48–1.29)	0.79 (0.44–1.41)	0.76 (0.43–1.36)
Quartile 4 (Q4)	2165	38	2.46	0.92 (0.57–1.46)	0.86 (0.46–1.59)	0.84 (0.45–1.56)
HBS [c]						
Quartile 1 (Q1)	1619	14	1.19	1.00	1.00	1.00
Quartile 2 (Q2)	2042	37	2.57	2.24 (1.21–4.15) *	2.38 (1.12–5.05)*	2.38 (1.12–5.06) *
Quartile 3 (Q3)	2168	44	2.89	2.48 (1.36–4.53) **	2.38 (1.14–5.00)*	2.39 (1.14–5.01) *
Quartile 4 (Q4)	2012	47	3.43	3.12 (1.72–5.68) ***	3.15 (1.50–6.63)**	3.31 (1.57–6.97) **
DQD [d]						
Quartile 1 (Q1)	1853	26	1.98	1.00	1.00	1.00
Quartile 2 (Q2)	1797	32	2.50	1.28 (0.76–2.15)	1.34 (0.75–2.37)	1.33 (0.75–2.36)
Quartile 3 (Q3)	2166	36	2.36	1.29 (0.78–2.13)	1.11 (0.62–2.01)	1.13 (0.63–2.04)
Quartile 4 (Q4)	2025	48	3.44	1.99 (1.23–3.23) **	2.19 (1.24–3.86) **	2.26 (1.28–4.00) **

Table 4. *Cont.*

Indicators	No (*n*)	Cases (*n*)	Incident Density (Cases per 1000 PYs)	HR (95%CI) [a]		
				Model 1	Model 2	Model 2
Cereals						
Score 0	1252	23	2.61	1.00	1.00	1.00
Score (−12)–(−7)	220	3	1.90	0.76 (0.23–2.51)	1.06 (0.31–3.62)	1.03 (0.30–3.53)
Score (−6)–(−1)	680	11	2.26	0.85 (0.41–1.74)	0.69 (0.29–1.65)	0.65 (0.27–1.55)
Score (1)–(6)	2192	31	2.01	0.74 (0.43–1.26)	0.55 (0.29–1.04)	0.54 (0.28–1.02)
Score (7)–(12)	3497	74	3.02	1.08 (0.67–1.72)	0.95 (0.56–1.61)	0.94 (0.55–1.59)
Vegetables						
Score 0	2982	56	2.67	1.00	1.00	1.00
Score (−6)–(−1)	4859	86	2.52	0.98 (0.70–1.37)	1.02 (0.68–1.53)	1.05 (0.70–1.57)
Fruits						
Score 0	354	4	1.62	1.00	1.00	1.00
Score (−6)–(−1)	7487	138	2.62	1.75 (0.65–4.73)	1.88 (0.59–6.04)	1.95 (0.61–6.28)
Dairy						
Score 0	25	1	5.84	1.00	1.00	1.00
Score (−6)–(−1)	7816	141	2.57	0.54 (0.07–3.82)	0.31 (0.04–2.30)	0.31 (0.04–2.26)
Soybeans						
Score 0	3575	63	2.52	1.00	1.00	1.00
Score (−6)–(−1)	4266	79	2.62	1.09 (0.78–1.53)	1.04 (0.69–1.57)	1.03 (0.68–1.54)
Meats						
Score 0	1544	23	2.12	1.00	1.00	1.00
Score (−4)–(−1)	2645	63	3.4	1.66 (1.03–2.68) *	1.13 (0.67–1.90)	1.08 (0.64–1.82)
Score (1)–(4)	3652	56	2.17	0.95 (0.59–1.55)	0.64 (0.37–1.12)	0.63 (0.36–1.09)
Fish/shrimps						
Score 0	194	2	1.47	1.00	1.00	1.00
Score (−4)–(−1)	7647	140	2.60	1.76 (0.44–7.10)	2.02 (0.28–14.50)	1.98 (0.28–14.20)
Eggs						
Score 0	1042	17	2.37	1.00	1.00	1.00
Score (−4)–(−1)	6547	122	2.64	1.01 (0.61–1.68)	0.78 (0.45–1.35)	0.79 (0.46–1.37)
Score (1)–(4)	252	3	1.69	0.65 (0.19–2.21)	0.52 (0.12–2.25)	0.50 (0.12–2.20)
Cooking oils						
Score 0	2814	28	1.39	1.00	1.00	1.00
Score (1)–(6)	5027	114	3.26	2.60 (1.72–3.94) ***	2.96 (1.75–5) ***	3.00 (1.77–5.07) ***
Alcoholic beverages						
Score 0	7353	130	2.51	1.00	1.00	1.00
Score (1)–(6)	488	12	3.64	1.62 (0.90–2.93)	1.30 (0.60–2.80)	1.35 (0.62–2.93)
Addible sugar						
Score 0	7757	141	2.59	1.00	1.00	1.00
Score (1)–(6)	84	1	1.74	0.81 (0.11–5.80)	0.84 (0.12–6.00)	0.81 (0.11–5.81)
Salt						
Score 0	3079	39	1.76	1.00	1.00	1.00
Score (1)–(6)	4762	103	3.13	2.04 (1.41–2.96) ***	1.98 (1.29–3.02) **	2.03 (1.33–3.10) **
Dietary variety						
Score 0	1195	17	1.97	1.00	1.00	1.00
Score (−12)–(−7)	337	7	2.85	1.58 (0.65–3.85)	5.24 (1.66–16.50) **	5.40 (1.70–17.20) **
Score (−6)–(−1)	6309	118	2.68	1.65 (0.99–2.76)	1.66 (0.94–2.95)	1.69 (0.95–3.01)

[a] Model 1: Adjusted for age only; Model 2: Model 1 + additionally adjusted for sex, area, ethnic group, education level, marriage status, economic level, smoking, diabetes, hypertension, dyslipidemia, and obesity status; Model 3: Model 2 + additionally adjusted for medication use and nutraceutical intake. [b] Quartile levels for the lower bound score (LBS): Q1, score 0–18; Q2, score 19–22; Q3, score 23–27; Q4, score 28–52. [c] Quartile levels for the higher bound score (HBS): Q1, score 0–6; Q2, score 7–12; Q3, score 13–17; Q4, score 18–33. [d] Quartile levels for the diet quality distance (DQD): Q1, score 9–29; Q2, score 30–34; Q3, score 35–40; Q4, score 41–68. * $p < 0.05$; ** $p < 0.01$; *** $p < 0.001$.

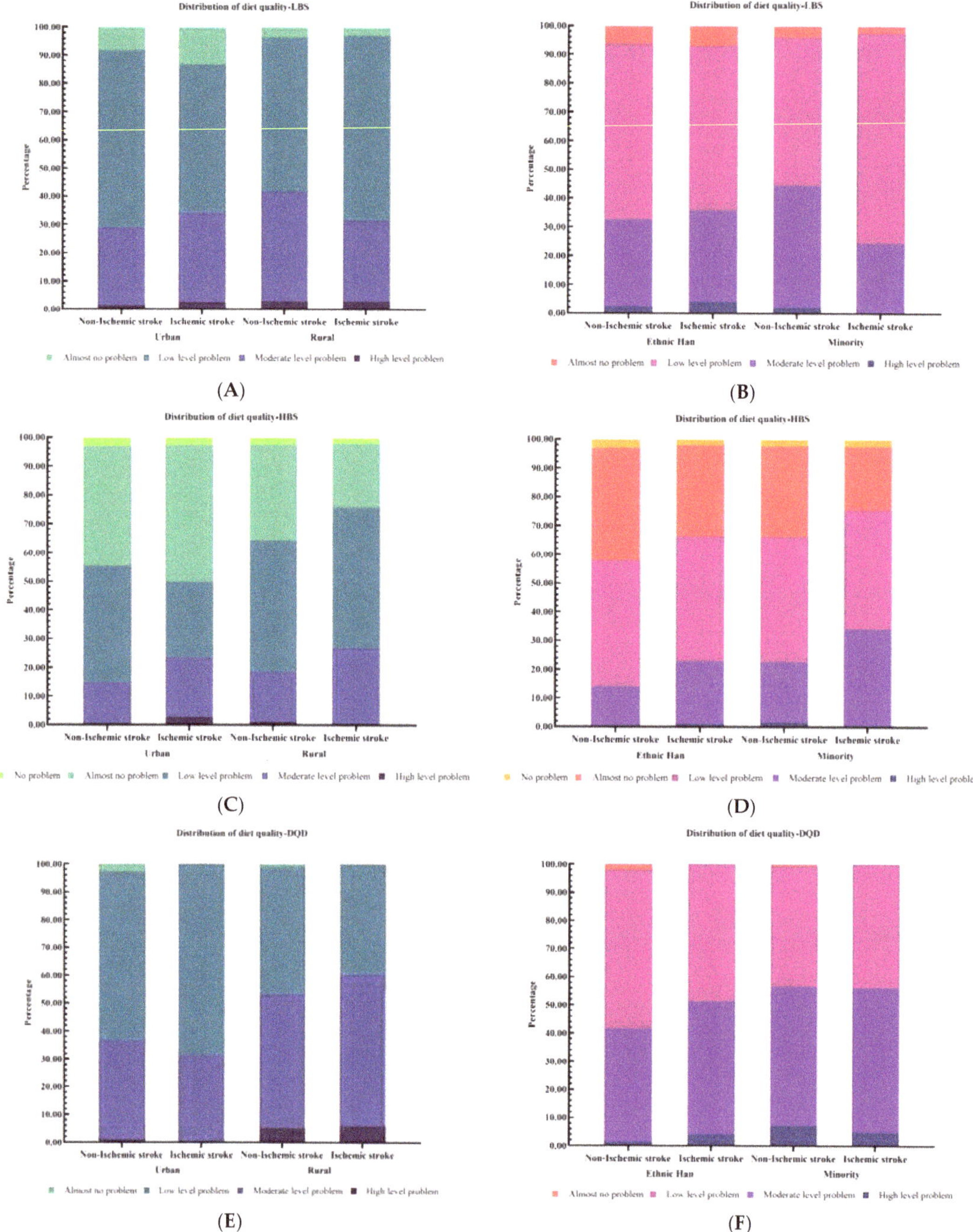

Figure 2. Distribution of the diet quality stratified by area (urban/rural) and ethnic group (ethnic Han/minority): (**A**,**B**) for the lower bound score (LBS); (**C**,**D**) for the higher bound score (HBS); (**E**,**F**) for the diet quality distance (DQD).

Among the thirteen components of DBI-16, both cooking oils and salt showed significant associations with ischemic stroke. Compared with the appropriate intake level (score = 0), excessive intake level (score 1–6) in cooking oils or salt added a 200% risk (HR: 3.00, 95%CI: 1.77–5.07, Model 3) and 103% risk (HR: 2.03, 95%CI: 1.33–3.10, Model 3) for ischemic stroke, respectively. Moreover, lower dietary variety (score −12 to −7) also promoted incident ischemic stroke (HR:5.40, 95%CI: 1.70–17.20, Model 3).

3.4. Stratified Analyses of the Association of Ischemic Stroke with Dietary Quality Indicators across Different Status of Comorbidities, Medication Use, and Nutraceutical Intake

After eliminating the role of comorbidities, medications, and nutraceuticals, HBS was still related to ischemic stroke among participants without diabetes or obesity, or those free of nutraceuticals or medications (Figure 3). In addition, the effect strengths of the associations between HBS with ischemic stroke were more evident in those with hypertension history (HR: 7.10, 95%CI: 2.71–19.9) and using medications (HR: 6.30, 95%CI: 1.44–18.6) than in total participants, comparing the highest quartile (Q4) of HBS to the lowest quartile (Q1) based on Model 3. However, HBS seemed to play less of a role in ischemic stroke among those with diabetes or intaking nutraceuticals. A similar pattern was also observed for DQD, with the increased HRs of 4.87 (95%CI: 1.66–14.20, Q4 vs. Q1, Model 3) and 6.00 (95%CI: 1.39–17.70, Q4 vs. Q1, Model 3) among those with hypertension history and using medications, respectively.

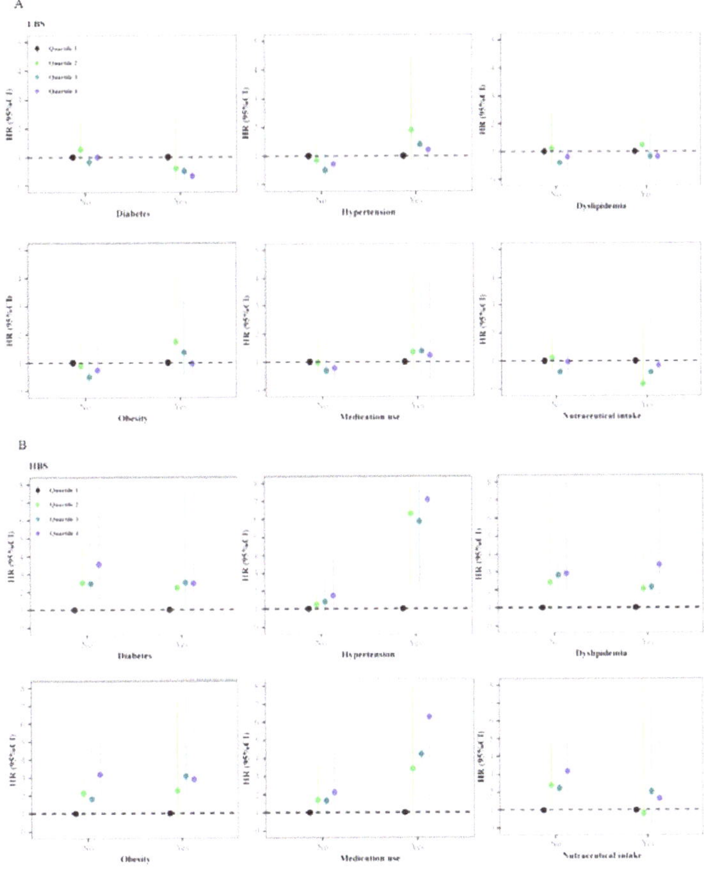

Figure 3. *Cont.*

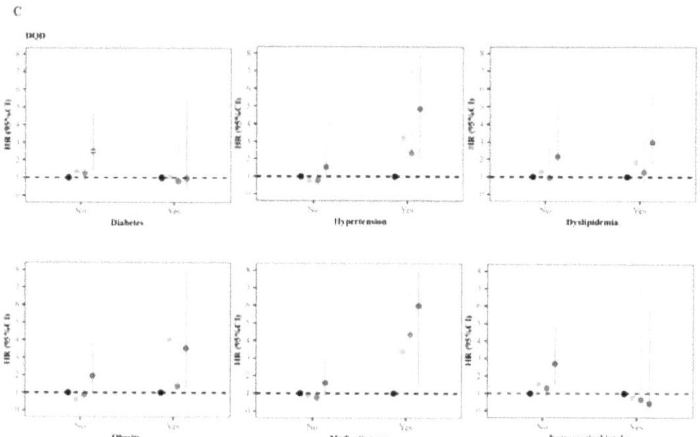

Figure 3. Adjusted hazard ratios (HRs) and 95% confidence intervals (95%CIs) for ischemic stroke associated with baseline dietary quality after stratified by the status of diabetes, hypertension, dyslipidemia, obesity, medication use and nutraceutical intake: (**A**) for the lower bound score (LBS); (**B**) for the higher bound score (HBS); (**C**) for the diet quality distance (DQD).

3.5. Stratified Analyses of the Associations of Ischemic Stroke with Dietary Quality Indicators across Baseline Demographic Factors

The multi-adjusted HRs for ischemic stroke by dietary quality indicators varied according to different demographic factors (Figure 4). The positive associations between HBS and ischemic stroke were seen only among participants with a baseline age of 60 years or more (HR: 4.70, 95%CI: 2.89–16.00, Q4 vs. Q1), female (HR: 2.94, 95%CI: 1.17–7.34, Q4 vs. Q1), rural residents (HR: 3.50, 95%CI: 1.57–7.81, Q4 vs. Q1) and the Ethnic Han (HR: 3.15, 95%CI: 1.36–7.32, Q4 vs. Q1), although there was no significant effect modification ($p_{interaction} > 0.05$) by age, sex, area, and ethnic group. When predicted by DQD, the risks for ischemic stroke elevated in females (HR: 2.68, 95%CI: 1.24–5.80, Q4 vs. Q1), rural residents (HR: 2.49, 95%CI: 1.26–4.90, Q4 vs. Q1), and the Ethnic Han (HR: 2.80, 95%CI: 1.36–5.77, Q4 vs. Q1). There was additionally a negative association between LBS and ischemic stroke in those aged more than 60 years (HR: 0.12, 95%CI: 0.03–0.55, Q3 vs. Q1).

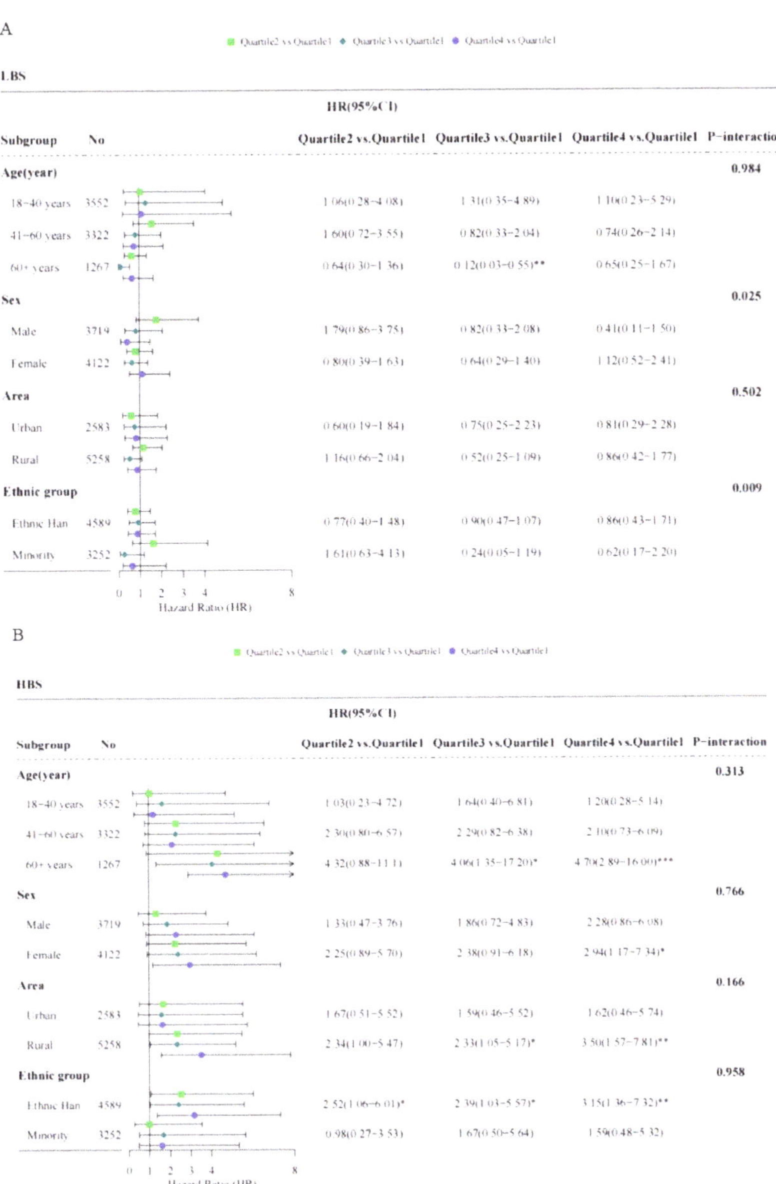

Figure 4. *Cont.*

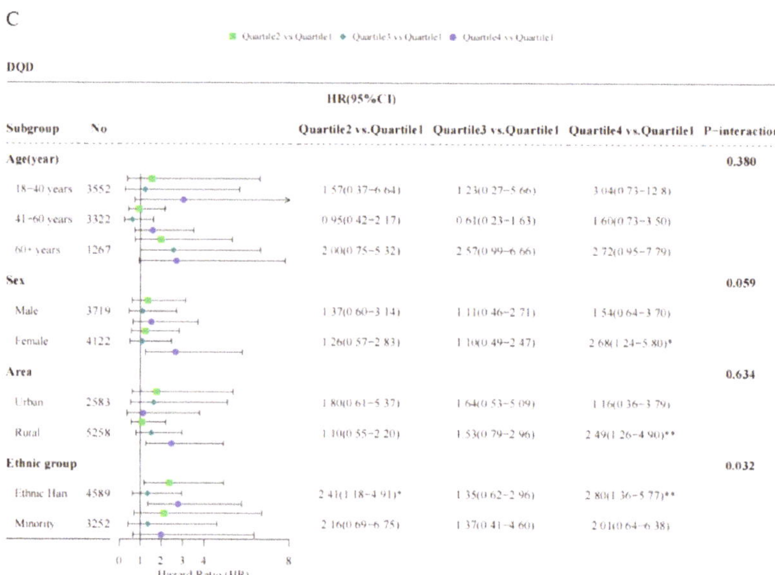

Subgroup	No	Quartile2 vs.Quartile1	Quartile3 vs.Quartile1	Quartile4 vs.Quartile1	P-interaction
Age(year)					0.380
18–40 years	3552	1.57(0.37–6.64)	1.23(0.27–5.66)	3.04(0.73–12.8)	
41–60 years	3322	0.95(0.42–2.17)	0.61(0.23–1.63)	1.60(0.73–3.50)	
60+ years	1267	2.00(0.75–5.32)	2.57(0.99–6.66)	2.72(0.95–7.79)	
Sex					0.059
Male	3719	1.37(0.60–3.14)	1.11(0.46–2.71)	1.54(0.64–3.70)	
Female	4122	1.26(0.57–2.83)	1.10(0.49–2.47)	2.68(1.24–5.80)*	
Area					0.634
Urban	2583	1.80(0.61–5.37)	1.64(0.53–5.09)	1.16(0.36–3.79)	
Rural	5258	1.10(0.55–2.20)	1.53(0.79–2.96)	2.49(1.26–4.90)**	
Ethnic group					0.032
Ethnic Han	4589	2.41(1.18–4.91)*	1.35(0.62–2.96)	2.80(1.36–5.77)**	
Minority	3252	2.16(0.69–6.75)	1.37(0.41–4.60)	2.01(0.64–6.38)	

Figure 4. Adjusted hazard ratios (HRs) and 95% confidence intervals (95%CIs) for ischemic stroke associated with baseline dietary quality after stratified by age, sex, area, and ethnic group: (**A**) for the lower bound score (LBS); (**B**) for the higher bound score (HBS); (**C**) for the diet quality distance (DQD); * $p < 0.05$; ** $p < 0.01$; *** $p < 0.001$.

4. Discussions

In this prospective cohort study conducted in Guizhou Province, Southwest China, we observed that the participants were exposed to dietary unbalance to different extents at baseline, mainly including the inadequate intakes of dairy, fish, fruits, eggs, vegetables, and soybeans, and the excessive intakes of cereals, cooking oils, salt, and meats. Our analyses further suggested that unfavorable dietary quality, including overall excessive consumption, high oils and salt diet, and low food diversity, may be a risk for ischemic stroke.

Several studies conducted in American, Swedish, and Chinese populations have estimated the effects of dietary quality on ischemic stroke. A recent study based on 73,890 women in Nurses' Health Study (NHS, 1984–2016), 92,352 women in NHSII (1991–2017), and 43,266 men in Health Professionals Follow-Up Study (1986–2012) in America revealed that the healthful plant-based dietary quality assessed by the Plant-based Diet Index (PDI) seemed to reduce the risk of ischemic stroke [22]. Similarly, a prospective cohort study of 26,547 Swedish aged 46 to 73 years found that less risk of incident ischemic stroke was related to the higher dietary quality, which was based on adherence to the Swedish nutrition recommendations [23]. By contrast, a large-scale prospective cohort study of 512,725 Chinese aged 30 to 79 years reported that less healthy dietary habits, which was defined as non-daily eating of vegetables, fruits, and eggs combined with eating daily or less than weekly, contributed to incident ischemic stroke [24].

China has experienced an ongoing transition of dietary patterns over the past decades, which mainly featured declines in the intakes of coarse and refined grains and vegetables, as well as increases in the intakes of animal-derived foods, with pork being most popular. Intakes of eggs, fish, and dairy have been consistently below recommended levels [25]. In addition, as the cooking methods markedly shifted from steaming and boiling to stir-fried and deep-fried, the daily consumption of cooking oils has been gradually increased from 18.2 g to 42.1 g from 1982 to 2012 [26]. People preferred to add a certain amount of salt to keep food from spoiling, such as pickles and salted fish, especially in rural or remote areas without a good condition for food storage. Data from China National Nutrition

Surveys (CHNS) also revealed that more than 55.9% and 71.8% of Chinese have consumed excessive cooking oil and salt, which were far above the recommended levels and strongly linked to increased risk of chronic diseases [27]. These cooking and eating behaviors are partly driven both by the great accessibility of cooking oils and salt and by their low price.

Compared to a previous study from Shanghai in East China and a cross-sectional study based on the CHNS, participants in this current study from Guizhou Province in Southeast China have a higher percentage of excessive intake of cooking oil and salt (score > 0) [28,29]. We also observed that the probabilities of occurring ischemic stroke for participants consuming the excessive level of cooking oil and salt were about appropriate two to three times as high as those consuming the moderate level. Cooking oil has been the main source of fat intake these decades. As expected, a diet with elevated fat or salt was associated with a significantly increased risk of hypertension and stroke, as reported previously [30,31]. A rat experiment found that a long-term administration of canola oil, sesame oil, or trans-fat led to marked dyslipidemia, fat accumulation, neuroinflammation, vascular lesion, and endothelial injury for rats, and thereby remarkably contributed to ischemic and hemorrhagic strokes [32]. Results from a gerbil animal model study indicated that a high-fat diet accelerates and exacerbates microgliosis and neuronal damage [33]. Although extra virgin olive oil (EVOO) or other oils rich in monounsaturated fat have been reported to bring some benefits to the cardiocerebral vascular system, but their consumption by the Chinese was relatively low [26]. In numerous epidemiology and clinical studies, excessive dietary sodium salt intake was linked to hypertension, which has been considered as the main risk factor for stroke [34]. A large prospective cohort study observed a significant linear association between calibrated urinary sodium excretion and stroke in patients with chronic kidney disease [35].

Given the potential effects of comorbidities, some medications for metabolic diseases, and some nutraceuticals, which may promote or decelerate the progress of cardiovascular diseases [36,37], we performed the same analyses in those with different status' of these conditions. The risk effects of excessive food intake (evaluated by HBS) and unbalanced food intake were more evident in those with hypertension history and taking medications for metabolic diseases, indicating that people with a high risk of stroke should pay more attention to the balance of food types and daily intake, reasonable diet is also one of the major measures to prevent hypertension [16]. However, we failed to observe any associations of ischemic stroke with diet quality among those with diabetes or intaking nutraceuticals, perhaps due to a smaller sample size in those subgroups.

The current study provides primary evidence that the risk of ischemic stroke among the residents from Guizhou Province in Southwest China may be partly due to the potential effects of dietary quality. The chief strengths lie in the use of a large population-based cohort, a prospective study design with more than six years of follow-up, a dietary assessment index suitable for Chinese people. The scores of DBI-16 are based on different levels of energy consumption for specific individuals recommended by the most recent Dietary Guidelines for Chinese residents, so the potential confounding effects of total energy intake may be appropriately controlled [38]. There also exist certain limitations. Firstly, dietary habits and socioeconomic characteristics were collected based on individual self-report, which might lead to recall bias. The food frequency questionnaire (FFQ) has been considered a convenient and widely-used dietary assessment tool, but this method is subject to less accuracy of quantification of food portions than the method of weighing, which might make some measurement errors [39]. Secondly, we just applied the proportion of condiment intakes at home to estimate the total daily intakes, inevitably ignoring the condiment intakes from eating outside. Thirdly, our study population was from Southeast China and the dietary assessment index was applicable to the Chinese, so the findings from the current study should be generalized with caution to other populations. In conclusion, considering a great disease burden caused by ischemic stroke in China, our study suggests that it is essential to adjust dietary habits and conduct dietary interventions, especially controlling risk factors in preventive mode better than therapeutic mode, because of a fairly short time to death or irreversible injuries after the onset of stroke.

5. Conclusions

Our results suggest that unfavorable dietary quality, including overall excessive consumption, excessive intake of cooking oils and salt, or under adequate dietary diversity, may increase the risk for ischemic stroke.

Supplementary Materials: The following are available online at: https://www.mdpi.com/article/10.3390/nu14030694/s1, Table S1: DBI-16 components and standard for scoring.

Author Contributions: Conceptualization, T.L., N.W. and C.F.; Data curation, X.S. and Y.C.; Formal analysis, Y.W. (Yingying Wang) and N.W.; Funding acquisition, T.L.; Investigation, Y.W. (Yiying Wang) and J.Z.; Methodology, Y.W. (Yingying Wang) and N.W.; Project administration, T.L. and C.F.; Supervision, C.F.; Writing—original draft, Y.W. (Yingying Wang); Writing—review & editing, X.S. and N.W. All authors have read and agreed to the published version of the manuscript.

Funding: This work was supported by Guizhou Province Science and Technology Support Program (Qiankehe [2018]2819).

Institutional Review Board Statement: The study was conducted according to the guidelines of the Declaration of Helsinki and approved by the Institutional Review Board (or Ethics Committee) of the Guizhou Center for Disease Control and Prevention (no. S2017-02).

Informed Consent Statement: Informed consent was obtained from all subjects that were involved in the study.

Data Availability Statement: The datasets for this manuscript will be made available upon request pending, further inquiries can be directed to the corresponding author Liu Tao, liutaombs@163.com and Na Wang, na.wang@fudan.edu.cn.

Conflicts of Interest: The authors declare that they have no known competing financial interest or personal relationship that could have appeared to influence the work reported in this paper.

References

1. Tsao, C.W.; Aday, A.W.; Almarzooq, Z.I.; Alonso, A.; Beaton, A.Z.; Bittencourt, M.S.; Boehme, A.K.; Buxton, A.E.; Carson, A.P.; Commodore-Mensah, Y.; et al. Heart Disease and Stroke Statistics-2021 Update: A Report From the American Heart Association. *Circulation* **2021**, *143*, e254–e743.
2. Wang, Y.-J.; Li, Z.-X.; Gu, H.-Q.; Zhai, Y.; Jiang, Y.; Zhao, X.-Q.; Wang, Y.-L.; Yang, X.; Wang, C.-J.; Meng, X.; et al. China Stroke Statistics 2019: A Report From the National Center for Healthcare Quality Management in Neurological Diseases, China National Clinical Research Center for Neurological Diseases, the Chinese Stroke Association, National Center for Chronic and Non-communicable Disease Control and Prevention, Chinese Center for Disease Control and Prevention and Institute for Global Neuroscience and Stroke Collaborations. *Stroke Vasc. Neurol.* **2020**, *5*, 211–239. [PubMed]
3. Li, Z.; Jiang, Y.; Li, H.; Xian, Y.; Wang, Y. China's response to the rising stroke burden. *BMJ* **2019**, *364*, l879. [CrossRef]
4. Wang, W.; Jiang, B.; Sun, H.; Ru, X.; Sun, D.; Wang, L.; Wang, L.; Jiang, Y.; Li, Y.; Wang, Y.; et al. Prevalence, Incidence, and Mortality of Stroke in China: Results from a Nationwide Population-Based Survey of 480 687 Adults. *Circulation* **2017**, *135*, 759–771. [CrossRef] [PubMed]
5. National Center for Cardiovascular Diseases. *Annual Report on Cardiovascular Health and Diseases in China 2019*; China Science Publishing House: Beijing, China, 2020. (In Chinese)
6. Lv, J.; Yu, C.; Guo, Y.; Bian, Z.; Yang, L.; Chen, Y.; Tang, X.; Zhang, W.; Qian, Y.; Huang, Y.; et al. Adherence to Healthy Lifestyle and Cardiovascular Diseases in the Chinese Population. *J. Am. Coll. Cardiol.* **2017**, *69*, 1116–1125. [CrossRef]
7. Chao, A.M.; Quigley, K.M.; Wadden, T.A. Dietary interventions for obesity: Clinical and mechanistic findings. *J. Clin. Investig.* **2021**, *131*, e140065. [CrossRef]
8. Ojo, O. Dietary Intake and Type 2 Diabetes. *Nutrients* **2019**, *11*, 2177. [CrossRef]
9. Badimon, L.; Chagas, P.; Chiva-Blanch, G. Diet and Cardiovascular Disease: Effects of Foods and Nutrients in Classical and Emerging Cardiovascular Risk Factors. *Curr. Med. Chem.* **2019**, *26*, 3639–3651. [CrossRef] [PubMed]
10. Krebs-Smith, S.M.; Pannucci, T.E.; Subar, A.F.; Sharon, R.D.; Kirkpatrick, I.; Lerman, J.L.; Tooze, J.; Wilson, M.M.; Reedy, J. Update of the Healthy Eating Index: HEI-2015. *J. Acad. Nutr. Diet.* **2018**, *118*, 1591–1602. [CrossRef]
11. Cruz, R.D.; Park, S.-Y.; Shvetsov, Y.B.; Boushey, C.J.; Monroe, K.R.; Le Marchand, L.; Maskarinec, G. Diet Quality and Breast Cancer Incidence in the Multiethnic Cohort. *Eur. J. Clin. Nutr.* **2020**, *74*, 1743–1747. [CrossRef]

12. Turati, F.; Carioli, G.; Bravi, F.; Ferraroni, M.; Serraino, D.; Montella, M.; Giacosa, A.; Toffolutti, F.; Negri, E.; Levi, F.; et al. Mediterranean Diet and Breast Cancer Risk. *Nutrients* **2018**, *10*, 326. [CrossRef]

13. He, Y.N.; Fang, Y.H.; Xia, J. Update of the Chinese Diet Balance Index: DBI-16. *Acta Nutr. Sin.* **2018**, *40*, 526–530.

14. Chinese Nutrition Society. *Dietary Guidelines for Chinese Residents (2016)*; People's Medical Publishing House: Beijing, China, 2016. (In Chinese)

15. Tan, S.; Lu, H.; Song, R.; Wu, J.; Xue, M.; Qian, Y.; Wang, W.; Wang, X. Dietary quality is associated with reduced risk of diabetes among adults in Northern China: A cross-sectional study. *Br. J. Nutr.* **2021**, *126*, 923–932. [CrossRef]

16. Wang, X.; Liu, A.; Du, M.; Wu, J.; Wang, W.; Qian, Y.; Zheng, H.; Liu, D.; Nan, X.; Jia, L.; et al. Diet quality is associated with reduced risk of hypertension among Inner Mongolia adults in northern China. *Public Health Nutr.* **2020**, *23*, 1543–1554. [CrossRef] [PubMed]

17. Gao, X.; Tian, Z.; Zhao, D.; Li, K.; Zhao, Y.; Xu, L. Associations between Adherence to Four A Priori Dietary Indexes and Cardiometabolic Risk Factors among Hyperlipidemic Patients. *Nutrients* **2021**, *13*, 2179. [CrossRef] [PubMed]

18. Ma, R.C.W. Epidemiology of diabetes and diabetic complications in China. *Diabetologia* **2018**, *61*, 1249–1260. [CrossRef] [PubMed]

19. He, Y.N.; Zhai, F.Y.; Yang, X.G.; Ge, K.Y. The Chinese Diet Balance Index Revised. *Acta Nutr. Sin.* **2009**, *31*, 532–536.

20. Writing group of Chinese Guideline for Cardiometabolic Diseases Prevention; Editorial board of Chinese Journal of Cardiometabolic Diseases. Chinese Guideline for Cardiometabolic Diseases Prevention (2017). *Chin. J. Cardiometabolic Dis.* **2018**, *46*, 10–25.

21. National Health and Family Planning Commission of the People's Republic of China. *Health Standard of the People's Republic of China*; No. WS/T 428-2013; Criteria of Weight for Adults 2013; Standards Press of China: Beijing, China, 2013. (In Chinese)

22. Baden, M.Y.; Shan, Z.; Wang, F.; Li, Y.; Manson, J.E.; Rimm, E.B.; Willett, W.C.; Hu, F.B.; Rexrode, K.M. Quality of Plant-Based Diet and Risk of Total, Ischemic, and Hemorrhagic Stroke. *Neurology* **2021**, *96*, e1940–e1953. [CrossRef]

23. Johansson, A.; Drake, I.; Engström, G.; Acosta, S. Modifiable and Non-Modifiable Risk Factors for Atherothrombotic Ischemic Stroke among Subjects in the Malmö Diet and Cancer Study. *Nutrients* **2021**, *13*, 1952. [CrossRef]

24. Han, Y.; Hu, Y.; Yu, C.; Guo, Y.; Pei, P.; Yang, L.; Chen, Y.; Du, H.; Sun, D.; Pang, Y.; et al. Lifestyle, cardiometabolic disease, and multimorbidity in a prospective Chinese study. *Eur. Heart J.* **2021**, *42*, 3374–3384. [CrossRef]

25. Zhai, F.; Du, S.F.; Wang, Z.; Zhang, J.G.; Du, W.; Popkin, B.M. Dynamics of the Chinese diet and the role of urbanicity, 1991-2011. *Obes. Rev.* **2014**, *15* (Suppl. 1), 16–26. [CrossRef] [PubMed]

26. Huang, L.; Wang, Z.; Wang, H.; Zhao, L.; Jiang, H.; Zhang, B.; Ding, G. Nutrition transition and related health challenges over decades in China. *Eur J. Clin. Nutr.* **2021**, *75*, 247–252. [CrossRef] [PubMed]

27. Jiang, H.; Zhang, J.; Su, C.; Zhang, J.; Zhang, B.; Wang, H. Cooking oil and salt consumption among chinese adults aged 18-59 years in 2015. *Acta Nutr. Sin.* **2018**, *40*, 27–31.

28. Zang, J.; Yu, H.; Zhu, Z.; Lu, Y.; Liu, C.; Yao, C.; Bai, P.; Guo, C.; Jia, X.; Zou, S.; et al. Does the Dietary Pattern of Shanghai Residents Change across Seasons and Area of Residence: Assessing Dietary Quality Using the Chinese Diet Balance Index (DBI). *Nutrients* **2017**, *9*, 251. [CrossRef] [PubMed]

29. He, D.; Qiao, Y.; Xiong, S.; Liu, S.; Ke, C.; Shen, Y. Association between Dietary Quality and Prediabetes based on the Diet Balance Index. *Sci. Rep.* **2020**, *10*, 3190. [CrossRef]

30. Muto, M.; Ezaki, O. High Dietary Saturated Fat is Associated with a Low Risk of Intracerebral Hemorrhage and Ischemic Stroke in Japanese but not in Non-Japanese: A Review and Meta-Analysis of Prospective Cohort Studies. *J. Atheroscler. Thromb.* **2018**, *25*, 375–392. [CrossRef]

31. Arsang-Jang, S.; Mansourian, M.; Mohammadifard, N.; Khosravi, A.; Oveis-Gharan, S.; Nouri, F.; Sarrafzadegan, N. Temporal trend analysis of stroke and salt intake: A 15-year population-based study. *Nutr. Neurosci.* **2021**, *24*, 384–394. [CrossRef]

32. Guo, H.; Ban, Y.-H.; Cha, Y.; Kim, T.-S.; Lee, S.-P.; An, E.S.; Choi, J.; Seo, D.W.; Yon, J.-M.; Choi, E.-K.; et al. Comparative effects of plant oils and trans-fat on blood lipid profiles and ischemic stroke in rats. *J. Biomed. Res.* **2017**, *31*, 122–129.

33. Seo, W.J.; Ahn, J.H.; Lee, T.-K.; Kim, B.; Lee, J.-C.; Park, J.H.; Yoo, Y.H.; Shin, M.C.; Cho, J.H.; Won, M.-H.; et al. High fat diet accelerates and exacerbates microgliosis and neuronal damage/death in the somatosensory cortex after transient forebrain ischemia in gerbils. *Lab. Anim. Res.* **2020**, *36*, 28. [CrossRef]

34. Frisoli, T.M.; Schmieder, R.E.; Grodzicki, T.; Messerli, F.H. Salt and hypertension: Is salt dietary reduction worth the effort? *Am. J. Med.* **2012**, *125*, 433–439. [CrossRef]

35. Mills, K.T.; Chen, J.; Yang, W.; Appel, L.J.; Kusek, J.W.; Alper, A.; Delafontaine, P.; Keane, M.G.; Mohler, E.; Ojo, A.; et al. Sodium Excretion and the Risk of Cardiovascular Disease in Patients With Chronic Kidney Disease. *JAMA* **2016**, *315*, 2200–2210. [CrossRef] [PubMed]

36. Li, X.; Wu, C.; Lu, J.; Chen, B.; Li, Y.; Yang, Y.; Hu, S.; Li, J. Cardiovascular risk factors in China: A nationwide population-based cohort study. *Lancet Public Health* **2020**, *5*, e672–e681. [CrossRef]

37. Scicchitano, P.; Cameli, M.; Maiello, M.; Modesti, P.A.; Muiesan, M.L.; Novo, S.; Palmiero, P.; Saba, P.S.; Pedrinelli, R.; Ciccone, M.M. Nutraceuticals and dyslipidaemia: Beyond the common therapeutics. *J. Funct. Foods* **2014**, *6*, 11–32. [CrossRef]

38. Zhang, Q.; Qin, G.; Liu, Z.; Li, Z.; Li, J.; Varma, D.S.; Wan, Q.; Zhao, J.; Min, X.; Han, X.; et al. Dietary Balance Index-07 and the Risk of Anemia in Middle Aged and Elderly People in Southwest China: A Cross Sectional Study. *Nutrients* **2018**, *10*, 162. [CrossRef]

39. McKenzie, B.L.; Coyle, D.H.; Santos, J.A.; Burrows, T.; Rosewarne, E.; Peters, S.A.E. Investigating sex differences in the accuracy of dietary assessment methods to measure energy intake in adults: A systematic review and meta-analysis. Am. *J. Clin. Nutr.* **2021**, *113*, 1241–1255. [CrossRef] [PubMed]

Review

The Role of Diet and Interventions on Multiple Sclerosis: A Review

Panagiotis Stoiloudis, Evangelia Kesidou, Christos Bakirtzis, Styliani-Aggeliki Sintila, Natalia Konstantinidou, Marina Boziki and Nikolaos Grigoriadis *

2nd Department of Neurology, AHEPA Hospital, Faculty of Health Sciences, Aristotle University of Thessaloniki, 54636 Thessaloniki, Greece; stpan@windowslive.com (P.S.); bioevangelia@yahoo.gr (E.K.); bakirtzischristos@yahoo.gr (C.B.); lina17puma@yahoo.gr (S.-A.S.); nataliak95@gmail.com (N.K.); bozikim@auth.gr (M.B.)

* Correspondence: grigoria@med.auth.gr or ngrigoriadis@auth.gr; Tel.: +30-2310-994683 or +30-2310-994681 or +30-2310-994665

Abstract: Multiple sclerosis (MS) is a chronic autoimmune disease of the central nervous system (CNS) characterized by inflammation and neurodegeneration. The most prominent clinical features include visual loss and sensorimotor symptoms and mainly affects those of young age. Some of the factors affecting its pathogenesis are genetic and/or environmental including viruses, smoking, obesity, and nutrition. Current research provides evidence that diet may influence MS onset, course, and quality of life of the patients. In this review, we address the role of nutrition on MS pathogenesis as well as dietary interventions that show promising beneficial results with respect to MS activity and progression. Investigation with large prospective clinical studies is required in order to thoroughly evaluate the role of diet in MS.

Keywords: multiple sclerosis; diet; nutrition; gut–brain axis; gut microbiota

Citation: Stoiloudis, P.; Kesidou, E.; Bakirtzis, C.; Sintila, S.-A.; Konstantinidou, N.; Boziki, M.; Grigoriadis, N. The Role of Diet and Interventions on Multiple Sclerosis: A Review. *Nutrients* **2022**, *14*, 1150. https://doi.org/10.3390/nu14061150

Academic Editor: Panagiota Mitrou

Received: 28 January 2022
Accepted: 8 March 2022
Published: 9 March 2022

Publisher's Note: MDPI stays neutral with regard to jurisdictional claims in published maps and institutional affiliations.

1. Introduction

Multiple sclerosis (MS) is a chronic autoimmune disease of the central nervous system (CNS) characterized by loss of myelin and inflammation, leading to neurodegeneration. Clinical features mainly include visual loss and sensorimotor symptoms as well as more atypical features such as fatigue and mental/cognitive impairment. It affects mainly patients of young age and mostly women [1]. It is classified in three clinical forms: relapsing-remitting (RRMS), primary progressive (PPMS), and secondary progressive (SPMS) disease [2], characterized by varying degree of pathology across the spectrum of acute/chronic inflammation and/or neurodegeneration. Its prevalence varies, with Europe and North America reporting the highest prevalence. Diagnosis is based on the revised 2017 McDonald criteria [3]. For RRMS, a clinical attack and dissemination in time and space is required. For PPMS, there is a need for disability progression confirmed for at least one year and dissemination in space. The SPMS subtype requires disability progression following the initial diagnosis of RRMS [3]. Several disease modifying treatments (DMTs) are currently in use in order to manage the ongoing disease activity in an attempt to control relapses and disability progression [4].

The pathogenesis of MS remains complicated and multifactorial. Other than genetic, various environmental factors seem to play a role in the development of MS. Microbial and viral infections, smoking, vitamin D, sun exposure, obesity, and dietary habits may be relevant to its pathogenesis. Environmental factors not only affect the development of MS but also the disease course and progression. Conversely, physical exercise and healthy diet appear to have an anti-inflammatory effect and to, at least in part, ameliorate the disease course [1,5–8].

Nutrition and dietary factors affect the mechanisms of MS pathology, its development, and degree of activity [9]. Although studies have shown the important role of nutrition in MS, the current therapy is not combined with any specific nutritional or lifestyle recommendation [1].

The present article reviews the current literature concerning the association of nutrition with the pathogenesis of MS and dietary interventions that affect its course.

2. Mechanisms of MS Pathology—The Effect of Diet/Nutritional Factors

2.1. Neurodegeneration

It is already known that neurodegeneration is presented even at the earliest stages of the disease [5]. In experimental models, oxidative stress leads to mitochondrial dysfunction, causing cell membrane disruption and eventually neuronal cell death [9]. Dietary antioxidant factors can dampen oxidative stress and may help against chronic demyelination and neuronal or axonal damage [5]. Both oxidative and mitochondrial injury primarily disrupt the function of neurons and glia, causing disturbances in cellular communication [10].

2.1.1. Oxidative Stress

Oxidative injury is involved in both relapsing-remitting and progressive forms of MS [11]. Inflammatory cytokines, reactive oxygen species, and phagocytes lead to damage of myelin and axons. It is found that oxidative stress enhances inflammation and causes damage of the myelin, consequently leading to cell death. Clinically, the course of MS has been associated with inflammatory and oxidative stress mediators including cytokines such as IL-1β, IL-6, IL-17, TNF-α, and INF-γ [12].

Dietary antioxidant factors may regulate the activation of immune inflammatory cells, leading to the reduction in inflammatory and may also dampen oxidative stress, thus preventing chronic demyelination and axonal damage. Antioxidant factors such as curcumin, vitamin D, and fatty acids have been studied and seem to play a role in the regulation of oxidative stress [13]. Curcumin, derived from the plant *Curcuma longa* [12], has been advocated to inhibit proinflammatory cytokines [14]. In animal models of MS, curcumin was shown to reduce clinical severity and decrease CNS infiltration by inflammatory cells in mice. Curcumin possesses antioxidant and anti-inflammatory properties. Its anti-oxidant effects have been assessed in several neurodegenerative diseases including Alzheimer's disease (AD), Parkinson's disease (PD), and MS [15]. Another nutritional factor is melatonin, which is produced naturally by the pineal gland during the night. It is formed exogenously from tryptophan. Melatonin is mainly consumed from meat, oily fish such as salmon, eggs, milk, seeds, nuts, almonds, and soy products. Melatonin is suggested to regulate anti-oxidative defensive systems by stimulating the synthesis of superoxide dismutase and glutathione peroxidase, especially in patients with SPMS [16].

Vitamin D plays a significant role not only in calcium homeostasis and bone health, but also in immunomodulation and the reduction in oxidative stress. MS patients frequently exhibit vitamin D deficiency [1]. Studies report that low levels of vitamin D are associated with a higher risk for the development and relapse of MS [8,17]. Supplementation with vitamin D has been shown to have anti-inflammatory and immunomodulatory effects on MS pathogenetic mechanisms by inhibiting the production of CD4+ T cells, thus lowering the risk of MS and diminishing disease progression [18]. However, Bagur et al. reported in their systematic review that existing studies on the effect of vitamin D supplementation in MS are inconsistent with respect to EDSS, MRI lesions, overall functional status, and relapse rate [13]. It has been suggested that empirical replacement with high doses of vitamin D supplementation (at least 4000 IU/day orally) and for a prolonged period appears to be safe and is associated with low risk for adverse events, although available data are limited [12,19–21].

Vitamin A is a fat-soluble nutrient with a variety of functions in visual ability, skin, and immunity. Vitamin A includes retinoids and carotenoids, available in liver, milk, cheese, green leaves, oil, vegetables, and fruit. Association between the pathogenesis of MS and vitamin A remains undefined. Studies in animal models demonstrate a possible role of vitamin A in the modulation of immunity [22,23]. A negative correlation has been found between the development of MS and low levels of vitamin A in plasma [12]. A randomized controlled trial showed benefits in fatigue, depression, and cognitive status of MS patients supplemented with high doses of vitamin A (400 IU/day), which were considered safe and were not associated with adverse effects [24].

Fatty acids, especially omega-3 polyunsaturated fatty acids (PUFAs), are other antioxidant compounds that are associated with ameliorating neurodegeneration in MS. Intake of PUFAs consumed via fish, nuts, and seeds seems to be associated with protective effects against demyelination [5]. In animal models, PUFAs decrease inflammation, maintain immunomodulation and promote neuroprotection and remyelination [5]. Some studies have shown inconsistent results indicating the effect of PUFAs mainly against progression. In one study, association between PUFA intake and MS incidence seems to be non-significant. Conversely, one Swedish and one Australian study reported low incidence of MS in people following diets enriched in PUFAs [5,12,13,25,26]. Results from meta-analyses suggest that PUFAs may reduce the frequency of relapses, but are not effective against the progression of the disease [1,19]. In human studies, a low fat diet supplemented with PUFAs was associated with lower levels of disability assessed by EDSS, slight improvement in relapse rat, as well as improved quality of life [13,25]. Another study provided evidence of PUFA-related improvement with respect to specific markers linked with inflammation and/or neurodegeneration in patients with MS (for instance, matrix metallopeptidase-9 (MMP-9) rather than in quality of life, EDSS score, or fatigue [26].

Among PUFAs, α-linolenic acid (ALA) is associated with low incidence of MS. It can contribute to the immune pathway by decreasing markers of inflammation. Eicosapentaenoic acids (EPAs) and docosahexaenoic acids (DHAs) can also play a role in in decreasing MMP-9 levels in patients with MS [25]. Riccio et al. reported that fish oil supplementation enriched with omega-3 fatty acids have a beneficial effect in the inhibition of the expression and reduction in the levels of MMP-9 in MS patients [27]. Ramirez et al. reported the beneficial effects of fish oil containing high amounts of omega-3 PUFAs into protecting against inflammation and oxidative stress [25]. Omega-3 fatty acid supplementation results in the decrease in proinflammatory cytokines, free radicals, and as a result, improving the quality of life of patients with MS by decreasing relapse rates [25].

Polyphenols, which are included in vegetables, fruit, wine, and tea, have been proven to be beneficial, leading to modulation of the immune response and affecting the expression of genes encoding pro-survival proteins including antioxidant enzymes. Polyphenols can also enhance neuronal survival [28]. Studies have focused particularly on polyphenols such as resveratrol and ginkgo biloba. In animal studies, these compounds seemed to promote protection against oxidative stress, also protecting against demyelination and axonal injury [26]. Khalili et al. suggested that lipoic acid consumption by patients with MS results in the improvement of total antioxidant capacity [13].

Randomized clinical trials seem to confirm the efficacy of some of the compounds discussed above such as melatonin, vitamin D3, omega-3 PUFAs, and polyphenol compounds. However, further research is needed in order to understand the potential protective effects exerted by antioxidants on the cellular immunology of MS neurodegeneration [12].

2.1.2. Mitochondria—Energy Production

Mitochondrial injury or the accumulation of iron in the brain is also enhanced in the progressive phase of the disease [12]. In patients with MS, mitochondrial structural changes and enzyme activity increase ROS production and cause oxidative damage [12]. Among the other antioxidants, curcumin is especially reported to play a major role against free radicals. Curcumin may benefit patients with MS by binding transition metals and forming stable inactive complexes, especially with ferrous ions, protecting against neurodegeneration [29].

2.2. Immune System (Innate and Adaptive) Responses—Factors of Immune System Activation

Nutrients and special diets such as saturated and 'trans' fatty acids, α-lipoic acid, polyphenols, high-fat diet, and high-carbohydrate diet result in the modulation of the components of inflammatory cascade. Several studies have shown that saturated and 'trans' fatty acids and lipopolysaccharide (LPS) may upregulate the activity of proinflammatory compounds, promoting inflammation; on the other hand, calorie restriction, polyphenols, and Ω-3 PUFAs would exert the opposite effect [26]. The influence of diet on inflammatory and autoimmune processes in MS is highlighted, supporting the hypothesis of a close relationship between nutritional factors and the immune system responses that play a role in the pathogenesis of MS [26].

2.3. Proinflammatory Diet

Recent studies have highlighted the role of proinflammatory diets in the pathogenesis of MS. Fatty acids and polyphenols as well as diets high in carbohydrates and fats may induce inflammatory cascade [26]. Diet can induce the production of inflammatory factors such as tumor necrosis factor, interleukins, MMP9, prostaglandins, and leukotrienes, leading to inflammation and oxidative stress [26].

Swank et al. reported adverse effects of saturated fatty acids (SFAs) on the course of MS, emphasizing their proinflammatory character [9]. High intake of SFAs leads to a dysbiosis of gut microbiota. Additionally, the consumption of vegetable oils, which are enriched with trans fatty acids, is associated with gut inflammation and the upregulation of proinflammatory cells [30]. Red meat leads to the formation of nitrous compounds increasing chronic inflammation. Red meat also contains arachidonic acid, which participates in inflammatory pathways by activating Th17 cells [27]. Furthermore, a high consumption of sugar-sweetened beverages and refined cereals leads to the production of insulin, which, in this way, is responsible for the upregulation of synthesis and the production of arachidonic acid. High salt intake can induce the production of Th17 cells and proinflammatory cytokines [27]. Proteins contained in cow-milk may play a role in the mechanisms of pathogenesis of MS. Particularly, butyrophilin can induce EAE by mechanisms of molecular mimicry with myelin oligodendrocyte glycoprotein [27].

2.4. Gut Brain-Axis and MS

2.4.1. Gut Microbiota

The gut–brain axis represents a bidirectional communication system between the CNS and the gastrointestinal system that includes the CNS, the enteric nervous system, the autonomic nervous system, the immune system, and the gut microbiota [31,32]. The role of gut microbiota is crucial because of its impact on regulating and maintaining the normal function of the innate immune system [31,32]. From birth to adolescence, commensal microbiota infests the gastrointestinal system, remaining in a stable condition, a state called eubiosis [31,32]. However, in early stages of life, factors such as antibiotics, infections, or unhealthy dietary habits may lead to alterations of the relative distribution and frequency of commensal microbiota, thus also, at least in part, predisposing to gut colonization by pathogens, a state called dysbiosis. In dysbiosis, there is an increase in the number of pathogenic bacteria and decrease in their biodiversity, resulting in gastrointestinal and systemic inflammation, possibly leading to increased risk for local or systemic inflammatory disease [31,32].

Studies on the experimental model of MS show the possible association of gut microbiota with the severity of the disease, indicating a possible protective role as well as a role in inducing pathological mechanisms in the context of immune dysregulation in CNS autoimmunity. The presence of gut commensal microbiota is necessary for the occurrence of CNS autoimmunity [33]. In animal models, it has been shown that the gut is involved in the modulation of inflammation of the CNS and may orchestrate mechanisms of immune tolerance, thus protecting from the development of CNS autoimmunity [34]. Metagenomic studies addressing the gut microbiota composition by next generation sequencing (NGS) demonstrated microbial imbalance and differences in the relative composition of gut microbiota in MS patients compared to healthy individuals, linking the dysbiosis with possible MS pathogenesis [35,36]. In this respect, nutritional modification with a potential to modulate the gut commensal microbiota has been advocated as a strategy that may affect the development of the disease and/or alter the disease course [1,32,33,37,38].

Metabolism of nutrients, especially carbohydrates, the production of neurotransmitters and vitamins, and competition with other colonizing pathogens are some of the main physiological functions of gut microbiota [32]. Furthermore, gut microbiota is possibly associated with CNS homeostasis and development and also with neuroimmunological and neurodegenerative disease [32]. Diet comprises a main factor determining the synthesis and metabolism of gut microbiota, thus enabling the host to defend against pathogens. The role of gut microbiota is also significant for the regulation of the immune system by affecting the overall activation status of T cells and other cells of the innate and adaptive immunity. In particular, T regulatory cells and T helper cells type 2 may suppress the activation of the immune system [1,32,33,39]. Moreover, short-chain fatty acids (SCFAs) such as butyrate, derived from gut microbiota, promote anti-inflammatory processes by producing anti-inflammatory cytokines and by inhibiting the connection of leukocytes to epithelium [1,32,33,39]. It is believed that the consumption of a diet with high fiber intake may increase the production of butyrate, thus leading to improved outcomes in patients with CNS disorders [1,32,33,39]. Studies in animal models demonstrated a strong and important connection between the microbiota, butyrate production, and the CNS. Patients with MS have lower levels of SCFAs in feces as well as reduced frequency of SCFA-producing bacteria in the gut [1,32,33,39].

Nutrition and dietary interventions regulate the gut microbiota affecting its composition and its functionality [32]. Diets characterized by a high intake of fat, sugar, and animal protein may lead to the development of specific pathogenic bacteria species such as Bacteroidetes in the gut, which, in turn, may induce enteric inflammation, damage of the intestinal barrier and increase in cross-reactive cells of the adaptive immunity [31]. Moreover, diet-induced low biodiversity of gut microbiota is associated with metabolic changes as well as an increase in inflammation markers [31].

It has been proposed that a diet enriched with vegetables, a high amount of fiber combined with probiotics, vitamin D and vitamin A supplementation, and lipoic acid results in gut eubiosis. This leads to an increase in microbial diversity and microbe-associated anti-inflammatory mediators such as short chained fatty acids (SCFAs). Conversely, a diet rich in animal fat and trans-fatty acids also including sugar and salt intake, promotes gut dysbiosis and results in an increase in the presence of pro-inflammatory mediators and also in gut barrier and blood–brain barrier (BBB) permeability, resulting in CNS autoimmunity [40].

2.4.2. Effects of Pre- and Probiotics in Patients with Multiple Sclerosis

The impact of diet on gut microbiota has been experimentally studied through the observations of effects of pre- and probiotics in patients with autoimmune diseases [37]. Prebiotics are nonviable compounds of living microorganisms with an ability to beneficially manipulate the host's microbiota. Many fermentable carbohydrates have prebiotic effects. Non digestible oligosaccharides such as fructans and glycans, which are utilized by Bifidobacteria, are reported to have the most beneficial effects. In addition, oligosaccharides identified in dairy products are reported to act as prebiotics. Probiotics are mostly consumable live microorganisms such as Lactobacillus. Sources of probiotics are contained in food such as yogurt [38]. Kouchaki et al. reported improvement in EDSS scale and decrease in inflammatory markers in patients with MS who were treated with probiotic supplementation [39]. A number of studies have encouraged the use of pre- and probiotics in patients with MS due to their benefits in maintaining the homeostasis of the CNS, improving the intestinal microbial balance and regulating the composition of gut microbiota. In these studies, the combination of prebiotics and probiotics is highly recommended [27].

3. Comorbidities in MS as an Independent Factor of Pathology—The Effect of Diet and Nutrition

Results from recent studies have shown that the presence of factors that predispose toward cardio-vascular risk in MS patients not only increases the risk for higher disability in the context of MS, but is also associated with diagnostic delay in MS. In addition, vascular comorbidity is reported to be associated with higher risk of hospitalization for patients with MS [5]. Of these, hyperlipidemia and obesity are the most common comorbidities among patients with MS [5].

3.1. Hyperlipidemia

Hyperlipidemia has been reported to be a common comorbidity among patients with MS. The mechanism by which hyperlipidemia affects MS is currently unexplained, but it may involve fatty acid metabolic and inflammatory pathways that influence the regulation of gene expression and metabolism. Lipid composition of myelin and its morphology are affected during neuroinflammation. A high-fat diet is thought to promote neuroinflammation. Cholesterol and its components are associated with adverse disease effects on patients with MS. Moreover, it is hypothesized that cholesterol and its molecules could be markers of disease activity, markers of efficacy of treatment, or possible therapeutic targets [1]. It is also suggested that fatty acids are not only associated with neuroinflammation, but also with neurodegeneration, affecting the progression of MS [1]. High consumption of fatty acids, especially saturated fats, evidently leads to increased levels of blood LDL cholesterol. In particular, long chain fatty acids included in processed food influence the immune system by activating proinflammatory components, leading to T cell and macrophage activation and expression of inflammatory cytokines [5]. Swank et al. reported in an epidemiological study in patients with MS that low intake of saturated fat was associated with lower disability and mortality rate. However, no controlled randomized trials that have focused on saturated fat consumption exist, thus conclusions are difficult to make [5].

3.2. Obesity and Increased BMI

Patients with MS often consume a low-carbohydrate and high-lipid diet associated with abdominal obesity and higher body mass index (BMI). This condition leads to a pro-inflammatory status increasing levels of interleukin 6 (IL-6), TNF-alpha, and leptin, factors that are associated with MS pathogenesis [41]. Additionally, recent studies suggest that increased BMI and obesity play a major role in MS development and progression [42]. Obesity and MS can lead to altered adipokine release into the blood circulation. This activates inflammatory pathways and increases the infiltration of immune cells in the CNS. Moreover, increased levels of pro-inflammatory cytokines in MS release pro-inflammatory adipokines and disrupt adipokine pathway, creating a feedback loop [42]. Pathophysiologically, obesity affects MS by promoting chronic inflammation, altering the endocrine system by disturbing the secretion of adipokines and influencing the gut microbiota [42]. Moreover, an unbalanced diet contributes to an increase in BMI and abdominal obesity. BMI is affected by protein and lipid intake as well as carbohydrate intake. Excessive simple carbohydrate intake is related to obesity and being overweight as well as an increase in adipose tissue in the abdomen [43]. It has been reported that in MS patients, intake of lipid and protein favors abdominal obesity and increases BMI. This fact is associated with an increase in proinflammatory cytokines, promoting inflammation [43].

Overweight and obesity are associated with chronic inflammation of adipose tissue, which is accompanied by an altered secretion of adipokines. These adipokines are hormones and cytokines that regulate metabolic pathways [42]. Among others, leptin and adiponectin seem to be associated with the pathogenesis of MS. Leptin regulates energy and acute phase reactions, interfering in the activity and progression of experimental autoimmune encephalomyelitis (EAE) in mice or the pathogenesis of MS in humans [26]. Increased leptin in humans has been suggested to be related with proinflammatory conditions and autoimmunity. As a result, a diet associated with increased levels of leptin could possibly influence the balance between T cells, promoting inflammation and inducing cell-mediated autoimmunity [26]. In humans, hyperleptinemia has been correlated with pro-inflammatory conditions [44]. In patients with MS, expression of leptin receptors has been found to be significantly higher in the relapse phase than that observed in remission [45]. Piccio et al. suggested that caloric restriction, leading to a reduction in leptin levels, can reduce inflammation, demyelination, and axonal injury [46]. Adiponectin (APN) is another adipokine that exhibits anti-inflammatory activity on immune system cells [47]. Musabak et al. found that in patients with MS, the APN serum levels were lower than those of the healthy controls [48]. Their study found that low APN levels were associated with high risk for early onset of MS, especially in females, and also with increased disability and progression, predicted by higher EDSS score [48].

4. Dietary Patterns—Interventions and Effects on MS

Dietary approaches have been studied in order to improve outcomes in MS. However, several studies have provided at least some indications on the potential role of dietary habits in the course of MS. Among the most popular dietary interventions overall are the Mediterranean, the Paleolithic, the Swank, the McDougall, and the Hyberbolic-caloric restriction diets [1,31].

4.1. Mediterranean Diet

The Mediterranean diet consists of a high intake of fruit, vegetables, and whole grains including olive oil as the main source. It also includes a moderate amount of fish and dairy products and a low intake of red meat [5]. Phenols in olive oil are responsible for its anti-inflammatory effects, thus protecting the nervous system from oxidative stress. It seems that the Mediterranean diet reduces inflammatory markers [31], regulates predisposing factors of vascular pathology in the context of several autoimmune disorders [28,49], and regulates gut microbiota [28,49]. Thus, it has been suggested that the Mediterranean diet is associated with low risk for MS onset [5,30,32].

4.2. Paleolithic Diet

The Paleolithic diet is characterized by the consumption of leaf green vegetables, plant proteins, soy, and nuts excluding the consumption of dairy and processed food. Studies indicate that fatigue in MS patients following the Paleolithic diet was improved, although risk for nutritional deficiencies was increased [5,32].

4.3. Swank Diet

The Swank diet is based on limited saturated fat intake. Increased fruit, vegetable, and oil intake is encouraged. Swank observed that the risk of MS development was higher in people living in areas with a high consumption of fat [9]. This study showed that patients with MS who followed the Swank diet exhibited a lower risk for relapse, disease progression, and reduced mortality. Although possible underlying mechanisms have not been identified, it has been advocated that reduction in fat consumption may be related to protection against inflammation and demyelination [1,32].

4.4. McDougall Diet

The main caloric source of the McDougall diet is carbohydrates, based on consuming plants [5]. In addition, olive oil and animal products including eggs and dairy products are not preferred and restricted. Studies showed an association with lower fatigue in the group of MS patients who followed the McDougall diet [9,32]. Conversely, there was no significant effect on relapse rate, magnetic resonance imaging activity and disability [5,32].

4.5. Hyberbolic Diet-Caloric Restriction

It is believed that an increased amount of consumed calories is associated with inflammation, especially following meals [5,9]. As a result, caloric restriction has been suggested to reduce the risk of postprandial inflammation. Results from studies concerning caloric restriction revealed a reduction in oxidative stress in patients with relapsing and progressive types of MS, leading to a better quality of life [32]. Trials on the possible beneficial effect of intermittent fasting on relapse rate and progression of MS are currently ongoing [49,50]. Overall, following an anti-inflammatory diet, in terms of intermittent fasting, seems to regulate inflammation and protect against oxidative damage and progression of MS in experimental models [27]. Choi et al. reported in their study that an anti-inflammatory diet that included mainly caloric restriction diet showed promising results by reducing inflammation and promoting regeneration in animal models with experimented autoimmune encephalomyelitis [51].

4.6. Ketogenic Diet

The ketogenic diet is low in carbohydrates and high in fat [32]. It is characterized by the induction of ketones released in blood circulation that may exert anti-inflammatory effects [32]. Trials in MS patients revealed a possible improvement in quality of life, fatigue, and depression related to the ketogenic diet, although negative effects including deficiency of vitamins, weight loss, and gastrointestinal symptoms may also arise [49].

4.7. Gluten Free Diet

Current evidence does not show a significant effect of the gluten free diet in MS [49,50]. Few gluten-free interventions showed an improvement in EDSS, lesion activity, fatigue, and quality of life, however, existing studies pose limitations in terms of the high risk of bias and also the lack of controlled randomized trials [49,50].

5. Conclusions

Although a balanced diet involving high amount of fruit, vegetables, and low fat may not replace DMTs in controlling disease activity in MS, it may have an add on value in a more efficient management of the disease overall. Existing evidence indicates that nutrition and diet may play a role in MS pathogenesis and course. These factors may affect gut microbiota function, enzyme activity, and risk factors of vascular pathology in MS patients. At the moment, precise recommendations regarding a specific dietary plan in patients with MS do not exist. However, clinical and experimental studies provide indirect evidence that a balanced diet in combination with an overall healthy lifestyle is linked with an improvement in several clinical parameters as well as measurements of quality of life for patients with MS. Furthermore, we strongly suggest large, well-scheduled clinical trials containing both clinical and biochemical, molecular, metagenomic, and metabolomic technologies aiming to clarify the role of diet in MS management.

Author Contributions: Conceptualization, M.B. and N.G.; Writing—original draft preparation, P.S. and M.B.; Reviewing and editing, E.K., C.B., S.-A.S., N.K. and M.B.; Supervision, N.G. All authors have read and agreed to the published version of the manuscript.

Funding: This work was supported by an unrestricted fund provided by Genesis Pharma entitled "Educational Program in experimental and clinical neuroimmunology aiming in the study of autoimmune neurological diseases—NEURO" (ESO.0423006) via the Institute of Applied Biosciences at the Centre for Research and Technology Hellas (INAB ǀ CERTH).

Institutional Review Board Statement: Not applicable.

Informed Consent Statement: Not applicable.

Data Availability Statement: Not applicable.

Conflicts of Interest: The authors declare no conflict of interest.

References

1. Penesová, A.; Dean, Z.; Kollár, B.; Havranová, A.; Imrich, R.; Vlček, M.; Rádiková, Ž. Nutritional intervention as an essential part of multiple sclerosis treatment? *Physiol. Res.* **2018**, *67*, 521–533. [CrossRef]
2. Lublin, F.D.; Reingold, S.C.; Cohen, J.A.; Cutter, G.R.; Sørensen, P.S.; Thompson, A.J.; Wolinsky, J.S.; Balcer, L.J.; Banwell, B.; Barkhof, F.; et al. Defining the clinical course of multiple sclerosis: The 2013 revisions. *Neurology* **2014**, *83*, 278–286. [CrossRef]
3. Thompson, A.J.; Banwell, B.L.; Barkhof, F.; Carroll, W.M.; Coetzee, T.; Comi, G.; Correale, J.; Fazekas, F.; Filippi, M.; Freedman, M.S.; et al. Diagnosis of multiple sclerosis: 2017 revisions of the McDonald criteria. *Lancet Neurol.* **2018**, *17*, 162–173. [CrossRef]
4. Kołtuniuk, A.; Chojdak-Łukasiewicz, J. Adherence to Therapy in Patients with Multiple Sclerosis-Review. *Int. J. Environ. Res. Public Health* **2022**, *19*, 2203. [CrossRef]
5. Katz Sand, I. The Role of Diet in Multiple Sclerosis: Mechanistic Connections and Current Evidence. *Curr. Nutr. Rep.* **2018**, *7*, 150–160. [CrossRef]
6. Wu, H.; Zhao, M.; Yoshimura, A.; Chang, C.; Lu, Q. Critical Link Between Epigenetics and Transcription Factors in the Induction of Autoimmunity: A Comprehensive Review. *Clin. Rev. Allergy Immunol.* **2016**, *50*, 333–344. [CrossRef]
7. Marck, C.H.; Neate, S.L.; Taylor, K.; Weiland, T.; Jelinek, G. Prevalence of Comorbidities, Overweight and Obesity in an International Sample of People with Multiple Sclerosis and Associations with Modifiable Lifestyle Factors. *PLoS ONE* **2016**, *11*, e0148573. [CrossRef]
8. Dos Passos, G.R.; Sato, D.K.; Becker, J.; Fujihara, K. Th17 Cells Pathways in Multiple Sclerosis and Neuromyelitis Optica Spectrum Disorders: Pathophysiological and Therapeutic Implications. *Mediat. Inflamm.* **2016**, *2016*, 5314541. [CrossRef]
9. Mao, P.; Reddy, P.H. Is multiple sclerosis a mitochondrial disease? *Biochim. Biophys. Acta* **2010**, *1802*, 66–79. [CrossRef]
10. Pathak, D.; Berthet, A.; Nakamura, K. Energy failure: Does it contribute to neurodegeneration? *Ann. Neurol.* **2013**, *74*, 506–516. [CrossRef]
11. Lee, D.H.; Gold, R.; Linker, R.A. Mechanisms of oxidative damage in multiple sclerosis and neurodegenerative diseases: Therapeutic modulation via fumaric acid esters. *Int. J. Mol. Sci.* **2012**, *13*, 11783–11803. [CrossRef]
12. Miller, E.D.; Dziedzic, A.; Saluk-Bijak, J.; Bijak, M. A Review of Various Antioxidant Compounds and their Potential Utility as Complementary Therapy in Multiple Sclerosis. *Nutrients* **2019**, *11*, 1528. [CrossRef]
13. Bagur, M.J.; Murcia, M.A.; Jiménez-Monreal, A.M.; Tur, J.A.; Bibiloni, M.M.; Alonso, G.L.; Martínez-Tomé, M. Influence of Diet in Multiple Sclerosis: A Systematic Review. *Adv. Nutr.* **2017**, *8*, 463–472. [CrossRef]
14. Garcia-Alloza, M.; Borrelli, L.A.; Rozkalne, A.; Hyman, B.T.; Bacskai, B.J. Curcumin labels amyloid pathology in vivo, disrupts existing plaques, and partially restores distorted neurites in an Alzheimer mouse model. *J. Neurochem.* **2007**, *102*, 1095–1104. [CrossRef]
15. Miller, E.; Markiewicz, L.; Kabzinski, J.; Odrobina, D.; Majsterek, I. Potential of redox therapies in neurodegenerative disorders. *Front. Biosci. (Elite Ed.)* **2017**, *9*, 214–234. [CrossRef]
16. Rathnasamy, G.; Ling, E.A.; Kaur, C. Therapeutic implications of melatonin in cerebral edema. *Histol. Histopathol.* **2014**, *29*, 1525–1538.
17. Pierrot-Deseilligny, C.; Souberbielle, J.C. Vitamin D and multiple sclerosis: An update. *Mult. Scler. Relat. Disord.* **2017**, *14*, 35–45. [CrossRef]
18. Salzer, J.; Hallmans, G.; Nystrom, M.; Stenlund, H.; Wadell, G.; Sundstrom, P. Vitamin D as a protective factor in multiple sclerosis. *Neurology* **2012**, *79*, 2140–2145. [CrossRef]
19. Tredinnick, A.R.; Probst, Y.C. Evaluating the Effects of Dietary Interventions on Disease Progression and Symptoms of Adults with Multiple Sclerosis: An Umbrella Review. *Adv. Nutr.* **2020**, *11*, 1603–1615. [CrossRef]
20. Dobson, R.; Cock, H.R.; Brex, P.; Giovannoni, G. Vitamin D supplementation. *Pract. Neurol.* **2018**, *18*, 35–42. [CrossRef]
21. Rito, Y.; Torre-Villalvazo, I.; Flores, J.; Rivas, V.; Corona, T. Epigenetics in Multiple Sclerosis: Molecular Mechanisms and Dietary Intervention. *Cent. Nerv. Syst. Agents Med. Chem.* **2018**, *18*, 8–15. [CrossRef]
22. IOM (Institute of Medicine). *Dietary Reference Intakes for Vitamin A, Vitamin K, Arsenic, Boron, Chromium, Copper, Iodine, Iron, Manganese, Molybdenum, Nickel, Silicon, Vanadium, and Zinc*; The National Academies Press: Washington, DC, USA, 2001.
23. Xiao, S.; Jin, H.; Korn, T.; Liu, S.M.; Oukka, M.; Lim, B.; Kuchroo, V.K. Retinoic acid increases Foxp3+ regulatory T cells and inhibits development of Th17 cells by enhancing TGF-beta-driven Smad3 signaling and inhibiting IL-6 and IL-23 receptor expression. *J Immunol.* **2008**, *181*, 2277–2284. [CrossRef]
24. Bitarafan, S.; Saboor-Yaraghi, A.; Sahraian, M.-A.; Soltani, D.; Nafissi, S.; Togha, M.; Moghadam, N.B.; Roostaei, T.; Honarvar, N.M.; Harirchian, M.-H. Effect of Vitamin A Supplementation on fatigue and depression in Multiple Sclerosis patients: A Double-Blind Placebo-Controlled Clinical Trial. *Iran. J. Allergy Asthma Immunol.* **2016**, *15*, 13–19.
25. AlAmmar, W.A.; Albeesh, F.H.; Ibrahim, L.M.; Algindan, Y.Y.; Yamani, L.Z.; Khattab, R.Y. Effect of omega-3 fatty acids and fish oil supplementation on multiple sclerosis: A systematic review. *Nutr. Neurosci.* **2021**, *24*, 569–579. [CrossRef]
26. Esposito, S.; Bonavita, S.; Sparaco, M.; Gallo, A.; Tedeschi, G. The role of diet in multiple sclerosis: A review. *Nutr. Neurosci.* **2018**, *21*, 377–390. [CrossRef]
27. Riccio, P.; Rossano, R. Nutrition facts in multiple sclerosis. *ASN Neuro* **2015**, *7*, 1759091414568185. [CrossRef]

28. Santangelo, C.; Vari, R.; Scazzocchio, B.; De Sanctis, P.; Giovannini, C.; D'Archivio, M.; Masella, R. Anti-inflammatory Activity of Extra Virgin Olive Oil Polyphenols: Which Role in the Prevention and Treatment of Immune-Mediated Inflammatory Diseases? *Endocr. Metab. Immune Disord. Drug Targets* **2018**, *18*, 36–50. [CrossRef]

29. Lee, W.H.; Loo, C.Y.; Bebawy, M.; Luk, F.; Mason, R.S.; Rohanizadeh, R. Curcumin and its derivatives: Their application in neuropharmacology and neuroscience in the 21st century. *Curr. Neuropharmacol.* **2013**, *11*, 338–378. [CrossRef]

30. Hoare, S.; Lithander, F.; van der Mei, I.; Ponsonby, A.L.; Lucas, R.; Ausimmune Investigator, G. Higher intake of omega-3 polyunsaturated fatty acids is associated with a decreased risk of a first clinical diagnosis of central nervous system demyelination: Results from the Ausimmune Study. *Mult. Scler.* **2016**, *22*, 884–892. [CrossRef]

31. Altowaijri, G.; Fryman, A.; Yadav, V. Dietary Interventions and Multiple Sclerosis. *Curr. Neurol. Neurosci. Rep.* **2017**, *17*, 28. [CrossRef]

32. Fleck, A.K.; Schuppan, D.; Wiendl, H.; Klotz, L. Gut-CNS-Axis as Possibility to Modulate Inflammatory Disease Activity-Implications for Multiple Sclerosis. *Int. J. Mol. Sci.* **2017**, *18*, 1526. [CrossRef] [PubMed]

33. Berer, K.; Mues, M.; Koutrolos, M.; Al Rasbi, Z.; Boziki, M.; Johner, C.; Wekerle, H.; Krishnamoorthy, G. Commensal microbiota and myelin autoantigen cooperate to trigger autoimmune demyelination. *Nature* **2011**, *479*, 538–541. [CrossRef] [PubMed]

34. Berer, K.; Boziki, M.; Krishnamoorthy, G. Selective accumulation of pro-inflammatory T cells in the intestine contributes to the resistance to autoimmune demyelinating disease. *PLoS ONE* **2014**, *9*, e87876. [CrossRef] [PubMed]

35. Chen, J.; Chia, N.; Kalari, K.R.; Yao, J.Z.; Novotna, M.; Paz Soldan, M.M.; Luckey, D.H.; Marietta, E.V.; Jeraldo, P.R.; Chen, X.; et al. Multiple sclerosis patients have a distinct gut microbiota compared to healthy controls. *Sci. Rep.* **2016**, *6*, 28484. [CrossRef]

36. Jangi, S.; Gandhi, R.; Cox, L.; Li, N.; Von Glehn, F.; Yan, R.; Patel, B.; Mazzola, M.A.; Liu, S.; Glanz, B.L.; et al. Alterations of the human gut microbiome in multiple sclerosis. *Nat. Commun.* **2016**, *7*, 12015. [CrossRef]

37. Moles, L.; Otaegui, D. The Impact of Diet on Microbiota Evolution and Human Health. Is Diet an Adequate Tool for Microbiota Modulation? *Nutrients* **2020**, *12*, 1654. [CrossRef]

38. Lombardi, V.C.; De Meirleir, K.L.; Subramanian, K.; Nourani, S.M.; Dagda, R.K.; Delaney, S.L.; Palotás, A. Nutritional modulation of the intestinal microbiota; future opportunities for the prevention and treatment of neuroimmune and neuroinflammatory disease. *J. Nutr. Biochem.* **2018**, *61*, 1–16. [CrossRef]

39. Kouchaki, E.; Tamtaji, O.R.; Salami, M.; Bahmani, F.; Kakhki, R.D.; Akbari, E.; Tajabadi-Ebrahimi, M.; Jafari, P.; Asemi, Z. Clinical and metabolic response to probiotic supplementation in patients with multiple sclerosis: A randomized, double-blind, placebo-controlled trial. *Clin. Nutr.* **2017**, *36*, 1245–1249. [CrossRef]

40. Boziki, M.K.; Kesidou, E.; Theotokis, P.; Mentis, A.-F.A.; Karafoulidou, E.; Melnikov, M.; Sviridova, A.; Rogovski, V.; Boyko, A.; Grigoriadis, N. Microbiome in Multiple Sclerosis; Where Are We, What We Know and Do Not Know. *Brain Sci.* **2020**, *10*, 234. [CrossRef]

41. Drehmer, E.; Platero, J.L.; Carrera-Juliá, S.; Moreno, M.L.; Tvarijonaviciute, A.; Navarro, M.Á.; López-Rodríguez, M.M.; Ortí, J.E.d.L.R. The Relation between Eating Habits and Abdominal Fat, Anthropometry, PON1 and IL-6 Levels in Patients with Multiple Sclerosis. *Nutrients* **2020**, *12*, 744. [CrossRef]

42. Rijnsburger, M.; Djuric, N.; Mulder, I.A.; de Vries, H.E. Adipokines as Immune Cell Modulators in Multiple Sclerosis. *Int. J. Mol. Sci.* **2021**, *22*, 10845. [CrossRef] [PubMed]

43. de Oliveira, C.; de Mattos, A.B.; Biz, C.; Oyama, L.M.; Ribeiro, E.B.; do Nascimento, C.M. High-fat diet and glucocorticoid treatment cause hyperglycemia associated with adiponectin receptor alterations. *Lipids Health Dis.* **2011**, *10*, 11. [CrossRef] [PubMed]

44. Guerrero-Garcia, J.J.; Carrera-Quintanar, L.; Lopez-Roa, R.I.; Marquez-Aguirre, A.L.; Rojas-Mayorquin, A.E.; Ortuno-Sahagun, D. Multiple Sclerosis and Obesity: Possible Roles of Adipokines. *Mediat. Inflamm.* **2016**, *2016*, 4036232. [CrossRef] [PubMed]

45. Matarese, G.; Carrieri, P.B.; Montella, S.; De Rosa, V.; La Cava, A. Leptin as a metabolic link to multiple sclerosis. *Nat. Rev. Neurol.* **2010**, *6*, 455–461. [CrossRef] [PubMed]

46. Piccio, L.; Stark, J.L.; Cross, A.H. Chronic calorie restriction attenuates experimental autoimmune encephalomyelitis. *J. Leukoc. Biol.* **2008**, *84*, 940–948. [CrossRef] [PubMed]

47. Ouchi, N.; Walsh, K. A novel role for adiponectin in the regulation of inflammation. *Arter. Thromb. Vasc. Biol.* **2008**, *28*, 1219–1221. [CrossRef]

48. Musabak, U.; Demirkaya, S.; Genc, G.; Ilikci, R.S.; Odabasi, Z. Serum adiponectin, TNF-alpha, IL-12p70, and IL-13 levels in multiple sclerosis and the effects of different therapy regimens. *Neuroimmunomodulation* **2011**, *18*, 57–66. [CrossRef]

49. Evans, E.; Levasseur, V.; Cross, A.H.; Piccio, L. An overview of the current state of evidence for the role of specific diets in multiple sclerosis. *Mult. Scler. Relat. Disord.* **2019**, *36*, 101393. [CrossRef]

50. Thomsen, H.L.; Jessen, E.B.; Passali, M.; Frederiksen, J.L. The role of gluten in multiple sclerosis: A systematic review. *Mult. Scler. Relat. Disord.* **2019**, *27*, 156–163. [CrossRef]

51. Choi, I.Y.; Piccio, L.; Childress, P.; Bollman, B.; Ghosh, A.; Brandhorst, S.; Suarez, J.; Michalsen, A.; Cross, A.; Morgan, T.E.; et al. A Diet Mimicking Fasting Promotes Regeneration and Reduces Autoimmunity and Multiple Sclerosis Symptoms. *Cell Rep.* **2016**, *15*, 2136–2146. [CrossRef]

 *nutrients*

Brief Report

Fast Anxiolytic-Like Effect Observed in the Rat Conditioned Defensive Burying Test, after a Single Oral Dose of Natural Protein Extract Products

Thomas Freret [1,2,*], Stacy Largilliere [1], Gerald Nee [1], Melanie Coolzaet [1], Sophie Corvaisier [1] and Michel Boulouard [1]

[1] UNICAEN, INSERM, COMETE, Cyceron, CHU Caen, Normandie University, 14000 Caen, France; stacy.largilliere@unicaen.fr (S.L.); gerald.nee@unicaen.fr (G.N.); coolzaet.melanie@gmail.com (M.C.); sophie.corvaisier@unicaen.fr (S.C.); michel.boulouard@unicaen.fr (M.B.)
[2] Behavioral Research Platform, Normandie University, 14000 Caen, France
* Correspondence: thomas.freret@unicaen.fr; Tel.: +33-2315-66877

Abstract: Anxiety appears among the most frequent psychiatric disorders. During recent years, a growing incidence of anxiety disorders can be attributed, at least in part, to the modification of our eating habits. To treat anxiety disorders, clinicians use benzodiazepines, which unfortunately display many side effects. Herein, the anxiolytic-like properties of two natural products (αS1–casein hydrolysate and Gabolysat®) were investigated in rats and compared to the efficacy of benzodiazepine (diazepam). Thus, the conditioned defensive burying test was performed after a unique oral dose of 15 mg/kg, at two time-points (60 min and then 30 min post oral gavage) to show potential fast-onset of anxiolytic effect. Both natural products proved to be as efficient as diazepam to reduce the time rats spent burying the probe (anxiety level). Additionally, when investigated as early as 30 min post oral gavage, Gabolysat® also revealed a fast-anxiolytic activity. To date, identification of bioactive peptide, as well as how they interact with the gut–brain axis to sustain such anxiolytic effect, still remains poorly understood. Regardless, this observational investigation argues for the consideration of natural compounds in care pathway.

Keywords: fish protein hydrolysate; anxiety; burying test; rodent

Citation: Freret, T.; Largilliere, S.; Nee, G.; Coolzaet, M.; Corvaisier, S.; Boulouard, M. Fast Anxiolytic-Like Effect Observed in the Rat Conditioned Defensive Burying Test, after a Single Oral Dose of Natural Protein Extract Products. *Nutrients* **2021**, *13*, 2445. https://doi.org/10.3390/nu13072445

Academic Editor: Panagiota Mitrou

Received: 8 June 2021
Accepted: 14 July 2021
Published: 17 July 2021

Publisher's Note: MDPI stays neutral with regard to jurisdictional claims in published maps and institutional affiliations.

1. Introduction

Anxiety disorders are among the most prevalent and disabling psychiatric disorders worldwide [1]. Currently, on the market anxiolytic drugs are mostly benzodiazepine or benzodiazepine-like agents, which particularly target the GABAergic system. Unfortunately, their use is not harmless and requires caution and vigilance. Indeed, a life-threatening anxiety rebound effect may appear when withdrawal of benzodiazepine is too rapid [2–4]. Furthermore, benzodiazepines display significant side effects and drug interactions [5]. Risk-taking behavior have been observed under benzodiazepines prescription [6,7], together with impaired cognition, mobility and driving skills, as well as increased fall risk [8]. Additionally, if their consumption lasts for a long period (usually considered as 3 months), GABAergic medications might create a tolerance and dependence [5]. Finally, remote adverse effects have more recently been suspected, notably with an increased risk, to develop an Alzheimer's disease [9].

In view of the risks associated with benzodiazepine intake, intense preclinical and clinical research were carried out to identify/develop new therapeutic strategies for the management of anxiety disorders. Therefore, anxiolytic drugs free of GABAergic action (such as buspirone) or even anti-depressant drugs (such as Selective Serotonin/Norepinephrine Reuptake Inhibitors) have been proposed as a substitute. Although they might be used routinely to cure general anxiolytic disorders, they are unfortunately devoid of fast-acting effects. In addition, they may lack therapeutic efficacy and/or display important side effects. Thus, apart from only a few drug developments and/or repurposing of some antidepressant drugs, fast-acting, efficient and safe therapeutic alternatives are scarce [10].

Surprisingly, hope for a new therapeutic strategy could come from diet supplements. Indeed, they are increasingly considered as a natural alternative to treat—or to manage—anxiety disorders [11]. This assumption comes, at least in part, from an observed close relationship (that some would qualify as a causal relationship) between changes in eating habits (mostly in Western countries) and the occurrence of anxiety disorders (associated or not with sleep disturbances) [12]. As a consequence, increasing efforts have been made in the search for natural, non-chemical, diet supplements with anxiolytic-like effect. Notably, natural peptides extracted from animal proteins appear as promising candidates. Thus, several interesting results were reported from experiments conducted with the milk-derived αS1–casein hydrolysate or with the fish protein hydrolysate. In fact, under a chronic oral administration regimen, both hydrolysates demonstrated anxiolytic-type properties in rodents (both mice and rats [13–17]), but also in several other animal species such as cats and dogs [18–21]. A few investigating works in humans have also been published ([22,23], see also for review [24]). For instance, level of state anxiety—assessed by the STAI test—was reduced in young adult students aged 18 to 25 after 1 week of supplementation with fish protein hydrolysate [25]. Similarly, 30 days of daily administration of αS1–casein hydrolysate (150 mg) appeared efficient to relieve 63 women suffering from stress-related symptoms [23]. Of note, treatment efficiency was in this work evaluated through a newly constructed questionnaire on a mix of two previously published ones (Hamilton Anxiety scale and Ferreri Anxiety Rating Diagram). Nevertheless, other stress-related physiological parameters seem to benefit from oral intake regimen of hydrolysates. Indeed, a dampening effect in stress-induced variations was also observed on different physiological parameters, such as stress hormone level (corticosterone) or systolic blood pressure [22].

To further examine underlying mechanisms, several preclinical investigations have been performed and each reported an anxiolytic effect either with one or the other natural hydrolysates. To date, no study has compared similar experimental protocols for the onset/efficacy of different hydrolysates. In fact, as worthwhile it can be, such a comparison is not possible given the high level of disparities in terms of tested doses or administration regimen as well as behavioral test used. For instance, anxiolytic properties of fish protein hydrolysate were reported in the elevated plus maze and in the defensive burying test, as well as in the conditioned light extinction test. All those experiments were conducted using a chronic administration regimen, i.e., an oral gavage twice a day, for either 3 or 8 days, with doses ranging from 25 to 100 mg/kg [15,26]. In addition, when using even higher range doses (i.e., 300 and 1200 mg/kg during 5 days), biochemical experiments demonstrated that Gabolysat®, a proprietary fish protein hydrolysate, displayed a diazepam-like effect on stress responsiveness of the pituitary–adrenal system and sympathoadrenal activity (with a reduced adrenaline and noradrenaline level in a stress condition and increase brain hippocampal GABA content in a non-stress condition) [13]. Conversely, when anxiolytic properties of αS1–casein hydrolysate were investigated under an acute administration regimen, a dose effect study (from 5 to 50 mg/kg) demonstrated that the minimal effective dose (15 mg/kg) was as efficient as diazepam (3 mg/kg) in the conditioned burying test. This beneficial effect was then observed for 60 min after a single oral administration [16,17].

2. Materials and Methods

We aimed to assess and compare acute anxiolytic properties (i.e., after a single oral dose) of two compounds (Gabolysat® and αS1–casein hydrolysate), through a one trial burying behavioral test. This test was chosen given its high degree of face and construct validity, but also given its good pharmacological validation (predictive validity) [27]. Efficacy of the tested compound was compared to a reference drug (the most frequently consumed benzodiazepine on the market, i.e., diazepam).

Animals: All experiments were carried out in accordance with the European Communities Council Directive (63/2010) regarding the care and use of animals for experimental procedures, and they were approved by the local ethics committee (Comité d'Ethique NOrmandie en Matière d'EXpérimentation Animale, CENOMEXA, agreement number: 03-08-11/16/08-14). Aged of 8 weeks, male Wistar RjHan:WI rats were obtained from Janvier Labs (France) and pair-housed in Plexiglas cages with ad libitum access to food and water. A total number of 102 rats were necessary to perform displayed experiments. The animal facility was maintained on a controlled light–dark cycle (lights on from 7 a.m. to 7 p.m.), with a constant temperature (22 ± 2 °C). After 1 week acclimatization, animals were daily handled (2 × 5 min) during the following week to reduce the stress induced through the restraint required for oral gavage (Figure 1).

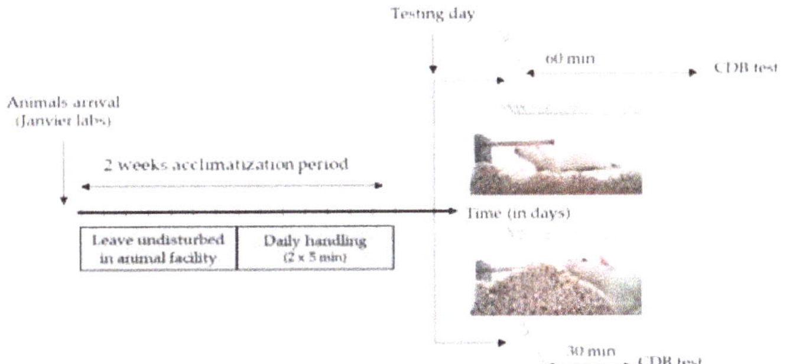

Figure 1. Experimental design. There was a 2 week acclimatization period, where animals were first left undisturbed for the first week, then handled twice daily in order to acclimate the animals to the experimenter. On the testing day, 60 or 30 min after an oral gavage, the Conditioned-Defense Burying (CDB) task was performed.

Group testing: αS1–casein hydrolysate and Gabolysat® were obtained from Ingredia® (Aras, France) and Laboratoire Dielen® (Cherbourg, France), respectively. Diazepam (Sigma, France) was used as the pharmacological benzodiazepine references substance. All compounds were solubilized in a saline (NaCl, 0.9%) mix with 1% Carboxymethylcellulose (Sigma Aldrich®), which was used as treatment for the control group. The dose of 15 mg/kg was selected according to previous related studies [13,17]. Each compound was orally administered in a volume of 2 mL/kg body weight.

Defensive probe-burying test: Animals were familiarized in pairs for 2 consecutive days (20 min session) with the testing room (dim lit) and apparatus which consisted in a Plexiglas acrylic cage 42.5 × 26.6 × 18.5 cm (Intelli-bio®, Seichamps, France) [28,29]. The floor of the cage was covered with 5 cm of bedding material (fine wood sawdust). The following day, each rat was individually placed in the testing cage, with an electrified probe (7 cm long) emerging from one of its walls 2 cm above the bedding material. The latency for first approach to the probe, reflecting level of exploration behavior, was then recorded to ensure no bias interpretation of the data. Once the rat touched the probe, it received an electric shock of 0.2 mA (constant current shocker). All tested rats, whatever group or experiments considered, touched the probe only once and, therefore, received only a single shock. Burying behavior (pushing the sawdust ahead with rapid alternating movements of the forepaws oriented to cover the electrified probe) has been directly related with the experimental anxiety levels [29]. This behavior was constantly recorded during a 15 min period and in 5 min time slices.

Statistical analysis: All graphs displayed results as the mean ± standard errors of mean (SEM). Statistical analyses were performed with Statview®. ANOVAs were performed, followed when appropriate by post-hoc group comparison tests. The *p*-value was set at 0.05. Bonferroni correction was applied for post-hoc multiple comparisons.

3. Results

3.1. Evaluation of Anxiolytic-Like Effect (60 min after Oral Gavage)

One-way ANOVA of elapsed time to first approach to the probe (probe contact latency) did not reveal any statistical group differences ($F_{(3,48)}$ = 0.301, p = 0.8245), thus, ensuring no locomotor and/or exploration behavior bias (Figure 2A). Furthermore, two-way ANOVA with a repeated measurement of burying time during the behavioral test did not reveal a group effect ($F_{(3,48)}$ = 1.655, p = 0.1892), but a significant time effect ($F_{(2,96)}$ = 27.773, p < 0.0001) and a group x time interaction ($F_{(6,96)}$ = 3.158, p = 0.0072) (Figure 2B). When focused on the first 5 min of the test, a significant group difference was revealed (one-way ANOVA, $F_{(3,48)}$ = 5.390 and p = 0.0028). In fact, the post-hoc test showed that all treated groups of rats (Diazepam, αS1–casein hydrolysate as well as Gabolysat®-treated animals) spent significantly less time burying the probe compared to the control group (p = 0.0162, 0.0159 and 0.0362, respectively). Thereafter (for the last 10 min of the test), probe-burying time for the control group massively dropped down (time effect: $F_{(2,24)}$ = 52.373 and p < 0.0001). No group difference can then be revealed for the last 10 min of the test (p > 0.05).

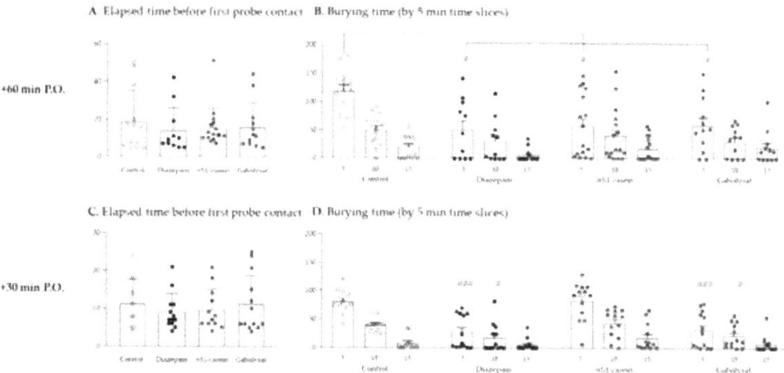

Figure 2. Conditioned behavioral test performed 60 min (**A,B**) and 30 min (**C,D**) after oral gavage. Figure displays the latency for first contact to probe, as a control measurement of motivation/locomotor activity for all animals' groups. Figure display probe-burying behavior during the test (by 5 min time slices), reflective of anxiety levels. A total of 102 rats were used. Groups sizes were 13, 11, 16 and 12 for control, diazepam, αS1–casein hydrolysate and Gabolysat®, respectively, when the test was performed 60 min after gavage (**A,B**); corresponding 13, 11, 13 and 13 when test was performed 30 min after gavage (**C,D**) (ANOVA with repeated measurement, post-hoc test: # p < 0.05 and ### p < 0.001 compared to respective control time).

3.2. Evaluation of Rapid (30 min after Oral Gavage) Anxiolytic-Like Effect

One-way ANOVA of elapsed time to first approach to the probe did not reveal any statistical group difference ($F_{(3,46)}$ = 0.376, p = 0.7705), thus, ensuring no locomotor biases (Figure 2C). Furthermore, two-way ANOVA with a repeated measurement of burying time during the behavioral test revealed a significant group ($F_{(3,46)}$ = 12.749, p < 0.0001) and time effect ($F_{(2,92)}$ = 75.708, p < 0.0001), as well as a group x time interaction ($F_{(6,92)}$ = 3.779, p = 0.0021) (Figure 2D). Whichever group considered, animals spent less and less time burying the probe during the behavioral test (Control group: $F_{(2,20)}$ = 53.993, p < 0.0001 and for diazepam, αS1–casein hydrolysate and Gabolysat®-treated group: $F_{(2,24)}$ = 7.490, 25.232 and 12.089, respectively with p < 0.001). The post-hoc tests showed that the global group difference relied on significantly lesser probe burying time for diazepam- and Gabolysat®-treated animals compared to the control group (p = 0.0001 and 0.0005, respectively). In addition, if we consider only the first or second 5 min-section of the test, one-way ANOVA of burying time revealed a significant group difference ($F_{(3,46)}$ = 12.981 and 4.556, with p < 0.0001 and p = 0.0071, respectively). The post-hoc tests showed that both diazepam-

and Gabolysat®-treated animals spent significantly less time burying the probe compared to the control group (from 0 to 5 min test: $p < 0.0001$ and $p = 0.0002$; for 5–10 min test: $p = 0.0150$ and 0.0463, respectively).

4. Discussion

We herein experimentally demonstrated acute anxiolytic-like properties of Gabolysat® and αS1–casein hydrolysate [17]. Orally given at a dose of 15 mg/kg and tested in the conditioned burying task 60 min later, both natural products were as efficient as diazepam (used as benzodiazepine reference pharmacological drug) to elicit anxiolytic activity. More interestingly, when the anxiolytic-effect was assessed even as soon as 30 min after the oral dose, a similar efficiency (still compared to diazepam) was observed for Gabolysat®. This last result demonstrates a fast onset of anxiolytic activity for Gabolysat®.

Diazepam is a well-known and frequently prescribed benzodiazepine, which already experimentally demonstrated its anxiolytic activity in the conditioned burying test. Thus, its anxiolytic activity was described either when administrated intra-peritoneal (doses ranging from 0.5 to 2 mg/kg) [29,30], or orally given (3 mg/kg) [16,17]. Despite any potential differences of laboratory experimental conditions (light/dark cycle, strain, testing room environment, …), an anxiolytic effect of diazepam is always found, as herein. The same holds true for αS1–casein hydrolysate. Indeed, when administered orally at 15 mg/kg and then tested 60 min after, αS1–casein hydrolysate demonstrated anxiolytic properties in the conditioned burying test. Such a result is in accordance with literature data [16,17], reinforcing our choice to use αS1–casein hydrolysate as a reference natural product having demonstrated anxiolytic activity. Here, we demonstrated for the first time, that Gabolysat®—a fish proteins extract—also has rapid anxiolytic properties, as testified by the reduced probe-burying time observed in animals with acute treatment. Additionally, one can note that, whichever treatment was considered, the animals' anxiety levels (behaviorally assessed through probe-burying time measurement) were similar (no statistical difference between treated animals' groups). Thus, within the sensitivity limits of the test, we observed similar anxiolytic properties of each of the natural compounds and the benzodiazepine.

In a rat, the maximum plasmatic concentration of diazepam *per os* is observed at 30 min [31]. Hence, the experiment was conducted again with a shorten elapsed time (30 min) after oral dose to attempt to observe a rapid anxiolytic effect of both natural compounds and benzodiazepine. Since the maximum plasma concentration of diazepam is observed 30 min after oral gavage, this was taken as the new elapsed time to evaluate anxiolytic properties. Thus, we demonstrated for the first time to the best of our knowledge, that diazepam—but also Gabolysat®—have rapid anxiolytic properties. This was testified by the decreased probe-burying behavior during the first 10 min of the test, performed only 30 min after oral dose. This last finding regarding the fast onset of Gabolysat® could draw clinicians' attention as it might find application in several disorders (such as sleep anxiety and insomnia) or even psychiatric pathologies, such as depression. Indeed, depression, for instance, displays a 57% rate to co-occur with one anxiety disorder. Antidepressants prescribed so far, have revealed to be moderately efficient. This is firstly because some patients do not respond to treatment, and then because the average time to achieve remission for those who are sensitive to it is approximately 7 weeks [32]. Thus, natural products such as Gabolysat®—provided they are devoid of toxicity—may offer a therapeutic adjuvant during the first phase of treatment. So far, no side effect was reported in rodents after chronic (5 days) oral treatment of high doses of Gabolysat® (dose ranging from 300 to 1200 mg/kg) [13].

Beyond toxicity, the mechanisms underlying anxiolytic activity observed after an oral dose of those two natural compounds still remains unknown. Given the different behavioral profile observed in our study, one might imagine that their mechanisms may slightly diverge. Among possible mechanisms, a direct action on central GABAergic and/or serotonergic transmission, as well as modulatory role on corticotrope axis have shown preliminary results [13,33]. However, one exciting hypothesis recently emerged from current ongoing research on microbiota–gut–brain axis functioning. Indeed, gut peptides were recently evoked as important regulators of microbiota–gut–brain signaling in health and stress-related psychiatric illnesses [34]. Further works are required to investigate how those two natural compounds might interact with microbiome and, as a consequence, on gut–brain axis functioning.

A major study limitation is the chosen route of administration. In fact, while closely related to human practice, oral gavage required animal restraint which is a stress-induced event. In addition, the technique might be harmful or induce pain (esophageal trauma, etc.), if the experimenter is not skilled in animal handling and restraint. All animals herein underwent handling/restraint by the same experimenter (with acknowledged expertise) and the animals' groups were counterbalanced across the session.

5. Conclusions

In conclusion, our results confirm the efficacy of αS1–casein hydrolysate to dampen anxiety level but they also demonstrate, for the first time, the efficacity of Gabolysat® to do so. In addition, this last compound also displays a fast onset of action affording a rapid and lasting anxiolytic action. Numerous natural products are marketed for behavior therapy, but very few have demonstrated any evidence of efficacy. Additionally, as both natural products were described as being devoid of any major side effects, they appear as powerful natural alternative solutions to benzodiazepine drugs and to their constraints of use.

Author Contributions: The authors' responsibilities were as follows: Conceptualization, M.B. and T.F.; formal analysis, investigation and data curation, S.L., M.C., G.N. and S.C.; writing—original draft preparation and editing, T.F. All authors have read and agreed to the published version of the manuscript.

Funding: This research received no external funding.

Institutional Review Board Statement: All experiments were carried out in accordance with the European Communities Council Directive (63/2010) regarding the care and use of animals for experimental procedures, and they were approved by the local ethics committee (Comité d'Ethique NOrmandie en Matière d'EXpérimentation Animale, CENOMEXA, agreement number: 03-08-11/16/08-14).

Informed Consent Statement: Not applicable.

Data Availability Statement: The data presented in this study are available on request from the corresponding author.

Acknowledgments: Diet complement αS1–casein hydrolysate and Gabolysat® was kindly given by INGREDIA and DIELEN, respectively. Additionally, authors wish to warmly thank Miss Nicole Turnbull for proofreading the manuscript, spelling corrections and editing English.

Conflicts of Interest: The authors declare no conflict of interest. DIELEN and INGREDIA had no role in the design of the study; in the behavioral analyses, or interpretation of data; in the writing of the manuscript; or in the decision to publish the results.

References

1. Murrough, J.W.; Yaqubi, S.; Sayed, S.; Charney, D.S. Emerging drugs for the treatment of anxiety. *Expert Opin. Emerg. Drugs* **2015**, *20*, 393–406. [CrossRef]
2. Bandelow, B.; Michaelis, S.; Wedekind, D. Treatment of anxiety disorders. *Dialogues Clin. Neurosci.* **2017**, *19*, 93–107.
3. Fluyau, D.; Revadigar, N.; Manobianco, B.E. Challenges of the pharmacological management of benzodiazepine withdrawal, dependence, and discontinuation. *Ther. Adv. Psychopharmacol.* **2018**, *8*, 147–168. [CrossRef]
4. Lader, M.; Kyriacou, A. Withdrawing Benzodiazepines in Patients with Anxiety Disorders. *Curr. Psychiatry Rep.* **2016**, *18*, 8. [CrossRef]
5. Votaw, V.R.; Geyer, R.; Rieselbach, M.M.; McHugh, R.K. The epidemiology of benzodiazepine misuse: A systematic review. *Drug Alcohol Depend.* **2019**, *200*, 95–114. [CrossRef] [PubMed]
6. Deakin, J.B.; Aitken, M.R.; Dowson, J.H.; Robbins, T.W.; Sahakian, B.J. Diazepam produces disinhibitory cognitive effects in male volunteers. *Psychopharmacology* **2004**, *173*, 88–97. [CrossRef] [PubMed]

7. Lane, S.D.; Tcheremissine, O.V.; Lieving, L.M.; Nouvion, S.; Cherek, D.R. Acute effects of alprazolam on risky decision making in humans. *Psychopharmacology* **2005**, *181*, 364–373. [CrossRef] [PubMed]

8. Olfson, M.; King, M.; Schoenbaum, M. The popularity of benzodiazepines, their advantages, and inadequate pharmacological alternatives-reply. *JAMA Psychiatry* **2015**, *72*, 624. [CrossRef]

9. Endres, J.; Graber, M.A.; Dachs, R. Benzodiazepines and Alzheimer disease. *Am. Fam. Physician* **2015**, *91*, 191–192. [PubMed]

10. Bandelow, B. Current and Novel Psychopharmacological Drugs for Anxiety Disorders. *Adv. Exp. Med. Biol.* **2020**, *1191*, 347–365. [CrossRef]

11. Fernandez-Rodriguez, M.; Rodriguez-Legorburu, I.; Lopez-Ibor Alcocer, M.I. Nutritional supplements in Anxiety Disorder. *Actas Esp. Psiquiatr.* **2017**, *45*, 1–7.

12. Godos, J.; Currenti, W.; Angelino, D.; Mena, P.; Castellano, S.; Caraci, F.; Galvano, F.; Del Rio, D.; Ferri, R.; Grosso, G. Diet and Mental Health: Review of the Recent Updates on Molecular Mechanisms. *Antioxidants* **2020**, *9*, 346. [CrossRef]

13. Bernet, F.; Montel, V.; Noel, B.; Dupouy, J.P. Diazepam-like effects of a fish protein hydrolysate (Gabolysat PC60) on stress responsiveness of the rat pituitary-adrenal system and sympathoadrenal activity. *Psychopharmacology* **2000**, *149*, 34–40. [CrossRef]

14. Dela Pena, I.J.I.; Kim, H.J.; de la Pena, J.B.; Kim, M.; Botanas, C.J.; You, K.Y.; Woo, T.; Lee, Y.S.; Jung, J.C.; Kim, K.M.; et al. A tryptic hydrolysate from bovine milk alphas1-casein enhances pentobarbital-induced sleep in mice via the GABAA receptor. *Behav. Brain Res.* **2016**, *313*, 184–190. [CrossRef] [PubMed]

15. Messaoudi, M.; Lalonde, R.; Nejdi, A.; Bisson, J.F.; Rozan, P.; Javelot, H.; Schroeder, H. The effects of Garum Armoricum®(GA) on elevated-plus maze and conditioned light extinction tests in rats. *Curr. Top. Nutraceutical Res.* **2008**, *6*, 41–45.

16. Messaoudi, M.; Lalonde, R.; Schroeder, H.; Desor, D. Anxiolytic-like effects and safety profile of a tryptic hydrolysate from bovine alpha s1-casein in rats. *Fundam. Clin. Pharmacol.* **2009**, *23*, 323–330. [CrossRef]

17. Violle, N.; Messaoudi, M.; Lefranc-Millot, C.; Desor, D.; Nejdi, A.; Demagny, B.; Schroeder, H. Ethological comparison of the effects of a bovine alpha s1-casein tryptic hydrolysate and diazepam on the behaviour of rats in two models of anxiety. *Pharmacol. Biochem. Behav.* **2006**, *84*, 517–523. [CrossRef] [PubMed]

18. Landsberg, G.M.; Mougeot, I.; Kelly, S.; Milgram, N.W. Assessment of noise-induced fear and anxiety in dogs: Modification by a novel fish hydrolysate supplemented diet. *J. Vet. Behav.* **2015**, *10*, 391–398. [CrossRef]

19. Makawey, A.; Iben, C.; Palme, R. Cats at the Vet: The Effect of Alpha-s1 Casozepin. *Animals* **2020**, *10*, 2047. [CrossRef] [PubMed]

20. Tynes, V.V.; Landsberg, G.M. Nutritional Management of Behavior and Brain Disorders in Dogs and Cats. *Vet. Clin. N. Am. Small Anim. Pract.* **2021**, *51*, 711–727. [CrossRef] [PubMed]

21. Beata, C.; Beaumont-Graff, E.; Coll, V.; Cordel, J.; Marion, M.; Massal, N.; Marlois, N.; Tauzin, J. Effect of alpha-casozepine (Zylkene) on anxiety in cats. *J. Vet. Behav.* **2007**, *2*, 40–46. [CrossRef]

22. Messaoudi, M.; Lefranc-Millot, C.; Desor, D.; Demagny, B.; Bourdon, L. Effects of a tryptic hydrolysate from bovine milk alphaS1-casein on hemodynamic responses in healthy human volunteers facing successive mental and physical stress situations. *Eur. J. Nutr.* **2005**, *44*, 128–132. [CrossRef] [PubMed]

23. Kim, J.H.; Desor, D.; Kim, Y.T.; Yoon, W.J.; Kim, K.S.; Jun, J.S.; Pyun, K.H.; Shim, I. Efficacy of alphas1-casein hydrolysate on stress-related symptoms in women. *Eur. J. Clin. Nutr.* **2007**, *61*, 536–541. [CrossRef] [PubMed]

24. Phing, C.H. A Review of the Potential Efficacy of Alpha-S1-Casein Tryptic Hydrolysate on Stress and Sleep Disorder. *Malays. J. Psychiatry* **2016**, *25*, 99–109.

25. Dorman, T.B.L.; Glaze, P.; Hogan, J.; Skinner, R.; Nelson, D.; Bowker, L.; Head, D. The effectiveness of Garum armoricum (Stabilium) in reducing anxiety in college students. *J. Adv. Med.* **1995**, *8*, 193–200. [CrossRef]

26. Messaoudi, M.; Nejdi, A.; Bisson, J.F.; Rozan, P.; Javadov, H.; Lalonde, R. Anxiolytic and antidepressant-like effects of Garum Armoricum® (GA), a blue ling fish protein autolysate in male wistar rats. *Curr. Top. Nutraceutical Res.* **2008**, *6*, 115–123.

27. De Boer, S.F.; Koolhaas, J.M. Defensive burying in rodents: Ethology, neurobiology and psychopharmacology. *Eur. J. Pharmacol.* **2003**, *463*, 145–161. [CrossRef]

28. Fucich, E.A.; Morilak, D.A. Shock-probe Defensive Burying Test to Measure Active versus Passive Coping Style in Response to an Aversive Stimulus in Rats. *Bio-protocol* **2018**, *8*, e2998. [CrossRef]

29. Treit, D.; Pinel, J.P.; Fibiger, H.C. Conditioned defensive burying: A new paradigm for the study of anxiolytic agents. *Pharmacol. Biochem. Behav.* **1981**, *15*, 619–626. [CrossRef]

30. Wilson, M.A.; Burghardt, P.R.; Ford, K.A.; Wilkinson, M.B.; Primeaux, S.D. Anxiolytic effects of diazepam and ethanol in two behavioral models: Comparison of males and females. *Pharmacol. Biochem. Behav.* **2004**, *78*, 445–458. [CrossRef]

31. Gong, W.; Liu, S.; Xu, P.; Fan, M.; Xue, M. Simultaneous Quantification of Diazepam and Dexamethasone in Plasma by High-Performance Liquid Chromatography with Tandem Mass Spectrometry and Its Application to a Pharmacokinetic Comparison between Normoxic and Hypoxic Rats. *Molecules* **2015**, *20*, 6901–6912. [CrossRef] [PubMed]

32. Gaynes, B.N.; Warden, D.; Trivedi, M.H.; Wisniewski, S.R.; Fava, M.; Rush, A.J. What did STAR*D teach us? Results from a large-scale, practical, clinical trial for patients with depression. *Psychiatr. Serv.* **2009**, *60*, 1439–1445. [CrossRef]

33. Mizushige, T.; Sawashi, Y.; Yamada, A.; Kanamoto, R.; Ohinata, K. Characterization of Tyr-Leu-Gly, a novel anxiolytic-like peptide released from bovine alphaS-casein. *FASEB J.* **2013**, *27*, 2911–2917. [CrossRef] [PubMed]

34. Lach, G.; Schellekens, H.; Dinan, T.G.; Cryan, J.F. Anxiety, Depression, and the Microbiome: A Role for Gut Peptides. *Neurotherapeutics* **2018**, *15*, 36–59. [CrossRef] [PubMed]

nutrients

Article

Poor Dietary Quality and Patterns Are Associated with Higher Perceived Stress among Women of Reproductive Age in the UK

Karim Khaled [1,*], Vanora Hundley [2] and Fotini Tsofliou [1,2]

[1] Department of Rehabilitation & Sport Sciences, Faculty of Health & Social Sciences, Bournemouth University, Bournemouth BH8 8GP, UK; ftsofliou@bournemouth.ac.uk
[2] Centre for Midwifery, Maternal & Perinatal Health, Faculty of Health & Social Sciences, Bournemouth University, Bournemouth BH8 8GP, UK; vhundley@bournemouth.ac.uk
* Correspondence: Khaledk@bournemouth.ac.uk

Abstract: The aim of this study was to investigate the association between stress and diet quality/patterns among women of reproductive age in UK. In total, 244 reproductive aged women participated in an online survey consisting of the European Prospective into Cancer and Nutrition food frequency questionnaire in addition to stress, depression, physical-activity, adiposity, and socioeconomic questions. An a-priori diet quality index was derived by assessing the adherence to Alternate Mediterranean Diet (aMD). A-posteriori dietary-patterns (DPs) were explored through factor analysis. Regression models were used to assess the predictors of the DPs. Participants mainly had medium (n = 113) aMD adherence. Higher stress levels were reported by participants with low aMD adherence. Participants with high aMD adherence were of normal BMI. Factor analysis revealed three DPs: fats and oils, sugars, snacks, alcoholic-beverages, red/processed meat, and cereals (DP-1), fish and seafood, eggs, milk and milk-products (DP-2), and fruits, vegetables, nuts and seeds (DP-3). Regression models showed that DP-1 was positively associated with stress (p = 0.005) and negatively with age (p = 0.004) and smoking (p = 0.005). DP-2 was negatively associated with maternal educational-level (p = 0.01) while DP-3 was negatively associated with stress (p < 0.001), BMI (p = 0.001), and white ethnicity (p = 0.01). Stress was negatively associated with healthy diet quality/patterns among reproductive aged women.

Keywords: perceived stress; psychological; stress; diet quality; dietary patterns; women; reproductive age; childbearing age; a-priori; a-posteriori

Citation: Khaled, K.; Hundley, V.; Tsofliou, F. Poor Dietary Quality and Patterns Are Associated with Higher Perceived Stress among Women of Reproductive Age in the UK. *Nutrients* 2021, 13, 2588. https://doi.org/10.3390/nu13082588

Academic Editor: Panagiota Mitrou

Received: 2 July 2021
Accepted: 26 July 2021
Published: 28 July 2021

Publisher's Note: MDPI stays neutral with regard to jurisdictional claims in published maps and institutional affiliations.

1. Introduction

Increased body weight before pregnancy is associated with higher risk of pregnancy complications [1–4] and of severe maternal morbidity and mortality [5]. A recent meta-analysis has indicated that a healthy diet is crucial to prevent increased weight gain before and during pregnancy and its related complications (e.g., gestational diabetes, preeclampsia, caesarean section delivery) [6].

There are several predictors of diet quality, one of which is perceived stress. Stress is increasing among people and has been associated with poorer diet quality among women of reproductive age [7–11]. However, most studies have focused on the association between stress and individual foods/food groups. For example, a higher level of psychological stress among women of reproductive age was found to be significantly associated with a greater consumption of fat in their diet [12–15]. Studies have also found that stress has been associated with decreased intake of fruits and vegetables among women of reproductive age [16,17]. Moreover, higher levels of perceived stress have been found to associate with increased consumption of sweets, fast foods, snacks, and saturated fats and decreased intake of fruits, vegetables, and unsaturated oils [18,19]. However, these studies assessed dietary intake through recall questionnaires that do not include a variety and wide range of food items and food groups; this might predispose participants to misreporting [8,20–25].

With respect to the evaluation of diet quality, most studies on stress and diet have included only the a-priori dietary approach (based on measuring adherence to diet index) [12,22,25]. To the best of our knowledge, no study on stress and diet have combined both the a-priori and a-posteriori (based on statistical techniques such as factor analysis) approaches which offer comprehensive insight and characterisation of the diet pattern specific to the population group under investigation. Additionally, the small sample sizes and the lack of representativeness in those studies mean that generalising the results is not possible. Not considering confounding factors such as sociodemographic characteristics and physical activity was also a major limitation of some studies [12,16,24,26,27]. Moreover, most studies on the association between stress and diet were conducted among the general adult population, however evidence is scarce among women of reproductive age (18–49 years old) [26,28,29].

In summary, studies on stress and diet in the literature have several limitations. Investigating the factors that affect diet quality in women of reproductive age has crucial importance especially because diet-related morbidity among these women had an increasing trend during the past years [30]. To our knowledge, this study is the first to examine the association between stress and diet quality among women of reproductive age in UK. The aim of this study is to investigate whether higher level of perceived stress is associated with lower diet quality among women of reproductive age in the UK.

2. Materials and Methods

This was a cross sectional study targeting women of reproductive age in the UK. The study used an online questionnaire survey developed and administered via the Bristol Online Surveys (BOS).

The sample was one of convenience and consisted of females of reproductive age (between 18 and 49 years old) who were students and staff at a UK University. There are varying definitions for reproductive age in the literature; the range for this study was chosen to reflect the majority of recent studies [31,32]. Participants were excluded if they were: males at birth, below 18 or above 49 years old, not students or staff, suffering from a chronic illness or disease such as: cancer, Crohn's disease, diabetes, heart disease, HIV/AIDS/multiple sclerosis, depression, asthma, COPD, cystic fibrosis, or mental health disorder, having any kind of food intolerance or food allergy, pregnant or breastfeeding, or were on any medication known to affect appetite or body weight or have undergone bariatric surgery. The sample size was calculated by applying the correlation sample-size method [33] with a power of 80%, and an α (significance level) of 0.05. A correlation coefficient of 0.18 was chosen for the power calculation and it was based on the lowest correlation coefficient r reported in studies about stress and diet quality in women of reproductive age [24,30]. This yielded a total sample of 240 participants.

Potential participants were targeted through posters and social media advertisements (e.g., twitter). Consent was ascertained on the landing page of the survey.

2.1. Methodological Measurements and Procedures

2.1.1. Diet

Diet quality and patterns were estimated via the European Prospective into Cancer and Nutrition food frequency questionnaire (EPIC FFQ) which has been previously validated among UK adult females [34,35]. The EPIC-FFQ consists of 130 food items and one additional question for milk (131 items). The questionnaire represents either individual food (51%), combination of between two and four individual foods (23%), or food types (26%) that are further described by examples of individual foods. The number and percentage of food types in the list are: vegetables, 25 (19%); fruit and fruit juices, 12 (9%); meats, poultry, fish and eggs, 18 (14%); breads, cereals and starches, 18 (14%); dairy foods and fats, 15 (11%); beverages, 10 (8%); sweets and confectionery items, 14 (11%) and miscellaneous foods, 19 (14%). The food list is associated with a set of nine frequency choices for consumption ranging from 'never or less than once a month' to '6 or more times per day'.

The questionnaire consists of two parts. Part 1, the main part, lists 130 food items. Part 2 includes a set of additional questions that determine further information on the type and brand of breakfast cereal and kind of fat used in frying, roasting, grilling or baking and the amount of visible fat on meat.

2.1.2. Dietary Data Analysis

The FETA software was used to analyse the EPIC FFQ data and calculate the grams/day of nutrients and food groups [36]. Eleven food groups (grams/day) were derived from the EPIC food frequency questionnaire data analysis which included fats and oils, sugars and snacks, cereals, alcoholic beverages, red and processed meat, fish and seafood, eggs, milk and milk products, fruits, vegetables, and nuts and seeds. Adherence to the Alternate Mediterranean Diet Index (aMED) was used to assess the a-priori approach for diet quality assessment. The aMED is an adjustment of the Mediterranean Diet Index, which is based on the evidence suggesting a protective effect of this diet on the risk of chronic diseases, and that was previously developed by Trichopoulou et al. [37]. The aMED is based on the consumption of nine food groups: vegetables (excluding potatoes), fruit, nuts, legumes, fish, whole grains, mono-unsaturated fatty acids to saturated fatty acids ratio, alcohol, and red and processed meat [38]. The a-posteriori approach was based on factor analysis that derived the dietary patterns of participants. The importance of factor analysis (a-posteriori approach) is that it characterises the sample's variation in dietary intake and provides a more meaningful description of the overall patterns and quality of the diet which complements the a-priori dietary approach [39].

2.1.3. Mental Health Indicators

Perceived stress was measured using the 14-item Perceived Stress Scale (PSS) [40]. PSS measures the level of psychological stress, thoughts, and feelings of each participant over the past month. The scale has been tested in several trials in adult populations and showed significant consistency with Cronbach's alpha = 0.75 and 0.85 [40]. The PSS is not a diagnostic tool; hence there are no cut-off points that determine if an individual is stressed. PSS was equally divided into two categories: low to medium level of stress (score = 0–27) and medium to high level of stress (score = 28–56) as per previous studies [11,41].

Depression was measured using the 21-item Beck Depression Inventory II (BDI-II) [42]. The BDI-II has become one of the most widely used measures to assess depressive symptoms and their severity in adolescents and adults [43]. The BDI-II is a 21-item self-report measure that taps major depression symptoms according to diagnostic criteria listed in the Diagnostic and Statistical Manual for Mental Disorders [44]. Since its publication, a number of studies have examined the validity and reliability of BDI-II across different populations and countries [45]. Results have consistently shown good internal consistency and test-retest reliability of the BDI-II in community [46,47] adolescent and adult clinical outpatients [48] as well as in adult clinical inpatients [49].

2.1.4. Physical and Socio-Demographic Characteristics

Physical activity was measured using the International Physical Activity Questionnaire (IPAQ) [50]. The IPAQ records the activity of participants of four intensity levels: vigorous-intensity activity such as aerobic, moderate-intensity activity such as leisure cycling, walking, and sitting [50].

Adiposity measures: weight in kg and height in cm were self-reported and body mass index (BMI) in kg/m^2 was estimated to classify body weight status [51]. BMI was calculated by dividing weight in kilograms over height squared in centimetres [51]. Previous papers stated that self-reported weight and height are acceptable for determining BMI [52].

Data on socioeconomic factors (age, education, income, race, and marital status) were collected to control for the influence of these confounding factors [53,54].

2.2. Statistical Analysis

IBM SPSS statistics version 25 (Chicago, IL, USA) was used for data analysis. The normality of the data was assessed by Shapiro–Wilk test. Descriptive data are presented as median and interquartile range (IQR) for data with non-normal distribution. Kruskal–Wallis test was used to compare continuous data among the aMD adherence categories (low, medium, high). Categorical data among the three aMD adherence categories were compared through Chi-squared test. The Bonferroni method was used to correct for multiplicity in data.

The normality of the food groups' data was assessed, and appropriate transformation was undertaken when high skewness was detected in the data. Kaiser–Meyer–Olkin (KMO) measure of sampling adequacy and Bartlett's test of sphericity were conducted to check the appropriateness of factor analysis. Results revealed a large KMO of 0.75 (>0.5) and a very significant Bartlett's test ($p < 0.00001$) with an approximate Chi-square of 832 and 55 degrees of freedom; therefore, factor analysis was deemed appropriate to use [55]. Additionally, the sample size of the present study ($n = 244$) is acceptable for conducting factor analysis [56,57]. To derive the number of factors from the food groups' data, a scree plot was generated showing the factors that have an eigenvalue >1. A varimax rotation was assigned to calculate factor loadings for each factor (dietary pattern) based on the assumption that factors were not correlated. Simple linear regression models carried out for each factor (dietary pattern) were revealed to investigate the association between that factor (dietary pattern) and the following variables separately: perceived stress, depression scores, BMI (kg/m^2), PA (Mets) and socioeconomic measures. The predictors with significant association were then included in a multiple linear regression model of the diet pattern along with the other significant predictors. Some categories of the socioeconomic measures were merged together before inclusion in the regression models due to their small size (e.g., marital status (single/divorced/widowed, living together/married), parity (never, once/two times or more), religion (no religion, Christian, other), education (No qualification/Certificate of Secondary School (CSE)/General Certificate of Secondary School (GCSE), A-level/higher education), ethnicity (white, other), smoking status (smoker, non-smoker), income (below the average, above the average), parents occupation (employee, other)).

3. Results

A sample of 252 women participated in the study, and after screening eight were excluded since they did not meet the eligibility criteria (e.g., food intolerance/food allergy/chronic disease). In total, the data of 244 women were included in the analysis of the present study. Overall, participants had an average age of 24 years, were mainly of white ethnicity, single, non-smokers, and their parental educational level was mainly O-level or GCSE examinations taken at 16 years. In addition, 47% of the total sample had a moderate level of physical activity (Tables 1 and 2).

The participants' characteristics are reported across the three categories of the Alternate Mediterranean Diet Scores (aMDS: low, medium, and high). The majority of the 244 participants had a medium adherence to aMD (46%), followed by 39% having low adherence, and only 15% of participants had high adherence to aMD.

There was a significant association between perceived stress and diet quality. Medium to high levels of stress were more likely to be reported by participants (73%) with a low adherence to aMDS (X^2 (2, $n = 244$) = 14.08, $p = 0.001$). Pairwise comparisons showed that stress was different between low and high aMD adherence ($p = 0.005$) and between low and medium aMD adherence categories ($p = 0.003$) but not between medium and high adherence categories ($p = 0.467$).

Table 1. Physical and mental characteristics of participants (n = 244).

Participants' Characteristics(N (%))	Total Sample	Alternate Mediterranean Diet Adherence Categories			p-Value
		Low aMDS (0–3)	Medium aMDS (4–6)	High aMDS (7–9)	
		95 (39)	113 (46)	36 (15)	
Physical and lifestyle characteristics					
Age (years) #	24.0 (21.0–32.0)	23.0 (21.0–29.0)	25.0 (21.5–32.0)	24.0 (20.3–35.0)	0.277
Age (years) * 18–24 25–34 35–49	124 (51) 77 (32) 43 (17)	54 (57) 26 27) 15 (16)	51 (45) 44 (39) 18 (16)	19 (53) 7 (19) 10 (28)	0.09
BMI (kg/m^2) #	23.7 (20.9–27.9)	26.1 (21.5–49.4)	23.7 (20.6–27.5)	21.9 (20.3–23.9)	0.093
BMI * Underweight Normal Weight Overweight/obese	14 (6) 120 (49) 108 (44)	4 (4) 38 (40) 52 (56)	7 (6) 56 (50) 50 (44)	3 (8) 26 (72) 6 (17)	0.005
Physical Activity (METs-h/wk) #	1429 (464.3–2824.5)	1159 (330.0–2615.0)	1440 (479.3–2886.3)	2380 (1325.5–3464.3)	0.336
Physical Activity level * Low (<600 MET minutes/week)	76 (31)	39 (41)	33 (29)	4 (11)	0.018
Moderate (>600 MET minutes/week)	114 (47)	39 (41)	55 (49)	20 (56)	
High (>3000 MET minutes/week)	54 (22)	17 (18)	25 (22)	12 (33)	
Mental Health Indicator					
Stress #	29 (22.0–33.0)	31 (26.0–34.0)	27 (22.0–27.0)	26.5 (18.0–31.8)	0.002
Stress * Low-Medium Medium-High	103 (42) 141 (58)	26 (27) 69 (73)	58 (51) 55 (49)	19 (53) 17 (47)	0.001
Depression #	5 (2.0–12.0)	5 (2.0–13.0)	5 (2.0–11.0)	5 (1.0–13.0)	0.926
Depression * Minimal (0–13) Mild (14–19) Moderate (20–28) Severe (29–63)	191 (78) 28 (11) 12 (5) 13 (5)	73 (77) 15 (16) 4 (4) 3 (3)	90 (80) 10 (9) 8 (7) 5 (4)	28 (78) 3 (8) 0 (0) 5 (14)	0.07

METs-h/wk: Metabolic equivalents of tasks-hours per week, BMI: body mass index., GCSE: General Certificate of Secondary Education, O-level: ordinary level. p-values were derived through a Chi-squared test of independence to display differences in physical activity, mental health indicators, and BMI of participants across the three Alternate Mediterranean diet (aMD) scores categories. The differences between median (IQR) of physical, mental health, and lifestyle characteristics were explored with Kruskal–Wallis test and post-hoc pairwise comparisons. * represents N (%). # represents median (IQR). aMDS: alternate Mediterranean Diet Score.

Table 2. Socio-demographic characteristics of participants (*n* = 244).

Participants' Characteristics (N (%))	Total Sample N (%)	Alternate Mediterranean Diet Adherence Categories			*p*-Value
		Low aMDS (0–3)	Medium aMDS (4–6)	High aMDS (7–9)	
		95 (39)	113 (46)	36 (15)	
Father's education					
No qualifications	23 (9)	8 (8)	11 (10)	4 (11)	
Certificate of Secondary education (CSE) taken at 14–16 years at a lower level than GCSE	57 (23)	28 (29)	25 (22)	4 (11)	
O-level or GCSE examinations taken at 16 years	71 (29)	23 (24)	35 (31)	13 (36)	0.626
A-level school examinations taken at 18 years	45 (18)	18 (19)	19 (17)	8 (22)	
Higher education	48 (20)	18 (19)	23 (20)	7 (19)	
Mother's education					
No qualifications	16 (7)	5 (5)	9 (8)	2 (6)	
Certificate of Secondary education (CSE) taken at 14–16 years at a lower level than GCSE	47 (19)	24 (25)	20 (18)	3 (8)	
O-level or GCSE examinations taken at 16 years	82 (34)	26 (27)	41 (36)	15 (42)	0.399
A-level school examinations taken at 18 years	44 (18)	17 (18)	18 (16)	9 (25)	
Higher education	55 (23)	23 (24)	25 (22)	7 (19)	
Father's occupation					
Working as an employee	92 (38)	35 (37)	42 (37)	15 (42)	
On a government sponsored training scheme	5 (2)	4 (4)	1 (1)	0 (0)	
Self-employed or freelance	68 (28)	28 (29)	30 (27)	10 (28)	
Working paid or unpaid for your own or your family's business	31 (13)	13 (14)	15 (13)	3 (8)	0.424
Doing any other kind of paid work	9 (4)	6 (6)	1 (1)	2 (6)	
Retired (whether receiving a pension or not)	36 (15)	8 (8)	22 (19)	6 (17)	
Long-term sick or disabled	3 (1)	1 (1)	2 (2)	0 (0)	
Mother's occupation					
Working as an employee?	101 (41)	38 (40)	43 (38)	20 (56)	
On a government sponsored training scheme	9 (4)	4 (4)	5 (4)	0 (0)	
Self-employed or freelance	25 (10)	6 (6)	14 (12)	5 (14)	
Working paid or unpaid for your own or your family's business	17 (7)	9 (9)	7 (6)	1 (3)	
Doing any other kind of paid work	20 (8)	11 (12)	6 (5)	3 (8)	0.266
Retired (whether receiving a pension or not)	34 (14)	8 (8)	22 (19)	4 (11)	
Looking after home or family	27 (11)	14 (15)	11 (10)	2 (6)	
Long-term sick or disabled	11 (5)	5 (5)	5 (4)	1 (3)	
Income per year					
<£13,000	119 (49)	43 (45)	54 (48)	22 (61)	
£13,000 to £33,800	99 (40)	45 (48)	47 (41)	7 (19)	0.047
>£33,800	26 (11)	7 (7)	12 (11)	7 (19)	

Table 2. *Cont.*

Participants' Characteristics (N (%))	Total Sample N (%)	Alternate Mediterranean Diet Adherence Categories			*p*-Value
		Low aMDS (0–3)	Medium aMDS (4–6)	High aMDS (7–9)	
		95 (39)	**113 (46)**	**36 (15)**	
Parents' annual income					
<£13,000	36 (15)	14 (15)	15 (13)	7 (19)	
£13,000 to £23,400	51 (21)	21 (22)	25 (22)	5 (14)	
>£23,400 to £33,800	69 (28)	31 (33)	33 (29)	5 (14)	0.432
>£33,800 to £52,000	47 (19)	16 (17)	20 (18)	11 (31)	
>£52,000	41 (17)	13 (14)	20 (18)	8 (22)	
Marital Status					
Single	176 (72)	67 (71)	84 (74)	25 (69)	
Married	43 (18)	16 (17)	18 (16)	9 (25)	
Divorced	17 (7)	9 (9)	8 (7)	0 (0)	0.46
Separated but still legally married	6 (2)	3 (3)	2 (2)	1 (3)	
Widowed	2 (1)	0 (0)	1 (1)	1 (3)	
Smoking					
Current Smoker	56 (23)	25 (26)	23 (20)	8 (22)	
Ex-smoker	27 (11)	10 (11)	11 (10)	6 (17)	0.47
Never smoked	161 (66)	60 (63)	79 (70)	22 (61)	
Religion					
No religion	104 (43)	36 (38)	53 (47)	15 (42)	
Christian	105 (43)	45 (47)	42 (37)	18 (50)	
Buddhist	7 (3)	2 (2)	5 (4)	0 (0)	
Hindu	9 (4)	5 (5)	3 (3)	1 (3)	0.437
Jewish	(0)	(0)	(0)	(0)	
Muslim	19 (8)	7 (7)	10 (9)	2 (6)	
Sikh	(0)	(0)	(0)	(0)	
Ethnicity					
Mixed/multiple ethnic groups	10 (4)	5 (5)	4 (4)	1 (3)	
White	177 (73)	61 (64)	82 (73)	34 (34)	
Asian/Asian British	35 (14)	18 (19)	16 (14)	1 (3)	0.231
Black/African/Caribbean/Black British	15 (6)	7 (7)	8 (7)	0 (0)	
Other ethnic group	7 (3)	4 (4)	3 (3)	0 (0)	
Parity					
Never	189 (77)	72 (76)	92 (81)	25 (69)	
Once	26 (11)	14 (15)	6 (5)	6 (17)	0.229
Two times or more	29 (12)	9 (9)	15 (13)	5 (14)	

aMDS: alternate Mediterranean Diet Score. GCSE: General Certificate of Secondary Education, O-level: ordinary level. *p*-values were derived through a Chi-squared test of independence to display differences in socio-demographics of participants across the three alternate Mediterranean diet scores (aMDS) categories.

BMI was also found to be different among aMD adherence groups (X^2 (4, n = 244) = 14.815, p = 0.005) (Table 1). Participants who had normal BMI were more likely to have high aMDS (72%) compared to those who were underweight (8%) and overweight/obese (17%). The physical activity level of participants differed across the three categories (X^2 (4, n = 244) = 11.92, p = 0.018). A higher percentage of participants with high aMDS adherence were engaging in moderate (56%) and high (33%) physical activity levels whereas those with low aMDS adherence were engaging in low (41%) and moderate (41%) physical activity levels.

Income per year showed a significant, but weak, association with adherence to aMDS (X^2 (10, n = 244) = 18.48, p = 0.047). However, adherence to aMD was not associated with any other socio-demographic characteristics (Table 2).

Factor Analysis

Figure 1 demonstrates a scree plot showing the number of factors (dietary patterns) retained from factor analysis. As shown in the scree plot, the number of factors (dietary patterns) with eigenvalue ≥ 1 is 3. The three factors (dietary patterns) explained 60% of the total variance in data.

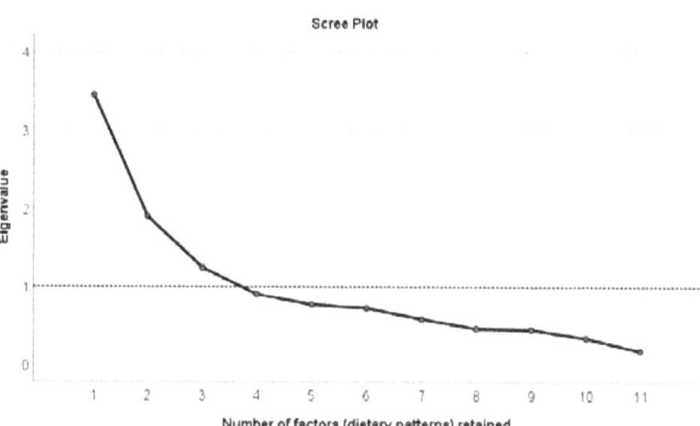

Figure 1. Scree plot showing the number of factors (dietary patterns) retained from factor analysis on the X-axis and the eigenvalues on Y-axis.

Table 3 demonstrates the 11 food groups with factor loadings for the three factors (dietary patterns). Coefficients with absolute value below 0.3 for each factor were suppressed, therefore five food groups were assigned to factor 1 (DP-1), three to factor 2 (DP-2), and three to factor 3 (DP-3). The first dietary pattern (DP-1) had high factor loadings for the following food groups: fats and oils, sugars and snacks, alcoholic beverages, cereals, and red and processed meat and was labelled "Western-style" dietary pattern. DP-2 had high factor loadings for food groups such as fish and seafood, eggs, and milk and milk products and was labelled "high-quality protein" dietary pattern. The third dietary pattern (DP-3) was labelled "vegetarian-like" dietary pattern with factor loadings high for fruits, vegetables, and nuts and seeds food groups.

Regression analysis that was used to examine the association between the three dietary patterns (DPs), which were derived from factor analysis, and all other variables indicated the following: In the first model, DP-1 was positively associated with stress ($p = 0.005$) and negatively with age ($p = 0.004$) and smoking ($p = 0.005$) (Table 4). DP-1 was common among young, smokers, and highly stressed women. Model 2 of the second dietary pattern showed that DP-2 was negatively associated with mother's educational level ($p = 0.019$) (Table 4). The second dietary pattern (DP-2) was common among participants who had mothers with lower educational level. The third dietary pattern (DP-3) was common among normal weight people who had low stress level and non-white. As shown in Table 4, DP-3 was negatively associated with stress ($p < 0.001$), BMI ($p = 0.001$), and ethnicity ($p = 0.013$).

Table 3. Orthogonally rotated (varimax) factor loadings for the 3 factors (dietary patterns) of the 11 food groups (grams/day).

11 Food Groups Derived from the European Prospective into Cancer and Nutrition (EPIC) Food Frequency Questionnaire	Factors (Dietary Patterns)		
	1	2	3
Fats and Oils (grams/day)	0.838		
Sugars and Snacks (grams/day)	0.738		
Cereals (grams/day)	0.712		
Alcoholic beverages (grams/day)	0.665		
Red and processed meat (grams/day)	0.553		
Fish and Seafood (grams/day)		0.821	

Table 3. *Cont.*

11 Food Groups Derived from the European Prospective into Cancer and Nutrition (EPIC) Food Frequency Questionnaire	Factors (Dietary Patterns)		
	1	2	3
Eggs (grams/day)		0.809	
Milk and milk products (grams/day)		0.518	
Fruits (grams/day)			0.750
Vegetables (grams/day)			0.747
Nuts and Seeds (grams/day)			0.619

Table 4. Multiple regression models showing the association between each a-posteriori-derived diet pattern and its predictor variables.

Model	Predictor	Coefficient Estimate	*p*-Value
1 (DP-1) "fats & oils, sugars & snacks, alcoholic beverages, red and processed meat, and cereals" DP	Intercept	0.419	<0.001
	Stress	0.003	0.005
	Physical activity (METs-h/wk)	−0.0000002	0.395
	BMI	0.002	0.062
	Age	−0.003	0.004
	Father's educational level (A-level/higher)	−0.027	0.107
	Mother's educational level (A-level/higher)	−0.006	0.713
	Ethnicity (white)	0.026	0.128
	Father's occupation (other)	0.015	0.369
	Mother's occupation (other)	0.022	0.174
	Smoking status (smoker)	−0.05	0.005
	Participant's income (above average)	0.026	0.098
2 (DP-2) "fish & seafood, eggs, and milk & milk products" DP	Intercept	0.441	<0.0001
	Stress	−0.002	0.14
	Depression	0.0001	0.676
	Mother's education (A-level/higher)	−0.038	0.019
	Father's occupation (other)	0.035	0.057
	Mother's occupation (other)	0.018	0.313
	Participant's income (above average)	0.033	0.069
3 (DP-3) "fruits, vegetables, and nuts & seeds" DP	Intercept	0.653	<0.001
	Stress	−0.005	<0.001
	Physical activity (METs-h/wk)	0.0000006	0.115
	BMI	−0.005	0.001
	Ethnicity (white)	−0.047	0.013
	Parent's income (above average)	0.023	0.184
	Smoking (smoker)	0.033	0.092

DP: dietary pattern. METs-h/wk: metabolic equivalents of tasks-hours per week. Model 1 of the first dietary pattern was based on the following formula: DP − 1 = β0 + β1 Stress + β2 Physical activity + β3 BMI + β4 Age + β5 Father's educational level + β6 Mother's educational level + β7 Ethnicity + β8 Father's occupation + β9 Mother's occupation + β10 Smoking status + β11 Participant's income. Model 2 of the second dietary pattern was based on the following formula: DP − 2 = β0 + β1 Stress + β2 Depression + β3 Mother's education + β4 Father's occupation + β5 Mother's occupation + β6 Participant's income. Model 3 of the third dietary pattern was based on the following formula: DP − 3 = β0 + β1 Stress + β2 Physical activity + β3 BMI + β4 Ethnicity + β5 Parent's income + β6 Smoking.

4. Discussion

This is the first study to investigate the association between perceived stress and diet quality/patterns among women of reproductive age in the UK. The association between stress and diet quality/patterns has recently gained the interest of health researchers, especially that diet is a main modifiable risk factor of obesity and many chronic diseases [21]. In the present study, diet quality/patterns analysis was used, rather than individual-nutrient assessment, because it allows the description of the whole diet of the population and is considered essential in understanding the relationship between dietary consumption and diet-related diseases [16]. Additionally, the association between stress and single nutrients is difficult to investigate since they are never consumed separately but rather within meals, and they metabolically interact with one another [16]. Our findings indicate that stress is associated with lower diet quality where 73% of participants who had low adherence to the alternate Mediterranean Diet (aMD) had a high stress level. Therefore, stress-coping strategies and programs for women of reproductive age should be implemented to prevent unhealthy eating habits and poor diet quality and their adverse health consequences.

Participants in this study were recruited from a university setting and included both students and employees (18–49 years old) to provide a more representative sample of reproductive aged women of this setting.

The a-priori assessment of diet quality indicated an overall medium adherence to the alternate Mediterranean Diet index (46% of the total sample). Similar results were found in the US where 43% of women of reproductive age ($n = 248$) had a moderate adherence to the Mediterranean Diet [58] and in the UK where most workplace females ($n = 426$) were moderate adherers to the Mediterranean Diet Index ($n = 346$) [59]. Similarly, our research team has previously assessed diet quality by the Mediterranean Diet Index, among 123 women of reproductive age in the UK and also reported an overall moderate adherence [60]; the alternate Mediterranean Diet Index was used in the present study because it has been considered more reflective of MD for non-Mediterranean countries such as the UK [61]. In this context, women of reproductive age should be supported with nutrition counselling and education, in addition to reproductive health care services, to further enhance their diet quality [62].

The a-posteriori dietary approach (Table 4) corroborated further the negative association between stress and healthy diet quality/patterns and offered additional dietary insight by highlighting the types of food groups that might contribute to this association. It showed that stress was positively associated with the Western-style dietary pattern (DP-1) consisting of fats and oils, sugar and snacks, alcoholic beverages, red/processed meat, and cereals ($p = 0.005$) and negatively with the vegetarian-like dietary pattern (DP-3) consisting of fruits, vegetables, nuts and seeds ($p < 0.001$). These findings agree with other studies targeting the association between stress and diet. For instance, El-Ansari et al. [16] assessed stress levels using the Perceived Stress Scale and nutritional habits through a 12-food item FFQ and found that among female university students in the UK, stress was significantly associated with poorer diet quality resembled by high intake of sugar, snacks, fat, and low intake of unsaturated fats, fruits, and vegetables. Additionally, Isasi et al. [25] found that stress was negatively linked with diet quality (Alternate-Healthy Eating Index 2010) among Hispanic/Latino females in the US. Similarly, Groesz et al. [63] targeted 561 females from the US and found that highly stressed females reported high consumption of unhealthy foods (fast food, sweets, etc.) and low consumption of whole grains, fruits, and vegetables as assessed via a food frequency questionnaire. In another study conducted among females across three countries (Germany, Poland, and Bulgaria), a positive association between stress and poor dietary patterns was reported [26]. Habhab et al. [64] also assessed the link between stress and food restraint and diet quality/patterns among 40 women of childbearing age via mixed-design analysis of variance and found that women with poorer diet quality had a high stress level. These findings were corroborated by our recent systematic review and meta-analysis [30] that was the first to examine the association between perceived stress and diet quality in women of reproductive age. The systematic literature review included 24 studies (8 had diet quality as the primary outcome and 16 assessed food frequency of consumption) with a total of 41,033 participants. Overall, the 16 studies on food intake and frequency of consumption ($n = 33,477$) found that stress was associated with high intake of fat, fast food, sweets, processed foods, and low intake of fruits, vegetables, whole grains, and legumes. The meta-analysis included the 8 studies on diet quality ($n = 7556$) and reported a significantly negative association between stress and diet quality (r = -0.35, $p < 0.001$, 95% CI (-0.56; -0.15)).

On the contrary, some studies reported different findings. For example, Richardson et al. [24] assessed stress through the 14-item PSS and diet quality via Healthy Eating Index-2010 among 101 childbearing aged women (aged 18–44 years) and found no association between stress and diet quality. Similarly, Ferranti et al. [65] found no association between perceived stress and diet quality among 433 females in the US who were university and health centre employees. The study assessed stress via the 14-item PSS and diet quality via a-priori approach using three diet quality indices: Alternate Healthy Eating Index, Mediterranean Diet Index, and Dietary Approach to Stop Hypertension Index. Two other studies in Egypt [66] and Iran [67] among women of childbearing age also found no significant association between diet quality and stress.

Discrepancies in findings between these studies and the present study might be explained by variations in sample sizes, diversity of the tools used to assess variables, and difference in the population from which the sample was taken. For instance, Richardson et al. [24] recruited 101 women and Widaman et al. [23] recruited 75 females. On the other hand, the present study recruited 244 participants. Secondly, most studies on the association between perceived stress and diet quality in women of reproductive age have used 24-h recalls as the dietary assessment tool [21–25] whereas the EPIC food frequency questionnaire, which measures a wide variety of food items and the frequency of consumption over the past one year, was used in the present study.

In understanding the stress/diet relationship, studies have argued that the association between perceived stress and dietary quality/patterns is bidirectional: psychological stress symptoms could be associated with behaviours that are considered "health-compromising" that put the individual at risk of health problems [16]. For example, in a group of female students, high stress levels were associated with weight dissatisfaction and other health-compromising behaviours such as alcohol intake, binge eating and smoking, skipping breakfast [68]. Stressed people tend to consume high energy-dense foods to taper down their stressful emotions [69]. Adam et al. [69] suggested that the important reason behind these eating behaviours resulting from negative emotions and stress is the lack of eating control. This is when the consumption of high caloric and palatable foods relates to satisfaction and reward and becomes comfort eating during the stressful periods [69]. However, the absence of significant association between perceived stress and diet quality/patterns in some studies does not support these views. The findings of these studies can be explained by the following coping strategies that are not related to food such as spirituality that could attenuate the effect of stress on dietary behaviour [70]. Although these studies show no significant associations between perceived stress and diet quality, some environmental factors including stress coping strategies, cultural food traditions, cognitive factors (such as the knowledge of nutrition), and the cost of food might contribute to the dietary pattern and quality and must be further studied. Another explanation of the stress/diet relationship is derived from the fact that perceived stress causes physiological changes (in addition to psychological changes) to the human body that trigger food craving [71]. Upon stress, the hypothalamus and central nervous system secrete the hormone cortisol into the bloodstream which leads, if in high circulating concentrations, to the formation and accumulation of visceral fat in the body [69]. Additionally, several studies have pointed out that elevated levels of stress cortisol can be associated with increased food intake [72,73]. This is because perceived stress, and elevated serum cortisol, stimulate the secretion of the gastric hormone Ghrelin that increases appetite and food craving [71].

Strengths and Limitations

The study has several strengths. To our knowledge, it is the first to assess the association between perceived stress and diet quality in women of reproductive age in the United Kingdom. Diet quality/patterns were assessed comprehensively through two approaches: the a-priori (hypothesis-driven) and the a-posteriori (data-driven) which gave robust results and clearer insight about the overall dietary quality/patterns of the study's participants. Another strength of this study is the fact that the tools used to assess all variables were validated and standardised, such as the Perceived Stress Scale to assess stress levels [40], Becks Depression Inventory II to assess depression [42], in addition to the anthropometric and socioeconomic questions [74–76]. Furthermore, while most studies on the association between stress and diet utilised dietary recalls to assess dietary intake, this study used the EPIC food frequency questionnaire which is considered a gold standard dietary assessment tool [34,35].

On the other hand, there are several limitations which are worth acknowledging. The cross-sectional design of the study made it hard to draw and generalise a definitive conclusion about the association between perceived stress and diet quality/patterns among women of reproductive age. Additionally, the convenience sample that was selected from a population of a UK university setting and consequently might not be representative of the general population of women of reproductive age. Although all variables were measured using validated and standardised tools, they have been self-reported by participants, which might have caused inaccuracy in the results. For example, anthropometric measures would be better estimated using advanced and more accurate tools such as Dual-Energy X-ray Absorptiometry (DEXA) which measures the whole-body composition including weight, height, fat mass, and fat-free body mass [77]. Similarly, the Perceived Stress Scale, that was used to assess stress levels of participants, was self-reported and hence participants might not have accurately recalled the stressful situations that occurred over the past weeks. A more accurate measure of stress should be used in future studies such as blood or salivary cortisol [78]. Moreover, food intake biomarkers (such as urinary and blood samples), which objectively measure the nutritional intake of individuals, should complement the food frequency questionnaires and other self-reported measures of dietary intake [79].

5. Conclusions

In conclusion, the negative association between perceived stress and diet quality (in both a-priori and a-posteriori approaches) that was found among a sample of women of reproductive age in the present study is important and merits further investigation. The results of this study have implications for future interventions which should include not only dietary but also other behavioural aspects to support lifestyle changes among women of reproductive age. In other words, the interventions are complex; they are more than simply changing the diet alone. Women of reproductive age seem to eat depending on the level of stress and therefore dietary interventions need take that into consideration when applying it. Future randomised controlled trials with accurate measures should be implemented to further confirm this negative association.

Author Contributions: Conceptualization, K.K., F.T. and V.H.; methodology, K.K., F.T. and V.H.; formal analysis, K.K.; data curation, K.K.; writing—original draft preparation, K.K.; writing—review and editing, K.K., F.T. and V.H.; visualization, K.K., F.T. and V.H.; supervision, F.T. and V.H. All authors have read and agreed to the published version of the manuscript.

Funding: This research received no external funding.

Institutional Review Board Statement: The study was conducted according to the guidelines of the Declaration of Helsinki and approved by the Institutional Review Board (or Ethics Committee) of Bournemouth University (protocol code 22344) on 27 January 2020.

Informed Consent Statement: Informed consent was obtained from all subjects involved in the study.

Acknowledgments: The authors would like to thank Laya Awada for her help in data cleaning. As authors, who have used the FETA software for the analysis of the EPIC FFQ data in the present study, we are asked by FETA-software team to acknowledge the following: The EPIC-Norfolk study (DOI 10.22025/2019.110.1105.00004) has received funding from the Medical Research Council (MR/N003284/1 and MC-UU_112015/1) and Cancer Research UK (C864/A14136). We are grateful to all the participants who have been part of the project and to the many members of the study teams at the University of Cambridge who have enabled this research.

Conflicts of Interest: The authors declare no conflict of interest.

References

1. Durst, J.K.; Tuuli, M.G.; Stout, M.J.; Macones, G.A.; Cahill, A.G. Degree of obesity at delivery and risk of preeclampsia with severe features. *Am. J. Obstet. Gynecol.* **2016**, *214*, 651.e1–651.e5. [CrossRef]
2. Guelinckx, I.; Devlieger, R.; Beckers, K.; Vansant, G. Maternal obesity: Pregnancy complications, gestational weight gain and nutrition. *Obes. Rev.* **2008**, *9*, 140–150. [CrossRef] [PubMed]
3. Wei, Y.-M.; Yang, H.; Zhu, W.-W.; Liu, X.-Y.; Meng, W.-Y.; Wang, Y.-Q.; Shang, L.-X.; Cai, Z.-Y.; Ji, L.-P.; Wang, Y.-F.; et al. Risk of adverse pregnancy outcomes stratified for pre-pregnancy body mass index. *J. Matern. Neonatal Med.* **2015**, *29*, 2205–2209. [CrossRef] [PubMed]
4. Sun, Y.; Shen, Z.; Zhan, Y.; Wang, Y.; Ma, S.; Zhang, S.; Liu, J.; Wu, S.; Feng, Y.; Chen, Y.; et al. Effects of pre-pregnancy body mass index and gestational weight gain on maternal and infant complications. *BMC Pregnancy Childbirth* **2020**, *20*, 1–13. [CrossRef]
5. Lisonkova, S.; Muraca, G.; Potts, J.; Liauw, J.; Chan, W.-S.; Skoll, A.; Lim, K.I. Association Between Prepregnancy Body Mass Index and Severe Maternal Morbidity. *JAMA* **2017**, *318*, 1777–1786. [CrossRef] [PubMed]
6. International Weight Management in Pregnancy (i-WIP) Collaborative Group. Effect Of Diet And Physical Activity Based Interventions In Pregnancy On Gestational Weight Gain And Pregnancy Outcomes: Meta-Analysis Of Individual Participant Data From Randomised Trials. *BMJ* **2017**, *358*, j3991.
7. Ng, D.M.; Jeffery, R.W. Relationships Between Perceived Stress and Health Behaviors in a Sample of Working Adults. *Health Psychol.* **2003**, *22*, 638–642. [CrossRef]
8. Tseng, M.; Fang, C. Stress is Associated with Unfavorable Patterns of Dietary Intake Among Female Chinese Immigrants. *Annal. Behav. Med.* **2011**, *41*, 324–332. [CrossRef]
9. Bayram, N.; Bilgel, N. The prevalence and socio-demographic correlations of depression, anxiety and stress among a group of university students. *Soc. Psychiatry Psychiatr. Epidemiol.* **2008**, *43*, 667–672. [CrossRef] [PubMed]
10. Bauer, K.W.; Hearst, M.O.; Escoto, K.; Berge, J.; Neumark-Sztainer, D. Parental employment and work-family stress: Associations with family food environments. *Soc. Sci. Med.* **2012**, *75*, 496–504. [CrossRef] [PubMed]
11. Laugero, K.D.; Falcon, L.M.; Tucker, K. Relationship between perceived stress and dietary and activity patterns in older adults participating in the Boston Puerto Rican Health Study. *Appetite* **2011**, *56*, 194–204. [CrossRef] [PubMed]
12. Vidal, E.J.; Alvarez, D.; Martinez-Velarde, D.; Vidal-Damas, L.; Yuncar-Rojas, K.A.; Julca-Malca, A.; Bernabe-Ortiz, A. Perceived stress and high fat intake: A study in a sample of undergraduate students. *PLoS ONE* **2018**, *13*, e0192827. [CrossRef] [PubMed]
13. Wichianson, J.R.; Bughi, S.A.; Unger, J.; Spruijt-Metz, D.; Nguyen-Rodriguez, S.T. Perceived stress, coping and night-eating in college students. *Stress Health* **2009**, *25*, 235–240. [CrossRef]
14. Nastaskin, R.S.; Fiocco, A.J. A survey of diet self-efficacy and food intake in students with high and low perceived stress. *Nutr. J.* **2015**, *14*, 1–8. [CrossRef]
15. Roohafza, H.; Sarrafzadegan, N.; Sadeghi, M.; Rafieian-Kopaei, M.; Sajjadi, F.; Khosravi-Boroujeni, H. The Association Be-tween Stress Levels And Food Consumption Among Iranian Population. *Arch. Iran. Med.* **2013**, *16*, 145–148.
16. El Ansari, W.; Adetunji, H.; Oskrochi, R. Food and Mental Health: Relationship between Food and Perceived Stress and Depressive Symptoms among University Students in the United Kingdom. *Central Eur. J. Public Health* **2014**, *22*, 90–97. [CrossRef]
17. Papier, K.; Ahmed, F.; Lee, P.; Wiseman, J. Stress and dietary behaviour among first-year university students in Australia: Sex differences. *Nutrients* **2015**, *31*, 324–330. [CrossRef]
18. O'Connor, D.B.; O'Connor, R.C. Perceived changes in food intake in response to stress: The role of conscientiousness. *Stress Health* **2004**, *20*, 279–291. [CrossRef]
19. Tryon, M.; Carter, C.; DeCant, R.; Laugero, K. Chronic Stress Exposure May Affect the Brain's Response to High Calorie Food Cues And Predispose To Obesogenic Eating Habits. *Physiol. Behav.* **2013**, *120*, 233–242. [CrossRef]
20. Hwang, J.; Lee, S.; Kim, S.; Chung, H.; Kim, W. Psychological Distress Is Associated with Inadequate Dietary Intake In Viet-namese Marriage Immigrant Women In Korea. *J. Am. Diet. Assoc.* **2010**, *110*, 779–785. [CrossRef] [PubMed]
21. Fowles, E.R.; Bryant, M.; Kim, S.; Walker, L.O.; Ruiz, R.J.; Timmerman, G.M.; Brown, A. Predictors of Dietary Quality in Low-Income Pregnant Women. *Nurs. Res.* **2011**, *60*, 286–294. [CrossRef] [PubMed]
22. Fowles, E.R.; Stang, J.; Bryant, M.; Kim, S. Stress, Depression, Social Support, and Eating Habits Reduce Diet Quality in the First Trimester in Low-Income Women: A Pilot Study. *J. Acad. Nutr. Diet.* **2012**, *112*, 1619–1625. [CrossRef] [PubMed]
23. Widaman, A.; Witbracht, M.G.; Forester, S.M.; Laugero, K.D.; Keim, N.L. Chronic Stress is Associated with Indicators of Diet Quality in Habitual Breakfast Skippers. *J. Acad. Nutr. Diet.* **2016**, *116*, 1776–1784. [CrossRef]
24. Richardson, A.S.; Arsenault, J.E.; Cates, S.C.; Muth, M.K. Perceived stress, unhealthy eating behaviors, and severe obesity in low-income women. *Nutr. J.* **2015**, *14*, 1–10. [CrossRef]
25. Isasi, C.; Parrinello, C.; Jung, M.; Carnethon, M.; Birnbaum-Weitzman, O.; Espinoza, R.; Penedo, F.; Perreira, K.; Schneiderman, N.; Sotres-Alvarez, D.; et al. Psychosocial Stress Is Associated with Obesity and Diet Quality in Hispanic/Latino Adults. *Annal. Epidemiol.* **2015**, *25*, 84–89. [CrossRef]
26. Mikolajczyk, R.T.; El Ansari, W.; Maxwell, A.E. Food consumption frequency and perceived stress and depressive symptoms among students in three European countries. *Nutr. J.* **2009**, *8*, 31. [CrossRef]
27. González, A.M.; Cruz, S.Y.; Ríos, J.L.; Pagán, I.; Fabián, C.; Betancourt, J.; Rivera-Soto, W.T.; González, M.J.; Palacios, C. Alcohol consumption and smoking and their associations with socio-demographic characteristics, dietary patterns, and perceived academic stress in Puerto Rican college students. *Puerto Rico Health Sci. J.* **2013**, *32*, 82–88.

28. Fulkerson, J.E.; Sherwood, N.E.; Perry, C.L.; Neumark-Sztainer, D.; Story, M. Depressive symptoms and adolescent eating and health behaviors: A multifaceted view in a population-based sample. *Prev. Med.* **2004**, *38*, 865–875. [CrossRef]

29. Liu, C.; Xie, B.; Chou, C.-P.; Koprowski, C.; Zhou, D.; Palmer, P.; Sun, P.; Guo, Q.; Duan, L.; Sun, X.; et al. Perceived stress, depression and food consumption frequency in the college students of China seven cities. *Physiol. Behav.* **2007**, *92*, 748–754. [CrossRef] [PubMed]

30. Khaled, K.; Tsofliou, F.; Hundley, V.; Helmreich, R.; Almilaji, O. Perceived stress and diet quality in women of reproductive age: A systematic review and meta-analysis. *Nutr. J.* **2020**, *19*, 1–15. [CrossRef] [PubMed]

31. Morema, E.N.; Atieli, H.E.; Onyango, R.O.; Omondi, J.H.; Ouma, C. Determinants of Cervical screening services uptake among 18–49 year old women seeking services at the Jaramogi Oginga Odinga Teaching and Referral Hospital, Kisumu, Kenya. *BMC Health Serv. Res.* **2014**, *14*, 1–7. [CrossRef]

32. Proportion of Women of Reproductive Age (Aged 15–49 Years) Who Have Their Need for Family Planning Satisfied with Modern Methods (SDG 3.7.1). Available online: https://www.who.int/data/maternal-newborn-child-adolescent-ageing/indicator-explorer-new/mca/proportion-of-women-of-reproductive-age-(aged-15--49-years)-who-have-their-need-for-family-planning-satisfied-with-modern-methods (accessed on 22 July 2021).

33. Hulley, S.; Cummings, S.; Browner, W.; Grady, D.; Newman, T. *Designing Clinical Research*, 4th ed.; Wolters Kluwer/Lippincott Williams & Wilkins: Philadelphia, PA, USA, 2013.

34. Verkasalo, P.K.; Appleby, P.N.; Allen, N.E.; Davey, G.; Adlercreutz, H.; Key, T.J. Soya intake and plasma concentrations of daidzein and genistein: Validity of dietary assessment among eighty British women (Oxford arm of the European Prospective Investigation into Cancer and Nutrition). *Br. J. Nutr.* **2001**, *86*, 415–421. [CrossRef]

35. Welch, A.A.; Luben, R.; Khaw, K.T.; Bingham, S.A. The CAFE computer program for nutritional analysis of the EPIC-Norfolk food frequency questionnaire and identification of extreme nutrient values. *J. Hum. Nutr. Diet.* **2005**, *18*, 99–116. [CrossRef]

36. Mulligan, A.A.; Luben, R.; Bhaniani, A.; Parry-Smith, D.J.; O'Connor, L.; Khawaja, A.; Forouhi, N.; Khaw, K.-T. A new tool for converting food frequency questionnaire data into nutrient and food group values: FETA research methods and availability. *BMJ Open* **2014**, *4*, e004503. [CrossRef] [PubMed]

37. Trichopoulou, A.; Costacou, T.; Bamia, C.; Trichopoulos, D. Adherence to a Mediterranean Diet and Survival in a Greek Population. *N. Engl. J. Med.* **2003**, *348*, 2599–2608. [CrossRef] [PubMed]

38. Fung, T.; McCullough, M.; Newby, P.; Manson, J.; Meigs, J.; Rifai, N.; Willett, W.; Hu, F. Diet-Quality Scores and Plasma Concentrations Of Markers Of Inflammation And Endothelial Dysfunction. *Am. J. Clin. Nutr.* **2005**, *82*, 163–173. [CrossRef] [PubMed]

39. Zhao, J.; Li, Z.; Gao, Q.; Zhao, H.; Chen, S.; Huang, L.; Wang, W.; Wang, T. A review of statistical methods for dietary pattern analysis. *Nutr. J.* **2021**, *20*, 1–18. [CrossRef]

40. Cohen, S.; Kamarck, T.; Mermelstein, R. A Global Measure of Perceived Stress. *J. Health Soc. Behav.* **1983**, *24*, 385. [CrossRef]

41. Tavolacci, M.P.; Ladner, J.; Grigioni, S.; Richard, L.; Villet, H.; Dechelotte, P. Prevalence and association of perceived stress, substance use and behavioral addictions: A cross-sectional study among university students in France, 2009–2011. *BMC Public Health* **2013**, *13*, 724. [CrossRef]

42. Beck, A.; Steer, R.; Brown, G. *Manual for the Beck Depression Inventory-II*, 2nd ed.; Psychological Corporation: San Antonio, TX, USA, 1996.

43. García-Batista, Z.E.; Guerra-Peña, K.; Cano-Vindel, A.; Herrera-Martínez, S.X.; Medrano, L.A. Validity and reliability of the Beck Depression Inventory (BDI-II) in general and hospital population of Dominican Republic. *PLoS ONE* **2018**, *13*, e0199750. [CrossRef] [PubMed]

44. American Psychiatric Association. *Diagnostic and Statistical Manual of Mental Disorders*, 4th ed.; American Psychiatric Association: Washington, DC, USA, 2011.

45. Wang, Y.-P.; Gorenstein, C. Psychometric properties of the Beck Depression Inventory-II: A comprehensive review. *Rev. Bras. Psiquiatr.* **2013**, *35*, 416–431. [CrossRef] [PubMed]

46. Gomes-Oliveira, M.H.; Gorenstein, C.; Neto, F.L.; Andrade, L.H.; Wang, Y.P. Validation of the Brazilian Portuguese Version of the Beck Depression Inventory-II in a community sample. *Rev. Bras. Psiquiatr.* **2012**, *34*, 389–394. [CrossRef] [PubMed]

47. Segal, D.L.; Coolidge, F.L.; Cahill, B.S.; O'Riley, A.A. Psychometric Properties of the Beck Depression Inventory—II (BDI-II) Among Community-Dwelling Older Adults. *Behav. Modif.* **2008**, *32*, 3–20. [CrossRef]

48. Grothe, K.B.; Dutton, G.R.; Jones, G.; Bodenlos, J.; Ancona, M.; Brantley, P.J. Validation of the Beck Depression Inventory-II in a Low-Income African American Sample of Medical Outpatients. *Psychol. Assess.* **2005**, *17*, 110–114. [CrossRef] [PubMed]

49. Subica, A.M.; Fowler, J.C.; Elhai, J.D.; Frueh, B.C.; Sharp, C.; Kelly, E.L.; Allen, J.G. Factor structure and diagnostic validity of the Beck Depression Inventory–II with adult clinical inpatients: Comparison to a gold-standard diagnostic interview. *Psychol. Assess.* **2014**, *26*, 1106–1115. [CrossRef] [PubMed]

50. Craig, C.L.; Marshall, A.; Sjöström, M.; Bauman, A.E.; Booth, M.L.; Ainsworth, B.E.; Pratt, M.; Ekelund, U.; Yngve, A.; Sallis, J.F.; et al. International Physical Activity Questionnaire: 12-Country Reliability and Validity. *Med. Sci. Sports Exerc.* **2003**, *35*, 1381–1395. [CrossRef]

51. World Health Organization. *Obesity: Preventing and Managing the Global Epidemic: Report of a WHO Consultation*; WHO: Geneva, Switzerland, 2000.

52. Nikolaou, C.K.; Hankey, C.R.; Lean, M.E.J. Accuracy of on-line self-reported weights and heights by young adults. *Eur. J. Public Health* **2017**, *27*, 898–903. [CrossRef]

53. Martorell, R.; Kettel Khan, L.; Hughes, M.; Grummer-Strawn, L. Obesity in Women From Developing Countries. *Eur. J. Clin. Nutr.* **2000**, *54*, 247–252. [CrossRef]

54. Monteiro, C.A.; Conde, W.L.; Lu, B.; Popkin, B. Obesity and inequities in health in the developing world. *Int. J. Obes.* **2004**, *28*, 1181–1186. [CrossRef]

55. Matore, E.; Khairani, A.; Adnan, R. Exploratory Factor Analysis (EFA) For Adversity Quotient (AQ) Instrument among Youth. *J. Crit. Rev.* **2019**, *6*, 234–242.

56. MacCallum, R.; Widaman, K.; Zhang, S.; Hong, S. Sample Size in Factor Analysis. *Psychol. Methods* **1999**, *4*, 84–99. [CrossRef]

57. de Winter, J.; Dodou, D.; Wieringa, P.A. Exploratory Factor Analysis with Small Sample Sizes. *Multivar. Behav. Res.* **2009**, *44*, 147–181. [CrossRef] [PubMed]

58. Boghossian, N.S.; Yeung, E.H.; Mumford, S.L.; Zhang, C.; Gaskins, A.J.; Wactawski-Wende, J.; Schisterman, E.; for the BioCycle Study Group. Adherence to the Mediterranean diet and body fat distribution in reproductive aged women. *Eur. J. Clin. Nutr.* **2013**, *67*, 289–294. [CrossRef]

59. Papadaki, A.; Wood, L.; Sebire, S.J.; Jago, R. Adherence to the Mediterranean diet among employees in South West England: Formative research to inform a web-based, work-place nutrition intervention. *Prev. Med. Rep.* **2015**, *2*, 223–228. [CrossRef] [PubMed]

60. Khaled, K.; Hundley, V.; Almilaji, O.; Koeppen, M.; Tsofliou, F. A Priori and a Posteriori Dietary Patterns in Women of Childbearing Age in the UK. *Nutrients* **2020**, *12*, 2921. [CrossRef] [PubMed]

61. Chen, G.-D.; Dong, X.-W.; Zhu, Y.-Y.; Tian, H.-Y.; He, J.; Chen, Y.-M. Adherence to the Mediterranean diet is associated with a higher BMD in middle-aged and elderly Chinese. *Sci. Rep.* **2016**, *6*, 25662. [CrossRef] [PubMed]

62. Nestle, M. Obesity. Halting the obesity epidemic: A public health policy approach. *Public Health Rep.* **2000**, *115*, 12–24. [CrossRef] [PubMed]

63. Groesz, L.M.; McCoy, S.; Carl, J.; Saslow, L.; Stewart, J.; Adler, N.; Laraia, B.; Epel, E. What is eating you? Stress and the drive to eat. *Appetite* **2012**, *58*, 717–721. [CrossRef] [PubMed]

64. Habhab, S.; Sheldon, J.; Loeb, R. The Relationship Between Stress, Dietary Restraint, And Food Preferences in Women. *Appetite* **2009**, *52*, 437–444. [CrossRef]

65. Ferranti, E.P.; Dunbar, S.B.; Higgins, M.; Dai, J.; Ziegler, T.R.; Frediani, J.; Reilly, C.; Brigham, K.L. Psychosocial factors associated with diet quality in a working adult population. *Res. Nurs. Health* **2013**, *36*, 242–256. [CrossRef]

66. El Ansari, W.; Berg-Beckhoff, G. Nutritional Correlates of Perceived Stress among University Students in Egypt. *Int. J. Environ. Res. Public Health* **2015**, *12*, 14164–14176. [CrossRef]

67. Valipour, G.; Esmaillzadeh, A.; Azadbakht, L.; Afshar, H.; Hassanzadeh, A.; Adibi, P. Adherence To The DASH Diet In Re-lation To Psychological Profile Of Iranian Adults. *Eur. J. Nutr.* **2015**, *56*, 309–320. [CrossRef] [PubMed]

68. Stice, E.; Presnell, K.; Spangler, D. Risk Factors For Binge Eating Onset In aAdolescent Girls: A 2-Year Prospective Investiga-tion. *Health Psychol.* **2002**, *21*, 131–138. [CrossRef] [PubMed]

69. Adam, T.; Epel, E. Stress, Eating And The Reward System. *Physiol. Behav.* **2007**, *91*, 449–458. [CrossRef]

70. Musgrave, C.F.; Allen, C.E.; Allen, G.J. Spirituality and Health for Women of Color. *Am. J. Public Health* **2002**, *92*, 557–560. [CrossRef] [PubMed]

71. Azzam, I.; Gilad, S.; Limor, R.; Stern, N.; Greenman, Y. Ghrelin Stimulation by Hypothalamic–Pituitary–Adrenal Axis Acti-vation Depends On Increasing Cortisol Levels. *Endo. Connect.* **2017**, *6*, 847–855. [CrossRef]

72. Herhaus, B.; Ullmann, E.; Chrousos, G.; Petrowski, K. High/low cortisol reactivity and food intake in people with obesity and healthy weight. *Transl. Psychiatry* **2020**, *10*, 1–8. [CrossRef]

73. Hewagalamulage, S.; Lee, T.; Clarke, I.; Henry, B. Stress, Cortisol, And Obesity: A Role For Cortisol Responsiveness in Iden-tifying Individuals Prone To Obesity. *Domes. Anim. Endocrinol.* **2016**, *56*, S112–S120. [CrossRef]

74. Flint, E.; Webb, E.; Cummins, S. Change in commute mode and body-mass index: Prospective, longitudinal evidence from UK Biobank. *Lancet Public Health* **2016**, *1*, e46–e55. [CrossRef]

75. Prady, S.L.; Pickett, K.E.; Croudace, T.; Fairley, L.; Bloor, K.; Gilbody, S.; Kiernan, K.E.; Wright, J. Psychological Distress during Pregnancy in a Multi-Ethnic Community: Findings from the Born in Bradford Cohort Study. *PLoS ONE* **2013**, *8*, e60693. [CrossRef]

76. Bowman, S.A. A comparison of the socioeconomic characteristics, dietary practices, and health status of women food shoppers with different food price attitudes. *Nutr. Res.* **2006**, *26*, 318–324. [CrossRef]

77. Shepherd, J.A.; Ng, B.K.; Sommer, M.J.; Heymsfield, S.B. Body composition by DXA. *Bone* **2017**, *104*, 101–105. [CrossRef] [PubMed]

78. Hellhammer, D.; Wüst, S.; Kudielka, B. Salivary Cortisol as A Biomarker in Stress Re-search. *Psychoneuroendocrinology* **2009**, *34*, 163–171. [CrossRef] [PubMed]

79. Picó, C.; Serra, F.; Rodríguez, A.; Keijer, J.; Palou, A. Biomarkers of Nutrition and Health: New Tools For New Approaches. *Nutrients* **2019**, *11*, 1092. [CrossRef] [PubMed]

nutrients

Article

Study of Physical Fitness, Bone Quality, and Mediterranean Diet Adherence in Professional Female Beach Handball Players: Cross-Sectional Study

Alejandro Martínez-Rodríguez [1,2,*], Javier Sánchez-Sánchez [3,*], María Martínez-Olcina [1], Manuel Vicente-Martínez [4], Laura Miralles-Amorós [1] and Juan Antonio Sánchez-Sáez [5]

[1] Department of Analytical Chemistry, Nutrition and Food Science, Faculty of Sciences, Alicante University, 03690 Alicante, Spain; maria.martinezolcina@ua.es (M.M.-O.); lma52@alu.ua.es (L.M.-A.)
[2] Alicante Institute for Health and Biomedical Research (ISABIAL Foundation), 03010 Alicante, Spain
[3] Department of Sport Sciences, School of Sport Sciences, Universidad Europea de Madrid, 28670 Madrid, Spain
[4] Faculty of Health Science, Miguel de Cervantes European University, 47012 Valladolid, Spain; mvmartinez11006@alumnos.uemc.es
[5] GDOT Research Group, Faculty of Sport, Universidad Católica de Murcia, 30107 Murcia, Spain; jasanchez419@ucam.edu
* Correspondence: amartinezrodriguez@ua.es (A.M.-R.); javier.sanchez2@universidadeuropea.es (J.S.-S.)

Abstract: (1) Background: Beach handball is a relatively new type of sport, derived from team handball. The purpose of the study was to evaluate the physical fitness of elite players of this sport by studying some variables of sports performance, including strength, endurance and power, and dietary habits, and to assess bone ultrasonographic variables. (2) Methods: 33 beach handball players have participated in this research; 18 juniors (age: 16.7 ± 0.50) and 15 seniors (age: 24.8 ± 4.71). The athletes' strength was evaluated using the Handgrip Test on the dominant hand, the height of jump was evaluated by a counter-jump on a contact platform, and velocity, agility, and resistance by the Yo-Yo test. The broadband ultrasound attenuation (BUA) and the sound of speed (SOS) through the calcaneus were also measured. The Mediterranean diet adherence (KIDMED) was the questionnaire used to evaluate eating habits. In the statistical analysis, descriptions and correlations were made between the study variables. (3) Results: Both in the case of the dynamometric hand strength test ($p < 0.05$) and in the lower extremity power test ($p < 0.01$), senior players presented significantly higher values compared to junior players (35.1 ± 3.84 vs. 31.8 ± 3.37 and 35.1 ± 6.89 vs. 28.5 ± 5.69 with the dynamometry and Abalakov tests, respectively). However, no differences were observed in the variables by playing position. Significant correlations between different variables have been established, highlighting negative correlations between BMI and weight with the Abalakov Jump Test and positive correlations between Yo-Yo and BUA, and, between BMI and BUA. (4) Conclusions: Older and trained players are in better physical fitness; high weight and BMI have a negative influence on power, agility, speed, and endurance. In general, adherence to the Mediterranean diet is moderate and it seems evident that there is a beneficial influence of beach handball on bone condition, as measured by ultrasound. However future research should be carried on, including dual-energy x-ray absorptiometry assessments and food intake registers for a whole week.

Keywords: women's health; bone mineral density; diet; beach sports; osteoporosis

Citation: Martínez-Rodríguez, A.; Sánchez-Sánchez, J.; Martínez-Olcina, M.; Vicente-Martínez, M.; Miralles-Amorós, L.; Sánchez-Sáez, J.A. Study of Physical Fitness, Bone Quality, and Mediterranean Diet Adherence in Professional Female Beach Handball Players: Cross-Sectional Study. *Nutrients* **2021**, *13*, 1911. https://doi.org/10.3390/nu13061911

Academic Editor: Panagiota Mitrou

Received: 19 May 2021
Accepted: 31 May 2021
Published: 2 June 2021

Publisher's Note: MDPI stays neutral with regard to jurisdictional claims in published maps and institutional affiliations.

1. Introduction

Beach handball (BH) is relatively a new sport specialty. It has led to arousing the interest of the scientific community in the sport, analyzing the different specific performance variables of BH players (e.g., physical and physiological demands or body composition and anthropometric profile) [1–7] resulting from the particularities of this discipline (e.g., in-flight throw, spin shot, shoot-out, attack in numerical superiority, or defense in numerical inferiority) [8] and caused by the attack–defense actions of the game cycle [9]. Moreover, the competition format of a BH tournament —two to three games per day for two to three days—makes it necessary to design specific training exercises that respond to the actual physiological needs required by the players [7].

In relation to the physiological demands (internal load), BH, like other team sports (e.g., ice hockey or soccer [10,11]), requires intermittent efforts [7]. In this sense, this sport discipline is considered to be high-intensity because most of the playing time, the player is over 80% of the HRMAX [2,4,7,12]. Other characteristics to consider because they affect the internal load of the athlete would be the kinematic variables: (a) distance: 445.6 ± 156.3 to 669 ± 155 m per set, and (b) speed: 4.2 ± 0.6 to 5.7 ± 1.2 m/s [4,7,12].

There are different validated instruments and tests to determine performance variables in team sports, and they can provide the necessary data to assess the physical and physiological profile of the athlete. One of these tests is the Yo-Yo Intermittent Recovery Test Level 1 [13], widely applied in recent decades by researchers in indoor handball [14–20]. Another of the tests, the use of which is widespread, is the Abalakov Jump Test, used to determine the lower body power and fatigue of athletes [21–24]. In this case, we used the research of Sánchez-Sáez et al. [7] as a reference study of BH, which determined that there were no significant decreases in jumping ability during the tournament in female beach handball players ($p > 0.05$).

Regarding muscle strength, the Handgrip Test is used to evaluate the musculoskeletal fitness of the upper extremities [3] and muscle fatigue [7] in BH players. This variable could be considered of vital importance when establishing the performance profile of the BH player since the grip of the ball should be propitiated with sufficient strength both in the control and mastery of the ball, and at the time of execution of the passes or throws. To perform all the actions above, the implementation of an adequate high-performance training program is essential, which includes good eating habits, such as the ones offered by the Mediterranean diet (MD) [25]. In this way, in addition to improving physical capacities, the risk of suffering sports injuries, such as bone fractures, will be reduced [26].

Regarding bone quality, it has been observed that in addition to being determined by body composition, genetic/ethnic factors, and hormonal status, it is also related to lifestyle behaviors, such as diet and physical activity, which influence bone gain during growth and bone loss later in life [27]. From a mechanical point of view, changes in bone mineral density (BMD) can be made by inducing mechanical stimuli with loading forces or external loads during skeletal muscle contractions. For this reason, athletes practicing sports, such as beach handball, that require high muscular forces (strength training) or general high impacts (sprinting and jumping), are regularly recommended to work on the improvement of bone mass and density [28]. It appears that exercise activities that combine the tension produced by intense muscle contractions and the mechanical stimulus of ground reaction forces are considered better for bone stimulation [29–31] than sports such as swimming, water polo, or rowing, which do not show measurable osteogenic benefits [32].

In fact, it has been shown that an optimal nutritional status can enhance the effects of training, speeding up recovery and optimizing body composition [33]. The dietary culture of the Spanish population is within the framework of the Mediterranean diet [34–36]. This one is characterized by being a balanced diet that provides sufficient calories in the right proportions through high consumption of vegetables, legumes, fruits, nuts, cereals, and olive oil, moderate consumption of fish, eggs, and dairy products, preferably yogurt or cheese, and a lower intake of meat and less consumption of animal fats [37].

Therefore, the main objective of this research was to evaluate the physical fitness of elite female BH players through the study of the determinant variables of sports performance, including strength, endurance and power, and their dietary habits, and to assess bone ultrasonographic variables. This research provides relevant information for coaches and physical trainers when planning the season and training sessions.

2. Materials and Methods

Figure 1 presents a brief scheme of this section.

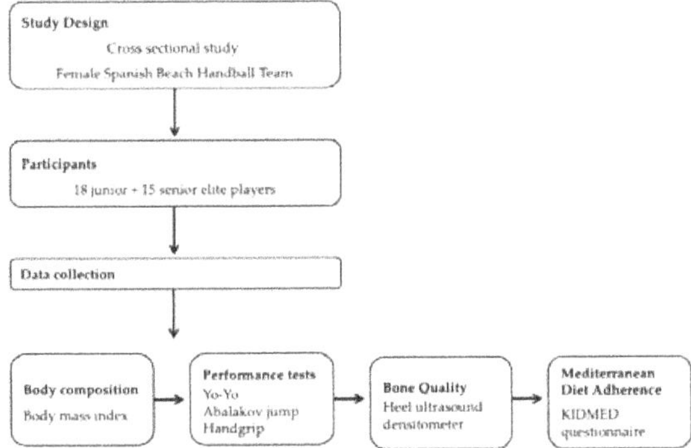

Figure 1. Material and methods scheme.

2.1. Study Design

A descriptive, cross-sectional study was conducted to analyze the influence of age category and playing position on the results of performance tests, adherence to the Mediterranean diet, and bone mineral density in female BH players. The research involved the best international players from Spain of this sport modality, who represent elite BH players from all over the world. The Declaration of Helsinki guidelines (revised in Hong Kong in September 1989 and in Edinburgh in 2000) and the recommendations of Good Clinical Practice of the EEC (Document 111/3976/88 of July 1990) were followed for all the procedures carried out. Approval of the research was granted by the Human Research Ethics Committee of the University of Alicante (Spain) (UA-2019-04-09).

2.2. Participants and Eligibility Criteria

A total of 33 female beach handball players participated in the study; 54% were juniors and 45% were seniors. For the distribution of the sample by playing position, 6 goalkeepers, 10 wings, 8 specialists, 6 pivots, and 3 defenders participated. All the players who were part of the current Spanish BH national team participated. The following exclusion criteria were considered: the presence of chronic diseases, injury during the training camp preventing the performance of any test, and non-compliance with the informed consent. However, no player refused to participate, and no athlete was excluded; every athlete gave written informed consent. No financial compensation was given to the participants for their collaboration. For those who were minors, consent was given by their parents or legal guardians. The players' anonymity was always preserved.

2.3. Data Collection

2.3.1. Body Composition

Anthropometric measurements included standing body height (stadiometer accuracy of 0.1 cm; Holtain, Crosswell, Crymych, Pembs, UK) and body mass (0.1 kg; Tanita BF683W scale, Munich, Germany). With the weight and height data, the body mass index BMI (kg/m^2) was calculated for each of the players.

2.3.2. Performance Tests

The test known as Yo-Yo IR1 was performed in accordance with the procedures described by Krustrup et al. [38]. The test consisted of 20 m runs performed at increasing speeds with 10 s of recovery between runs until the individual became exhausted. Previously, the players performed a warm-up of 5 min of low intensity running, followed by a 5 min warm-up of running at medium-high intensity. The test was terminated when the participant did not reach the first line in time (objective evaluation) twice in a row, or if she felt that she would not be able to complete another sprint (subjective evaluation). The total distance covered was recorded as the "score" of the test. The reliability of the Yo-Yo IR1 is demonstrated by a coefficient of variation (CV) of 3.6% and an ICC of 0.94% [38].

The Abalakov Jump Test was used; participants performed 3 countermovement's with 30 s rests between jumps [39] performed on a stable surface. All the players had to stand upright and perform a 90° knee flexion followed by the fastest possible extension, with the goal of reaching the highest achievable jump height. An optical (infrared) data collection system (Optojump Next Microgate, Bolzano, Italy) was used to calculate the Abalakov jump height. Of the 3 results, the best one was used for statistical analysis.

A hand-held dynamometer (Takei, Tokyo, Japan) was used with the arm at right angles and the elbows at the sides of the body. The instrument was adjusted, its base rested on the first metacarpal and the handle rested on the middle of the participant's four fingers. All players performed the test with their dominant hand. Their maximal isometric effort was maintained for 5 s. The test was performed twice, with a 1 min rest between trials, and the highest value being the one to be used in the subsequent analysis.

2.3.3. Bone Quality—Ultrasound Measurement

The bilateral calcanei of each subject were measured using a heel ultrasound densitometer (Achilles EXP II, GE Healthcare, Chicago, IL, USA). To assess bone health, quantitative ultrasound (QUS) measurements are widely used since they are noninvasive measurements, and are also less expensive and simpler than laboratory techniques (e.g., dual-energy X-ray absorptiometry—DXA). Quality assurance was performed using a specific dummy before the first measurement. Along with a coupling medium to ensure good contact, ultrasound gel was applied. During each ultrasonographic assessment, the broadband ultrasound attenuation (BUA) and the speed of sound (SOS) were directly measured. The BUA and SOS measurements, and their derived parameters (rigidity), are seriously associated with fracture risk, while bone quality is the determinant of both bone strength and bone fragility. Therefore, these variables can be used to refer to bone quality [40,41]. The calcaneus stiffness index was calculated as follows [28]: Calcaneus stiffness (A.U.) = $(0.67 \cdot \text{BUA} + 0.28 \cdot \text{SOS}) - 420$

2.3.4. Mediterranean Diet Adherence (KIDMED Questionnaire)

The KIDMED questionnaire was published in 2004 to evaluate adherence to the Mediterranean diet (MD) in children and adolescents [42]. The questionnaire constitutes 16 questions, 12 of which represent a positive score in relation to the adherence to the MD, and the other 4 represent a negative score. Positive answers to questions involving greater adherence to the diet are worth +1 point. Positive answers to questions involving less dietary adherence are worth −1 point [43]. Depending on the scores obtained in the questionnaire, the population was divided into 2 groups: (1) poor or average adherence, and (2) excellent adherence.

2.4. Statistical Analysis

The Jamovi statistical program (version 1.6.15, Sydney, Australia) was used for data analysis. First, the descriptive data (mean and standard deviation) were calculated. To assess the normality of the descriptive statistics (mean ± standard deviation) and inferential analysis, the Kolmogorov–Smirnov test was performed. To determine homogeneity, the Levene test was used to evaluate the normality of the data ($p > 0.05$). To calculate the differences between the age groups (junior vs. senior), as well as for the analysis of the different variables according to playing position, analysis of covariance (ANCOVA) with Bonferroni correction was used, controlling for the effect of BMI. The effect size was also calculated using a partial eta-squared ($\eta p2$) considering <0.25, 0.26–0.63, and >0.63 as small, medium, and large effect sizes, respectively [44].

Correlations between the performance variables (Abalakov, Handgrip, Yo-Yo), the KIDMED total score, bone variables, weight, and BMI (kg/m^2) were determined utilizing Pearson's product-moment correlation coefficient (r), with 95% confidence intervals (CI).

3. Results

A total of 33 female BH players (18 junior and 15 senior) of the Spanish Nationality Team participated in this study. Table 1 shows the basic anthropometric and demographic characteristics of the sample. The mean height and body mass were 167 ± 4.90 cm and 62.4 ± 7.29 kg for the junior females, and 169 ± 5.31 cm and 64.9 ± 7.87 kg for the senior players. There were only significant differences in the variable of age ($p < 0.001$), being obviously greater in the senior players. In the rest of the variables, there were no significant differences between the junior and senior players. However, as expected, the senior players showed higher values for weight and height.

Table 1. Demographic and basic anthropometric characteristics of professional female beach handball players.

Demographic and Anthropometric Variables	Junior (n = 18)			Senior (n = 15)		
	Mean	±	SD	Mean	±	SD
Age	16.7	±	0.50	24.8	±	4.71
Height (cm)	167	±	4.90	169	±	5.31
Weight (kg)	62.4	±	7.29	64.9	±	7.87
BMI (kg/m^2)	22.5	±	2.28	22.8	±	2.75

SD = Standard deviation; BMI: Body Mass Index.

The variables of adherence to the Mediterranean diet, sports performance, and bone mineral density are presented in Table 2. Except for the KIDMED total score, the senior players had higher scores for all the variables. In addition, the senior female players present significantly higher values in the case of the dynamometric hand strength test (Handgrip; $p < 0.05$) and the lower limb power test (Abalakov; $p < 0.01$).

Table 2. Descriptive statistics for the KIDMED, performance tests, and bone mineral density.

Study Variables	Total (N = 33)						ANCOVA Comparison (Adjusting for BMI)			
	Junior (n = 18)			Senior (n = 15)			Mean Difference	SE	t	*p* Value
Performance Test	Mean ± SD			Mean ± SD						
Yo-Yo (m)	382	±	81.4	414	±	109	−36.4	31.2	−1.17	0.253
Abalakov jump (cm)	28.5	±	5.69	35.1	±	6.89	−7.03	1.84	−3.83	<0.001
Handgrip (kg)	31.8	±	3.37	35.1	±	3.84	−3.11	1.20	−2.59	0.015
Bone Quality (Ultrasounds)										
BUA (dB/MHz)	137	±	9.55	139	±	11.7	−2.04	3.47	−0.601	0.552
SOS (m/s)	1666	±	37.2	1664	±	42.6	2.44	13.9	0.176	0.861
Stiffness (A.U)	139	±	13.7	138	±	17.9	−0.72	5.31	−0.135	0.894
Mediterranean Diet Adherence										
KIDMED	7.33	±	1.61	6.27	±	2.05	1.14	0.601	1.89	0.068

SD = Standard deviation; Yo-Yo IR: intermittent recovery test; KIDMED: Mediterranean Diet Quality Index; BUA: Broadband ultrasound attenuation; SOS: Speed of sound; BMI: Body Mass Index; N = total number of the sample; n = number of players on each team; t = t student; mean differences were significant at $p < 0.05$.

The results in Table 3 also indicate differences by position for the BMI, KIDMED total score, performance test, and BDM variables. Post hoc analysis revealed that there were no differences in any of the variables studied because of the playing position.

Table 3. Position-related differences in variables.

Study Variables	Goalkeepers (n = 6)			Wings (n = 10)			Specialist (n = 8)			Pivots (n = 6)			Defenders (n = 3)			ANCOVA (Adjusted by BMI)		
	Mean	±	SD	Mean	±	SD	Mean	±	SD	Mean	±	SD	Mean	±	SD	F	*p*	ηp2
Body Composition																		
BMI (kg/m^2)	24.9	±	3.60	22.2	±	2.50	22.3	±	1.45	21.8	±	1.72	21.8	±	1.31			
Performance Test																		
Yo-Yo (m)	313	±	46.8	412	±	103	405	±	104	400	±	63.2	480	±	80.0	1.18	0.343	0.153
Abalakov jump (cm)	30.9	±	6.06	30.7	±	8.73	29.6	±	4.73	34.6	±	7.73	34.1	±	8.34	0.821	0.523	0.108
Handgrip (kg)	35.3	±	5.85	31.6	±	3.48	33.6	±	3.21	32.8	±	3.40	35.2	±	2.28	0.751	0.566	0.100
Bone Quality (Ultrasounds)																		
BUA (dB/MHz)	142	±	10.6	141	±	9.47	139	±	9.84	134	±	8.20	128	±	17.1	0.969	0.441	0.126
SOS (m/s)	1660	±	36.1	1673	±	38.9	1666	±	32.1	1659	±	43.7	1657	±	74.1	0.317	0.864	0.045
Stiffness (A.U)	138	±	17.1	143	±	15.1	139	±	11.6	133	±	12.6	129	±	32.2	0.522	0.720	0.072
Mediterranean Diet Adherence																		
KIDMED	7.67	±	1.63	6.80	±	2.20	6.25	±	1.83	6.67	±	1.75	7.33	±	2.08	0.273	0.893	0.039

SD = Standard deviation; BMI: Body Mass Index; Yo-Yo: intermittent recovery test; KIDMED: Mediterranean Diet Quality Index; BUA: Broadband ultrasound attenuation; SOS: Speed of sound; n = number of players of each position; np2 = partial eta (effect size); mean differences were significant at $p < 0.05$.

The correlations between the different variables, separated by age category (junior vs. senior) are presented in Table 4. In the case of junior players, statistically significant negative correlations were observed between BMI and the Abalakov test results ($p < 0.05$), and between weight and the Abalakov ($p < 0.05$). However, significant positive correlations were also observed between the Yo-Yo and Abalakov test results ($p < 0.05$) and between the Yo-Yo test results and SOS values. For the senior players, statistically significant positive relationships were observed between BMI and BUA ($p < 0.05$) and between KIDMED and BMI ($p < 0.05$). However, between the results of the KIDMED and Abalakov ($p < 0.05$), and between BMI and the Abalakov test ($p < 0.05$), the correlations were negative.

Table 4. Female beach handball players' correlations.

		SENIOR								
		Weight	**BMI**	**Yo-Yo**	**Abalakov**	**Handgrip**	**BUA**	**SOS**	**Stiffness**	**KIDMED**
	Weight	n/a	0.868 *	−0.383	−0.493	0.320	0.333	−0.024	0.111	0.605 *
	BMI	0.856 **	n/a	−0.439	−0.512 *	0.359	0.661 *	0.343	0.316	0.524 *
	Yo-Yo	−0.359	−0.386	n/a	0.232	−0.101	−0.216	0.247	0.074	0.068
	Abalakov	−0.471 *	−0.630 *	0.653 *	n/a	0.133	−0.134	−0.011	−0.069	−0.477 *
JUNIOR	Handgrip	0.338	0.314	−0.089	−0.340	n/a	0.108	0.062	0.093	0.044
	BUA	0.242	0.057	−0.045	−0.196	−0.059	n/a	0.464 *	0.777 **	−0.079
	SOS	−0.046	0.029	0.602 *	0.011	−0.035	0.298	n/a	0.915 **	0.001
	Stiffness	0.078	0.049	0.438	−0.083	−0.054	0.681 *	0.894 **	n/a	−0.035
	KIDMED	0.051	0.186	0.174	−0.085	0.059	−0.335	0.057	−0.113	n/a

BUA: Broadband ultrasound attenuation; SOS: Speed of sound; BMI: Body Mass Index; Yo-Yo: intermittent recovery test; KIDMED: Mediterranean Diet Quality Index; n/a: not applicable; * = mean differences were significant at $p < 0.05$; ** mean differences were significant at $p < 0.01$. Gray color: numerical representation of the senior category. White color: numerical representation of the junior category.

4. Discussion

The main purpose of this investigation was to assess the physical fitness (through different physical tests), bone mass quality, and adherence to the Mediterranean diet of top-level female beach handball players. The present manuscript studied the differences by age group and playing position, and the correlations between the included variables.

Regarding physical fitness, the senior players were generally heavier and taller than the junior players. Moreover, the seniors scored higher in both the Abalakov jump and Handrip tests. Therefore, the senior players had better lower body power than the junior players ($p < 0.001$). The same occurred with the variable of musculoskeletal fitness of the upper extremities, measured by Handgrip since the senior players had shown higher values ($p = 0.015$). This coincides with previous research conducted with other samples of handball players [25,45]. It seems that the reasons, as demonstrated in female volleyball players [46], are that these differences may be directly related to neuromechanical adaptations because of the repeated actions taking place in training and matches. However, it has also been shown that the results of these tests improved, and therefore, improving physical abilities (jumping, reaction speed, and running speed), when a Plyometric Training Program training was performed [21,22]. This data is of special interest for all the coaches.

In the Yo-Yo Test, there were no significant differences between the values obtained for the junior vs. senior players. One of the reasons why this test was used was because it has been shown [47] that the Yo-Yo Test contributes specifically to the handball performance of elite players, which means that this test, among others, measures important qualities for success in handball. If the results achieved in the present study are compared to those of [20,47], much lower values are observed; the main reason for this distinction is the surface since, in this study, the players did the tests on sand.

Furthermore, some researchers underline [47] the importance of aerobic and anaerobic qualities for success in handball, supporting the emphasis on the development of prolonged intermittent running ability, sprinting qualities, and repeated sprinting ability in elite handball players.

In addition, for correlations, negative correlations were observed in the junior players between BMI and the Abalakov jump and between weight and the Abalakov jump. This means that lower weight and lower BMI are related to higher jump heights and therefore, more power in the lower limbs. In previous studies [48] conducted on female handball players, these physical characteristics were already correlated. As already corroborated [49], it seems that an excess of body mass has a negative role in jumping. For the same reason, the senior female players also have negative correlations between BMI and the Abalakov jump. However, significant positive correlations have been observed in the junior players between the results of the Yo-Yo and Abalakov jump tests, and therefore, those players who present greater power in the upper limbs also present greater endurance, agility, and speed.

In terms of bone health, previous studies [50] have demonstrated an osteogenic effect of high-impact exercise (such as BH) and weight training on BMD by DXA. Additionally, a study using DXA and QUS measurements by Lehtonen-Veromaa et al. [51] showed that both femoral neck BMD and heel QUS parameters increased in the following order: control, runners, and gymnasts. In many of the studies performed, in which they have compared an athlete population and a control group, such as the case of female runners [28] or powerlifters [52], it has been observed that athletes who perform a specific sport modality in which there are both impact movements and strength exercises, the values of SOS, BUA, and Stiffness were higher than the control group. Moreover, if the values obtained in the present study of SOS, BUA, and Stiffness are compared for both the junior and senior players, the values of all three parameters are higher than those of female runners [51,53], both long and short distance [28], gymnasts [51], and powerlifters [52]. Other studies [54] also relate higher values to a higher number of hours devoted to training. Therefore, it seems that the fact that beach handball is played on a surface such as sand, and that there are repeated impacts after jumps, turns, and sprints is favored by the development and bone quality of the growing skeleton.

It should be noted that this is the first study to research the correlation between bone variables (BUA, SOS, and Stiffness) and performance results using different tests that measure strength, endurance, and power. Positive relationships have been observed between SOS values and Yo-Yo test results in the junior players; those players who have managed to run a longer distance (m) are those who have higher values for the speed of sound. These results indicate that athletes' training endurance, speed, and agility may exert progressive adaptation on the calcaneal bone, tending to produce superior elastic bone strength (this correlates with mineral and protein contents). Furthermore, in the case of the senior players, a positive correlation can be observed between the BMI and BUA values; therefore, those players who have a higher BMI have higher values for broadband ultrasound attenuation. One of the possible reasons could be that players who are taller and heavier have higher bone mineral density, and therefore, a lower risk of fractures.

Regarding the nutrition of the players, positive correlations have been established in senior players between KIDMED and BMI, and thus, those players with higher scores have higher BMI values. Negative correlations have also been found in the senior players between the KIDMED score and the Abalakov jump results; the better the KIDMED score, the lower the Abalakov jump test results, and therefore, lower power. It should be noted that the KIDMED measures the quality of the diet in terms of adherence to the Mediterranean diet; however, it is not able to measure the amount of food eaten by the players. As previously observed [25], adherence to the Mediterranean diet in female beach handball players is moderate.

It has been previously observed that both correct hydration and nutrition are essential for better performance in beach handball [55,56]. Because this sport is performed in hot environments, female athletes are at greater risk of dehydration and hyperthermia, hindering overall physiological function and cognitive and athletic performance. In addition, sports performance and recovery from exercise have been found to be enhanced by optimal nutrition. Appropriate food choices should be made, in addition to considering sufficient energy intake during both high intensity and long duration training periods, with the goal of maintaining body weight and health and maximizing the effects of training [33]. There is an evident need for athletes to improve their dietary habits, so the intervention of dieticians/nutritionists in multidisciplinary teams is necessary and essential.

This research has some limitations. Firstly, body composition variables were measured by anthropometry and bone quality by ultrasound. Although both methods are related to bone densitometry (dual-energy X-ray absorptiometry), it was not possible to use this method, which is considered the "gold standard". In addition, as previously indicated, the quality of the food was evaluated, but it was not possible to assess the quantity. It would be interesting to be able to make 7-day records with a photographic report, to be able to assess both quality and quantity. Due to the fact that there are not many sports played on sand, there are tests, such as the Yo-Yo, that are not validated for this surface, which is why the players presented lower values. Therefore, it would be interesting to validate this test on a sand surface. The Abalakov jump test had to be performed on a regular surface to be able to use the contact platform correctly because the sand was very unstable.

Future research will consider the mentioned limitations. Researchers in the field are invited to provide more specific information regarding the evaluation of physical fitness, body composition, and dietary habits of professional beach handball players, as research is still scarce, as well as regarding the validation of specific tests for this sport on its usual playing surface, sand.

5. Conclusions

The results of this cross-sectional study in female athletes from the Spanish BH indicate that older players performed better in the Handgrip and Abalakov jump tests. Correlation analysis highlighted the relationship of weight and poor nutrition with worse results in jumping, agility, power, and endurance tests. Regarding bone mass, it was observed that female handball players have higher values of SOS, BUA, and Stiffness than athletes of other sports disciplines, confirming that sports involving impact, jumps, and sprints have a higher level of bone mineral density.

Author Contributions: Conceptualization, A.M.-R., J.S.-S. and J.A.S.-S.; data curation, M.M.-O., L.M.-A., and M.V.-M.; formal analysis, A.M.-R., L.M.-A., and M.M.-O.; investigation, A.M.-R., J.S.-S., M.V.-M. and J.A.S.-S.; methodology, A.M.-R., J.S.-S., and J.A.S.-S.; resources, A.M.-R., J.S.-S., M.V.-M. and J.A.S.-S.; software, M.M.-O., M.V.-M. and A.M.-R.; validation, A.M.-R., J.S.-S., and J.A.S.-S.; writing—original draft, A.M.-R., M.M.-O., and J.A.S.-S.; writing—review and editing, A.M.-R., J.S.-S., and J.A.S.-S. All authors have read and agreed to the published version of the manuscript.

Funding: This research received no external funding.

Institutional Review Board Statement: The study was conducted according to the guidelines of the Declaration of Helsinki and approved by the University Human Research Ethics Committee of the University of Alicante (Spain) (UA-2019-04-09).

Informed Consent Statement: Informed consent was obtained from all subjects involved in the study.

Data Availability Statement: The data presented in this study is available on request from the corresponding author. The data are not publicly available due to is personal health information.

Acknowledgments: To the players and coaches of the Spanish Beach Handball National Team, to the Royal Spanish Handball Federation and to the European Institute of Exercise and Health (EIEH) of Alicante University for their selfless collaboration in this research.

Conflicts of Interest: The authors declare no conflict of interest.

References

1. Gutiérrez-Vargas, R.; Gutiérrez-Vargas, J.C.; Ugalde-Ramírez, J.A.; Rojas-Valverde, D. Kinematics and thermal sex-related responses during an official beach handball game in Costa Rica: A pilot study. *Arch. Med. Deport.* **2019**, *36*, 13–18.
2. Lara Cobos, D. La respuesta cardiaca durante la competición de balonmano playa femenino. *Apunt. Med. L'Esport* **2011**, *46*, 131–136. [CrossRef]
3. Lemos, L.F.; Oliveira, V.C.; Duncan, M.J.; Ortega, J.P.; Martins, C.M.; Ramirez-Campillo, R.; Sanchez, J.S.; Nevill, A.M.; Nakamura, F.Y. Physical fitness profile in elite beach handball players of different age categories. *J. Sports Med. Phys. Fitness* **2020**, *60*, 1544–1550. [CrossRef]
4. Pueo, B.; Jimenez-Olmedo, J.M.; Penichet-Tomas, A.; Ortega Becerra, M.; Espina Agullo, J.J. Analysis of time-motion and heart rate in elite male and female beach handball. *J. Sports Sci. Med.* **2017**, *16*, 450–458. [PubMed]
5. Pueo, B.; Espina-Agullo, J.J.; Selles-Perez, S.; Penichet-Tomas, A. Optimal body composition and anthropometric profile of world-class beach handball players by playing positions. *Sustainability* **2020**, *12*, 6789. [CrossRef]
6. Reigal, R.E.; Vázquez-Diz, J.A.; Morillo-Baro, J.P.; Hernández-Mendo, A.; Morales-Sánchez, V. Psychological profile, competitive anxiety, moods and self-efficacy in beach handball players. *Int. J. Environ. Res. Public Health* **2020**, *17*, 241. [CrossRef] [PubMed]
7. Sánchez-Sáez, J.A.; Sánchez-Sánchez, J.; Martínez-Rodríguez, A.; Felipe, J.L.; García-Unanue, J.; Lara-Cobos, D. Global positioning system analysis of physical demands in elite women's beach handball players in an official spanish championship. *Sensors* **2021**, *21*, 850. [CrossRef]
8. Rules of the Game—International Handball Federation—Flipbook by l FlipHTML5. Available online: https://fliphtml5.com/ufty/matt (accessed on 21 April 2021).
9. Lara Cobos, D.; Sánchez Sáez, J.A.; Morillo Baro, J.P.; Sánchez Malia, J.M. Beach Handball Game Cycle. *Rev. Int. Deport. Colect.* **2018**, *34*, 89–100.
10. Roczniok, R.; Stanula, A.; Maszczyk, A.; Mostowik, A.; Kowalczyk, M.; Fidos-Czuba, O.; Zając, A. Physiological, physical and on-ice performance criteria for selection of elite ice hockey teams. *Biol. Sport* **2016**, *33*, 43–48. [CrossRef]
11. Fernandes-da-Silva, J.; Castagna, C.; Teixeira, A.S.; Carminatti, L.J.; Guglielmo, L.G.A. The peak velocity derived from the Carminatti Test is related to physical match performance in young soccer players. *J. Sports Sci.* **2016**, *34*, 2238–2245. [CrossRef] [PubMed]
12. Zapardiel, J.C.; Asín-Izquierdo, I. Conditional analysis of elite beach handball according to specific playing position through assessment with GPS. *Int. J. Perform. Anal. Sport* **2020**, *20*, 118–132. [CrossRef]
13. Bangsbo, J.; Iaia, F.M.; Krustrup, P. The Yo-Yo intermittent recovery test: A useful tool for evaluation of physical performance in intermittent sports. *Sports Med.* **2008**, *38*, 37–51. [CrossRef]
14. Hermassi, S.; Aouadi, R.; Khalifa, R.; Van Den Tillaar, R.; Shephard, R.J.; Chelly, M.S. Relationships between the Yo-Yo intermittent recovery test and anaerobic performance tests in adolescent handball players. *J. Hum. Kinet.* **2015**, *45*, 197–205. [CrossRef]
15. Hermassi, S.; Bragazzi, N.L.; Majed, L. Body Fat Is a Predictor of Physical Fitness in Obese Adolescent Handball Athletes. *Int. J. Environ. Res. Public Health* **2020**, *17*, 8428. [CrossRef] [PubMed]
16. Hermassi, S.; Chelly, M.S.; Michalsik, L.B.; Sanal, N.E.M.; Hayes, L.D.; Cadenas-Sanchez, C. Relationship between fatness, physical fitness, and academic performance in normal weight and overweight schoolchild handball players in Qatar State. *PLoS ONE* **2021**, *16*, e0246476. [CrossRef] [PubMed]
17. Hermassi, S.; Laudner, K.; Schwesig, R. The Effects of Circuit Strength Training on the Development of Physical Fitness and Performance-Related Variables in Handball Players. *J. Hum. Kinet.* **2020**, *71*, 191–203. [CrossRef] [PubMed]
18. Souhail, H.; Castagna, C.; Mohamed, H.Y.; Younes, H.; Chamari, K. Direct validity of the yo-yo intermittent recovery test in young team handball players. *J. Strength Cond. Res.* **2010**, *24*, 465–470. [CrossRef] [PubMed]
19. Michalsik, L.B.; Madsen, K.; Aagaard, P. Match performance and physiological capacity of female elite team handball players. *Int. J. Sports Med.* **2014**, *35*, 595–607. [CrossRef]

20. Moss, S.L.; McWhannell, N.; Michalsik, L.B.; Twist, C. Anthropometric and physical performance characteristics of top-elite, elite and non-elite youth female team handball players. *J. Sports Sci.* **2015**, *33*, 1780–1789. [CrossRef]

21. Hammami, M.; Gaamouri, N.; Suzuki, K.; Shephard, R.J.; Chelly, M.S. Effects of Upper and Lower Limb Plyometric Training Program on Components of Physical Performance in Young Female Handball Players. *Front. Physiol.* **2020**, *11*, 1028. [CrossRef]

22. Hammami, M.; Hermassi, S.; Gaamouri, N.; Aloui, G.; Comfort, P.; Shephard, R.J.; Chelly, M.S. Field Tests of Performance and Their Relationship to Age and Anthropometric Parameters in Adolescent Handball Players. *Front. Physiol.* **2019**, *10*, 1124. [CrossRef]

23. Santos, D.A.; Matias, C.N.; Monteiro, C.P.; Silva, A.M.; Rocha, P.M.; Minderico, C.S.; Sardinha, L.B.; Laires, M.J.; Santos, D.A. Magnesium intake is associated with strength performance in elite basketball, handball and volleyball players. *Magnes. Res.* **2011**, *24*, 215–224.

24. Wagner, H.; Sperl, B.; Bell, J.W.; Von Duvillard, S.P. Testing Specific Physical Performance in Male Team Handball Players and the Relationship to General Tests in Team Sports. *J. Strength Cond. Res.* **2019**, *33*, 1056–1064. [CrossRef]

25. Martínez-Rodríguez, A.; Martínez-Olcina, M.; Hernández-García, M.; Rubio-Arias, J.; Sánchez-Sánchez, J.; Lara-Cobos, D.; Vicente-Martínez, M.; Carvalho, M.J.; Sánchez-Sáez, J.A. Mediterranean diet adherence, body composition and performance in beach handball players: A cross sectional study. *Int. J. Environ. Res. Public Health* **2021**, *18*, 1–14.

26. Ramírez-Vélez, R.; Ojeda-Pardo, M.L.; Correa-Bautista, J.E.; González-Ruíz, K.; Navarro-Pérez, C.F.; González-Jiménez, E.; Schmidt-RioValle, J.; Izquierdo, M.; Lobelo, F. Normative data for calcaneal broadband ultrasound attenuation among children and adolescents from Colombia: The FUPRECOL Study. *Arch. Osteoporos.* **2016**, *11*, 1–10. [CrossRef]

27. Babaroutsi, E.; Magkos, F.; Manios, Y.; Sidossis, L.S. Lifestyle factors affecting heel ultrasound in Greek females across different life stages. *Osteoporos. Int.* **2005**, *16*, 552–561. [CrossRef] [PubMed]

28. Lara, B.; Salinero, J.J.; Gutiérrez, J.; Areces, F.; Abián-Vicén, J.; Ruiz-Vicente, D.; Gallo-Salazar, C.; Jiménez, F.; Del Coso, J. Influence of endurance running on calcaneal bone stiffness in male and female runners. *Eur. J. Appl. Physiol.* **2016**, *116*, 327–333. [CrossRef] [PubMed]

29. Nikander, R.; Kannus, P.; Rantalainen, T.; Uusi-Rasi, K.; Heinonen, A.; Sievänen, H. Cross-sectional geometry of weight-bearing tibia in female athletes subjected to different exercise loadings. *Osteoporos. Int.* **2010**, *21*, 1687–1694. [CrossRef] [PubMed]

30. Guadalupe-Grau, A.; Fuentes, T.; Guerra, B.; Calbet, J.A.L. Exercise and bone mass in adults. *Sports Med.* **2009**, *39*, 439–468. [CrossRef] [PubMed]

31. Ju, Y.-I.; Sone, T.; Ohnaru, K.; Tanaka, K.; Yamaguchi, H.; Fukunaga, M. Effects of Different Types of Jump Impact on Trabecular Bone Mass and Microarchitecture in Growing Rats. *PLoS ONE* **2014**, *9*, e107953. [CrossRef]

32. Gómez-Bruton, A.; González-Agüero, A.; Gómez-Cabello, A.; Matute-Llorente, A.; Casajús, J.A.; Vicente-Rodríguez, G. The effects of swimming training on bone tissue in adolescence. *Scand. J. Med. Sci. Sports* **2015**, *25*, e589–e602. [CrossRef]

33. Rodriguez, N.R.; DiMarco, N.M.; Langley, S. Position of the American Dietetic Association, Dietitians of Canada, and the American College of Sports Medicine: Nutrition and athletic performance. *J. Am. Diet. Assoc.* **2009**, *109*, 509–527.

34. Serra-Majem, L.; Tomaino, L.; Dernini, S.; Berry, E.M.; Lairon, D.; de la Cruz, J.N.; Bach-Faig, A.; Donini, L.M.; Medina, F.X.; Belahsen, R.; et al. Updating the mediterranean diet pyramid towards sustainability: Focus on environmental concerns. *Int. J. Environ. Res. Public Health* **2020**, *17*, 8758. [CrossRef] [PubMed]

35. Willett, W. Mediterranean dietary pyramid. *Int. J. Environ. Res. Public Health* **2021**, *18*, 4568. [CrossRef] [PubMed]

36. Fernandez, M.L.; Raheem, D.; Ramos, F.; Carrascosa, C.; Saraiva, A.; Raposo, A. Highlights of current dietary guidelines in five continents. *Int. J. Environ. Res. Public Health* **2021**, *18*, 2814. [CrossRef] [PubMed]

37. Alacid, F.; Vaquero-Cristóbal, R.; Sánchez-Pato, A.; Muyor, J.M.; López-Miñarro, P.Á. Adhesión a la dieta mediterránea y relación con los parámetros antropométricos de mujeres jóvenes kayakistas. *Nutr. Hosp.* **2014**, *29*, 121–127. [PubMed]

38. Krustrup, P.; Mohr, M.; Amstrup, T.; Rysgaard, T.; Johansen, J.; Steensberg, A.; Pedersen, P.K.; Bangsbo, J. The Yo-Yo intermittent recovery test: Physiological response, reliability, and validity. *Med. Sci. Sports Exerc.* **2003**, *35*, 697–705. [CrossRef]

39. Bosco, C.; Luhtanen, P.; Komi, P.V. A simple method for measurement of mechanical power in jumping. *Eur. J. Appl. Physiol. Occup. Physiol.* **1983**, *50*, 273–282. [CrossRef]

40. Fonseca, H.; Moreira-Gonçalves, D.; Coriolano, H.J.A.; Duarte, J.A. Bone quality: The determinants of bone strength and fragility. *Sports Med.* **2014**, *44*, 37–53. [CrossRef]

41. Marín, F.; González-Macías, J.; Díez-Pérez, A.; Palma, S.; Delgado-Rodríguez, M. Relationship between bone quantitative ultrasound and fractures: A meta-analysis. *J. Bone Miner. Res.* **2006**, *21*, 1126–1135. [CrossRef]

42. Altavilla, C.; Caballero-Perez, P. An update of the KIDMED questionnaire, a Mediterranean Diet Quality Index in children and adolescents. *Public Health Nutr.* **2019**, *22*, 2543–2547. [CrossRef]

43. Galan-Lopez, P.; Sánchez-Oliver, A.J.; Ries, F.; González-Jurado, J.A. Mediterranean diet, physical fitness and body composition in sevillian adolescents: A healthy lifestyle. *Nutrients* **2019**, *11*, 2009. [CrossRef]

44. Richardson, J.T.E. The analysis of 2 × 2 contingency tables-Yet again. *Stat. Med.* **2011**, *30*, 890. [CrossRef]

45. Gabrys, T.; Stanula, A.; Gupta, S.; Szmatlan-Gabrys, U.; Benešová, D.; Wicha, Ł.; Baron, J. A comparative study on the performance profile of under-17 and under-19 handball players trained in the sports school system. *Int. J. Environ. Res. Public Health* **2020**, *17*, 7979. [CrossRef] [PubMed]

46. Kitamura, K.; Pereira, L.A.; Kobal, R.; Abad, C.C.C.; Finotti, R.; Nakamura, F.Y.; Loturco, I. Loaded & unloaded jump performance of top-level volleyball players from different age categories. *Biol. Sport* **2017**, *34*, 273–278. [PubMed]

47. Hermassi, S.; Hoffmeyer, B.; Irlenbusch, L.; Fieseler, G.; Noack, F.; Delank, K.S.; Gabbett, T.J.; Souhaiel Chelly, M.; Schwesig, R. Relationship between the handball-specific complex-test and intermittent field test performance in professional players. *J. Sports Med. Phys. Fitness* **2018**, *58*, 8–16. [PubMed]

48. Saavedra, J.M.; Kristjánsdóttir, H.; Einarsson, I.; Guðmundsdóttir, M.L.; Þorgeirsson, S.; Stefansson, A. Anthropometric characteristics, physical fitness, and throwing velocity in elite women's handball teams. *J. Strength Cond. Res.* **2018**, *32*, 2294–2301. [CrossRef]

49. Nikolaidis, P.T.; Gkoudas, K.; Afonso, J.; Clementesuarez, V.J.; Knechtle, B.; Kasabalis, S.; Kasabalis, A.; Douda, H.; Tokmakidis, S.; Torres-Luque, G. Who jumps the highest? Anthropometric and physiological correlations of vertical jump in youth elite female volleyball players. *J. Sports Med. Phys. Fitness* **2017**, *57*, 802–810.

50. Yung, P.S.; Lai, Y.M.; Tung, P.Y.; Tsui, H.T.; Wong, C.K.; Hung, V.W.Y.; Qin, L.; Lai, Y.; Hong, T. Effects of weight bearing and non-weight bearing exercises on bone properties using calcaneal quantitative ultrasound. *Br. J. Sports Med.* **2005**, *39*, 547–551. [CrossRef] [PubMed]

51. Lehtonen-Veromaa, M.; Möttönen, T.; Nuotio, I.; Heinonen, O.J.; Viikari, J. Influence of physical activity on ultrasound and dual-energy X-ray absorptiometry bone measurements in peripubertal girls: A cross-sectional study. *Calcif. Tissue Int.* **2000**, *66*, 248–254. [CrossRef]

52. Jawed, S.; Horton, B.; Masud, T. Quantitative Heel Ultrasound Variables in Powerlifters and Controls. *Br. J. Sports Med.* **2001**, *35*, 274–275. [CrossRef] [PubMed]

53. Brahm, H.; Ström, H.; Piehl-Aulin, K.; Mallmin, H.; Ljunghall, S. Bone metabolism in endurance trained athletes: A comparison to population-based controls based on DXA, SXA, quantitative ultrasound, and biochemical markers. *Calcif. Tissue Int.* **1997**, *61*, 448–454. [CrossRef] [PubMed]

54. Scheffler, C.; Gniosdorz, B.; Staub, K.; Rühli, F. Skeletal robustness and bone strength as measured by anthropometry and ultrasonography as a function of physical activity in young adults. *Am. J. Hum. Biol.* **2014**, *26*, 215–220. [CrossRef]

55. Wilson, T.E. Renal sympathetic nerve, blood flow, and epithelial transport responses to thermal stress. *Auto. Neurosci. Basic Clin.* **2017**, *204*, 25–34. [CrossRef] [PubMed]

56. Cheuvront, S.N.; Kenefick, R.W. Dehydration: Physiology, assessment, and performance effects. *Compr. Physiol.* **2014**, *4*, 257–285.

Article

Changes in Bone Mineral Density and Serum Lipids across the First Postpartum Year: Effect of Aerobic Fitness and Physical Activity

Erin M. Kyle [1], Hayley B. Miller [2], Jessica Schueler [1], Michelle Clinton [1], Brenda M. Alexander [3], Ann Marie Hart [4] and D. Enette Larson-Meyer [1,2,*]

[1] Department of Family and Consumer Sciences, University of Wyoming, Laramie, WY 82071, USA; erin.kyle.728@gmail.com (E.M.K.); jesschueler@gmail.com (J.S.); michelle_felts@live.com (M.C.)
[2] Department of Human Nutrition, Foods and Exercise, Virginia Tech, Blacksburg, VA 24061, USA; hayley@vt.edu
[3] Department of Animal Science, University of Wyoming, Laramie, WY 82071, USA; balex@uwyo.edu
[4] School of Nursing, University of Wyoming, Laramie, WY 82071, USA; annmhart@uwyo.edu.com
* Correspondence: enette@vt.edu; Tel.: +1-540-231-1025

Abstract: This study evaluated the changes in bone mineral density (BMD) and serum lipids across the first postpartum year in lactating women compared to never-pregnant controls, and the influence of physical activity (PA). The study also explored whether N-telopeptides, pyridinoline, and deoxypyridinoline in urine serve as biomarkers of bone resorption. A cohort of 18 initially lactating postpartum women and 16 never pregnant controls were studied. BMD (dual energy X-ray absorptiometry), serum lipid profiles, and PA (Baecke PA Questionnaire) were assessed at baseline (4–6 weeks postpartum), 6 months, and 12 months. Postpartum women lost 5.2 ± 1.4 kg body weight and BMD decreased by 1.4% and 3.1% in the total body and dual-femur, respectively. Furthermore, BMDdid not show signs of rebound. Lipid profiles improved, with increases in high-density lipoprotein-cholesterol (HDL-C) and decreases in low-density lipoprotein cholesterol (LDL-C) and the cholesterol/HDL-C ratio at 12 months (vs. baseline). These changes were not influenced by lactation, but the fall the Cholesterol/HDL-C ratio was influenced by leisure-time ($p = 0.051$, time X group) and sport ($p = 0.028$, time effect) PA. The decrease in BMD from baseline to 12 months in total body and dual femur, however, was greater in those who continued to breastfeed for a full year compared to those who stopped at close to 6 months. Urinary markers of bone resorption, measured in a subset of participants, reflect BMD loss, particularly in the dual-femur, and may reflect changes bone resorption before observed changes in BMD. Results provide support that habitual postpartum PA may favorably influence changes in serum lipids but not necessarily BMD. The benefit of exercise and use of urinary biomarkers of bone deserves further exploration.

Keywords: bone density; bone resorption markers; HDL-C; LDL-C; lipid profile; cardiovascular disease; exercise; aerobic fitness

Citation: Kyle, E.M.; Miller, H.B.; Schueler, J.; Clinton, M.; Alexander, B.M.; Hart, A.M.; Larson-Meyer, D.E. Changes in Bone Mineral Density and Serum Lipids across the First Postpartum Year: Effect of Aerobic Fitness and Physical Activity. *Nutrients* 2022, 14, 703. https://doi.org/10.3390/nu14030703

Academic Editor: Maria Luz Fernandez

Received: 11 November 2021
Accepted: 2 February 2022
Published: 8 February 2022

1. Introduction

Pregnancy and lactation are characterized by a variety of physiological changes in the mother, including changes in bone mineral density (BMD) and lipid profiles. Extensive research has found significant decreases in BMD [1–3] and increases in plasma cholesterol and triglyceride concentrations throughout the gestational period [4–6]. While maternal adaptation, including increased absorption of intestinal calcium [7] and changes in lipid metabolism [8], help ensure supply of calcium and lipid to the fetus and placenta for steroid hormone synthesis during pregnancy, the persistence of reduced BMD and elevated serum lipids and triglycerides (TG) during the postpartum period can become risk factors for lactation-induced osteoporosis [9] or cardiovascular disease [10].

According to the Institute of Medicine, lactating women provide two-to-three times more calcium to their infant through breast milk during the first 6 months of breastfeeding than during the entirety of pregnancy [11]; this results in significant alterations in maternal calcium metabolism to maintain serum calcium within the normal range [7,12]. While increased intestinal calcium absorption helps accommodate these demands during pregnancy, the primary mechanisms of calcium conservation during lactation include renal resorption via the distal tubules, skeletal demineralization, and bone resorption [7]. Additionally, studies have found that neither calcium intake nor supplementation influences lactation-induced BMD loss [13–15], highlighting that these changes are physiologically driven. Previous research has found inconsistent results regarding the impact of lactation on BMD loss and its subsequent recovery that may vary by skeletal site and length of lactation. Several studies observed significant decreases in BMD from baseline in women lactating at least 3 to 6 months [1–3,16–19] with those who breastfed more than 4 to 6 months experiencing greater bone loss than those that weaned earlier [1–3,16,20,21]. In longitudinal studies with mostly young, healthy, Caucasian mothers, BMD was shown to decrease ~2 to 7.5% during the first 4–6 months of lactation in the lumbar spine and/or hip [2,3,13,16,18–20,22–24], 0 to 5% in the forearm radius [13,16,19,22,24], and 0 to 3% of the total body BMD [2,3,13,18,19,24]. Although most BMD loss within the first 6 months has been observed to approach complete recovery following weaning [2,22,25–30], some studies have found that lactation past 6 months is associated with only partial recovery [1,16,19,20,31], which suggests that extended lactation can delay the return of BMD to baseline levels. As a result, limited research has focused on the impact of exercise on lactation-related bone loss, with several studies supporting an association between exercise and reduced bone loss [32–34] and others reporting no significant difference [17,20,35]. To our knowledge, no studies have used emergent urinary markers of bone resorption such as n-telopeptides (NTX) [36], pyridinoline (PYD), or deoxypyridinoline (DPYD) [37] to help better understand the dynamic changes in bone that occur postpartum.

Changes in serum cholesterol and lipid profiles may also be of concern postpartum. Lipid metabolism adapts during pregnancy to transport adequate cholesterol to the placenta and developing fetus, which supports steroid hormone synthesis and fetal nervous system development. These changes promote maternal fat accumulation during the first two trimesters and enhance the breakdown of fat depots during the third trimester, resulting in hypercholesteremia [8]. While it is well-recognized that cholesterol and TG are elevated during pregnancy due to the aforementioned changes in lipid metabolism, independent of dietary patterns [38], less is known during lactation. The persistence of atherogenic lipid concentrations postpartum has been investigated as a potential predictor of cardiovascular disease risk with research suggesting that lactation may help mitigate elevated plasma cholesterol and TG in postpartum mothers [39,40]. Several studies found that total serum cholesterol and triglyceride concentrations significantly decrease with at least 2 to 6 months of lactation [39,40]. More specifically, a study conducted by Kallio et al. found that total serum cholesterol, LDL-C and HDL-C, and TG all returned to baseline concentrations after one year of exclusive lactation [39]. Limited research has explored the effect of exercise on lipid profiles. For example, a randomized controlled trial by Lovelady et al. [41] found modest increases in HDL-C in women assigned to an aerobic exercise intervention in comparison to sedentary controls but found no changes in other lipid values. Another study that investigated the impact of both diet and exercise on cardiovascular risk factors and weight loss during lactation found no effect of exercise on blood lipids independent of an energy-restricted dietary intervention intended for weight loss [42].

Currently, there is a lack of cohesive research evaluating BMD and/or lipid profile changes across lactation and during postpartum and the potential influence of physical activity, exercise and aerobic fitness during this period. Therefore, the present study aimed to investigate changes in BMD and serum cholesterol and TG across the first postpartum year in lactating women relative to age-matched never-pregnant controls We specifically tested the hypotheses that (1) BMD would decrease during the first 6 months of lactation, followed by at least a partial rebound after weaning; and (2) that previously elevated serum cholesterol and triglyceride concentrations would significantly decrease by 12 months of lactation; Secondary objectives were to evaluate whether habitual physical activity assessed by questionnaire and aerobic fitness would positively influence changes in BMD and serum lipid profile during the 12 months postpartum and explore the use of urinary NTX, PYD, and DPYD as markers of bone resorption during lactation.

2. Materials and Methods

This analysis contains longitudinal data collected as part of a study that evaluated the effect of appetite-regulating hormones on body weight retention in lactating mothers [43] and the presence of appetite-regulation hormones in breast milk [44]. Participants were recruited through flyers posted within the community, university, and doctors' offices. Eligibility criteria included age over 18, singleton birth, 4–6 weeks postpartum, and intentions to fully breastfeed for one year. Participants were excluded if they smoked, had pregnancy complications (e.g., gestational diabetes, preeclampsia), had preexisting kidney, liver, hormonal, stomach, intestine, lung, heart, or blood disease, were taking prescription or over-the-counter medications or herbal supplements, or had a history of anxiety, depression, disordered eating, alcoholism or substance abuse. Twenty-four healthy primiparous women and 20 never-pregnant controls were initially enrolled and provided written, informed consent. Of these, 18 postpartum women and 16 never-pregnant controls completed the one year follow up and were used in the current analysis.

2.1. Laboratory Measurements

Measurements were taken at baseline (4 to 6 weeks postpartum), and at 6 and 12 months postpartum in participants and at corresponding time points in control women. At each visit, maternal anthropometrics and body composition/bone density analysis (by dual energy X-ray absorptiometry scan, DXA) were performed, and blood was collected for analysis of serum lipids. Urine was collected in a subset of postpartum ($n = 14$) and control ($n = 8$) women for analysis of urinary markers of bone resorption Physical activity questionnaires were collected at all time points, and a maximal oxygen uptake test (VO_{2max}) to assess aerobic fitness was conducted at 12 months. Study visits were scheduled in the follicular phase of the menstrual cycle (1–9 days after the start of menstruation) for never-pregnant control women and for postpartum women who had resumed their menstrual cycles at the 6 and 12 month visits.

2.2. Assessment of Anthropometrics and Bone Mineral Density

Participants were measured without shoes and in light clothing. Height was measured using a stadiometer (Invicta Plastics, Leicester, UK) with weight measured on a digital scale (Tanita, Tokyo, Japan). A Gulick tape measure was used to measure maternal hip and waist circumference. Hip circumference was defined as a horizontal measure taken at the maximum circumference of the buttocks [45]. Waist circumference was defined as the narrowest part of the torso above the umbilicus and below the xiphoid process [45]. Bone density and body composition were assessed by DXA (Lunar Prodigy, GE Healthcare, Fairfield, CT). Participants were placed in the supine position, and instructed to lay still on the X-ray table while the fan-beam scanner made a series of transverse scans from head to toe in (0.6- to 1.0-cm) intervals. Three scans were performed: total body, lumbar spine (L1–L4), and dual femur (hips; femoral neck and trochanter). Scans were analyzed utilizing manufacturer-provided software (encore Software v13.6) and standardized for the evaluation of adults. Standardized t-scores, matched against a healthy 30-year female population were recorded for all sites when calculated by the software.

2.3. Serum Lipid Profiles

At each testing point, blood for analysis of serum triglycerides (TG), total cholesterol, high-density lipoprotein cholesterol (HDL-C), low-density lipoprotein cholesterol (LDL-C) and very low-density lipoprotein cholesterol (VLDL-C) was collected into serum gel tubes and centrifuged between 30 and 120 min of draw time and kept refrigerated until analysis. Analysis was performed by a commercial laboratory (Regional West Laboratories, Scottsbluff, NE, USA) using standardized procedures. Specifically, TG, total cholesterol and HDL-C were measured by spectrophotometry and LDL-C and VLDL-C were calculated using the Friedewald equation.

2.4. Markers of Bone Resorption

Urine samples were collected in the morning (first-morning void) after an overnight fast for analysis of markers of bone resorption including NTX, a general marker of bone-resorption, PYD, a general marker of collagen degradation, and DPYD, a sensitive and specific marker of bone-specific collagen degradation/bone resorption [37] in a subset of postpartum (n = 14) and control (n = 8) women. Urine samples were immediately frozen (without preservatives) and later analyzed for the aforementioned markers using chemiluminescent immunoassay (NTX, Mayo Clinic Laboratories, Rochester, MN, USA) and high-performance liquid chromatography (PYD, DPYD, ARUP Laboratories, Salt Lake City, UT, USA) by a commercial laboratory (Regional West, Scottsbluff, NE, Powered by Mayo Clinic Laboratories). Values were normalized to urinary creatinine to account for the variation in urinary concentrations between individuals. Urinary creatinine concentration was analyzed via enzymatic colorimetric assay (Mayo Clinic Laboratories, Rochester, MN, USA).

2.5. Assessment of Habitual Exercise and Aerobic Fitness

Habitual PA was assessed using the Baecke Physical Activity Questionnaire at each time point. The Baecke questionnaire is a validated assessment tool that categorizes PA into sport, work, and leisure-time activity, and has been shown to be a reliable predictor of habitual PA [46]. Cardiovascular fitness was determined during a walking VO_{2max} test on the treadmill as previously described [43]. Briefly, after a 2 min warm up, the test began with participants walking at 1.1 m/s at 0% grade for one minute. Speed and/or grade was increased each minute in increments appropriate for non-athletic subjects of varying fitness levels to achieve maximal effort within 12–15 min. Heart rate (HR) and rating of perceived exertion (RPE; modified Borg scale) were recorded at the end of the third, sixth, and ninth stages, and at every stage following the ninth. The test was terminated when the participant reached volitional exhaustion. Cardiorespiratory data were collected at 20 s intervals using a computerized system (PARVO, Sandy, UT, USA), O_2 and CO_2 analyzers, and a 5 L mixing chamber. To determine if VO_{2max} had been attained, at least two of the following criteria had to be satisfied; Plateau in VO_2, HR within 10 beats of age-predicted max HR ($208 - (0.07 \times age)$) [47], or respiratory exchange ratio (RER) greater than 1.1 and RPE of $\geq$ 18.

2.6. Menstrual and Lactation Logs

Participants were instructed to keep records of menstrual cycles and lactation frequency throughout the study. Postpartum women were encouraged to breastfeed for a minimum of one year; however, the frequency and duration of lactation were recorded using lactation logs. Menstrual cycle logs were used to determine a return of menses at 6 and 12 months. This information was used to facilitate scheduling of follow-up visits during the follicular phase. Lactation duration and menstrual cycle status were used as group variables in statistical analysis as explained below.

2.7. Statistical Analysis

Data were analyzed using SPSS software (Version 26; SPSS Inc., Chicago, IL, USA). Values are reported as mean $\pm$ SEM plus data range for all variables except those determined to be highly skewed (skewness <-1 or >1); skewed variables are reported using the median with data range. Differences between body mass, body composition, bone mineral density, serum lipids, and habitual PA between lactating and never pregnant controls were analyzed using independent sample t-tests. Changes in BMD, urinary bone resorption markers (subset), and serum lipids across time were evaluated by repeated-measures ANOVA to test for time (baseline, 6 month, and 12 month) and time X group (postpartum vs. control) effects for these and other key variables including PA during the 12-month study period. Paired *t*-tests corrected for multiple comparisons were used to determine differences between baseline and 6 months and baseline and 12 months when a significant time effect was observed. PA and aerobic fitness were added to ANOVA models of postpartum women as a cofactor for variables that changed over time. Independent sample *t*-tests were used to evaluate change in key variables from baseline to 6 months and baseline to 12 months in postpartum women who continued to lactate compared to those who discontinued lactation and to compare women who had a return of their menstrual cycle by 6 months compared to those who did not. Associations between habitual PA and VO_{2max} and the change in BMD and urinary markers were evaluated using Pearson product moment correlations. Statistical significance was set at $p \leq 0.05$ unless otherwise specified.

3. Results

3.1. Missing Data

At baseline, hip and waist circumferences were missed for one lactating participant due to recording error. Blood for analysis of lipid profile was missed in 1 control and 2 lactating participants. All data at the six-month collection were missed for one control participant with a scheduling conflict. VO_{2max} data at 12 months could not be obtained from 2 postpartum participants, 1 due to being diagnosed with ataxia oculomotor apraxia type II and one due to computer malfunction. Complete total body t-scores were missing for 2 control and 4 lactating women, complete dual femur t-scores were missing for 4 lactating and control women, and complete lumbar spine t-scores were unavailable for 4 control and 7 lactating women. Incomplete t-scores were present in control and postpartum women who were <21 years old, absent due to a scheduling conflict, or when t-scores were not computed by DXA software.

3.2. Body Mass and Body Composition

The 18 postpartum women who completed the study reported weighing 63.7 ± 1.9 kg before pregnancy, gaining 16.0 ± 1.1 kg during pregnancy and delivering term babies which weighed 3.3 ± 0.1 kg at birth. The characteristics of body composition for the postpartum and never-pregnant participants at each time point are shown in Table 1.

At baseline, differences were not detected in age or anthropometric variables except for waist circumference and percent body fat, which was higher ($p < 0.001$) in the postpartum group as previously reported [43]. During the 1-year follow-up, lactating women lost 5.2 ± 1.4 kg and experienced reductions in body fat and waist circumference, whereas the control women remained relatively weight stable (Table 1). The majority of loss in weight, waist circumference, and body fat occurred during the first six months postpartum. At 12 months, body composition did not differ between groups ($p < 0.05$).

3.3. Bone Mineral Density and Markers of Bone Resorption

3.3.1. BMD at Dual-Femur, Spine and Total Body

Bone mineral density at all sites for control and postpartum women is summarized in Table 2 and Figure 1. BMD of total body, dual femur or spine did not differ between postpartum and control women at baseline ($p = 0.62$, $p = 0.58$, and $p = 0.42$, respectively). At baseline, two women in the control group and two in the lactating group had evidence of osteopenia (t-score between -1 and -2.5) in the dual femur, and one of the same controls had evidence of osteopenia in the spine. During the one year follow up, total BMD decreased by $1.4 \pm 0.5\%$ and hip BMD decreased by $3.1 \pm 0.9\%$ in the postpartum group with changes of $+0.2 \pm 0.3\%$ and $-2.2 \pm 1.3\%$ in controls. Although average spine BMD decreased by $2.0 \pm 2.3\%$ in the postpartum and 2.2 ± 2.9 in controls (Figure 1), these changes were highly variable among individuals in both groups and did not differ by time or time X group ($p = 0.19$ and $p = 0.96$, respectively).

Table 1. Anthropometric Characteristics in Lactating ($n = 18$) and Control ($n = 16$) Women.

		Baseline	6 Months	12 Months	*p* Value
Age (years)	Control	26.4 ± 1.4 (19–38)	-	-	-
	Lactating	27.9 ± 1.5 (19–38)	-	-	
Height (cm)	Control	169.0 ± 1.4 (160.0–178.5)	-	-	-
	Lactating	167.7 ± 1.8 (154.1–179.2)	-	-	
Weight (kg)	Control	68.0 ± 2.2 (55.4–86.1)	69.1 ± 2.3 (55.6–87.4) [a]	67.9 ± 2.5 (54.2–92.2)	* *p* = 0.019
	Lactating	70.8 ± 2.2 (52.8–92.3)	66.8 ± 2.3 (48.5–85.2) [1]	65.6 ± 2.3 (48.2–86.6) [2]	
BMI (kg/m^2)	Control	23.9 ± 0.7 (19.5–30.0)	24.2 ± 0.7 (20.3–30.2) [a]	23.7 ± 0.8 (19.0–32.2)	* *p* = 0.026
	Lactating	25.0 ± 0.8 (20.3–32.9)	23.6 ± 0.9 (19.0–32.0) [1]	23.0 ± 0.9 (18.4–33.1) [2]	
Waist Circumference (cm)	Control	81.5 (68.0–101.0)	81.1 (72.5–104.0) [a]	80.3 (29.5–106.0)	* *p* = 0.027
	Lactating	93.5 (77.0–107.0) [b]	83.8 (32.0–109.4)	80.5 (71.5–105.4) [2]	
Hip Circumference (cm)	Control	99.0 ± 1.9 (84.5–110.0)	102.3 ± 1.4 (92.7–110.6) [a]	97.7 ± 4.4 (37.5–115.0)	* † NS
	Lactating	104.5 (85.5–118.0) [b]	98.8 (37.5–117.2)	99 (86.6–120.7)	
Body Fat (%)	Control	34.2 ± 1.8 (18.0–44.8)	36.0 ± 1.5 (26.1–44.1) [a]	34.5 ± 2.0 (15.4–46.4)	* *p* = 0.007
	Lactating	38.9 ± 1.4 (26.5–47.0)	36.0 ± 1.7 (24.4–52.3) [1]	34.5 ± 1.9 (22.5–52.9) [2]	

Results reported as mean ± SEM (range); [a] missing data due to scheduling conflict ($n = 1$) or [b] recording error ($n = 1$); * significant time X group interaction by repeated measures ANOVA; † significant time effect by repeated measures ANOVA; [1] significant difference by paired *t*-test 6 months vs. baseline ($p < 0.025$); [2] significant difference by paired *t*-test 12 months vs. baseline ($p < 0.025$); NS = no significant time or time X group effect by repeated measures ANOVA.

Table 2. Bone Mineral Density in Lactating ($n = 18$) and Control ($n = 16$) Women.

Bone Density		Baseline	6 Months	12 Months	*p* Value
Total Body (g/cm^2)	Control	1.17 ± 0.02 (1.07–1.27)	1.16 ± 0.02 (1.06–1.27) [a]	1.17 ± 0.02 (1.07–1.28)	* *p* = 0.011
	Lactating	1.16 ± 0.01 (1.08–1.31)	1.15 ± 0.01 (1.07–1.24)	1.14 ± 0.01 (1.04–1.23) [2]	
Spine (g/cm^2)	Control	1.17 ± 0.04 (0.94–1.41)	1.15 ± 0.04 (0.82–1.39) [a]	1.14 ± 0.03 (0.93–1.40)	* † NS
	Lactating	1.20 ± 0.03 (0.99–1.34)	1.18 ± 0.03 (0.96–1.46)	1.17 ± 0.01 (0.81–1.40)	
Dual Femur (g/cm^2)	Control	1.06 ± 0.03 (0.85–1.33)	1.05 ± 0.03 (0.84–1.24) [a]	1.04 ± 0.03 (0.84–1.19)	† *p* = 0.014
	Lactating	1.03 ± 0.02 (0.87–1.21)	0.99 ± 0.02 (0.82–1.13) [1]	1.00 ± 0.03 (0.85–1.15) [2]	
T-scores Total Body	Control	0.59 ± 0.19 (−0.5 – 1.7) [c]	0.56 ± 0.21 (−0.8 – 1.8) [ac]	0.54 ± 0.20 (−0.7 – 1.9)	† *p* = 0.041
	Lactating	0.49 ± 0.17 (−0.5 – 2.3) [d]	0.29 ± 0.16 (−0.7 – 1.4) [bc1]	0.22 ± 0.13 (−0.6 – 1.3)	
T-scores Spine	Control	0.41 ± 0.28 (−2.1 – 1.8) [c]	0.28 ± 0.30 (−2.0 – 1.6) [ad]	0.31 ± 0.32 (−2.2 – 2.3) [d]	* † NS
	Lactating	0.35 ± 0.17 (−0.9 – 1.2) [e]	0.16 ± 0.26 (−1.2 – 2.1) [be1]	0.18 ± 0.22 (−1.7 – 1.7) [c]	
T-scores Dual Femur	Control	0.45 ± 0.23 (−1.3 – 1.7) [c]	0.29 ± 0.26 (−1.4 – 1.8) [ad]	0.06 ± 0.22 (−1.3 – 1.4) [d]	† *p* = 0.009
	Lactating	0.25 ± 0.20 (−1.1 – 1.6) [d]	−0.18 ± 0.17 (−1.2 – 1.0) [bc1]	−0.12 ± 0.17 (−1.1 – 1.1)[2]	

Results reported as mean ± SEM (range); [a] missing data due to scheduling conflict ($n = 1$) or [b] ($n = 2$); [c] missing t-score ($n = 1$) or [d] ($n = 2$) or [e] ($n = 3$); * significant time X group interaction by repeated measures ANOVA; † significant time effect by repeated measures ANOVA; [1] significant difference by paired *t*-test 6 months vs. baseline ($p < 0.025$); [2] significant difference by paired *t*-test 12 months vs. baseline ($p < 0.025$); NS = not significant.

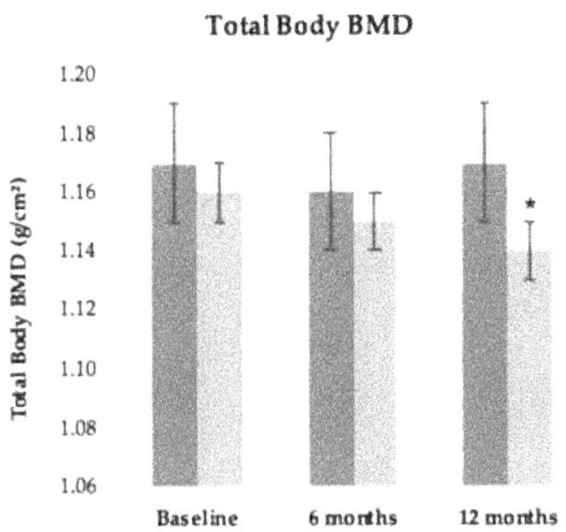

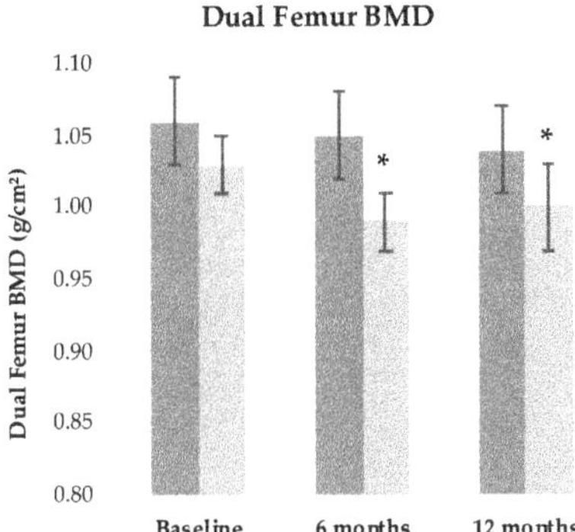

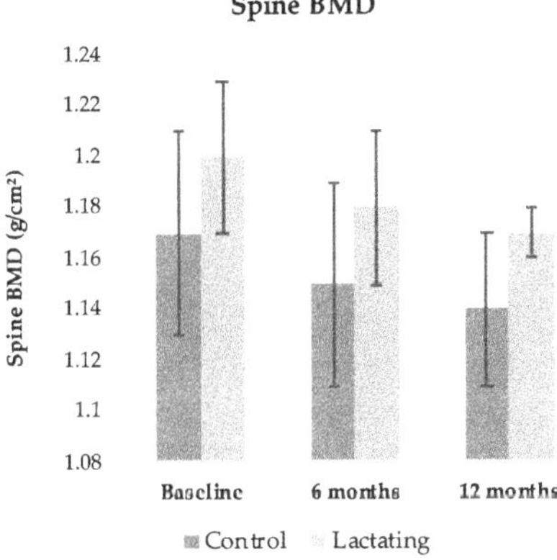

Figure 1. The change in total body (panel A), dual-femur panel B) and lumbar spine (panel C) BMD in lactating verses never-pregnant controls. BMD in total body (time X group interaction, $p = 0.011$) and dual femur (time effect, $p = 0.014$) decreased from 4 to 6 weeks (baseline) to 12 months postpartum. No differences across time were observed for BMD of the spine. * $p < 0.025$ vs. baseline by paired *t*-test.

3.3.2. Biochemical Markers of Bone Resorption

Markers of bone resorption for control and postpartum women are summarized in Table 3. At baseline, urinary PYD and DPYD concentrations were higher in the postpartum compared to the control women ($p < 0.001$) with trends for a higher NTX ($p = 0.07$) and lower DPYD/PYD ratio ($p < 0.01$). All markers changed across time with a significant time X group interaction only for PYD, DPYD, and the PYD/DPYD ratio (Table 3). NTX, PYD and DPYD concentrations at baseline were predictive of greater absolute (data not shown) and percent deficits in dual-femur BMD from baseline to 6 months ($r = 0.49$, 0.52 and 0.62, respectively, $p < 0.05$) in the full group ($n = 22$) with similar patterns in the postpartum group only ($r = 0.41$, $p = 0.15$; $r = 0.56$ and 0.74, $p < 0.05$, respectively).

Table 3. Urinary Biochemical Markers of Bone Resorption in a Subset of Lactating ($n = 8$) and Control ($n = 5$) Women.

		Baseline	6 Months	12 Months	p Value
NTX (nM BCE/mM creatinine)	Control	67.1 ± 11.1 (25–121)	54.3 ± 6.7 (26–76)	46.6 ± 3.4 (31–57)	† $p = 0.007$
	Lactating	101.1.6 ± 14.0 (34–219)	101.1 ± 8.0 (66–156)	73.8 ± 7.0 (28–124)[2]	
PYD (µmol/mol creatinine)	Control	46.5 ± 4.5 (34.3–72.8)	47.7 ± 3.7 (34.6–64.4)	50.7 ± 3.7 (38–69.6)	* $p = 0.000$
	Lactating	151.3 ± 11.8 (101.5–266.7)	80.4 ± 5.5 (59.4–119.9)[1]	69.0 ± 4.4 (43.4–100.7)[2]	
DPYD (µmol/mol creatinine)	Control	13.9 ± 1.4 (8.8–20.9)	14.4 ± 1.4 (8.4–18.9)	14.8 ± 1.3 (9.0–20.6)	* $p = 0.021$
	Lactating	34.3 ± 2.7 (23.2–64.4)	26.9 ± 2.0 (18.4–37.8)[1]	23.1 ± 1.7 (14.2–32.7)[2]	
DPYD/PYD Ratio	Control	0.31 ± 0.01 (0.25–0.42)	0.31 ± 0.01 (0.23–0.34)	0.29 ± 0.01 (0.24–0.32)	* $p = 0.000$
	Lactating	0.23 ± 0.01 (0.17–0.36)	0.33 ± 0.01 (0.25–0.44)[1]	0.33 ± 0.01 (0.27–0.42)[2]	

NTX, N-telopeptide PYD, pyridinoline; DPYD, deoxypyridinoline. Results reported as mean ± SEM (range); * significant time X group interaction by repeated measures ANOVA; † significant time effect by repeated measures ANOVA; [1] significant difference by paired *t*-test 6 months vs. baseline ($p < 0.025$); [2] significant difference by paired *t*-test 12 months vs. baseline ($p < 0.025$); NS = no significant time or group effect by repeated measures ANOVA.

3.4. Serum Lipids

Fasting serum TG and cholesterol concentrations at baseline, 6 months, and 12 months are summarized in Table 4. At baseline, no differences were observed between cholesterol ($p = 0.42$), HDL-C ($p = 0.33$), LDL-C ($p = 0.12$), VLDL-C ($p = 0.60$) or triglyceride concentrations ($p = 0.57$) nor the Cholesterol/HDL-C ratio ($p = 0.33$) between lactating and control women. However, one woman in the control group and 4 in the postpartum group had serum cholesterol concentration of ≥200 mg/dL. Four and 6 control women and 6 and 9 postpartum women had LDL-C and HDL-C concentrations, respectively, that were out of the optimal ranges of <100 mg/dL and >50 mg/dL [48]. During the first postpartum year, women experienced an overall improvement in lipid profile with significant increases in HDL-C concentration and decreases in the cholesterol/HDL-C ratio by 6 months and decreases in LDL-C concentration at 12 months. The control women were observed to have improvements in total and HDL-C during the study period. Changes in VLDL-C and triglyceride concentrations were not observed in either group.

3.5. Aerobic Fitness and Physical Activity

Reported habitual PA at baseline, 6 months, and 12 months and VO_{2max} at 12 months are shown in Table 5. Total PA and PA associated with work, sport and leisure-time indices did not differ between lactating and control women at baseline and did not change in either group over time. Therefore, the total Baecke score and the work, sport, and leisure-time indices at all time points were averaged and used for further analysis. All women except one postpartum participant met the criteria for an acceptable VO_{2max} test at 12 months. VO_{2max} was highly variable among control (19.9 to 97.5 percentile) and lactating women (15.1 to 95.7 percentile) [45] but was not different between groups. VO_{2max} was found to correlate with the Baecke sport index (r = 0.523, $p < 0.01$) but not with the work (r = 0.002), leisure-time (r = −0.13) or total Baecke score (r = −0.20) ($p > 0.05$). Results were not affected by the participant who did not have an acceptable VO_{2max} test.

3.6. Effect of Habitual Exercise, Lactation and Return of Menses

3.6.1. Habitual exercise and Aerobic Fitness

In postpartum women, measures of PA including VO_{2max}, total Baecke score, or work, sport or leisure-time indices did not influence change in BMD across the first postpartum year. The decrease in the Cholesterol/HDL-C ratio was positively influenced by PA during leisure-time (time X group effect ($p = 0.051$, time X group) and sport ($p = 0.028$, time effect). No other interactions were observed.

Table 4. Cholesterol and Triglyceride (TG) Concentrations in Lactating ($n = 18$) and Control ($n = 16$) Women.

		Baseline	6 Months	12 Months	*p* Value
Cholesterol (mg/dL)	Control	163.1 ± 6.5 (108.0–208.0) [b]	186.3 ± 7.5 (154.0–253.0) [a1]	183.8 ± 8.4 (133.0–270.0) [2]	* $p = 0.002$
	Lactating	171.6 ± 8.0 (115.0–227.0) [b]	165.6 ± 5.8 (130.0–204.0)	157.2 ± 5.7 (113.0–206.0)	
HDL-C (mg/dL)	Control	51.1 ± 3.6 (32.0–73.0) [b]	62.3 ± 4.7 (37.0–109.0) [a1]	59.3 ± 3.4 (32.0–89.0) [2]	† $p = 0.000$
	Lactating	46.7 ± 2.6 (29.0–68.0) [b]	53.7 ± 2.3 (31.0–68.0) [1]	52.3 ± 2.1 (39.0–71.0)	
LDL-C (mg/dL)	Control	91.7 ± 6.3 (53–156) [b]	103.2 ± 6.6 (67.0–170.0) [a]	104.3 ± 9.3 (44.0–212.0)	* $p = 0.010$
	Lactating	106.4 ± 6.8 (70.0–165.0) [b]	95.6 ± 4.8 (69.0–124.0)	89.7 ± 4.1 (58–122) [2]	
VLDL-C (mg/dL)	Control	19 (10.0–49.0) [b]	21 (9.0–36.0) [a]	19 (10.0–45.0)	* † NS
	Lactating	18.6 ± 2.5 (10–47) [b]	16.4 ± 1.8 (9–39)	15.2 ± 2.0 (9–47)	
TG (mg/dL)	Control	96 (52.0–244.0) [b]	103 (46.0–179.0) [a]	93.5 (51.0–227.0)	* † NS
	Lactating	76.1 (50–234) [b]	71 (43–195)	64.5 (47–234)	
Cholesterol/HDL-C Ratio	Control	3.1 (2.1–6.5) [b]	2.8 (2.1–6.0) [a1]	2.9 (2.0–7.1)	† p = 0.001
	Lactating	3.5 (2.7–6.7) [b]	3 (2.4–6.0) [1]	3 (2.4–3.8) [2]	

Results reported as mean ± SEM (range); [a] missing data due to scheduling conflict; [b] missing blood sample collection ($n = 1$ control and 2 lactating); * significant time X group interaction by repeated measures ANOVA; † significant time effect by repeated measures ANOVA; [1] significant difference by paired *t*-test 6 months vs. baseline ($p < 0.025$); [2] significant difference by paired *t*-test 12 months vs. baseline ($p < 0.025$); NS = no significant time or time X group effect by repeated measures ANOVA.

3.6.2. Lactation Duration

Nine of the 18 postpartum women (50%) continued to breastfeed for the full postpartum year while nine had stopped before ($n = 1$) or shortly after the 6 month ($n = 8$) visit. Those who continued to lactate for 12 months had a greater fall in BMD in the total body (0.0311 ± 0.0092 g/cm^2) and hip (0.0542 ± 0.0083 g/cm^2) than those who had stopped lactating (0.0028 ± 0.0048 g/cm^2 and 0.0117 ± 0.0138 g/cm^2, respectively ($p < 0.05$)). Differences were not detected in the spine ($p = 0.638$). There was also no evidence of BMD re-bound at 12 months in any site. In the subset with urinary bone resorption markers, those who continued to lactate ($n = 8$) had less of a fall in NTX (4.8 ± 10.6 vs. 57.3 ± 21.6 nM BCE/mM creatinine), PYD (63.5 ± 12.2 vs. 107.4 ± 12.7 µmol/mol creatinine), and DPYD (5.8 ± 2.7 vs. 18.5 ± 3.4 µmol/mol creatinine) than those ($n = 6$) who stopped earlier. There were no differences observed for the change in serum lipid profile.

Table 5. Reported Physical Activity and Aerobic Fitness in Lactating (n = 18) and Control (n = 16) Women.

		Baseline	6 Months	12 Months	*p* Value
Work Index	Control	2.5 ± 0.1 (1.6–3.5)	2.4 ± 0.1 (1.6–3.1) [a]	2.6 ± 0.2 (1.4–3.9)	* † NS
	Lactating	2.2 ± 0.2 (0.0–3.1)	2.3 ± 0.1 (1.6–3.3)	2.3 ± 0.1 (1.6–3.3)	
Sport Index	Control	3.7 ± 0.2 (2.3–4.5)	3.2 ± 0.2 (2.0–4.8) [a]	3.4 ± 0.2 (2.0–4.5)	* † NS
	Lactating	2.4 ± 0.2 (1.0–3.5)	2.5 ± 0.2 (1.3–3.8)	2.7 ± 0.2 (1.5–4.5)	
Leisure Index	Control	3.6 ± 0.4 (2.5–10.0)	3.7 ± 0.3 (2.3–7.3) [a]	3.6 ± 0.4 (2.8–8.9)	* † NS
	Lactating	2.9 ± 0.2 (1.5–6.5)	2.8 ± 0.1 (2.0–3.5)	2.8 ± 0.1 (2.0–3.8)	
Total Baecke Score	Control	9.8 ± 2.2 (7.3–16.8)	9.3 ± 1.6 (6.5–12.5.0) [a]	9.6 ± 2.2 (6.1–15.6)	* † NS
	Lactating	7.5 ± 0.4 (3.5–11.3)	7.6 ± 0.2 (5.4–10.3)	7.8 ± 0.3 (5.9–10.0)	
VO_{2max} (ml/kg/min)	Control	-	-	37.9 ± 1.7 (24.2–50.9)	-
	Lactating	-	-	37.5 ± 1.5 (29.1–48.7) [b]	

Results reported as mean ± SEM (range); [a] missing data due scheduling conflict (n = 1); [b] unable to complete aerobic fitness test = 2); * = significant time X group interaction by repeated measures ANOVA; † = significant time effect by repeated measures ANOVA; NS = no significant time or time X group effect by repeated measures ANOVA.

3.6.3. Return of Menses

Six of the 18 postpartum women (33%) had resumed menses by six months postpartum whereas 12 (67%) resumed menses after six months, with 10 of the 12 not experiencing menses by study end. The change in hip BMD from baseline to 12 months was positively influenced by the return of menses in that those who had a return of menses before 6 months had a dampened fall in BMD in the hip (0.0047 ± 0.0148 g/cm^2)) compared to those who didn't who experienced a greater fall in BMD (0.04709 ± 0.00994 g/cm^2) (p = 0.028). Resumption of menstrual cycle did not impact other areas of BMD or the change in serum lipid profile.

4. Discussion

Physiological changes that occur following pregnancy and during lactation can have an impact on bone health and cardiovascular risk. The current study evaluated the longitudinal changes in BMD and serum lipids and TG across the first postpartum year in lactating women relative to age-matched never pregnant controls. While results confirm previous, somewhat inconsistent findings that observed decreases in BMD during lactation, they further highlight that BMD rebound does not necessarily occur by 12 months in those continuing to breastfeed, and that BMD loss occurs in never-pregnant controls regardless of PA patterns. Results also add to the currently limited studies evaluating serum lipids during postpartum and with lactation that demonstrate general improvements in HDL-C, LDL-C and the cholesterol/HDL-C ratio independent of habitual PA. Additionally notable was our exploratory observation in a subset of participants that urinary biomarkers of bone resorption, namely NTX, DPYD and PYD, may help better understand changes in BMD in the postpartum period. Overall these findings contribute to understanding of whether persistent reductions in BMD and elevated serum lipids postpartum increase risk for osteoporosis and CVD later in life.

Our results specifically demonstrate the impact of breastfeeding on BMD in the dual femur, lumbar spine, and total body, and support the concept that skeletal demineralization and bone resorption during lactation support the calcium demands of milk production [10,49]. We chose to examine total body BMD, which overall contains a high content of cortical bone, as well as the dual femur (hip) and lumbar spine that are rich in trabecular bone. We found that the lactation-induced effect on maternal bone is particularly pronounced in the total body [13,18,24,50,51] and hip [13,16,17,20,22–24,35,50,52–54]. The calcium demand of producing breast milk, however, should theoretically have a greater influence on trabecular bone because it has a higher turnover rate and is more metabolically active than cortical bone [55]. In a 24-month longitudinal study, Hopkinson and colleagues [18] found that women who breastfed for a longer duration experienced loss in bone mineral accretion in cortical-rich sites but that trabecular-rich sites were the first to recover from lactation-induced BMD loss. Several other well-controlled studies support these findings [1,2,16,23,31].

The current study found that lactating women had an average decrease of 1.4% in total body BMD from early (4–6 months) to 12 months postpartum, which was significantly greater than the average total body BMD loss observed in the control group. The magnitude of these lactation-induced changes in total body BMD is in agreement with previous studies which have reported deficits ranging from 0.86 to 3% within the first 6 months of lactation [2,13,18,19,24,50,53] but not in alignment with others that found no changes in BMD during the first 3 months postpartum [35,49] and post-weaning [53]. Similar to the current study, DXA was employed to measure BMD in the majority of previous studies. However, variations in breastfeeding frequency, duration, time of weaning and sample size may explain the discrepancy in results. To determine the clinical significance of this loss, we further evaluated t-scores, which standardizeresults to those of healthy 30-year-old female adults. Despite the significant reduction across the 12-months postpartum, osteopenia risk (t-score <-1 and >-2.5) [56] was not immediately increased except in the spine of two postpartum participants who had baseline t-scores in the lower range (0.7 and 0.9) that dropped into the osteopenic range during the study (-1.2 and -1.3). The two lactating and two control women who were at risk at baseline based on dual femur t-scores remained at risk.

In addition, our results show an average decrease of 3.1% in dual femur BMD in the postpartum group, which agrees with previous studies that found deficits ranging from 2 to 7% [2,3,13,17,20,35,52–54]. Unlike BMD of total body, the decrease in dual femur BMD occurred in both groups over the year of study follow up with the controls experiencing an average decrease of 2.1%. These results align with existing research in postpartum women that found initial recovery to originate in trabecular-rich sites, such as the dual femur, by 12 months [1,2,16,18,23,31]. Contrary to the majority of previous studies, we did not find a significant decline in spine BMD [13,16,18,20,22–24,35,49–52,54,57,58], which may be due to notable variation among individual women in both the postpartum and control groups. There is also the possibility early postpartum changes in spine BMD (i.e., over the first 4 to 6 weeks), before our baseline assessment, were missed. By comparison, bone loss occurs at an annual rate of -1.8 to -2.5% in lumbar spine and -1.0 to -1.7% in the hip in the perimenopausal period [59] and may be as high as -3.3% in the spine and -2.0% in the femoral neck following menopause [60].

Although there was a slight average increase (1.01%) in dual-femur BMD during the 6- to 12-month period, this change was not statistically significant and could not be considered a rebound of BMD, as has been reported by many previous studies [61]. According to a systematic review by Grizzo et al. [61], the majority of BMD loss occurs during the first 6 months of lactation with several studies demonstrating complete or almost complete recovery to baseline levels at all skeletal sites measured post-weaning. However, other studies only showed partial recovery to pre-pregnancy BMD levels, especially in those who lactated for more than 6 months [1,16,18–20,31]. A well-controlled study by Cross et al. [49] found that spinal BMD lost during lactation was regained approximately 3 months after weaning, regardless of the duration of breastfeeding. Other research suggests that recovery of BMD is site specific and takes as long as 6 to 7 months post-weaning for sites such as the hip [18,62]. Given that 50 percent of our participants were still lactating at 12 months, a rebound in BMD by study end may not be expected. The mechanism of bone recovery after lactation is unknown, but reestablishment of ovarian hormones leading to resumption of ovulation and decreased prolactin (PRL) production following weaning may influence recovery [63]. Not surprisingly, the fall in BMD from baseline to 12 months in the total body and dual femur was greater in those who continued to breastfeed for a full year compared to those who stopped lactating earlier, which at least in the hip was related to the return of menstrual function and estrogen cycling in women who ceased breastfeeding. Additionally, we found no evidence supporting a beneficial effect of habitual PA (either from sport, work or leisure-time) or aerobic fitness on change in postpartum BMD or the propensity for BMD rebound. From previous studies, a beneficial effect of physical activity is inconsistently observed. Previous observational studies by Little et al. [35] and Sowers et al. [20] reported no association between exercise and dampened BMD loss over 12 months in lactating women who participated in self-selected exercise. In contrast, two more recent randomized controlled trials found that those assigned to an exercise intervention group (which included both resistance and aerobic exercise training) experienced less BMD loss in the lumbar spine, but not the total body or hip, than those in the control group [32,33].

A secondary purpose of the current study was to evaluate the usefulness of urinary biochemical markers of bone resorpton in postpartum, lactating women. We showed, in a subset of participants, that all urinary markers including NTX, PYD and DPYD decreased from early to 12 months postpartum. NTX is a urinary amino-terminal cross-linking telopeptide of type I collagen whereas urinary DPYD is derived from proteolytic hydrolysis of collagen found in bone, and PYD is a less specific marker of bone as well as cartilage, tendon, and blood vessels [37]. As these markers are byproducts of bone remodeling that can be detected in urine, the reduced urinary concentrations across ~one to six and 12 months postpartum is indicative of decreased bone resorption across this same time period. The increase in the DPYD/PYD ratio, however, suggests that a proportion of this reduced resorption may be from collagen that is not necessarily specific to bone. Additionally, our finding that baseline NTX, PYD, and DPYD concentrations were predictive of changes in dual-femur BMD during the first six months postpartum suggest that these markers could be useful in future studies of pregnant and postpartum women, or at other times when younger women may be at risk for increased bone resorption. Urinary markers are thought to be a more cost-effective and obtainable measures of BMD and are more reflective of current bone status as compared to DXA, which provides only a static snapshot of BMD [36]. Thus, in clinical settings like the current, where BMD values show no observable signs of rebound, urinary markers may be useful to further explore the changes in skeletal demineralization and bone resorption in association with weaning and/or the return of menses that occur before observed changes in BMD. Use of these markers, however, need to be further evaluated for sensitivity and specificity as not all findings in our control women, including the fall in NTX concentration in light of stable PYD and DPYD, can be easily explained.

A second major objective of this study was to investigate changes in cholesterol and lipid profiles in the year following childbirth and throughout lactation. While hypercholesterolemia is common during gestation due to changes in lipid metabolism [8], a persistent elevation of atherogenic lipid concentrations postpartum may increase cardiovascular disease risk with research suggesting that lactation may help mitigate elevated plasma cholesterol and TG in postpartum mothers [39,40,64]. Little is known about the effect of postpartum exercise. Our results are in partial support of both our initial hypothesis and limited previous research that found a reduction in both total serum cholesterol and LDL-C following at least two to six months of lactation [10,40,65]. While we observed decreases of

−8.4% and −15.7% in total serum cholesterol and LDL-C concentrations, respectively, in the postpartum group, only the drop in LDL-C was significant and not until 12 months postpartum. The changes in total cholesterol were highly variable among both postpartum and control women, as were the changes in TG which did not change throughout the study in either group. This is in contrast with existing research that demonstrated decreases in TG during lactation [10,40,65] and could be attributed to the presence of confounding variables in both the postpartum and control groups which include variations in dietary intake, body weight and alcohol consumption. The current study, however, did observe a significant improvement in HDL-C and cholesterol/HDL-C ratio which was evident by 6 months. There is limited research surrounding the effects of exercise on expediting the return of cholesterol and TG to baseline levels. An earlier study by Lovelady et al. [41] found a slight increase in HDL-C in lactating women assigned to an exercise intervention when compared to the non-exercising control group. Our data support this, indicating that both Leisure-time and Sport PA Indices were influential on the change in the cholesterol/HDL-C ratio over the first postpartum year.

Although our study provides insight into the changes in BMD and serum lipid profile during the first postpartum year, it is limited by a small sample size, absence of pre-pregnancy data and a non-lactating postpartum group, time frame of follow up, and methodological concerns of assessing PA by self-report questionnaire. The majority of previous studies utilized sample sizes between 6 and 139 lactating women and between 9 and 98 controls. Our sample size was on the lower end, which did not allow us to simultaneously account for all potentially important cofounders (lactation duration, menstrual cycle status, seasonality, etc.) in our repeated measures analysis models. Additionally, the inclusion of a non-lactating postpartum group would have allowed for teasing out the effect of lactation vs. childbearing on BMD and serum lipids. Unfortunately, recruitment of a non-breastfeeding control group proved impossible because of the high breastfeeding initiation rates of our state (>80%) (https://www.cdc.gov/breastfeeding/data/reportcard.htm accessed on 24 January 2022). Most importantly, however, a longer duration of follow-up which included testing pre-pregnancy, in earlier postpartum (several days after delivery), and as far out as 18 or 24 months postpartum would have allowed us to better capture true changes in BMD and serum lipids postpartum and determine the pattern of these changes and presence of a true rebound to pre-pregnancy values. Finally, our assessment of PA was limited because it addressed PA in general, albeit from sport, work and leisure-time, and did not specifically address exercise intensity or differentiate between weight bearing vs. non-weight bearing activity. Questionnaires, such as the Baecke PA Questionnaire, can be subject to recall and social bias and while VO_{2max} is commonly used to assess cardiovascular fitness, it is still influenced by genetics and does not account for how exercise habits may have changed after giving birth. It is also important to consider that our participant population consisted of relatively healthy and active women living in a university town, which may have reduced the amount of variation necessary to detect a difference and limit the generality of our findings.

5. Conclusions

Results of the present study confirm previous investigations which observed decreases in BMD during lactation and highlight that BMD rebound does not necessarily occur by 12 months in women who continue to breastfeed into the latter part of the first postpartum year. They, however, provide evidence that the overall lipid profile improves from early to late postpartum, with significant increases in HDL-C concentration and decreases in the cholesterol/HDL-C ratio by 6 months and decreases in LDL-C concentration at 12 months. Breastfeeding status was observed to influence the change in BMD in the dual femur and total body, whereas reported physical activity during sport and leisure heightened the change in the Cholesterol/HDL ratio. The potential benefits of exercise and use of urinary biomarkers of bone deserves further exploration during and following pregnancy and lactation.

Author Contributions: Conceptualization, D.E.L.-M.; methodology, J.S., B.M.A., A.M.H. and D.E.L.-M.; validation, J.S., B.M.A. and D.E.L.-M.; formal analysis, E.M.K., H.B.M. and D.E.L.-M.; investigation, E.M.K., J.S., M.C., B.M.A., A.M.H. and D.E.L.-M.; resources, D.E.L.-M. and B.M.A.; data curation, E.M.K. and J.S.; writing—original draft preparation, E.M.K., H.B.M., D.E.L.-M. and M.C.; writing—review and editing, J.S., B.M.A. and A.M.H.; visualization, E.M.K., M.C. and D.E.L.-M.; supervision, D.E.L.-M.; project administration, D.E.L.-M. Funding acquisition, D.E.L.-M. All authors have read and agreed to the published version of the manuscript.

Funding: This research was supported by an Institutional Development Award (IDeA) from the National Institute of General Medical Sciences of the National Institutes of Health under Grant # 2P20GM103432. Its contents are solely the responsibility of the authors and do not necessarily represent the official views of NIH.

Institutional Review Board Statement: The study was conducted according to the guidelines of the Declaration of Helsinki, and approved by the Institutional Review Board of the University of Wyoming (Approved 24 March 2009; "Role of Ghrelin and PYY in postpartum body weight regulation and presence in human milk; Protocol #s not assigned").

Informed Consent Statement: Written, informed consent was obtained from all subjects involved in the study.

Data Availability Statement: The data presented in this study are available on reasonable request from the corresponding author. The data are not publicly available due to privacy concerns.

Acknowledgments: The authors would like to acknowledge the contributions of time and effort from the women who participated in this study. We would like to thank the research assistants who facilitated the data collection; Kathy Austin, Ed VanKirk, and the late Susan N. Lescznske.

Conflicts of Interest: The authors declare no conflict of interest.

References

1. More, C.; Bettembuk, P.; Bhattoa, H.P.; Balogh, A. The Effects of Pregnancy and Lactation on Bone Mineral Density. *Osteoporos. Int.* **2001**, *12*, 732–737. [CrossRef]
2. Møller, U.K.; Við Streym, S.; Mosekilde, L.; Rejnmark, L. Changes in bone mineral density and body composition during pregnancy and postpartum. A controlled cohort study. *Osteoporos. Int.* **2012**, *23*, 1213–1223. [CrossRef]
3. Karlsson, C.; Obrant, K.J.; Karlsson, M. Pregnancy and Lactation Confer Reversible Bone Loss in Humans. *Osteoporos. Int.* **2001**, *12*, 828–834. [CrossRef]
4. Saarelainen, H.; Laitinen, T.; Raitakari, O.T.; Juonala, M.; Heiskanen, N.; Lyyra-Laitinen, T.; Viikari, J.S.; Vanninen, E.; Heinonen, S. Pregnancy-related hyperlipidemia and endothelial function in healthy women. *Circ. J.* **2006**, *70*, 768–772. [CrossRef]
5. Diareme, M.; Karkalousos, P.; Theodoropoulos, G.; Strouzas, S.; Lazanas, N. Lipid profile of healthy women during pregnancy. *J. Med. Biochem.* **2009**, *28*, 152–160. [CrossRef]
6. Bao, W.; Dar, S.; Zhu, Y.; Wu, J.; Rawal, S.; Li, S.; Weir, N.L.; Tsai, M.Y.; Zhang, C. Plasma concentrations of lipids during pregnancy and the risk of gestational diabetes mellitus: A longitudinal study. *J. Diabetes* **2018**, *10*, 487–495. [CrossRef]
7. Kovacs, C.S. Maternal Mineral and Bone Metabolism During Pregnancy, Lactation, and Post-Weaning Recovery. *Physiol. Rev.* **2016**, *96*, 449–547. [CrossRef]
8. Herrera, E.; Ortega-Senovilla, H. Maternal lipid metabolism during normal pregnancy and its implications to fetal development. *Clin. Lipidol.* **2010**, *5*, 899–911. [CrossRef]
9. Salari, P.; Abdollahi, M. The influence of pregnancy and lactation on maternal bone health: A systematic review. *J. Fam. Reprod Health* **2014**, *8*, 135–148.
10. Mankuta, D.; Elami-Suzin, M.; Elhayani, A.; Vinker, S. Lipid profile in consecutive pregnancies. *Lipids Health Dis* **2010**, *9*, 58. [CrossRef]
11. Institute of Medicine (US) Committee to Review Dietary Reference Intakes for Vitamin D and Calcium. The National Academies Collection: Reports funded by National Institutes of Health. In *Dietary Reference Intakes for Calcium and Vitamin D*; Ross, A.C., Taylor, C.L., Yaktine, A.L., Del Valle, H.B., Eds.; National Academies Press (US): Washington, DC, USA, 2020.
12. Prentice, A. Micronutrients and the bone mineral content of the mother, fetus and newborn. *J. Nutr.* **2003**, *133*, 1693s–1699s. [CrossRef]
13. Laskey, M.A.; Prentice, A.; Hanratty, L.A.; Jarjou, L.M.; Dibba, B.; Beavan, S.R.; Cole, T.J. Bone changes after 3 mo of lactation: Influence of calcium intake, breast-milk output, and vitamin D-receptor genotype. *Am. J. Clin. Nutr.* **1998**, *67*, 685–692. [CrossRef]
14. Specker, B.L.; Tsang, R.C.; Ho, M.; Miller, D. Effect of vegetarian diet on serum 1,25-dihydroxyvitamin D concentrations during lactation. *Obstet. Gynecol.* **1987**, *70*, 870–874.
15. Moser, P.B.; Reynolds, R.D.; Acharya, S.; Howard, M.P.; Andon, M.B. Calcium and magnesium dietary intakes and plasma and milk concentrations of Nepalese lactating women. *Am. J. Clin. Nutr.* **1988**, *47*, 735–739. [CrossRef]

16. Affinito, P.; Tommaselli, G.A.; di Carlo, C.; Guida, F.; Nappi, C. Changes in bone mineral density and calcium metabolism in breastfeeding women: A one year follow-up study. *J. Clin. Endocrinol. Metab.* **1996**, *81*, 2314–2318.
17. Drinkwater, B.L.; Chesnut, C.H. Bone density changes during pregnancy and lactation in active women: A longitudinal study. *Bone Miner.* **1991**, *14*, 153–160. [CrossRef]
18. Hopkinson, J.M.; Butte, N.F.; Ellis, K.; Smith, E.O. Lactation delays postpartum bone mineral accretion and temporarily alters its regional distribution in women. *J. Nutr.* **2000**, *130*, 777–783. [CrossRef]
19. Brembeck, P.; Lorentzon, M.; Ohlsson, C.; Winkvist, A.; Augustin, H. Changes in cortical volumetric bone mineral density and thickness, and trabecular thickness in lactating women postpartum. *J. Clin. Endocrinol. Metab.* **2015**, *100*, 535–543. [CrossRef]
20. Sowers, M.; Corton, G.; Shapiro, B.; Jannausch, M.L.; Crutchfield, M.; Smith, M.L.; Randolph, J.F.; Hollis, B. Changes in bone density with lactation. *JAMA* **1993**, *269*, 3130–3135. [CrossRef]
21. Laskey, M.A.; Prentice, A. Bone mineral changes during and after lactation. *Obstet. Gynecol.* **1999**, *94*, 608–615. [CrossRef]
22. Polatti, F.; Capuzzo, E.; Viazzo, F.; Colleoni, R.; Klersy, C. Bone mineral changes during and after lactation. *Obstet. Gynecol.* **1999**, *94*, 52–56.
23. Krebs, N.F.; Reidinger, C.J.; Robertson, A.D.; Brenner, M. Bone mineral density changes during lactation: Maternal, dietary, and biochemical correlates. *Am. J. Clin. Nutr.* **1997**, *65*, 1738–1746. [CrossRef]
24. Kalkwarf, H.J.; Specker, B.L.; Bianchi, D.C.; Ranz, J.; Ho, M. The effect of calcium supplementation on bone density during lactation and after weaning. *N. Engl. J. Med.* **1997**, *337*, 523–528. [CrossRef]
25. Bezerra, F.F.; Mendonça, L.M.; Lobato, E.C.; O'Brien, K.O.; Donangelo, C.M. Bone mass is recovered from lactation to postweaning in adolescent mothers with low calcium intakes. *Am. J. Clin. Nutr.* **2004**, *80*, 1322–1326. [CrossRef]
26. Akesson, A.; Vahter, M.; Berglund, M.; Eklöf, T.; Bremme, K.; Bjellerup, P. Bone turnover from early pregnancy to postweaning. *Acta Obs. Gynecol. Scand.* **2004**, *83*, 1049–1055. [CrossRef]
27. Glerean, M.; Furci, A.; Galich, A.M.; Fama, B.; Plantalech, L. Bone and mineral metabolism in primiparous women and its relationship with breastfeeding: A longitudinal study. *Medicine* **2010**, *70*, 227–232.
28. Laskey, M.A.; Price, R.I.; Khoo, B.C.C.; Prentice, A. Proximal femur structural geometry changes during and following lactation. *Bone* **2011**, *48*, 755–759. [CrossRef]
29. Costa, M.L.; Krupa, F.G.; Rehder, P.M.; Sousa, M.H.; Costa-Paiva, L.; Cecatti, J.G. Forearm bone mineral density changes during postpartum and the effects of breastfeeding, amenorrhea, body mass index and contraceptive use. *Osteoporos. Int.* **2012**, *23*, 1691–1698. [CrossRef]
30. Méndez, R.O.; Gallegos, A.C.; Cabrera, R.M.; Quihui, L.; Zozaya, R.; Morales, G.G.; Valencia, M.E.; Méndez, M. Bone mineral density changes in lactating adolescent mothers during the first postpartum year. *Am. J. Hum. Biol* **2013**, *25*, 222–224. [CrossRef]
31. Kent, G.N.; Price, R.I.; Gutteridge, D.H.; Allen, J.R.; Rosman, K.J.; Smith, M.; Bhagat, C.I.; Wilson, S.G.; Retallack, R.W. Effect of pregnancy and lactation on maternal bone mass and calcium metabolism. *Osteoporos. Int.* **1993**, *3* (Suppl. S1), 44–47. [CrossRef]
32. Lovelady, C.A.; Bopp, M.J.; Colleran, H.L.; Mackie, H.K.; Wideman, L. Effect of exercise training on loss of bone mineral density during lactation. *Med. Sci. Sports Exerc.* **2009**, *41*, 1902–1907. [CrossRef]
33. Colleran, H.L.; Hiatt, A.; Wideman, L.; Lovelady, C.A. The Effect of an Exercise Intervention During Early Lactation on Bone Mineral Density During the First Year Postpartum. *J. Phys. Act. Health* **2019**, *16*, 197–204. [CrossRef]
34. Ebina, A.; Sawa, R.; Kondo, Y.; Murata, S.; Saito, T.; Isa, T.; Tsuboi, Y.; Torizawa, K.; Matsuda, N.; Ono, R. Daily physical activity is associated with increased sonographically measured bone status during lactation. *Womens Health* **2020**, *16*, 1745506519900582. [CrossRef]
35. Little, K.D.; Clapp, J.F., 3rd. Self-selected recreational exercise has no impact on early postpartum lactation-induced bone loss. *Med. Sci. Sports Exerc.* **1998**, *30*, 831–836.
36. Ganesan, G.R.; Vijayaraghavan, P.V. Urinary N-telopeptide: The New Diagnostic Test for Osteoporosis. *Surg. J.* **2019**, *5*, e1–e4. [CrossRef]
37. Vasikaran, S.; Eastell, R.; Bruyere, O.; Foldes, A.J.; Garnero, P.; Griesmacher, A.; McClung, M.; Morris, H.A.; Silverman, S.; Trenti, T.; et al. Markers of bone turnover for the prediction of fracture risk and monitoring of osteoporosis treatment: A need for international reference standards. *Osteoporos. Int.* **2011**, *22*, 391–420. [CrossRef]
38. Bartels, Ä.; O'Donoghue, K. Cholesterol in pregnancy: A review of knowns and unknowns. *Obs. Med.* **2011**, *4*, 147–151. [CrossRef]
39. Kallio, M.J.; Siimes, M.A.; Perheentupa, J.; Salmenperä, L.; Miettinen, T.A. Serum cholesterol and lipoprotein concentrations in mothers during and after prolonged exclusive lactation. *Metabolism* **1992**, *41*, 1327–1330. [CrossRef]
40. Qureshi, I.A.; Xi, X.R.; Limbu, Y.R.; Bin, H.Y.; Chen, M.I. Hyperlipidaemia during normal pregnancy, parturition and lactation. *Ann. Acad. Med. Singap.* **1999**, *28*, 217–221.
41. Lovelady, C.A.; Nommsen-Rivers, L.A.; McCrory, M.A.; Dewey, K.G. Effects of exercise on plasma lipids and metabolism of lactating women. *Med. Sci. Sports Exerc.* **1995**, *27*, 22–28. [CrossRef]
42. Brekke, H.K.; Bertz, F.; Rasmussen, K.M.; Bosaeus, I.; Ellegård, L.; Winkvist, A. Diet and exercise interventions among overweight and obese lactating women: Randomized trial of effects on cardiovascular risk factors. *PLoS ONE* **2014**, *9*, e88250. [CrossRef] [PubMed]
43. Larson-Meyer, D.E.; Schueler, J.; Kyle, E.; Austin, K.J.; Hart, A.M.; Alexander, B.M. Do Lactation-Induced Changes in Ghrelin, Glucagon-Like Peptide-1, and Peptide YY Influence Appetite and Body Weight Regulation during the First Postpartum Year? *J. Obes.* **2016**, *2016*, 7532926. [CrossRef] [PubMed]

44. Larson-Meyer, D.E.; Schueler, J.; Kyle, E.; Austin, K.J.; Hart, A.M.; Alexander, B.M. Appetite-Regulating Hormones in Human Milk: A Plausible Biological Factor for Obesity Risk Reduction? *J. Hum. Lact.* **2020**, *37*, 603–614. [CrossRef]
45. Pescatello, L.S.; American College of Sports Medicine. *ACSM's Guidelines for Exercise Testing and Prescription*, 9th ed.; Wolters Kluwer/Lippincott Williams & Wilkins Health: Philadelphia, PA, USA, 2014; p. 373.
46. Baecke, J.A.; Burema, J.; Frijters, J.E. A short questionnaire for the measurement of habitual physical activity in epidemiological studies. *Am. J. Clin. Nutr.* **1982**, *36*, 936–942. [CrossRef] [PubMed]
47. Tanaka, H.; Monahan, K.D.; Seals, D.R. Age-predicted maximal heart rate revisited. *J. Am. Coll. Cardiol.* **2001**, *37*, 153–156. [CrossRef]
48. Birtcher, K.K.; Ballantyne, C.M. Measurement of Cholesterol. *Circulation* **2004**, *110*, e296–e297. [CrossRef]
49. Cross, N.A.; Hillman, L.S.; Allen, S.H.; Krause, G.F. Changes in bone mineral density and markers of bone remodeling during lactation and postweaning in women consuming high amounts of calcium. *J. Bone Min. Res.* **1995**, *10*, 1312–1320. [CrossRef]
50. Ritchie, L.D.; Fung, E.B.; Halloran, B.P.; Turnlund, J.R.; Van Loan, M.D.; Cann, C.E.; King, J.C. A longitudinal study of calcium homeostasis during human pregnancy and lactation and after resumption of menses. *Am. J. Clin. Nutr.* **1998**, *67*, 693–701. [CrossRef]
51. Kalkwarf, H.J.; Specker, B.L. Bone mineral loss during lactation and recovery after weaning. *Obstet. Gynecol.* **1995**, *86*, 26–32. [CrossRef]
52. Lopez, J.M.; Gonzalez, G.; Reyes, V.; Campino, C.; Diaz, S. Bone turnover and density in healthy women during breastfeeding and after weaning. *Osteoporos. Int.* **1996**, *6*, 153–159. [CrossRef]
53. Malpeli, A.; Mansur, J.L.; De Santiago, S.; Villalobos, R.; Armanini, A.; Apezteguia, M.; Gonzalez, H.F. Changes in bone mineral density of adolescent mothers during the 12-month postpartum period. *Public Health Nutr.* **2010**, *13*, 1522–1527. [CrossRef]
54. Kolthoff, N.; Eiken, P.; Kristensen, B.; Nielsen, S.P. Bone mineral changes during pregnancy and lactation: A longitudinal cohort study. *Clin. Sci.* **1998**, *94*, 405–412. [CrossRef]
55. Clarke, B. Normal bone anatomy and physiology. *Clin. J. Am. Soc. Nephrol.* **2008**, *3* (Suppl. S3), S131–S139. [CrossRef]
56. Lems, W.F.; Raterman, H.G.; van den Bergh, J.P.W.; Bijlsma, H.W.J.; Valk, N.K.; Zillikens, M.C.; Geusens, P. Osteopenia: A diagnostic and therapeutic challenge. *Curr. Osteoporos. Rep.* **2011**, *9*, 167–172. [CrossRef]
57. Hayslip, C.C.; Klein, T.A.; Wray, H.L.; Duncan, W.E. The effects of lactation on bone mineral content in healthy postpartum women. *Obstet. Gynecol.* **1989**, *73*, 588–592.
58. Honda, A.; Kurabayashi, T.; Yahata, T.; Tomita, M.; Takakuwa, K.; Tanaka, K. Lumbar bone mineral density changes during pregnancy and lactation. *Int. J. Gynaecol. Obstet. Off. Organ Int. Fed. Gynaecol. Obstet.* **1998**, *63*, 253–258. [CrossRef]
59. Finkelstein, J.S.; Brockwell, S.E.; Mehta, V.; Greendale, G.A.; Sowers, M.R.; Ettinger, B.; Lo, J.C.; Johnston, J.M.; Cauley, J.A.; Danielson, M.E.; et al. Bone mineral density changes during the menopause transition in a multiethnic cohort of women. *J. Clin. Endocrinol. Metab.* **2008**, *93*, 861–868. [CrossRef]
60. Sowers, M.R.; Zheng, H.; Jannausch, M.L.; McConnell, D.; Nan, B.; Harlow, S.; Randolph, J.F., Jr. Amount of bone loss in relation to time around the final menstrual period and follicle-stimulating hormone staging of the transmenopause. *J. Clin. Endocrinol. Metab.* **2010**, *95*, 2155–2162. [CrossRef]
61. Grizzo, F.M.F.; Alarcão, A.C.J.; Dell' Agnolo, C.M.; Pedroso, R.B.; Santos, T.S.; Vissoci, J.R.N.; Pinheiro, M.M.; Carvalho, M.D.B.; Pelloso, S.M. How does women's bone health recover after lactation? A systematic review and meta-analysis. *Osteoporos. Int.* **2020**, *31*, 413–427. [CrossRef]
62. Dobnig, H.; Kainer, F.; Stepan, V.; Winter, R.; Lipp, R.; Schaffer, M.; Kahr, A.; Nocnik, S.; Patterer, G.; Leb, G. Elevated parathyroid hormone-related peptide levels after human gestation: Relationship to changes in bone and mineral metabolism. *J. Clin. Endocrinol. Metab.* **1995**, *80*, 3699–3707. [CrossRef]
63. Kalkwarf, H.J.; Specker, B.L. Bone mineral changes during pregnancy and lactation. *Endocrine* **2002**, *17*, 49–53. [CrossRef]
64. Butte, N.F.; Hopkinson, J.M. Body composition changes during lactation are highly variable among women. *J. Nutr.* **1998**, *128*, 381S–385S. [CrossRef] [PubMed]
65. Alvarez, J.J.; Montelongo, A.; Iglesias, A.; Lasunción, M.A.; Herrera, E. Longitudinal study on lipoprotein profile, high density lipoprotein subclass, and postheparin lipases during gestation in women. *J. Lipid Res.* **1996**, *37*, 299–308. [CrossRef]

nutrients

Article

Estimating Dietary Intake from Grocery Shopping Data—A Comparative Validation of Relevant Indicators in Switzerland

Jing Wu [1], Klaus Fuchs [2,*], Jie Lian [1], Mirella Lindsay Haldimann [3], Tanja Schneider [4,*], Simon Mayer [1], Jaewook Byun [5], Roland Gassmann [6], Christine Brombach [6] and Elgar Fleisch [7,8]

[1] Interaction- and Communication-based Systems, Institute of Computer Science, University of St. Gallen, 9000 St. Gallen, Switzerland; jing.wu@unisg.ch (J.W.); jie.lian@unisg.ch (J.L.); simon.mayer@unisg.ch (S.M.)
[2] ETH AI Center, ETH Zurich, Swiss Federal Institute of Technology in Zurich, 8092 Zurich, Switzerland
[3] D-One Solutions AG, 8003 Zurich, Switzerland; mirella.haldimann@d-one.ai
[4] Technology Studies, School of Humanities and Social Sciences, University of St. Gallen, 9000 St. Gallen, Switzerland
[5] Department of Software, Sejong University, Seoul 05006, Korea; jwbyun@sejong.ac.kr
[6] Institute of Food and Beverage Innovation, Zurich University of Applied Sciences, 8820 Waedenswil, Switzerland; garo@zhaw.ch (R.G.); broc@zhaw.ch (C.B.)
[7] Information Management, Department of Management, Technology, and Economics, Swiss Federal Institute of Technology in Zurich, 8092 Zurich, Switzerland; efleisch@ethz.ch
[8] Technology Management, Institute of Technology Management, University of St. Gallen, 9000 St. Gallen, Switzerland
* Correspondence: fuchsk@ethz.ch (K.F.); tanja.schneider@unisg.ch (T.S.); Tel.: +41-78-858-7037 (K.F.); +41-71-224-2929 (T.S.)

Citation: Wu, J.; Fuchs, K.; Lian, J.; Haldimann, M.L.; Schneider, T.; Mayer, S.; Byun, J.; Gassmann, R.; Brombach, C.; Fleisch, E. Estimating Dietary Intake from Grocery Shopping Data—A Comparative Validation of Relevant Indicators in Switzerland. *Nutrients* **2021**, *14*, 159. https://doi.org/10.3390/nu14010159

Academic Editor: Panagiota Mitrou

Received: 4 September 2021
Accepted: 6 December 2021
Published: 29 December 2021

Publisher's Note: MDPI stays neutral with regard to jurisdictional claims in published maps and institutional affiliations.

Abstract: In light of the globally increasing prevalence of diet-related chronic diseases, new scalable and non-invasive dietary monitoring techniques are urgently needed. Automatically collected digital receipts from loyalty cards hereby promise to serve as an objective and automatically traceable digital marker for individual food choice behavior and do not require users to manually log individual meal items. With the introduction of the General Data Privacy Regulation in the European Union, millions of consumers gained the right to access their shopping data in a machine-readable form, representing a historic chance to leverage shopping data for scalable monitoring of food choices. Multiple quantitative indicators for evaluating the nutritional quality of food shopping have been suggested, but so far, no comparison has validated the potential of these alternative indicators within a comparative setting. This manuscript thus represents the first study to compare the calibration capacity and to validate the discrimination potential of previously suggested food shopping quality indicators for the nutritional quality of shopped groceries, including the Food Standards Agency Nutrient Profiling System Dietary Index (FSA-NPS DI), Grocery Purchase Quality Index-2016 (GPQI), Healthy Eating Index-2015 (HEI-2015), Healthy Trolley Index (HETI) and Healthy Purchase Index (HPI), checking if any of them performs differently from the others. The hypothesis is that some food shopping quality indicators outperform the others in calibrating and discriminating individual actual dietary intake. To assess the indicators' potentials, 89 eligible participants completed a validated food frequency questionnaire (FFQ) and donated their digital receipts from the loyalty card programs of the two leading Swiss grocery retailers, which represent 70% of the national grocery market. Compared to *absolute* food and nutrient intake, correlations between density-based *relative* food and nutrient intake and food shopping data are stronger. The FSA-NPS DI has the best calibration and discrimination performance in classifying participants' consumption of nutrients and food groups, and seems to be a superior indicator to estimate nutritional quality of a user's diet based on digital receipts from grocery shopping in Switzerland.

Keywords: food shopping quality indicators; FSA-NPS DI; dietary intake; diet monitoring; digital receipts

1. Introduction

The globally increasing prevalence of diet-related chronic diseases, including obesity, diabetes and certain types of cancers, represents a growing burden for affected patients and health-care systems alike [1–4]. Due to societal trends, such as urbanization and the transformation of food systems toward more processed and convenience food items, dietary patterns around the world show an increase in consumed (added) sugar, sodium, saturated fats, and calorific energy [5]. These increases elevate the risks of diet-related chronic diseases [6–8]. Besides the uptake of food items of low nutritional quality, the global demand for meat, fish, and exotic fruits throughout the year takes a significant toll on the planetary ecosystem as well [1]. To counter the global trend toward unhealthy and unsustainable food choices, novel automatic and scalable monitoring tools are needed that have the potential to help deal with the burden of unhealthy dietary patterns, monitor and eventually improve food choices [9].

1.1. Conventional Diet Monitoring Approaches

To analyze an individual's dietary pattern, dietitians and researchers typically rely on self-reported and laborious dietary assessment approaches, such as seven-day weighed food diaries or records, 24 h recalls and food frequency questionnaires (FFQs) [10–13]. Although individual dietary data collected in these ways have an accepted accuracy, the required manual transcription is prone to recall biases and high attrition rates, particularly when the data collection is conducted over longer periods of time [9,14–16]. The European Food Safety Authority (EFSA) promoted the use of software-based applications for FFQs, 24 h recalls and food diaries to reduce the labor intensity of data collection [17]. Even so, the user attrition and adoption in such self-tracking applications remain challenging. For instance, manual diet logging apps (e.g., *MyFitnessPal*) are only actively used by 8% of smartphone users [18]. In addition, collecting data on individual dietary behavior can be expensive, especially within a validated setup, e.g., conducting FFQs under supervision or assessing sodium excretion from 24 h urine collection or measuring micro-nutrient content from blood samples.

Given that contemporary diet monitoring techniques suffer from short-lived retention, memory bias, low acceptance and expensive costs, it is hard to conduct longitudinal studies with a strong statistical power or even population-wide diet monitoring via self-reporting dietary monitoring methods only. To allow for more inclusive and continuous food choice monitoring, new scalable and automated food choice monitoring tools to substitute or complement contemporary dietary monitoring approaches are needed.

1.2. Digital Receipts

With the proliferation of digital payments and loyalty cards, digital receipts from grocery shopping can serve as a novel, scalable, automatically and continuously self-updating monitoring tool for food choices [19,20]. Digital receipts can be seen as machine-readable, electronic substitutes for their contemporary paper-based printed counterparts. They are expected to be adopted around the globe over the next decade, as they promise significant advantages with regard to environmental footprint [21] and mitigating tax evasion [22,23], offering superior advantages and transparency for consumers [24].

Food shopping comprises a major part of people's shopping in supermarkets. Assuming that consumers eat a majority of the groceries which they buy with their loyalty cards, there exists the possibility to infer individual eating behavior from the household shopping records. Compared to conventional diet monitoring tools which focus on collecting individual dietary intake data, digital receipts do not require the laborious active logging of every single meal, and all historic purchase records are available instantly. In addition to data on purchased quantities and types of selected food items, information about expenditures and corresponding timestamps is available. Compared to contemporary food monitoring approaches, the multi-dimensional digital receipt dataset thus provides new possibilities to explore further aspects of participants' food choice behavior. These facets can include their favorite stores and shopping habits, their price sensitivity, their preferences for specific brands, categories and flavors, or even their desire to purchase seasonally, regional or sustainable products, all potentially relevant avenues for diet-related interventions.

Although representing partial data in a household context, digital receipts can serve as a fully automated diet monitoring proxy for estimating individual food intake. An important distinction between digital receipts and the aforementioned contemporary diet monitoring tools is that digital receipts usually represent the shopping behavior of a household rather than of an individual. This is because households, e.g., a family or a shared flat, tend to share groceries and loyalty cards. Assessing household level food shopping data to infer insights on individual food intake behavior results in lower but still acceptable accuracy, compared to individual dietary monitoring tools (e.g., FFQs, food diaries, bio-samples, such as blood or urine sampling). The challenges are manifold. First, food shopping data are incomplete, and do not consider process factors such as food preparation, out-of-home consumption, delayed consumption and food waste [25,26]. Still, the purchases from supermarkets usually represent the vast majority of dietary intake. For example, in the case of sodium, which is the most often consumed via eating packaged food products shopped from supermarkets, an estimation of 80% of its intake originates from supermarket purchases [27]. Second, the conversion from household-level food shopping data to individual diet behavior is challenging. Nevertheless, previous studies have repeatedly demonstrated that applying statistical methods on the partial food shopping data allows the inference of absolute dietary intake as well as relative distributions of individual food choices [28,29]. Food purchases captured by receipts correlate with and can predict individual dietary patterns [20,28,30–33], demonstrating moderate to strong agreement [20,31,32,34–37]. The dietary calorific intake of a person correlates with the amount of shopped energy-dense food products [34,35,37], or that obesity could be detected via analyzing food shopping patterns [31]. More concretely, by assessing their relative distribution in terms of weight, calories or expenditures, household-level food shopping data can be converted to individual level dietary intake estimates. These weight-based, expenditure-weighted or calorie-weighted approaches thus implicitly assume that each person in the household consumes similar proportions of the purchased food categories. While this assumption might be justified for single and smaller households, it might lack validity for very large households. Still, given the literature in the field, digital receipts can be assessed as a scalable, non-invasive proxy for individual dietary intake behavior.

In the past, research in this field has been strongly limited by the restricted access to digital receipt data. This is because considerable efforts are involved in collecting product data and sample sizes are often small. A key aspect of monitoring food choices via digital receipts is collecting a user's electronic shopping history via loyalty cards, which previously was often done manually via collecting printed paper receipts, or taking stock of home inventory of shopped food items [19,28,29,38,39]. Because of the introduction of the General Data Protection Regulation (GDPR) [40], millions of customers in the European Union recently gained the right to access their digital receipts from loyalty cards. Consequentially, researchers may obtain these data with their study subjects' consents. Therefore, product shopping, including food shopping recorded on loyalty cards, is now retrievable from data processors (i.e., retailers and loyalty card providers). By itself, a digital receipt does not contain any nutritional details of the purchased food products. To this end, food product composition databases, which were mandated by the regulation on mandatory declaration of food information for food items sold online (EU)1169 [41], allow the data fusion of digital receipts and nutritional information on the shopped food products.

To objectively determine whether a shopping history record of a certain user indicates a healthy behavior, validated quantitative indicators are needed to offer a normative, reliable and interpretable basis for comparison between users, households, regions and retailers.

1.3. Proposed Food Shopping Quality Indicators

Multiple quantitative food shopping indicators have been suggested to evaluate the nutritional quality of food and beverage shopping records, including the Grocery Purchase Quality Index-2016 (GPQI) [28] (designed and validated within the United States (US)), the Healthy Trolley Index (HETI) [29] (designed and validated within Australia) and the Healthy Purchase Index (HPI) [42] (designed and validated within France). These three suggested indicators all assign the observed food items in the shopping data into different food group categories, thereby allowing to estimate how balanced each respective user's shopping history is. Taking their respective local dietary guidelines as the golden standards, these indicators might not necessarily be directly 'transferable' to all regions, but are valid in their respective geographies. In addition, the expenditure share of the respective food groups was used in the calculation of these three indices in this study's digital receipt dataset from Switzerland to guarantee the compliance with respect to each indicator's corresponding index guidelines. In absence of food composition data, price-based indicators might be an interim solution as suggested by the HETI, HPI and GPQI indices. However, if available, shopped quantities in grams or milliliters or energy in kilocalories (kcal) or kilojoules (kJ) would give a more accurate representation of the dietary impact from the evaluated shopping records than price-based estimations. Additionally, traditional diet indices can be used to evaluate food shopping behavior. The Healthy Eating Index-2010 (HEI-2010) scores derived from food shopping data showed moderate agreement and minimal bias with HEI-2010 scores from 24 h recalls [43]. Thus, using food shopping data to calculate Healthy Eating Index-2015 (HEI-2015) [44], the latest version of HEI, might be a feasible way of assessing nutritional quality from digital receipts. Finally, the Food Standards Agency Nutrient Profiling System Dietary Index (FSA-NPS DI) [45] can also be utilized to assess food shopping quality. Instead of assigning food to different food groups and evaluating how compliant the baskets are to a certain dietary guideline, FSA-NPS DI takes all food items into account, weighs them based on the calorific contribution and focuses on the overall nutritional quality of the entire basket, in this case, series of shopping baskets over the study's observation period.

Despite the existence of multiple shopping indices, a published assessment validating the calibration and discrimination potential of these indicators in a comparative environment does not yet exist. The calibration capacity of a model, or an indicator in this case, can be reflected by how agreeable the prediction and actual outcomes are. The discrimination capacity evaluates how well a model can perform in separating cases with and without certain outcomes [46,47]. The main objective of this study is to compare the calibration and discrimination ability of the aforementioned quantitative shopping indices, namely FSA-NPS DI, GPQI, HEI-2015, HETI and HPI. Digital receipts that were automatically captured from loyalty cards in Switzerland are used as the data for the calculation of the five indicators. The results derived from the validated FFQs were taken as the objective measurement and were what the food shopping quality indicators were used to calibrate and discriminate. The hypothesis is that some of them perform significantly better than the others in a general situation. The results should equip researchers, practitioners, and policymakers with the insights required to select one among the existing indices. The conclusions might be relevant for researchers and practitioners who work on designing novel food shopping quality indicators, monitoring systems or interventions in this domain.

2. Materials and Methods

This manuscript describes the first study comparing the calibration capacity and validating the discrimination potential of multiple previously suggested food shopping quality indicators for the nutritional quality of shopped groceries, including FSA-NPS DI, GPQI, HEI-2015, HETI and HPI. To be more specific, the calibration capacity indicates how closely the food shopping quality indicators calculated from digital receipts and the dietary intake reflected by the FFQ results are correlated. The discrimination capacity shows how well the food shopping quality indicators can distinguish people with different levels of dietary intake. In the following, the digital receipt integration, food composition database and study design are introduced.

2.1. Digital Receipt Integration

The digital receipt infrastructure was implemented in Switzerland due to the availability of digital receipts from the loyalty card systems of the two leading supermarket chains. To support the comparative analysis of the suggested quantitative food shopping quality indicators, a technical setup was designed and implemented to allow the collection of receipts from users who consented to participate in the study. The study was deployed on the Bitsaboutme (BAM) online platform (see https://bitsabout.me/, accessed on 15 December 2021). BAM is a GDPR-compliant data marketplace service located in Switzerland, which allows users to request their own personal user data from data controllers and store their data in an encrypted data vault. Just as data sources such as social media or messaging services process personal data, financial transactions from bank accounts and digital receipts are also considered as personal data by the GDPR. Hence, users of the BAM service are able to retrieve their digital receipts from the two leading Swiss loyalty card providers, namely Migros Cumulus and Coop Supercard. In this regard, Switzerland can be considered an ideal region to validate quantitative food shopping quality indicators, as just these two leading Swiss grocery chains, i.e., Migros and Coop, represent a sales share of 70% [48,49]. Additionally, the Swiss consumers can be considered frequent users of loyalty cards. Taking Coop as an example, around 80% of Coop's annual sales were achieved with Supercard customers [50]. In this respect, Switzerland is comparatively ahead of multiple countries. In regions other than Switzerland, digital receipts are also likely to be implemented and adopted in the future due to the steadily increasing acceptance of digital payment methods, such as credit cards and mobile payment, involving an expected transition from paper to digital receipts. Thus, we believe that the results and implications from carrying out this study in Switzerland can potentially be generalized toward other regions as well.

Once users decide to donate their digital receipts to the study via the BAM service, they need to agree to multiple opt-in consent forms before their historic and future receipts can be integrated into the study (see Figure 1). To join the study, users had to opt in at least four times before their digital receipts would become part of the sample analyzed in this study. First, prospective study participants needed to be enrolled into at least one of the two Swiss loyalty card systems. Consequentially, users needed to accept the terms and conditions of at least one or even both of the loyalty card providers and opt-in toward collecting digital receipts in a digital form. Second, prospective participants needed to join the BAM service before they could participate in the study. Therefore, they needed to agree to the terms and conditions of the BAM service so that the service could retrieve their personal data from data controllers on their behalf. Third, users who already collected digital receipts from their loyalty cards needed to consent to the BAM service retrieving their digital receipts on their behalf from the loyalty card system providers directly. In one case, i.e., Migros Cumulus, this is done by linking the corresponding online account, similar to using a Facebook connect to hand over user data. In the other case, i.e., Coop Supercard, a user needs to share email-based digital receipts with the BAM service. Only then can a user's up-to-two-year historic and new digital receipts, i.e., those that are created every time a user buys groceries and uses the loyalty card(s) at the supermarket checkout from now on, be automatically imported into the BAM service and stored in a standardized form in the personal BAM data vault. Finally, a prospective user had to join our study, which was displayed toward eligible users on the BAM service platform, and to donate their digital receipts to the study.

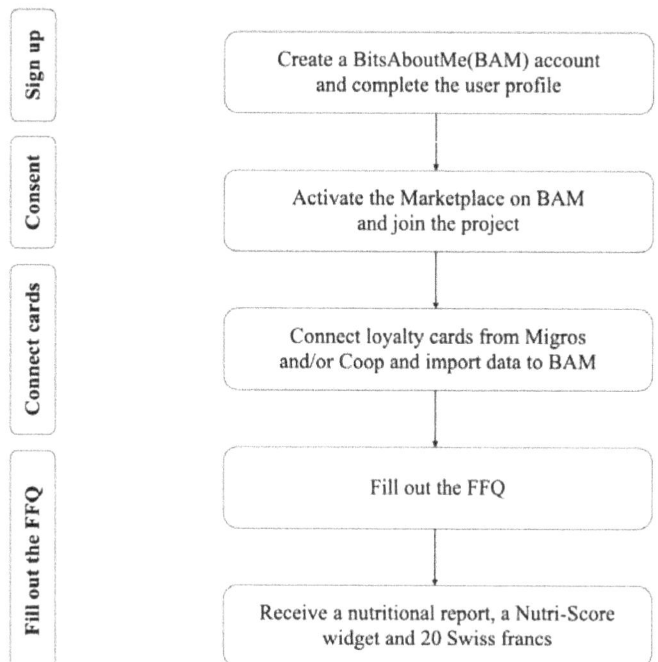

Figure 1. Participant on-boarding flow.

The study and its consent form were approved by the Ethics Committee of Swiss Federal Institute of Technology in Zurich (ETH Zurich) with the protocol code 2019-N-134 on 15 October 2019, prior to the launch of the study. In particular, the study protocol and the consent form on the BAM service required the anonymization of the donated digital receipt data. Concretely, no directly identifying personal data such as names, email addresses, phone numbers or loyalty card identifiers were shared by the BAM service with the study. To ensure the anonymity of the data donors, even the shopping locations and time of day were removed from the digital receipt dataset before the analyses conducted in this manuscript. Thus, each receipt in the final digital receipt dataset donated by the N = 464 users of the BAM service who participated in the study (see Figure 2) only contained a randomized study identifier of the respective user, the day of the year of the shopping, the shopped quantity amount, the identifier of the food item that was bought, the price, and potentially applied discounts. The possibility of re-identifying individual consumers from a maliciously acquired copy of the anonymized dataset was communicated to prospective study participants in the consent form and considered as an acceptable study risk, given the contribution of the study.

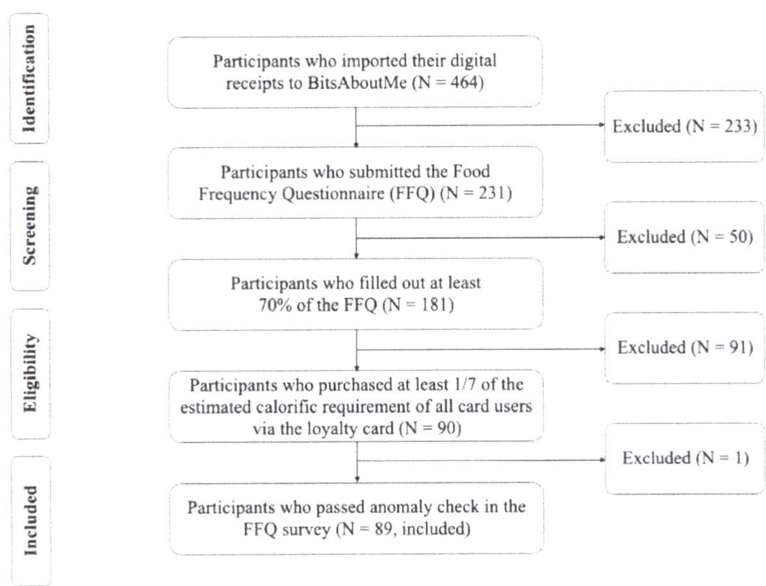

Figure 2. The procedure of excluding ineligible participants.

2.2. Food Composition Database

When using digital receipts as proxies for food choices, a common challenge is mapping food products captured via printed or digital receipts to nutrient information [20,51]. To conduct the comparisons of food shopping quality indicators, the anonymized digital receipt dataset was enriched with data about the nutritional composition of shopped food items, as digital receipts per se usually do not contain such information. The authors of this study leveraged an existing food composition database containing detailed information about over 50,000 grocery products frequently sold and consumed in Switzerland [52,53]. Driven by the recent mandates for online food nutrition databases [41], there are now trusted, curated databases (such as GS1 trustbox) as well as crowd-sourced databases (e.g., OpenFoodFacts) available to retrieve detailed nutritional information on products sold in a retail environment. This information becomes particularly useful when combined with a consumer's shopping history.

In food composition databases, products sold in retail environments are usually identified via their global trade item number (GTIN). The GTIN is a globally unique product identifier distributed by GS1, a globally operating non-profit standards organization. Unfortunately, paper-based or digital receipts usually only contain a product's name in terms of identifiers. Identifying a product's GTIN from a digital receipt in fact represents one of the key challenges in digital-receipt-based monitoring. Therefore, mapping the product names to GTIN is necessary. To ensure high data quality and correct product mapping, the authors decided to conduct the product matching manually. Since both retailers, i.e., Migros and Coop, do not include the GTIN within their digital receipt formats, a heuristic was applied to correctly identify the most frequently purchased food products. In the context of this study, a total of N = 464 users were invited to donate their shopping data. These data were used to identify the most frequently occurring food products. In total, 65,391 different products that were bought using the two loyalty card systems from the two leading Swiss supermarkets were observed by assessing the entire shopping history of the N = 464 users. In total, 5950 product article identifiers from the digital receipts were mapped to corresponding GTINs. For each of the matched products, its attributes such as nutritional details (e.g., calorific energy and macro-nutrients such as protein, carbohydrate, sugar, fat, saturated fat, dietary fiber, and micro-nutrients, such as sodium, all per 100 g or ml of product; 1 g corresponds to 1 ml), its logistical data (e.g., product size in grams, kilograms, milliliters or liters), product images, allergens, and ingredients were made available for the analysis. These mapped articles correspond to 4951 of the most frequently occurring products. This is because all coupons and all non-food items were labeled identically, i.e., the study does not differentiate between different types of coupons or non-food items. These 4951 labeled product items represent most of the frequently bought food items. Since the two supermarkets also feature non-food items (e.g., plastic bags, napkins, and toilet paper) and some food items cannot be identified (e.g., 'Menu 1', and 'Lunch Menu'), some frequently occurring products are not identified in this study. Nevertheless, the overall matching ratio, i.e., the proportion of identified products in the shopping history of the eligible users in the four-week observation period is 69.6%. This shows that in order to capture a majority of the products purchased in digital receipts in Switzerland, less than 10% of the products must be identified in the corresponding food composition database.

The study setup described in this manuscript not only allowed for the post hoc analysis of nutritional quality of shopped food items from digital receipts within this study, but also allowed study participants to analyze the nutritional quality of their food shopping after joining the study (see Figure 3). After joining the study on the BAM service website successfully, participants gained access to a new widget that shows users the aggregated Nutri-Score [54] of their recent food shopping. For displaying and faster processing for the user experience on the BAM website, we simplified the Nutri-Score framework by using a weight-based five-letter system (A = 0.5, B = 1.5, C = 2.5, D = 3.5, and E = 4.5). For each basket, the weight-weighted average of all products was calculated and displayed on a chart, as shown in the middle of Figure 3. A careful review demonstrated that the original framework and the simplified framework yielded very similar results. The simplified version was chosen for simplicity. The analysis was provided via an application programming interface (API) and demonstrated the potential of assessing the nutritional quality of digital receipts for tailored interventions to consumers, aiming at supporting healthy food choices.

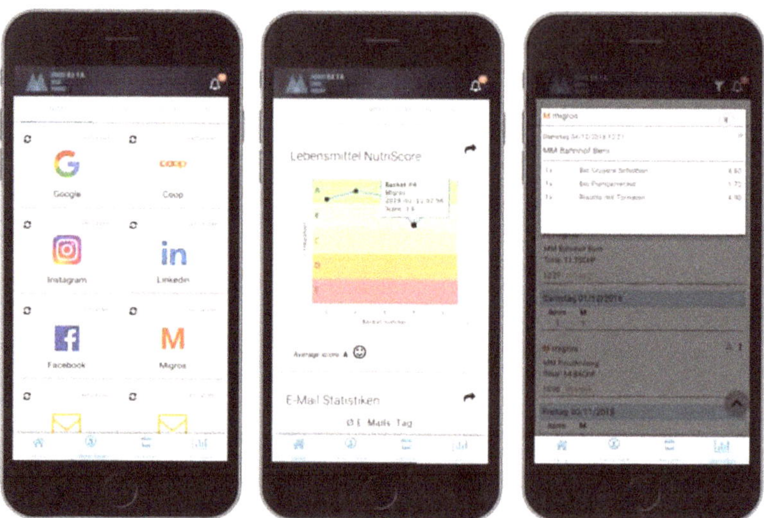

Figure 3. Enriching digital receipts with food composition data and displaying the weight-averaged Nutri-Scores of aggregated baskets.

2.3. Food Frequency Questionnaire

After having donated their digital receipts, prospective study participants were encouraged to also complete a previously validated FFQ in order to collect objective measurement data on individual diets for the validation of the food shopping quality indicators. Multiple alternative FFQs were identified and evaluated for the study. A systematic literature research and the assessment of a meta platform that compares available FFQs (see https://www.nutritools.org/, accessed on 15 December 2021) led to the comparison of five FFQs and one web-mediated 24 h recall tool. Namely, the VioCare FFQ (See https://vioscreen.com/, accessed on 15 December 2021), the Ernährungerhebung FFQ provided by Zurich University of Applied Science (ZHAW) (see https://r2n.ernaehrungserhebung.ch/, accessed on 15 December 2021) [55], the DHQ3 FFQ (See https://www.dhq3.org/login/, accessed on 15 December 2021), the Block FFQ (see https://nutritionquest.com/login/, accessed on 15 December 2021), and the MiniMeal-Q FFQ (see https://www.nutritools.org/tools/199, accessed on 15 December 2021) were compared. Aspects such as cost, required time, Swiss aptitude, accuracy, validation and setup effort were structurally assessed. Finally, the web-mediated FFQ (see https://r2n.ernaehrungserhebung.ch/, accessed on 15 December 2021) provided by ZHAW was selected (see Figure 4), primarily due to its previous validation in Switzerland, which was also the focus region of this study [55], as well as its easy-to-use web-based administration. The FFQ-mediated questionnaire used to validate the digital receipts can be found here (link: https://gitlab.ethz.ch/jingwu/shopping-index-comparison/-/blob/master/User_Survey_and_Food_Frequency_Questionnaire__1_.pdf, accessed on 15 December 2021). The FFQ instrument was validated prior to our study and the validation study also took place in Switzerland in 2017 [55]. The validation of the FFQ allows for inferring that its dietary intake estimates correlate well with actual individual dietary intake. Thus, demonstrating the correlation between digital receipts and the FFQ allows for inferring that the digital receipts correlate with actual individual food intake. The study data, its framework and resources are in line with other validation studies in the field (i.e., HETI, GPQI, HPI). In addition, Prof. Christine Brombach, who also supervised the FFQ validation study, served as a co-author in the study, ensuring a high validity of the conducted analyses.

The chosen FFQ took an average of around 30 min for each of the N = 181 users who filled out at least 70% of the FFQ. The technical link between the BAM service and the web-based FFQ was realized via personalized links. Participants who agreed to the study consent on the BAM service received an email from BAM and were then invited via a personalized link that included a pseudonymous user identifier, linking to a user's digital receipts as well as the user's FFQ responses. Table 1 shows the daily food and nutrient intake of participants, based on the self-reported FFQ results.

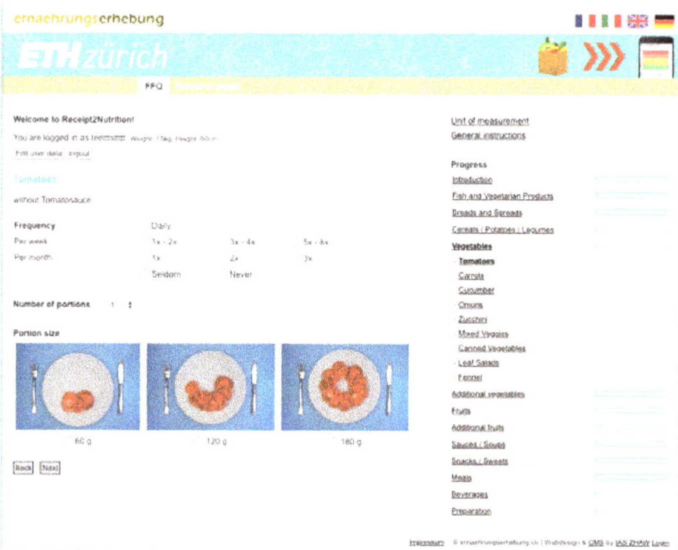

Figure 4. Web-mediated food frequency questionnaire (FFQ).

Table 1. *Absolute* individual daily nutritional intake, N = 89.

Category	Mean	Standard Deviation
Unit: portions/day		
Meat and meat products	1.17	1.20
Vegetables and salad	2.55	1.71
Fruits	1.38	1.16
Whole grain products	0.32	0.33
Sweets, salty snacks, sugar-sweetened beverages, alcohol	2.93	1.95
Unit: grams/day		
Sodium	2.1	1.5
Dietary fibers	27.1	14.3
Saturated fatty acids	37.5	26.0
Added sugar	10.4	8.52

An overview of the FFQ questions and potential answers can be found online (see https://gitlab.ethz.ch/food-coach/shopping-index-comparison, accessed on 15 December 2021). All donated data, i.e., digital receipts as well as the responses, were anonymized prior to the analyses conducted in this study. After successfully finishing these tasks, all participants received a nutritional assessment report based on their FFQ (see Figure 4) and an automatically self-updating Nutri-Score widget based on their recently shopped grocery baskets, displayed within the BAM service website (see Figure 3). In addition, participants received a financial compensation of CHF 20 (Swiss francs) (i.e., USD 21.80 (United States dollars), 28 June 2021) for completing the study.

2.4. Study Participants

The presented study was deployed on the BAM service online platform (see https://bitsabout.me/, accessed on 15 December 2021) and was advertised together with an invitation link through a variety of channels, including BAM's marketplace, BAM's newsletter, social media and billboards on the local university campus as well as university shuttle buses. This approach of using BAM's marketplace and the university network ensured to address younger as well as more mature households, as the socio-demographic characteristics on the BAM website represent more mature consumers (see Table 2).

Table 2. Demographic summary of participants.

Sample	Count (%)
Gender	
Male	68 (76.4)
Female	21 (23.6)
Other	0 (0.0)
Age [yrs]	
18–29	29 (32.6)
30–39	29 (32.6)
40–49	18 (20.2)
> 50	13 (14.6)
Body Mass Index [kg/m^2]	
Underweight (<18.5)	2 (2.3)
Normal ($\geq$18.5 and <25.0)	55 (61.8)
Overweight ($\geq$25.0 and <30.0)	22 (24.7)
Obese ($\geq$30)	10 (11.2)
Total	89 (100.0)

Participants were recruited from 22 December 2018 to 10 June 2021. Participants used their own devices (e.g., laptops and/or mobile phones) to enroll in the study. The relatively high barriers to join the study, as shown in Figure 1, led to a slow uptake of participants. Hence, the participant recruitment was conducted on a rolling basis of over two years in order to collect the required sample size for a robust comparison of the suggested food shopping quality indicators.

To ensure completeness and comparability between the FFQ data and the digital receipt data, participants were required to meet several eligibility criteria, which are illustrated in Figure 2. In total, N = 464 participants joined the study via accepting the study consent form that was displayed on the BAM service website. Out of those, N = 231 followed the email-based invitation to also start the web-based FFQ (i.e., email response rate of 50.2%). Fifty participants were excluded because they did not complete the FFQ (i.e., drop out rate 21.6%). To ensure that food shopping recorded on loyalty cards is representative of actual food intake, eligible users should shop at least 1/7 of the estimated calorific requirement of all people who share the loyalty cards with participants, within the four weeks preceding their FFQs. The estimated energy intake of a person who is $\geq$13 years old was 2250 kcal/day [56], irrespective of gender. That of a child (<13 years old) was estimated to be 0.575 [57] times the energy requirement of a person who is older than 13, i.e., 2250 kcal/day, irrespective of gender. The four-week time window was defined by the FFQ, which was validated to estimate a participant's typical diet on a four-week basis [55]. Consequently, 91 participants were excluded, as they did not meet the criteria (exclusion rate 50.2%, i.e., 91 from 181 users who completed their FFQ and donated digital receipts). Finally, one participant self-reported his/her own BMI to be 109.8 kg/m^2 (height, 88 cm; weight, 85 kg) and was therefore excluded. The final dataset included 89 users with a completed FFQ and an acceptable amount of shopped products captured by digital receipts in the four weeks prior to the FFQ.

On average, an eligible participant was 36.2 years of age (standard deviation (SD): 10.3) and had an average BMI of 24.4 (SD: 3.8). Compared to the general Swiss population, the recruited sample (see Table 2) has a higher male to female ratio (76.4% in the sample vs. 49.2% in Switzerland), a lower average BMI (24.4 in the sample vs. 25.3 in Switzerland), and a lower average age (36.2 years in the sample vs. 43.1 in Switzerland). With regard to the observed shopping behavior, participants in the final dataset spent on average a total of CHF 230.30 (SD: CHF 175.60) on an average of 39.9 kg (standard deviation: 32.1

kg) grocery products over the four weeks before finishing the FFQ. The average number of adults and children who were sharing their loyalty cards are 1.7 (SD: 1.0) and 0.5 (SD: 0.9) respectively (see Table 3). In terms of individual diet as determined by the FFQ, the N = 89 users on average consumed 1.17 (SD: 1.2) portions of meat per day (see Table 1). Similarly, they consumed on average 2.55 (SD: 1.71) portions of vegetables and 1.38 (SD: 1.16) portions of fruits per day. Hence, the average participant did not reach the publicly recommended three portions of vegetables and two portions of fruits per day [58]. The participants consumed on average 0.32 (SD: 0.33) portions of whole grains per day. Finally, the participants consumed an average of 2.93 (SD: 1.95) portions of sweets per day. The fact that the Swiss population consumes too many sweets has been observed in the annual food intake study MenuCH for many years [59,60]. In terms of nutritional intake, the N = 89 participants on average consumed 2.1 g (SD: 1.5) of sodium, 27.1 (SD: 14.3) grams of dietary fibers, 37.5 (SD: 26.0) grams of saturated fatty acids and an estimated amount of 10.4 (SD: 8.5) grams of added sugar per day. Compared to the annual food intake study MenuCH, these values seem representative of typical dietary intake and in line with the observations for the dietary behavior of the Swiss population [59,60].

Table 3. Food shopping characteristics of the study sample, N = 89.

Characteristics of Observed Food Shopping Behavior [a]	Mean (SD [b])
Household	
Adults sharing the loyalty card(s)	1.7 (1.0)
Children sharing the loyalty card(s)	0.5 (0.9)
Food shopping quantity identified via digital receipts	
Amount spent in Swiss francs (CHF)	230.30 (175.60)
Amount spent in United States dollars (USD) [c]	250.28 (190.83)
Weight [d] of shopped food products in kg	39.9 (32.1)

[a] This is based on the food shopping data in the four weeks before finishing FFQs. [b] SD: standard deviation. [c] Conversion CHF/USD on date: 30 June 2021. [d] Conversion for liquids: 1 g = 1 mL.

2.5. Validation and Comparison of Food Shopping Quality Indicators

To assess the calibration and discrimination capacity of the alternative food shopping quality indicators, five different indicators, namely FSA-NPS DI, GPQI, HEI-2015, HETI and HPI, were selected and calculated. Their respective calibration and discrimination performance for individual N = 89 dietary behavior was computed. We selected the indicators for multiple reasons. First, all five indicators were defined to represent a quantitative indication of an individual's general dietary habitual patterns. Second, the indicators can all be calculated using the digital receipts mandated by GDPR [40] as well as the food information, as mandated for products sold online by EU1169 [41]. Therefore, these indicators have the potential to support millions of consumers in the European Union in terms of monitoring nutritional quality using digital receipts. Finally, the five indicators have not yet been cross-validated in the same geographic region. While the GPQI and HEI-2015 were defined in the US, HETI was defined in Australia, and the HPI was defined in France. Further, the FSA-NPS DI was defined in the United Kingdom (UK) and later adopted in France. As argued earlier, Switzerland represents a suitable study context for such a validation study.

All five indicators were be calculated on a four-week basis, as the FFQ was previously validated over the course of a four-week time period [55]. Receipts whose timestamps were within the four weeks prior to the completion of the FFQ were used for calculation. The original definitions of the FSA-NPS DI, GPQI, HETI and HPI were followed and not adapted in terms of calculation. As discussed in the previous chapter, the HEI-2015 [44] was adopted to digital receipts, similar to how the Healthy Eating Index-2010 (HEI-2010) was adapted to food shopping [43]. As defined in their publications, the GPQI, HEI-2015, HETI and HPI are primarily based on health-relevant food groups. The HEI-2015 also includes some nutrients but not the other three food shopping quality indicators. Thus, for their calculation, the authors included only items belonging to the relevant food groups used in their respective definitions. When calculating the FSA-NPS DI, all shopped food items were included.

Results from the FFQ were taken as the objective measurements. We followed the definition of the Diet Quality Index [61], a Swiss dietary index, and aggregated food to five groups, namely meat and meat products; vegetables and salad; fruits; wholegrain products; and sweets, salty snacks, sugar-sweetened beverages and alcohol. On the nutrient level, we retrieved the consumption of sodium, dietary fiber and saturated fatty acids directly, using the food composition database. As added sugar content is not mandated in Europe [41], we estimated it based on an established approach proposed by Louie et al. [62].

To validate the indicators with the nutritional intake as determined by the FFQ, certain conditions were followed to ensure a coherent process. First, digital receipts that could not be identified (e.g., rare products) and therefore have missing nutritional data, were discarded. To assess the number of portions eaten by each individual user as determined by the FFQ, or shopped as determined by the digital receipts, the following considerations were agreed upon. First, to calculate the number of consumed meat portions, dried meat portions were defined as 30 g per portion, while non-dried meat as 120 g per portion. Second, to assess the number of fruit portions shopped or eaten, 30 g of dried fruits were defined as one portion, while 120 g of fresh or frozen fruits or 200 mL of fruit juices were defined as one portion. Third, 30 g of dried vegetables accounted for one portion, while 120 g of fresh or frozen vegetables or 200 mL of vegetable juices were defined as one portion. The number of portions of whole grains eaten or shopped were defined as follows: 100 g for bread, 30 g for cereals or flakes, 60 g for crisp bread and 30 g for cereals. In this study, whole grains were defined as grain products that include at least 5 g of dietary fiber per 100 g of product [63–65]. Finally, for sweets, the definitions of one portion were set to: 17 g for chocolate, 20 g for cocoa products and jams, 30 g for bonbons, cereal bars and cookies, 50 g for pudding and ice creams, 120 g for sweet cakes and similar. A detailed overview of the portion sizes used in the data processing for the FFQ and digital receipts can be found in the datasets that were published together with this manuscript.

Based on the obtained data from the participants' FFQs and digital receipts, we assessed the *calibration capacity* and the *discrimination capacity* of the considered food shopping quality indicators. To evaluate the *calibration capacity* of the models, Pearson correlation coefficients between food shopping quality indicators and *absolute/relative* nutritional intake were calculated. In this context, higher correlation coefficients represent higher calibration capacity. More specifically, a high correlation of a food shopping quality indicator with a health-relevant nutrient or food group underlines a valid calibration of the food shopping quality indicator for the nutrient or food group respectively.

To evaluate the *discrimination capacity* of the proposed food shopping quality indicators, the nutritional intake of three compliance tertiles were compared. Similar to validation studies of single food shopping quality indicators that divided their sample into three segments, tertile T1 maps to the lowest compliance to a given indicator, and T3 maps to the highest compliance. For the assessment of the discrimination capacity of the indicators, their potential to discriminate all three tertiles T1–T3, as well as their ability to distinguish pairwise differences between each combination of the tertiles were evaluated. In total, four statistical tests, namely one Kruskal–Wallis test for the three tertile comparisons (T1, T2, and T3) and three Mann–Whitney U tests for the pairwise comparisons (T1 vs. T2; T2 vs. T3; and T1 vs. T3, respectively) were conducted for each combination of an indicator with a nutritional intake category as determined by the FFQ. A significantly different result is counted as one point, while a non-significant result is counted as zero point. The total points of each indicator were obtained by summing up the points across all comparisons where the respective indicator was involved. Therefore, more points represent better discrimination capacity, and the maximum achievable number of points is 36.

3. Results

We obtained the following results on the *calibration* and *discrimination* capacities of the investigated indicators.

3.1. Calibration Capacity: Correlations between Food Shopping Quality Indicators and Nutritional Facts

As shown in Table 4, food shopping quality indicators and individual *absolute* daily nutritional intake were in general weakly to moderately correlated. Among all food shopping quality indicators, only FSA-NPS DI was weakly correlated with the highest number (4) of food and nutrient categories, i.e. the Pearson's correlation coefficient was between 0.1 and 0.3. The strongest correlation was observed between HEI-2015 and *absolute* dietary fibers. Generally, the correlations were significantly stronger when using individual daily density-based *relative* nutritional intake, as shown in Table 5. The strongest correlation was observed between FSA-NPS DI and *relative* intake of dietary fibers, with a Pearson's r of 0.500. Compared to the other four indicators, the FSA-NPS DI was correlated more strongly with the nutritional facts on both *absolute* and *relative* scales, demonstrating the highest correlation coefficient for four (six) out of the nine dimensions for *absolute* (*relative*) dietary intake captured from FFQs. On the other hand, the correlations between HETI, HPI and nutritional facts were in general weaker than those among other food shopping quality indicators and nutritional facts, no matter on the *absolute* or *relative* scale. On average, the five food shopping indicators calibrate fruits and dietary fibers the best, on both *absolute* or *relative* scales.

Table 4. Pearson correlation coefficients between food shopping quality indicators and *absolute* individual daily nutritional intake, N = 89.

Indicators [a]	- FSA-NPS DI [b]	GPQI [c]	HEI-2015 [d]	HETI [e]	HPI [f]
Unit: portions/day					
Meat and meat products	−0.246 *	0.000	−0.083	−0.099	−0.060
Vegetables and salad	**0.235 ***	0.140	0.190	0.181	0.177
Fruits	0.239 *	0.215 *	0.254 *	0.274 **	**0.288 ***
Wholegrain products	0.161	0.184	**0.322 ***	0.232 *	0.135
Sweets, salty snacks, sugar-sweetened beverages, alcohol	−0.111	−0.036	**−0.124**	−0.026	−0.002
Unit: grams/day					
Sodium	**−0.121**	0.050	0.023	0.072	0.027
Dietary fibers	0.312 **	0.178	**0.329 ***	0.296 **	0.173
Saturated fatty acids	**−0.144**	0.033	0.006	0.034	−0.008
Added sugar	−0.101	**−0.197**	−0.193	−0.137	0.138
Points	**4**	1	3	0	1

[a] The highest absolute value is marked bold to indicate the best calibrated food shopping quality indicator for each nutrient or food group. [b] - FSA-NPS DI: Inverted Food Standards Agency Nutrient Profiling System Dietary Index. We inverted the original FSA-NPS DI scores to make them directly comparable to other food shopping quality indicators. The higher the inverted FSA-NPS DI, the healthier the food shopping. [c] GPQI: Grocery Purchase Quality Index-2016. [d] HEI-2015: Healthy Eating Index-2015. [e] HETI: Healthy Trolley Index. [f] HPI: Healthy Purchase Index. * $p < 0.05$. ** $p < 0.01$.

Table 5. Pearson correlation coefficients between food shopping quality indicators and *relative* individual daily nutritional intake, N = 89.

Indicators [a]	- FSA-NPS DI [b]	GPQI [c]	HEI-2015 [d]	HETI [e]	HPI [f]
Unit: portions/1000kcal					
Meat and meat products	**−0.359 ***	−0.061	−0.241	−0.240 *	−0.090
Vegetables and salad	**0.321 **	0.136	0.191	0.108	0.133
Fruits	**0.354 ***	0.195 *	0.238 *	0.197	0.245 *
Wholegrain products	0.231	0.101	**0267 *	0.193	0.080
Sweets, salty snacks, sugar-sweetened beverages, alcohol	−0.097	−0.068	**−0.197**	−0.139	0.069
Unit: g/1000 kcal					
Sodium	**−0.244 *	−0.092	−0.178	−0.055	−0.045
Dietary fibers	**0.500 ***	0.126	0.342 **	0.235 *	0.139
Saturated fatty acids	**−0.367 ***	−0.125	−0.228 *	−0.177	−0.143
Added sugar	−0.093	**−0.329 **	−0.316 **	−0.306 **	−0.259 *
Points	**6**	1	2	0	0

[a] The highest absolute value is marked bold to indicate the best calibrated food shopping quality indicator for each nutrient or food group. [b] - FSA-NPS DI: Inverted Food Standards Agency Nutrient Profiling System Dietary Index. We inverted the original FSA-NPS DI scores to make them directly comparable to other food shopping quality indicators. The higher the inverted FSA-NPS DI, the healthier the food shopping. [c] GPQI: Grocery Purchase Quality Index-2016. [d] HEI-2015: Healthy Eating Index-2015. [e] HETI: Healthy Trolley Index. [f] HPI: Healthy Purchase Index. * $p < 0.05$. ** $p < 0.01$. *** $p < 0.001$.

3.2. Discrimination Capacity: Comparisons of Nutritional Facts across Compliance Tertiles

Tables 6 and 7 show the assessment results of the indicators' *discrimination capacity*, differentiating *absolute* and *relative* nutritional intake, respectively. Regarding the *absolute* food category intake as shown in Table 6, all indicators were able to differentiate participants' intake of fruits to a certain degree. On the contrary, no indicator managed to differentiate participants' added sugar intake, which is not declared on products in Switzerland, but was only indirectly estimated using products' category affiliation and sugar content [62], in contrast to the other nutrients and food groups. In general, the FSA-NPS DI outperformed the other indicators in differentiating participants' *absolute* nutritional intake, as 17 out of the 36 statistical tests were significant ($p < 0.05$). It was the only indicator that was capable of differentiating participants' intake of sweets, salty snacks, sugar-sweetened beverages, alcohol, sodium and saturated fatty acids well. All these aspects are important health-influencing factors. When it comes to differentiating the *relative* food category intake, as shown in Table 7, all indicators, except the HPI, performed better compared to distinguishing *absolute* food intake, yielding a higher number of significant results. Moreover, the FSA-NPS DI outperformed the other indicators, demonstrating 19 significant results out of 36 statistical tests. HEI-2015 and HETI performed equally well when the nutritional intake was assessed on a *relative* basis, having the same number of significant results. However, the corresponding details were different. For instance, HEI-2015 was able to differentiate the *relative* sodium intake, while HETI was not. Note that when considering specific food categories or nutrients, the best performance might not always be achieved by the same indicator. On average, the five food shopping indicators discriminate fruits and dietary fibers robustly on both *absolute* or *relative* scales.

Table 6. Discrimination potential of relevant food shopping quality indicators to differentiate health-relevant individual daily nutritional intake [a] behavior (*absolute*, i.e., weight-based), N = 89.

Indicator	- FSA-NPS DI [b]	GPQI [c]	HEI-2015 [d]	HETI [e]	HPI [f]
Unit: portions/day					
Meat and meat products	●	○	◑	◑	○
Vegetables and salad	○	○	◔	○	○
Fruits	◔	◕	●	◕	◕
Wholegrain products	◕	○	◕	◕	◑
Sweets, salty snacks, sugar-sweetened beverages, alcohol	●	○	○	○	○
Unit: grams/day					
Sodium	●	○	○	○	○
Dietary fiber	◕	○	●	◕	◑
Saturated fatty acids	●	○	○	○	○
Added sugar	○	○	○	○	○
Points	**17**	2	11	10	5

[a] The products contained in each food category can be found on https://gitlab.ethz.ch/food-coach/shopping-index-comparison. [b] - FSA-NPS DI: Inverted Food Standards Agency Nutrient Profiling System Dietary Index. We inverted the original FSA-NPS DI scores to make them directly comparable to other food shopping quality indicators. The higher the inverted FSA-NPS DI, the healthier the food shopping. [c] GPQI: Grocery Purchase Quality Index-2016. [d] HEI-2015: Healthy Eating Index-2015. [e] HETI: Healthy Trolley Index. [f] HPI: Healthy Purchase Index. ○ No test was significant. ◔ The Kruskal–Wallis test showed significant differences between all three tertiles of the indicator score distribution ($p < 0.05$). ● The Mann–Whitney U test between the 2nd and the 3rd tertiles was significant ($p < 0.05$). ◑ The Mann–Whitney U test between the 1st and the 3rd tertiles was significant ($p < 0.05$). ◕ The Mann–Whitney U test between the 1st and the 2nd tertiles was significant ($p < 0.05$). ◔◕◑● Multiple above-mentioned statistical tests were significant ($p < 0.05$).

To confirm the ability of the FSA-NPS DI to distinguish users with different dietary habits, the characteristics of the highest, medium and lowest tertiles of the inverted FSA-NPS DI were assessed in Tables 8 and 9, which shows the *absolute* and *relative*, i.e., per 1000 kcal, food intake of the entire study sample and three tertiles as measured by the FSA-NPS DI. The inverted FSA-NPS DI was used to make it more comparable to other food shopping quality indicators. The higher the inverted FSA-NPS DI, the healthier the food shopping. T3 has the shopping baskets of the healthiest nutritional quality. The median and interquartile ranges (IQRs) were reported because of the non-normal distribution of the inverted FSA-NPS DI. On both *absolute* and *relative* scales, T1 consumed significantly more meat, less fruit and less dietary fiber, compared to T2 and T3.

Table 7. Discrimination potential of relevant food shopping quality indicators to differentiate health-relevant individual daily nutritional intake [a] behavior (*relative*, i.e., calorie-adjusted), N = 89.

Indicator	- FSA-NPS DI [b]	GPQI [c]	HEI-2015 [d]	HETI [e]	HPI [f]
Unit: portions/1000 kcal					
Meat and meat products	◐	○	◕	◐	○
Vegetables and salad	◑	○	◕	○	○
Fruits	◑	◐	◐	○	◑
Wholegrain products	◐	◕	◐	◕	○
Sweets, salty snacks, sugar-sweetened beverages, alcohol	○	○	○	○	◑
Unit: g/1000 kcal					
Sodium	◐	○	◑	○	○
Dietary fibers	◐	◑	◑	◕	○
Saturated fatty acids	◐	◑	○	◕	○
Added sugar	○	○	◑	◑	○
Points	**19**	7	12	12	3

[a] The products contained in each food category can be found on https://gitlab.ethz.ch/food-coach/shopping-index-comparison. [b] - FSA-NPS DI: Inverted Food Standards Agency Nutrient Profiling System Dietary Index. We inverted the original FSA-NPS DI scores to make them directly comparable to other food shopping quality indicators. The higher the inverted FSA-NPS DI, the healthier the food shopping. [c] GPQI: Grocery Purchase Quality Index-2016. [d] HEI-2015: Healthy Eating Index-2015. [e] HETI: Healthy Trolley Index. [f] HPI: Healthy Purchase Index. ○ No test was significant. ◐ The Kruskal–Wallis test showed significant differences among all three tertiles ($p < 0.05$). ◑ The Mann–Whitney U test between the 2nd and the 3rd tertiles was significant ($p < 0.05$). ◐ The Mann–Whitney U test between the 1st and the 3rd tertiles was significant ($p < 0.05$). ◑ The Mann–Whitney U test between the 1st and the 2nd tertiles was significant ($p < 0.05$). ◕ Multiple above-mentioned statistical tests were significant ($p < 0.05$).

Table 8. Median and interquartile range (IQR) of the *absolute* nutritional intake across the tertiles of the inverted Food Standards Agency Nutrient Profiling System Dietary Index (-FSA-NPS DI) [a].

	-FSA-NPS DI Score Tertile							
	Overall (N = 89)	T1 (N = 30)	T2 (N = 29)	T3(N = 30)				
	Median (IQR)	Median (IQR)	Median (IQR)	Median (IQR)	p	$p_{T1\text{-}T2}$	$p_{T1\text{-}T3}$	$p_{T2\text{-}T3}$
Unit: portions/day								
Meat and meat products	1.02 (1.22)	1.28 (0.95)	1.06 (1.15)	0.37 (1.07)	<0.001 ***	0.744	<0.001 ***	0.003 **
Vegetables and salad	2.30 (1.99)	1.96 (1.73)	2.16 (2.14)	2.51 (1.60)	0.135	0.128	0.064	0.722
Fruits	1.17 (1.28)	0.74 (1.20)	1.16 (1.44)	1.45 (1.15)	0.063	0.200	0.018 *	0.336
Wholegrain products	0.25 (0.45)	0.09 (0.49)	0.25 (0.47)	0.33 (0.32)	0.049 *	0.367	0.012 *	0.185
Sweets, salty snacks, sugar-sweetened beverages, alcohol	2.53 (2.24)	2.58 (1.62)	3.12 (2.29)	1.97 (1.83)	0.038 *	0.471	0.030 *	0.032 *
Unit: grams/day								
Sodium	1.87 (1.00)	1.91 (0.78)	2.03 (1.39)	1.45 (0.98)	0.017 *	0.529	0.022 *	0.011 *
Dietary fibers	22.20 (17.30)	17.35 (10.48)	19.90 (23.30)	31.00 (17.80)	0.018 *	0.084	0.006 **	0.262
Saturated fatty acids	31.70 (17.10)	34.95 (15.08)	36.30 (17.70)	27.90 (16.23)	0.022 *	0.970	0.011 *	0.028 *
Added sugar	8.01 (7.82)	7.69 (6.81)	9.34 (9.20)	6.49 (7.41)	0.444	0.897	0.304	0.252

[a] -FSA-NPS DI: Inverted Food Standards Agency Nutrient Profiling System Dietary Index. We inverted the original FSA-NPS DI scores to make them directly comparable to other food shopping quality indicators. The higher the inverted FSA-NPS DI, the healthier the food shopping. p The Kruskal–Wallis test was performed to compare all three tertiles. $p_{T1\text{-}T2}$, $p_{T1\text{-}T3}$, $p_{T2\text{-}T3}$ The Mann–Whitney U tests were performed to compare pairwise tertiles. * $p < 0.05$. ** $p < 0.01$. *** $p < 0.001$.

Table 9. Median and interquartile range (IQR) of the *relative* (i.e., per 1000 kcal of) nutritional intake across the tertiles of the inverted Food Standards Agency Nutrient Profiling System Dietary Index (-FSA-NPS DI) [a].

	-FSA-NPS DI tertile							
	Overall (N = 89)	T1 (N = 30)	T2 (N = 29)	T3 (N = 30)				
	Median (IQR)	Median (IQR)	Median (IQR)	Median (IQR)	p	$p_{T1\text{-}T2}$	$p_{T1\text{-}T3}$	$p_{T2\text{-}T3}$
Unit: portions/1000 kcal								
Meat and meat products	0.57 (0.57)	0.75 (0.35)	0.68 (0.55)	0.31 (0.54)	<0.001 ***	0.611	<0.001 ***	0.002 **
Vegetables and salad	1.19 (1.06)	1.02 (0.96)	1.25 (0.97)	1.50 (1.55)	0.040 *	0.120	0.014 *	0.321
Fruits	0.60 (0.63)	0.49 (0.63)	0.59 (0.59)	0.82 (0.67)	0.008 **	0.190	0.002 **	0.080
Wholegrain products	0.13 (0.22)	0.06 (0.15)	0.15 (0.20)	0.22 (0.24)	0.007 **	0.299	0.002 **	0.043
Sweets, salty snacks, sugar-sweetened beverages, alcohol	1.50 (0.89)	1.59 (0.86)	1.49 (0.88)	1.36 (0.91)	0.165	0.779	0.115	0.097
Unit: g/1000 kcal								
Sodium	1.07 (0.36)	1.13 (0.25)	1.20 (0.40)	0.93 (0.36)	0.003	0.897	0.003 **	0.004 **
Dietary fibers	13.23 (8.21)	10.52 (4.11)	13.73 (5.43)	18.78 (11.08)	<0.001 ***	0.057	<0.001	0.012 *
Saturated fat	19.54 (6.27)	20.61 (3.68)	19.61 (5.14)	17.01 (5.74)	0.002 **	0.190	<0.001 ***	0.020 **
Added sugar	4.72 (2.64)	4.42 (2.89)	4.72 (3.43)	4.82 (2.86)	0.851	0.767	0.631	0.688

[a] -FSA-NPS DI: Inverted Food Standards Agency Nutrient Profiling System Dietary Index. We inverted the original FSA-NPS DI scores to make them directly comparable to other food shopping quality indicators. The higher the inverted FSA-NPS DI, the healthier the food shopping. p The Kruskal–Wallis test was performed to compare all three tertiles. $p_{T1\text{-}T2}$, $p_{T1\text{-}T3}$, $p_{T2\text{-}T3}$ The Mann–Whitney U tests were performed to compare pairwise tertiles. * $p < 0.05$. ** $p < 0.01$. *** $p < 0.001$.

4. Discussion

4.1. Summary

This manuscript represents the first ever quantitative validation study on the *calibration* and *discrimination* ability of previously suggested food shopping quality indicators, including FSA-NPS DI, GPQI, HEI-2015, HETI and HPI. In total, digital receipts from two loyalty card systems and validated FFQs from N = 89 individuals in Switzerland were collected to validate five indicators that were previously developed and validated in separate regions in the world.

As shown in Tables 4–5, the assessed indicators correlate weakly to moderately with *absolute* and slightly more strongly with *relative* individual daily nutritional intake. The following reasons might explain why the correlations are only weak to moderate. First, there is a time lag between food shopping and eating. For instance, a person bought fruits on 31 January but consumed them in the first week of February. Second, inter-day and inter-week variances can be high and are commonly present. For example, a person can eat pizza on a certain day and skip lunch on the next day. A person can also fast for some time for different reasons. Physical activity also influences how much energy a person needs [66]. Third, only 69.6% of receipts were successfully matched. With higher product detection rates, the correlation of digital-receipt-based purchase indicators and individual dietary intakes might even be higher. Fourth, the relationship of food purchases and individual dietary intake are moderated by hard-to-assess behavioral factors, such as food waste, food processing, and out-of-home eating. Still, the observed correlations are in line with the literature and acceptable in comparison with bio-sampling (e.g., sodium excretion or blood sampling) and even 24 h recalls [67]. Although the purchase indicators have a lower accuracy than contemporary diet intake assessment methods, they are more cost-effective and scalable. Thus, they are still qualified as either stand-alone or complementary indicators for large-scale diet monitoring and long-term dietary behavior change interventions.

As shown in Tables 4–7, the performance of indicators was generally better in calibrating and discriminating the calorie-weighted *relative* individual dietary intake than the *absolute* individual dietary intake. This could be because when both using *relative* scales, the gap between individual calorie-weighted food intake and household expenditure- or calorie-weighted food shopping is narrower. The FSA-NPS DI has the best *calibration capacity* and *discrimination capacity*, followed by HEI-2015. The relatively inferior performance of GPQI, HETI and HPI might be due to the following reasons. First, calorie-weighted approaches likely perform better than expenditure-weighted approaches in the nutrition context. This is intuitive since inflation, prices and promotions moderate the amounts consumed from food products that are similar in their nutritional composition but different in price. Even differences in currency exchange rates moderate the performance of expenditure-weighted indicators, as the GPQI, HETI and HPI are designed and calibrated in the US, Australia and France respectively by evaluating shares of wallet (in USD or Australian dollars or in Euros). A transformation to other currency regions is likely to inhibit the performance of such expenditure-based indicators. Hence, weight-based or calorie-based weighted indicators are likely to correlate better with individual food intake, as they measure the representation of food choices better from a dietary perspective. Therefore, if available, data on consumed products' respective weights and their calorific impact should be taken into account when designing purchase quality indicators.

The study does not give reason to believe that the geographic context in which an indicator was designed limits its performance abroad. While the FSA-NPS DI was developed in the UK, the GPQI and HEI-2015 were defined in the US, HETI was defined in Australia, and the HPI was defined in France. Since this validation study was conducted in Switzerland, it might be surprising that although Switzerland is closer to France, where the HPI was developed, than to the US, where the HEI-2015 was developed, HEI-2015 performed better than HPI in calibrating and discriminating the dietary intake of a Swiss study sample. In contrast, the FSA-NPS DI, which was developed in the UK, performed best. Hence, no structural dominance of European indicators could be observed. These results therefore indicate that food shopping quality indicators are indeed transferable to a new region, such as Switzerland and can, in fact, give a relatively reliable indication of the nutritional quality of individual diets by assessing their digital receipts. The FSA-NPS DI consistently outperformed the other food shopping quality indicators in calibrating and differentiating participants' nutritional intake, on both *absolute* and *relative* density-based scales. Hence, a validation of the FSA-NPS DI within digital receipts outside Europe could be an interesting foray for future research.

The results suggest that considering both food groups and nutrients, particularly fruits and dietary fibers, might be important in the design of food shopping quality indicators. The GPQI, HETI and HPI do not include any nutrient categories in their definitions. On the contrary, the HEI-2015 explicitly includes fatty acids, sodium, added sugars and saturated fats, and the FSA-NPS DI includes sodium, saturated fats, sugar and dietary fiber. These two indicators have better performance in calibrating and discriminating actual dietary intake. As shown in Tables 4–7, dietary fiber can generally be calibrated and discriminated the best by the food shopping indicators. Therefore, dietary fiber seems to be an important factor to consider in designing and choosing relevant purchase indicators. In terms of relevant food groups, all indicators can calibrate and discriminate fruits well. These results support the important roles of fruits in a healthy diet and in the design of food shopping indicators. Th European Commission also recommends its member states to track their consumers' intake of fruits within their European Core Health Indicators (ECHIM) list of health-relevant policy indicators [68].

4.2. Contribution

This study has multiple contributions to research and practice. First, this study represents the first validation of previously suggested food shopping quality indicators from different regions in the world within a thorough comparative quantitative assessment. Second, the consistent out-performance of the FSA-NPS DI in nutritional intake calibration and discrimination demonstrates that it could be a good choice for general purpose of use, particularly when developing tools for long-term diet monitoring and intervention. Third, the study found that fruits and dietary fibers are better calibrated and discriminated by food shopping indicators. This suggests the important roles of fruit and dietary fibers in a healthy diet. It might be useful to consider these two aspects when designing a food shopping indicators. Added sugar intake was not well captured by the assessed food shopping quality indicators. This finding suggests that there is still room for the development of different purpose-specific food shopping quality indicators (e.g., estimating the risk for diabetes type two by estimating intake levels of added sugar and carbohydrate quality). Fourth, researchers, practitioners and designers for food choice monitoring systems and behavior change interventions should take away that these indicators capture the calorie-adjusted *relative* nutritional intake better than *absolute* dietary intake. Finally, the study demonstrates that regulatory mandates such as the GDPR [40] and the mandate for nutritional declaration for food products sold online (EU)1169 [41] can pave the way for novel tools for scalable, non-invasive monitoring of food choices and tailored digital behavior change interventions, using digital receipts and food composition databases.

4.3. Limitations

This study possesses limitations. We select for discussion the ones which warrant special attention. First, the number of participants restricts the generalizability of the results. Compared with the actual composition of the Swiss population [69,70], the female participation ratio, the average age, and the mean BMI of the present sample were lower. Multiple factors could have led to the sample not being representative. First, most users on the BAM platform are male. Second, participants were mainly health-conscious individuals in or beyond tertiary education, considering where our advertisements were placed (e.g., university shuttle buses and billboards). In addition, the study is biased toward loyalty card holders of the leading two retailers in Switzerland since they are the only group that is able to participate in the study. Compared to non-card-holders, they might have slightly different eating or food shopping habits.

Regarding the FFQ, although the FFQ was validated previously in Switzerland, the tool has certain limitations, such as potentially missing food items in the questionnaire or the fact that vitamin and mineral supplements, which can be very important for vegetarians, are not covered in the FFQ. All these might lead to reporting biases, which are commonly present [71].

The five food shopping quality indicators were calculated using *relative* expenditure share or portion share per 1000 kcal. This implicitly assumed that each individual consumes the same proportion of food across all food categories, regardless of gender and age. This assumption could be biased. For instance, if a participant is a vegan but he or she lives with meat eaters, we assume that this participant consumes meat as well if there is meat in the receipts. Moreover, the food waste proportion in different food categories tends to be different. As the goal is to check if digital receipts are able to provide indications about dietary behavior, but not to predict actual food intake, these problems are rather mild for now.

Furthermore, inaccurate or missing information in the food composition database forms another limitation of the study. The manual matching process between article identifiers in digital receipts and their corresponding GTINs is challenging and can sometimes lead to ambiguous results. To mitigate this limitation, we invested a significant effort in the food composition database and manually mapped the most frequent 5950 products' article identifiers, which is far beyond the top 3000 products mapped and identified in similar studies.

Lastly, the amount of added sugar captured by the digital receipts seems low compared to the sugar content, which might be due to different food category definitions. As the declaration of added sugar is not mandated, we used an estimation based on sugar content and category affiliation of shopped products [62]. While the food categorization frameworks between studies might differ and were defined in regions other than Switzerland, this effect might have caused an underestimation of added sugar in this study. Coherent category definitions seem pivotal, as we found that the performance results of the GPQI and HPI indicators are very sensitive to changes in the definition of refined grains, for example.

4.4. Future Work

First and foremost, it is essential for the future of this and related studies to recruit a larger and more representative sample. In addition, data quality of digital receipts and food composition databases is another important factor that constructs gaps in research and practice. The authors have been and will continue being dedicated to increasing the matching ratio of digital receipts. To overcome the limitations of the FFQ, bio-sampling (e.g., sodium excretion or blood sampling) could be added to offset the potential of reporting biases, but it means higher costs as well. To capture information that is not in receipts, conducting a survey about other aspects of food behavior, such as food waste, might be necessary. To ensure the reproducibility of similar studies in the future, researchers should adhere to established food classification schema, such as FoodEx2 (revision 2) [72]. Although FoodEx2 is only defined for the European context, such classification schema can be very helpful in transferring concepts to other regions as well. Ultimately, a global food classification schema would be helpful to reproduce the suggested food shopping quality indicators internationally.

To support the extensive efforts required for the correct mapping of digital receipts and food composition databases, machine-learning based algorithms [73] can support the (semi-)automatic correction of errors in food composition databases [74] or the correct identification of products present in digital receipts (e.g., word2vec) [75,76].

This study also calls for the integration of product identifiers into digital receipt standards. Currently, retailers are not required to integrate relevant product identifiers, such as the GTIN, into their digital receipt structures. Therefore, the identification of text-based product identifiers within food product composition databases requires lots of manual effort. It would, therefore, be beneficial for the development of scalable digital receipt-based food choice monitoring and interventions if regulators would extend the right for data portability as mandated by the GDPR [40] and mandate the use of product identifiers within digital receipt standards.

In addition to the FSA-NPS DI, future research could assess the potential of superior food shopping indicators for estimating or even predicting the nutritional intake and health states of consumers. Finally, future regulatory mandates, such as the expected GDPR 2.0 in the European Union, could enforce the use of standardized, real-time APIs to further ease the barriers for data portability of personal datasets including shopping records.

5. Conclusions

In this study, we present the first comparison of the calibration and the discrimination capacities of shopping indicators FSA-NPS DI, GPQI, HEI-2015, HETI and HPI on a dataset that was obtained from 89 participants and included responses to an FFQ and real grocery shopping data using digital receipts collected from two loyalty card systems in Switzerland. Our results show that, overall, the surveyed indicators correlate only weakly to moderately with *absolute* and slightly more strongly with *relative* individual daily nutritional intake. All indicators are weakly to moderately suitable to differentiate health-relevant daily nutritional intake behavior using digital receipts. Among these indicators, the FSA-NPS DI in general outperforms the other indicators in our dataset, having the best calibration and discrimination ability, thus contributing an empirically validated guideline regarding the selection of shopping indicators. This is counterintuitive, as the FSA-NPS DI was primarily designed to discriminate the basket quality, but not the compliance to certain dietary guidelines nor to certain food group intake. The relatively inferior calibration performance of the GPQI, HETI and HPI might be because they were designed and validated by evaluating shares of wallet (in Australian dollars or in euros) in the US, Australia and France, rather than assessing nutrients or food groups by their weight or calorie contribution. This might make it inappropriate to translate them to other regions, as the prices and retail market dynamics in the US, Australia and France might differ from those in other countries. Adopting these indicators might require a re-calibration within other regions. In the future, we plan to enlarge the sample size, improve the data quality, and rerun the analysis using machine learning frameworks to estimate individual dietary intake deficits from digital receipts more accurately. In addition, we plan a further investigation into the reasons behind the differences in the calibration and discrimination capacities of the surveyed indicators in the future.

Author Contributions: Conceptualization, K.F., J.W. and J.L.; methodology, J.W., K.F., J.L.; formal analysis, J.W.; software, J.L., M.L.H.; resources, K.F., E.F., C.B., R.G.; data curation, K.F.; writing—original draft preparation, J.W.; writing—review and editing, all authors; supervision, S.M., T.S., E.F.; project administration, K.F.; funding acquisition, K.F., J.B., S.M., T.S. All authors have read and agreed to the published version of the manuscript.

Funding: This research was funded by the Swiss National Science Foundation under project grant number 188402 (see http://p3.snf.ch/project-188402, accessed on 15 December 2021). The Korean collaborator of this study was financed by the National Research Foundation of Korea (NRF) under project grant number 2019K1A3A1A14012376.

Institutional Review Board Statement: The study was conducted according to the guidelines of the Declaration of Helsinki, and approved by the Institutional Review Board (or Ethics Committee) of the Ethics Commission office, Stampfenbachstrasse 52/56, STE H 26, 8092 Zürich, of ETH Zurich (protocol code 2019-N-134, approved on 15 October 2019).

Informed Consent Statement: Informed consent was obtained from all subjects involved in the study.

Data Availability Statement: The dataset used in this study, including the anonymized FFQ and anonymized digital receipt datasets, can be found here: https://gitlab.ethz.ch/food-coach/shopping-index-comparison , accessed on 15 December 2021.

Acknowledgments: First, we would like to thank BitsaboutMe AG, Bollwerk 4, 3011 Bern, Switzerland, for providing the digital receipt infrastructure for this study. Second, we also would like to thank former Bachelor and Master students for supporting the data collection of product data used in this study: Lucia Caiata, Sebastian von der Mark, Sven Brunner, and Tristan Breyer. Third, we would like to thank Zurich University of Applied Science (ZHAW) for providing the web-mediated food frequency questionnaire for the study. Fourth, we would like to thank the food composition database operators GS1 trustbox, Codecheck and Openfoodfacts for offering their product datasets for research purposes. Fifth, we would like to thank the European parliament for the introduction of the GDPR, which significantly reduces the effort in collecting digital receipts for researchers and practitioners, and allows for the development of novel data-driven tools for scalable monitoring of food choices via loyalty cards. Last but not least, we would like to thank all study participants for the donation of their FFQ and digital receipt data, without which this study would not have been possible.

Conflicts of Interest: The authors declare no conflict of interest.

Sample Availability: https://gitlab.ethz.ch/food-coach/shopping-index-comparison.

Abbreviations

The following abbreviations are used in this manuscript:

API	Application Programming Interface
BAM	BitsaboutMe
B2C	Business2Consumer
CHF	Swiss franc
EFSA	European Food Safety Authority
ETH Zurich	Swiss Federal Institute of Technology in Zurich

FFQ	Food Frequency Questionnaire
FSA-NPS DI	Food Standards Agency Nutrient Profiling System Dietary Index
GDPR	General Data Protection Regulation
GPQI	Grocery Purchase Quality Index-2016
GTIN	Global Trade Item Number
HEI-2010	Healthy Eating Index-2010
HEI-2015	Healthy Eating Index-2015
HETI	Healthy Trolley Index
HPI	Healthy Purchase Index
kcal	kilocalories
kJ	kilojoules
UK	United Kingdom
US	United States
USD	United States dollar

References

1. Saeedi, P.; Petersohn, I.; Salpea, P.; Malanda, B.; Karuranga, S.; Unwin, N.; Colagiuri, S.; Guariguata, L.; Motala, A.A.; Ogurtsova, K.; et al. Global and regional diabetes prevalence estimates for 2019 and projections for 2030 and 2045: Results from the International Diabetes Federation Diabetes Atlas. *Diabetes Res. Clin. Pract.* **2019**, *157*, 107843. [CrossRef]

2. Hales, C.M.; Fryar, C.D.; Carroll, M.D.; Freedman, D.S.; Ogden, C.L. Trends in obesity and severe obesity prevalence in US youth and adults by sex and age, 2007–2008 to 2015–2016. *JAMA* **2018**, *319*, 1723–1725. [CrossRef]

3. Roy, A.; Praveen, P.A.; Amarchand, R.; Ramakrishnan, L.; Gupta, R.; Kondal, D.; Singh, K.; Sharma, M.; Shukla, D.K.; Tandon, N.; et al. Changes in hypertension prevalence, awareness, treatment and control rates over 20 years in National Capital Region of India: Results from a repeat cross-sectional study. *BMJ Open* **2017**, *7*, e015639. [CrossRef]

4. Popkin, B.M.; Corvalan, C.; Grummer-Strawn, L.M. Dynamics of the double burden of malnutrition and the changing nutrition reality. *Lancet* **2020**, *395*, 65–74. [CrossRef]

5. Hawkes, C.; Harris, J.; Gillespie, S. Changing diets: Urbanization and the nutrition transition. In *IFPRI Book Chapters*; International Food Policy Research Institute (IFPRI): Washington, DC, USA, 2017; pp. 34–41. doi:10.2499/9780896292529_04. [CrossRef]

6. Zhen, S.; Ma, Y.; Zhao, Z.; Yang, X.; Wen, D. Dietary pattern is associated with obesity in Chinese children and adolescents: Data from China Health and Nutrition Survey (CHNS). *Nutr. J.* **2018**, *17*, 1–9. [CrossRef]

7. Shi, Z.; Taylor, A.W.; Riley, M.; Byles, J.; Liu, J.; Noakes, M. Association between dietary patterns, cadmium intake and chronic kidney disease among adults. *Clin. Nutr.* **2018**, *37*, 276–284. [CrossRef]

8. Riddle, M.C.; Herman, W.H. The cost of diabetes care—An elephant in the room. *Diabetes Care* **2018**, *41*, 929–932. [CrossRef]

9. Fuchs, K.; Haldimann, M.; Vuckovac, D.; Ilic, A. Automation of Data Collection Techniques for Recording Food Intake: A Review of Publicly Available and Well-Adopted Diet Apps. In Proceedings of the 9th International Conference on Information and Communication Technology Convergence, ICT Convergence Powered by Smart Intelligence, Jeju, Korea, 17–19 October 2018; pp. 58–65. doi:10.1109/ICTC.2018.8539468. [CrossRef]

10. Fallaize, R.; Forster, H.; Macready, A.L.; Walsh, M.C.; Mathers, J.C.; Brennan, L.; Gibney, E.R.; Gibney, M.J.; Lovegrove, J.A. Online dietary intake estimation: Reproducibility and validity of the Food4Me food frequency questionnaire against a 4-day weighed food record. *J. Med. Internet Res.* **2014**, *16*, e190. [CrossRef]

11. Geelen, A.; Souverein, O.W.; Busstra, M.C.; de Vries, J.H.; van't Veer, P. Comparison of approaches to correct intake—Health associations for FFQ measurement error using a duplicate recovery biomarker and a duplicate 24 h dietary recall as reference method. *Public Health Nutr.* **2015**, *18*, 226–233. [CrossRef]

12. Denova-Gutiérrez, E.; Ramírez-Silva, I.; Rodríguez-Ramírez, S.; Jiménez-Aguilar, A.; Shamah-Levy, T.; Rivera-Dommarco, J.A. Validity of a food frequency questionnaire to assess food intake in Mexican adolescent and adult population. *Salud Pública De México* **2016**, *58*, 617–628. [CrossRef]

13. Carlsen, M.H.; Lillegaard, I.T.; Karlsen, A.; Blomhoff, R.; Drevon, C.A.; Andersen, L.F. Evaluation of energy and dietary intake estimates from a food frequency questionnaire using independent energy expenditure measurement and weighed food records. *Nutr. J.* **2010**, *9*, 1–9. doi:10.1186/1475-2891-9-37. [CrossRef]

14. El Kinany, K.; Garcia-Larsen, V.; Khalis, M.; Deoula, M.M.S.; Benslimane, A.; Ibrahim, A.; Benjelloun, M.C.; El Rhazi, K. Adaptation and validation of a food frequency questionnaire (FFQ) to assess dietary intake in Moroccan adults. *Nutr. J.* **2018**, *17*, 1–12. [CrossRef] [PubMed]

15. Noor Hafizah, Y.; Ang, L.C.; Yap, F.; Nurul Najwa, W.; Cheah, W.L.; Ruzita, A.T.; Jumuddin, F.A.; Koh, D.; Lee, J.A.C.; Essau, C.A.; et al. Validity and reliability of a food frequency questionnaire (FFQ) to assess dietary intake of preschool children. *Int. J. Environ. Res. Public Health* **2019**, *16*, 4722. [CrossRef] [PubMed]

16. Lovell, A.; Bulloch, R.; Wall, C.R.; Grant, C.C. Quality of food-frequency questionnaire validation studies in the dietary assessment of children aged 12 to 36 months: A systematic literature review. *J. Nutr. Sci.* **2017**, *6*, e16, doi:10.1017/jns.2017.12. [CrossRef] [PubMed]

17. European Food Safety Authority. Guidance on the EU Menu Methodology. Available online: https://www.efsa.europa.eu/it/efsajournal/pub/3944 (accessed on 15 November 2021).

18. König, L.M.; Sproesser, G.; Schupp, H.T.; Renner, B. Describing the process of adopting nutrition and fitness apps: Behavior stage model approach. *J. Med. Internet Res.* **2018**, *20*, doi:10.2196/mhealth.8261. [CrossRef]

19. Tin, S.T.; Mhurchu, C.N.; Bullen, C. Supermarket sales data: Feasibility and applicability in population food and nutrition monitoring. *Nutr. Rev.* **2007**, *65*, 20–30. doi:10.1301/nr.2007.jan.20. [CrossRef] [PubMed]

20. Tran, L.; Brewster, P.; Chidambaram, V.; Hurdle, J. An innovative method for monitoring food quality and the healthfulness of consumers' grocery purchases. *Nutrients* **2017**, *9*, 457, doi:10.3390/nu9050457. [CrossRef]

21. Varghese, S. The UK's 11 Billion Yearly Receipts Are an Environmental Nightmare, Technical Report, Wired, UK, 6 November 2018. Available online: https://www.wired.co.uk/article/receipt-recycling-uk-thermal-paper-digital-receipt (accessed on 24 November 2021)

22. Caicedo, O. eReceipts 101: The What, Why, Where and How of B2C Tax Compliance, Solve Tax for Good, 2018. Available online: https://sovos.com/blog/vat/ereceipts-b2c-compliance/ (accessed on 24 November 2021)

23. Ernst and Young. Tax Administration Is Going Digital, Technical Report, EY, 2016. Available online: https://www.ey.com/Publication/vwLUAssets/EY-tax-administration-is-going-digital/ (accessed on 24 November 2021)

24. Johnson, M. The Big Picture On Digital Receipts, Technical Report, Techcrunch, 2014. Available online: https://techcrunch.com/2014/09/19/the-big-picture-on-digital-receipts/?guccounter=1&guce_referrer=aHR0cHM6Ly93d3cuZ29vZ2xlLmNvbS8&guce_referrer_sig=AQAAAFTuzjMV55oZaS3adEq-N6YLjl1qQGorvezsUvAYZlGaHUE7RSiRTY9sXQwAEfpUt8iixN6-MaGI_hJb7iTga3uqFnv-COLHdBna2CyGdW_dE1UBiHwOFhwc3A2vEcymJZtELtsxPzHiJkMf3P0Kel1okl-IaBb-yVQitzLbhCcM (accessed on 24 November 2021)

25. Trichopoulou, A.; Naska, A.; Group, D.I. European food availability databank based on household budget surveys: The Data Food Networking initiative. *Eur. J. Public Health* **2003**, *13*, 24–28. [CrossRef]

26. Orfanos, P.; Naska, A.; Rodrigues, S.; Lopes, C.; Freisling, H.; Rohrmann, S.; Sieri, S.; Elmadfa, I.; Lachat, C.; Gedrich, K.; et al. Eating at restaurants, at work or at home. Is there a difference? A study among adults of 11 European countries in the context of the HECTOR* project. *Eur. J. Clin. Nutr.* **2017**, *71*, 407–419. [CrossRef] [PubMed]

27. Chappuis, A.; Bochud, M.; Glatz, N.; Vuistiner, P.; Paccaud, F.; Burnier, M. *Swiss Survey on Salt Intake: Main Results*; Service de N´ephrologie et Institut Universitaire de M´edecine Sociale et Pr´eventive Centre Hospitalier Universitaire Vaudois (CHUV): Lausanne, Switzerland, 2011; pp. 1–32.

28. Brewster, P.J.; Guenther, P.M.; Jordan, K.C.; Hurdle, J.F. The Grocery Purchase Quality Index-2016: An innovative approach to assessing grocery food purchases. *J. Food Compos. Anal.* **2017**, *64*, 119–126. [CrossRef]

29. Taylor, A.; Wilson, F.; Hendrie, G.A.; Allman-Farinelli, M.; Noakes, M. Feasibility of a Healthy Trolley Index to assess dietary quality of the household food supply. *Br. J. Nutr.* **2015**, *114*, 2129–2137. [CrossRef]

30. Tang, W.; Aggarwal, A.; Liu, Z.; Acheson, M.; Rehm, C.D.; Moudon, A.V.; Drewnowski, A. Validating self-reported food expenditures against food store and eating-out receipts. *Eur. J. Clin. Nutr.* **2016**, *70*, 352–357. doi:10.1038/ejcn.2015.166. [CrossRef] [PubMed]

31. Ransley, J.; Donnelly, J.; Botham, H.; Khara, T.; Greenwood, D.; Cade, J. Use of supermarket receipts to estimate energy and fat content of food purchased by lean and overweight families. *Appetite* **2003**, *41*, 141–148. doi:10.1016/S0195-6663(03)00051-5. [CrossRef]

32. Nelson, M. Family food purchases and home food consumption: Comparison of nutrient contents. *Br. J. Nutr.* **1985**, *54*, 373–387. [CrossRef]

33. Martin, S.L.; Howell, T.; Duan, Y.; Walters, M. The feasibility and utility of grocery receipt analyses for dietary assessment. *Nutr. J.* **2006**, *5*, 6–12. doi:10.1186/1475-2891-5-10. [CrossRef]

34. Ransley, J.; Donnelly, J.; Khara, T.; Botham, H.; Arnot, H.; Greenwood, D.; Cade, J. The use of supermarket till receipts to determine the fat and energy intake in a UK population. *Public Health Nutr.* **2001**, *4*, doi:10.1079/PHN2001171. [CrossRef]

35. Raynor, H.A.; Polley, B.A.; Wing, R.R.; Jeffery, R.W. Is dietary fat intake related to liking or household availability of high- and low-fat foods? *Obes. Res.* **2004**, *12*, 816–823. doi:10.1038/oby.2004.98. [CrossRef]

36. Fulkerson, J.A.; Nelson, M.C.; Lytle, L.; Moe, S.; Heitzler, C.; Pasch, K.E. The validation of a home food inventory. *Int. J. Behav. Nutr. Phys. Act.* **2008**, *5*, 1–10. doi:10.1186/1479-5868-5-55. [CrossRef] [PubMed]

37. Eyles, H.; Jiang, Y.; Ni Mhurchu, C. Use of Household Supermarket Sales Data to Estimate Nutrient Intakes: A Comparison with Repeat 24-h Dietary Recalls. *J. Am. Diet. Assoc.* **2010**, *110*, 106–110. doi:10.1016/j.jada.2009.10.005. [CrossRef] [PubMed]

38. Ni Mhurchu, C.; Blakely, T.; Wall, J.; Rodgers, A.; Jiang, Y.; Wilton, J. Strategies to promote healthier food purchases: A pilot supermarket intervention study. *Public Health Nutr.* **2007**, *10*, 608–615. doi:10.1017/S136898000735249X. [CrossRef] [PubMed]

39. Ni Mhurchu, C.; Blakely, T.; Jiang, Y.; Eyles, H.C.; Rodgers, A. Effects of price discounts and tailored nutrition education on supermarket purchases: A randomized controlled trial. *Am. J. Clin. Nutr.* **2010**, *91*, 736–747. doi:10.3945/ajcn.2009.28742. [CrossRef]

40. The European Parliament and the Council of European Union. Regulation (EU) 2016/679. Available online: https://eur-lex.europa.eu/eli/reg/2016/679/oj (accessed on 17 June 2021).

41. Parliament, T.E.; The Council of the European Union. Regulation (EU) 1169/2011, 2011. Available online: https://eur-lex.europa.eu/LexUriServ/LexUriServ.do?uri=OJ:L:2011:304:0018:0063:en:PDF (accessed on 24 November 2021)

42. Tharrey, M.; Dubois, C.; Maillot, M.; Vieux, F.; Méjean, C.; Perignon, M.; Darmon, N. Development of the Healthy Purchase Index (HPI): A scoring system to assess the nutritional quality of household food purchases. *Public Health Nutr.* **2019**, *22*, 765–775. [CrossRef] [PubMed]

43. Appelhans, B.M.; French, S.A.; Tangney, C.C.; Powell, L.M.; Wang, Y. To what extent do food purchases reflect shoppers' diet quality and nutrient intake? *Int. J. Behav. Nutr. Phys. Act.* **2017**, *14*, 46. [CrossRef]

44. Krebs-Smith, S.M.; Pannucci, T.E.; Subar, A.F.; Kirkpatrick, S.I.; Lerman, J.L.; Tooze, J.A.; Wilson, M.M.; Reedy, J. Update of the healthy eating index: HEI-2015. *J. Acad. Nutr. Diet.* **2018**, *118*, 1591–1602. [CrossRef] [PubMed]

45. Julia, C.; Méjean, C.; Touvier, M.; Péneau, S.; Lassale, C.; Ducrot, P.; Hercberg, S.; Kesse-Guyot, E. Validation of the FSA nutrient profiling system dietary index in French adults—Findings from SUVIMAX study. *Eur. J. Nutr.* **2016**, *55*, 1901–1910. [CrossRef] [PubMed]

46. Collins, G.S.; de Groot, J.A.; Dutton, S.; Omar, O.; Shanyinde, M.; Tajar, A.; Voysey, M.; Wharton, R.; Yu, L.M.; Moons, K.G.; et al. External validation of multivariable prediction models: A systematic review of methodological conduct and reporting. *BMC Med. Res. Methodol.* **2014**, *14*, 1–11. [CrossRef] [PubMed]

47. Steyerberg, E.W.; Vickers, A.J.; Cook, N.R.; Gerds, T.; Gonen, M.; Obuchowski, N.; Pencina, M.J.; Kattan, M.W. Assessing the performance of prediction models: A framework for some traditional and novel measures. *Epidemiology* **2010**, *21*, 128. [CrossRef] [PubMed]

48. Swiss Confederation. Retail. Available online: https://www.eda.admin.ch/aboutswitzerland/en/home/wirtschaft/taetigkeitsgebiete/detailhandel.html (accessed on 23 November 2021).

49. Statista. Market Share of the Leading Companies in Retail in Switzerland in 2018. Available online: https://www.statista.com/statistics/787298/switzerland-market-share-of-food-retailers/ (accessed on 24 November 2021).

50. Schmid, C. Kundenkarten: Zweifel am Nutzen. Handelszeitung. 2004. Available online: https://www.handelszeitung.ch/unternehmen/kundenkarten-zweifel-am-nutzen (accessed on 1 November 2021)

51. Brinkerhoff, K.M.; Brewster, P.J.; Clark, E.B.; Jordan, K.C.; Cummins, M.R.; Hurdle, J.F. Linking supermarket sales data to nutritional information: An informatics feasibility study. *AMIA* **2011**, *2011*, 598–606.

52. Fuchs, K.; Zeltner, M.; Haldimann, M.; Ilic, A. Icon-based Digital Food Allergen Labels for Complementation of Text-based Declaration. In Proceedings of the 28th European Conference on Information Systems (ECIS 2020), Marrakech, Morocco, 15–17 June 2020.

53. Fuchs, K.; Barattin, T.; Haldimann, M.; Ilic, A. Towards Tailoring Digital Food Labels: Insights of a Smart-RCT on User-specific Interpretation of Food Composition Data (in submission). In Proceedings of the CHItaly19 (The 13th Biannual Conference of the Italian SIGCHI Chapter), Adova, Italy, 23–25 September 2019, doi:10.1145/3347448.3357171. [CrossRef]

54. Chantal, J.; Hercberg, S.; WHO. Development of a new front-of-pack nutrition label in France: The five-colour Nutri-Score. *Public Health Panor.* **2017**, *3*, 712–725.

55. Steinemann, N.; Grize, L.; Ziesemer, K.; Kauf, P.; Probst-Hensch, N.; Brombach, C. Relative validation of a food frequency questionnaire to estimate food intake in an adult population. *Food Nutr. Res.* **2017**, *61*, 1305193. [CrossRef]

56. The National Health Service. Understanding Calories, 2019. Available online: https://www.nhs.uk/live-well/healthy-weight/managing-your-weight/understanding-calories/ (accessed on 1 November 2021)

57. Jensen, O.M.; Wahrendorf, J.; Rosenqvist, A.; Geser, A. The reliability of questionnaire-derived historical dietary information and temporal stability of food habits in individuals. *Am. J. Epidemiol.* **1984**, *120*, 281–290. [CrossRef]

58. Swiss Society for Nutrition. Swiss Food Pyramid. Available online: http://www.sge-ssn.ch/ich-und-du/essen-und-trinken/ausgewogen/schweizer-lebensmittelpyramide/ (accessed on 15 December 2021).

59. Bochud, M.; Chatelan, A.; Blanco, J.M.; Beer-Borst, S. *Anthropometric Characteristics and Indicators of Eating and Physical Activity Behaviors in the Swiss Adult Population*; Technical Report; The Federal Office of Public Health and the Food Safety and Veterinary Office: Bern, Switzerland, 2014.

60. Bochud, M.; Chatelan, A.; Blanco, J.M.; Beer-Borst, S.M. *Anthropometric Characteristics and Indicators of Eating and Physical Activity Behaviors in the Swiss Adult Population: Results from menuCH 2014–2015*; Technical Report; The Federal Office of Public Health and the Food Safety and Veterinary Office: Bern, Switzerland, 2017.

61. Hagmann, D.; Siegrist, M.; Hartmann, C. Meat avoidance: Motives, alternative proteins and diet quality in a sample of Swiss consumers. *Public Health Nutr.* **2019**, *22*, 2448–2459. [CrossRef]

62. Louie, J.C.Y.; Moshtaghian, H.; Boylan, S.; Flood, V.M.; Rangan, A.; Barclay, A.; Brand-Miller, J.; Gill, T. A systematic methodology to estimate added sugar content of foods. *Eur. J. Clin. Nutr.* **2015**, *69*, 154–161. [CrossRef] [PubMed]

63. USDA. How Much Fiber Is in Whole Grains versus Refined Grains/Non Whole Grains? Available online: https://ask.usda.gov/s/article/How-much-fiber-is-in-whole-grains-versus-refined-grains-non-whole-grains (accessed on 20 June 2021).

64. FDA. Health Claim Notification for Whole Grain Foods. Available online: https://www.fda.gov/food/food-labeling-nutrition/health-claim-notification-whole-grain-foods (accessed on 20 June 2021).

65. USDA. Nutrients: Total Dietary Fiber (g). Available online: https://www.nal.usda.gov/sites/www.nal.usda.gov/files/total_dietary_fiber.pdf (accessed on 20 June 2021).

66. Kumanyika, S.K.; Obarzanek, E.; Stettler, N.; Bell, R.; Field, A.E.; Fortmann, S.P.; Franklin, B.A.; Gillman, M.W.; Lewis, C.E.; Poston, W.C.; et al. Population-based prevention of obesity: The need for comprehensive promotion of healthful eating, physical activity, and energy balance: A scientific statement from American Heart Association Council on Epidemiology and Prevention,

Interdisciplinary Committee for Prevention (formerly the expert panel on population and prevention science). *Circulation* **2008**, *118*, 428–464. [PubMed]

67. Meyer, H.E.; Johansson, L.; Eggen, A.E.; Johansen, H.; Holvik, K. Sodium and potassium intake assessed by spot and 24-h urine in the population-based Tromsø Study 2015–2016. *Nutrients* **2019**, *11*, 1619. [CrossRef] [PubMed]
68. Verschuuren, M.; Gissler, M.; Kilpeläinen, K.; Tuomi-Nikula, A.; Sihvonen, A.P.; Thelen, J.; Gaidelyte, R.; Ghirini, S.; Kirsch, N.; Prochorskas, R.; et al. Public health indicators for the EU: The joint action for ECHIM (European Community Health Indicators & Monitoring). *Arch. Public Health* **2013**, *71*, 1–7.
69. Federal Statistical Office. Population. Available online: https://www.bfs.admin.ch/bfs/en/home/statistics/population/effectif-change/population.html (accessed on 17 June 2021).
70. Federal Statistical Office. Overweight-Share of the Population Aged 15 and over Who Are Overweight (BMI of 25 or More)—In percent (2019). Available online: https://www.bfs.admin.ch/bfs/en/home/statistics/sustainable-development/monet-2030/indikatoren/overweight.assetdetail.10187695.html (accessed on 17 June 2021).
71. Sluik, D.; Geelen, A.; de Vries, J.H.; Eussen, S.J.; Brants, H.A.; Meijboom, S.; van Dongen, M.C.; Bueno-de Mesquita, H.B.; Wijckmans-Duysens, N.E.; van't Veer, P.; et al. A national FFQ for the Netherlands (the FFQ-NL 1.0): Validation of a comprehensive FFQ for adults. *Br. J. Nutr.* **2016**, *116*, 913–923. [CrossRef]
72. European Food Safety Authority. *The Food Classification and Description System FoodEx 2 (Revision 2)*; EFSA Supporting Publications: Parma, Italy, 2015.
73. Muenzberg, A.; Sauer, J.; Hein, A.; Roesch, N. Machine Learning and Context-Based Approaches to Get Quality Improved Food Data. In *Proceedings of Sixth International Congress on Information and Communication Technology*; Yang, X.S., Sherratt, S., Dey, N., Joshi, A., Eds.; Springer: Singapore, 2022; pp. 423–435.
74. Presser, K.; Hinterberger, H.; Weber, D.; Norrie, M. A scope classification of data quality requirements for food composition data. *Food Chem.* **2016**, *193*, 166–172. doi:doi:10.1016/j.foodchem.2015.03.084. [CrossRef]
75. Kweeri. Everything You Need to Know from Receipts, 2021. Available online: https://kweeri.io/ (accessed on 7 November 2021)
76. Maslova, O.; Da Cruz, A. Product labels classification from Receipt: Word2vec and CNN Approach, 2019. Available online: https://blog.aboutgoods-company.com/receipt-labels-classification-word2vec-and-cnn-approach/ (accessed on 1 November 2021)

MDPI

St. Alban-Anlage 66

4052 Basel

Switzerland

Tel. +41 61 683 77 34

Fax +41 61 302 89 18

www.mdpi.com

Nutrients Editorial Office

E-mail: nutrients@mdpi.com

www.mdpi.com/journal/nutrients